Global Maternal and Child Health: Medical, Anthropological, and Public Health Perspectives

Series Editor:

David A. Schwartz
Department of Pathology
Medical College of Georgia
Augusta University
Augusta, GA, USA

Global Maternal and Child Health: Medical, Anthropological, and Public Health Perspectives is a series of books that will provide the most comprehensive and current sources of information on a wide range of topics related to global maternal and child health, written by a collection of international experts.The health of pregnant women and their children are among the most significant public health, medical, and humanitarian problems in the world today. Because in developing countries many people are poor, and young women are the poorest of the poor, persistent poverty exacerbates maternal and child morbidity and mortality and gender-based challenges to such basic human rights as education and access to health care and reproductive choices. Women and their children remain the most vulnerable members of our society and, as a result, are the most impacted individuals by many of the threats that are prevalent, and, in some cases, increasing throughout the world. These include emerging and re-emerging infectious diseases, natural and man-made disasters, armed conflict, religious and political turmoil, relocation as refugees, malnutrition, and, in some cases, starvation. The status of indigenous women and children is especially precarious in many regions because of ethnic, cultural, and language differences, resulting in stigmatization, poor obstetrical and neonatal outcomes, limitations of women's reproductive rights, and lack of access to family planning and education that restrict choices regarding their own futures. Because of the inaccessibility of women to contraception and elective pregnancy termination, unsafe abortion continues to result in maternal deaths, morbidity, and reproductive complications. Unfortunately, maternal deaths remain at unacceptably high levels in the majority of developing countries, as well as in some developed ones. Stillbirths and premature deliveries result in millions of deaths annually. Gender inequality persists globally as evidenced by the occurrence of female genital mutilation, obstetrical violence, human trafficking, and other forms of sexual discrimination directed at women. Many children are routinely exposed to physical, sexual, and psychological violence. Childhood and teen marriages remain at undesirably high levels in many developing countries.

Global Maternal and Child Health: Medical, Anthropological, and Public Health Perspectives is unique in combining the opinions and expertise of public health specialists, physicians, anthropologists and social scientists, epidemiologists, nurses, midwives, and representatives of governmental and non governmental agencies to comprehensively explore the increasing challenges and potential solutions to global maternal and child health issues.

More information about this series at http://www.springer.com/series/15852

Series Editorial Advisory Board

Frontispiece: Waterloo cemetery in Sierra Leone – one of Freetown's safe burial areas for Ebola victims. As can be seen in this photograph, there are many newborns and infants interred here who died from Ebola virus disease. Source: Simon Davis/DFID

David A. Schwartz
Julienne Ngoundoung Anoko
Sharon A. Abramowitz
Editors

Pregnant in the Time of Ebola

Women and Their Children in the 2013-2015 West African Epidemic

Springer

Editors
David A. Schwartz
Department of Pathology
Medical College of Georgia
Augusta University
Augusta, GA, USA

Julienne Ngoundoung Anoko
University of Rene Descartes Paris V La Sorbonne
Paris, France

Sharon A. Abramowitz
Department of Anthropology
Rutgers University
New Brunswick, NJ, USA

ISSN 2522-8382 ISSN 2522-8390 (electronic)
Global Maternal and Child Health: Medical, Anthropological, and Public Health Perspectives
ISBN 978-3-319-97636-5 ISBN 978-3-319-97637-2 (eBook)
https://doi.org/10.1007/978-3-319-97637-2

Library of Congress Control Number: 2018959098

This Springer imprint is published by the registered company Springer Nature Switzerland AG
The registered company address is: Gewerbestrasse 11, 6330 Cham, Switzerland

This book is dedicated to all of the victims of the West African Ebola outbreak who lost their lives during this tragic event. In addition, it is dedicated to the staff and volunteers from governmental agencies, the military, nongovernmental and international aid organizations, universities, ministries of health, public health organizations, missionaries, and members of healthcare teams as well as individuals who put themselves in harm's way in order to deliver care and minister to the many thousands of casualties of Ebola virus disease and their families. It is also dedicated to the 881 physicians, nurses, technicians, and midwives who developed Ebola virus infection, and the 513 who died from this terrible disease, while selflessly providing their services in Guinea, Liberia, and Sierra Leone. This loss of these healthcare workers, some of whom were faculty members, professors, and teachers, will have an effect upon the already fragile medical training capacities in these countries. And finally, this book is dedicated to Ebola virus disease survivors, and especially to women and their children, who are the subject of this book. Women and children are the most vulnerable members of society, who, after experiencing this horrendous and life-threatening disease and remaining alive, must pick up the pieces of their lives and move forward into their post-epidemic world. The consequences of the West African Ebola virus epidemic will be long-lasting, occurring just as these countries were emerging from political violence and, in Liberia and Sierra Leone, years of armed conflict and turmoil. This tiny organism, the Ebola virus, has forever changed the lives and futures of not only the people of these countries and West Africa as a whole but those who came from abroad as part of the international response to assist in this time of need.

Foreword

Over four decades ago, in 1976, our lab in Antwerp, Belgium, received a package from Kinshasa, Zaire (now the Democratic Republic of the Congo): a blue thermos flask filled with a sample of blood from a Flemish nun who had fallen ill from a mysterious illness. This flask contained what was later called Ebola, a deadly virus that claimed the lives of more than 300 people in a remote area of Northern Zaire around the village of Yambuku. Subsequent outbreaks of the Ebola virus were usually brief and remained confined to rural communities and small towns in Central Africa. This pattern unexpectedly changed in 2014 when the virus started to spread across urban centers in West Africa, reminding us that we can never assume things will remain the same.

The West African Ebola outbreak was the worst in history, infecting more than 28,600 people and claiming more than 11,300 lives. The crisis exposed some of the major fault lines of society, demonstrating how a "perfect storm" of poverty, weak health systems, and political instability can fuel epidemics with tragic impacts. In the aftermath of Ebola, a window of opportunity has emerged for us to examine and address the interlocking vulnerabilities that led to an outbreak of this magnitude and consequence.

I therefore welcome this timely and thoughtful book that places vulnerability at the center. *Pregnant in the Time of Ebola: Women and Their Children in the 2013-2015 West African Epidemic* addresses the unequal impact of the outbreak on two of the most vulnerable groups in society, women and children, whose health was threatened whether they contracted the virus or not.

In previous outbreaks of Ebola, we saw that pregnant women were one of the populations most at risk to mortality from the virus, where approximately 90% of those infected died. The mortality rate for neonates was equally troubling, with no documented cases of infants surviving infection following delivery to an infected mother. As highlighted in Caluwaerts and Kahn's chapter, these worrying figures were accompanied by a lack of knowledge and clinical guidance on how to manage pregnant women during the outbreak, which further complicated the response.

Pregnant in the Time of Ebola offers an authoritative account of the West African outbreak but also provides policy-relevant advice on how to manage future epidemics in resource-poor settings. It does so by drawing on the insights of leading experts from West Africa and across the globe, with experience in the fields of public health, clinical medicine, anthropology and the social sciences, epidemiology, nursing, and midwifery. Throughout my professional career, I have too often seen epidemic response activities being carried out in silos with a lack of coordination, and even cooperation, between different disciplines and sectors. I therefore welcome this effort to present a multidisciplinary perspective on an issue that brings us all together: the desire to improve maternal and child health. As we saw during the Ebola outbreak, addressing the complex health challenges of our time is not only dependent on epidemiology and biomedicine but must also engage with the political, social, and cultural factors that influence and determine health.

Ebola was a profound tragedy for all of the affected families, communities, and countries. We, as the global health community, have a responsibility to capitalize on the collective memory of the crisis to ensure that we are better prepared for the next epidemic. *Pregnant in the Time of Ebola* ensures that we keep the most vulnerable in focus during preparedness efforts and provides important lessons for the future.

Peter Piot
London School of Hygiene and Tropical Medicine and
Handa Professor of Global Health
London, UK

Preface

The first recognized Ebola virus outbreak began in August 1976 from the small village of Yambuku, located in the Bumba Zone of the Équateur Region of the northwestern part of the Democratic Republic of Congo (DRC), then called Zaire. Mabalo Lokela, the 44-year-old headmaster of the Yambuku Catholic Mission School, became mysteriously ill with a febrile disease after returning from a trip to the Mobaye-Bongo zone in the northern Équateur Region. He visited the Yambuku Mission Hospital, a 120-bed hospital staffed by Belgian nuns and a skilled Zairean medical assistant—there were no physicians on the staff. The nuns agreed that he had malaria and administered an injection of quinine. About 5 days after returning to his village of Yalikonde, he became seriously ill and returned to the hospital. He later developed signs of a hemorrhagic fever of unknown cause and died on September 8th. As is the custom, his body was hugged by family and friends and prepared for burial. By early September, the mysterious disease had recurred and spread, partly by the reuse of unsterilized syringes and needles by nuns at the hospital, to dozens of other patients and their contacts, including pregnant women, most of whom died. It was eventually diagnosed as a new hemorrhagic fever. Baron Sir Peter Piot, who wrote the introduction to this book, was then a 27-year-old physician and microbiologist at the Institute of Medicine in Antwerp, Belgium, who was instrumental in discovering the new virus and leading efforts to contain it. A Belgian doctor sent blood samples from a nun infected with the mysterious disease to the Institute in glass vials that he placed in a blue thermos flask. It was transported from Zaire to Antwerp by a passenger on a commercial flight in hand luggage. Dr. Piot and his colleagues analyzed these materials, and upon seeing an electron micrograph from the blood of one of the nuns infected in the outbreak, he recalls…

> *We saw a gigantic worm like structure—gigantic by viral standards. It's a very unusual shape for a virus, only one other virus looked like that and that was the Marburg virus.*

It had been suggested by Dr. Pierre Sureau of the French Institut Pasteur to name the deadly virus after the village where it was discovered - Yambuku. But in order to avoid permanently stigmatizing Yambuku (as had occurred when naming the Lassa fever virus after the village of Lassa, Nigeria), it was suggested by Dr. Joel Bremen of the CDC to consider a different name. Dr. Karl Johnson, also of the CDC and leader of the investigative team, thought that naming the newly discovered virus after a river would be appropriate and avoid forever associating the village with the infection. After examining a map of the area, it appeared that the nearest river to the village of Yambuku was the Ebola River, Ebola being a distortion of the local *Ngbandi* word for black - *legbala*. Dr. Peter Piot wrote that "*It appeared suitably ominous*". This outbreak eventually infected 318 individuals who lived within a 60 km radius of the hospital, had a case fatality rate of 88% (280 deaths), and left only 38 serologically confirmed survivors. Among the 17 staff members of the hospital, all of whom helped treat patients with the disease, 13 developed the infection and 11 died.

During the same year, a separate outbreak of the Ebola virus, but caused a different species, Sudan ebolavirus, occurred approximately 500 miles away in Nzara, South Sudan, where it infected 284 persons and had a case fatality rate of 53%.

Up until 2013, multiple Ebola outbreaks had sporadically occurred in Africa, and they all shared several features in common. They were self-limited and confined to rural areas, had never reached epidemic proportions, and never extended to involve urban areas or cross national boundaries. During these occurrences, pregnant women and their infants remained at the highest risk for death. Although survival data for pregnant women and their infants were unavailable from most of these pre-2013 outbreaks, in two reported studies, the case fatality rates for pregnant women were approximately 90% and 100%, and 100% for fetuses and neonates. During this period, and despite the Ebola virus infecting 2387 persons, and killing 1590 of them, there remained no specific treatment for the infection.

In December 2013, the unexpected happened. In a region of Africa that had never seen Ebola virus, a 2-year-old boy from the remote village of Meliandou in the Nzérékoré Region of southern Guinea became ill with fever, black stools, and vomiting; he died 4 days later. Shortly thereafter, his pregnant mother, sister, and grandmother developed a similar illness with symptoms consistent with Ebola infection and succumbed to the infection, as did several midwives, traditional healers, and staff at a hospital in the city of Gueckedou who treated them. The initial suspicion was raised on January 24, 2014, when the head of the Meliandou health post informed district health officials of five cases of severe diarrhea with a rapidly fatal outcome. The disease entered the capital city of Conakry on February 1st, 2014. It eventually crossed into Liberia. On March 1st, the infection was confirmed as a filovirus by the Institut Pasteur in Lyon; the following day it was identified as Zaire ebolavirus - the same strain that had been identifed in the Yambuku outbreak. Liberia's first two cases of Ebola infection were confirmed on March 30th, occurring in Foya District of Lofa County, near the border with Guinea. By April 7th, Liberia reported 21 confirmed, probable, and suspected cases and 10 deaths from Ebola virus infection. In Sierra Leone, the first cluster of cases was identified in June. These were associated with the funeral of a respected traditional healer in the remote village of Sokoma in Kailahun District—she had become infected while treating persons with Ebola infection from neighboring Guinea. Eventually, 365 Ebola-related deaths were traced back to this funeral. The course of the West African Ebola epidemic—which eventually involved endemic or imported cases in 10 countries; resulted in 11,310 deaths and killed large numbers of health-care workers in Liberia, Guinea, and Sierra Leone; resulted in at least 16,000 children who lost at least one parent or caregiver; and left approximately 17,000 survivors, many of whom have post-infectious symptoms—is now part of history.

Because of the historical vulnerability and extremely high case fatality rates among pregnant women and their children during Ebola epidemics, as well as problems related to the availability of medical and supportive care of women and infants during the West African epidemic, stigmatization of women both with and without Ebola infection, the effects of the epidemic on children and their lives, and the important roles of health-care providers, anthropologists, and other social scientists in providing services to pregnant women and children during the epidemic, Julienne, Sharon, and I decided to prepare this book. We believed that in the event that another outbreak of Ebola virus were to occur, this collection of shared knowledge and experiences from the 66 authors would be highly useful and, potentially, help to save lives. Unfortunately, and unexpectedly, we did not have long to wait.

In May, 2018, while this book was in still production, an outbreak of Ebola virus occurred in the Équateur Province of northwestern DR Congo—it was the ninth outbreak of Ebola to occur in that country. On May 8th, officials reported that 17 persons had died from Ebola virus infection near Bikoro, a small market town lying on Lake Tumba south of Mbdanka, and near the Republic of the Congo. The index case was a police officer, who died at the Ikoko-Ipenge health facility. Following his funeral, 11 members of his family developed the infection, and 7 who had cared for him or attended his funeral died. As the numbers of cases increased, on May 17th the first case was reported from Mbandaka, the capital city of Équateur province. It was the first time that Ebola virus had entered a city in the DRC and reawakened fears of what had occurred when the infection reached urban areas during the West African epidemic. To make matters worse, Mbandaka is a busy port city on the Congo River with a population of over 1 million persons, and it was feared that the virus could spread via river traffic to the capital city of Kinshasa, a city of approximately 11 million, as well as to Brazzaville, the capital city of the Republic of the Congo, both of which lie on the Congo River. The World Health Organization feared that the outbreak could spread across national borders to nine other countries as well, and Ebola virus deaths were being reported among health-care workers, evoking recent memories of the West African epidemic. Fortunately, during this outbreak, the recently developed live-attenuated vaccine to Ebola virus, recombinant vesicular stomatitis virus-Zaire Ebola virus or rVSV-ZEBOV, was available. This live-attenuated vaccine had been previously tested during the West African epidemic. Ring fence vaccinations were rapidly implemented across the affected areas—using this method, contacts of those infected, followed by contacts of those contacts, were vaccinated, as were health-care workers, laboratory personnel, surveillance workers, and people involved with burials. Unfortunately, pregnant and lactating women were not permitted to receive the potentially life-saving vaccine. During the West African epidemic, pregnant women and children had not been permitted to receive experimental antiviral drugs or vaccines—this despite the fact that there has never been a mother-infant pair that survived Ebola infection and that the only neonate to ever survive Ebola infection had received experimental treatments from Médecins Sans Frontières including ZMapp and the broad-spectrum antiviral GS-5734. The epidemic was declared over on July 24th—it had resulted in 54 confirmed or suspected cases and 33 deaths with a case fatality rate of 61%.

Tragically just 1 week later, Ebola returned to a different region of the DRC. A woman from Mangina, a town in North Kivu district in the northeastern part of the country, had been seen at a local health center on July 19th, 2018 for a heart condition. Following her discharge, she died at home on July 25th having symptoms of a hemorrhagic fever. Members of her family subsequently developed the same symptoms, dying soon afterwards. An investigation revealed an additional six cases, and following confirmation of the disease as Ebola virus, an outbreak was declared on August 1st. The area of this epidemic was especially challenging from the standpoint of epidemiological surveillance, medical intervention, and control. North Kivu is densely populated, borders Uganda to the east and Rwanda to the south, and is an active conflict zone. The Kivu conflict had been ongoing since 2004, and currently more than 100 armed groups operate in this region. Violence and crime are common, and there are intensive military operations ongoing—the administrative center of the district, Beni, is under military rule. The so-called red zones are inaccessible to public health workers due to fighting and the risk of kidnapping. Armed rebels have killed dozens of villagers and prevented health teams from reaching some areas. Vaccination was begun on August 8th, and the Congolese government authorized the use of the experimental drug mAb114—the first time that the NIH-developed monoclonal antibody has been used during an active outbreak. As the infection spread, the district of Ndindi in Beni city became the major focus of the epidemic. The WHO officials have commented that responders were reporting a higher-than-expected number of illnesses in women and children, accounting for 58% of affected persons. On September 4th the city of Butembo, with a population of almost one million people, reported its first fatality in the Ebola outbreak. As of November 22nd, 2018, there were 399 cases of Ebola infection (352 confirmed and 47 probable), including 228 deaths (181 confirmed and 47 probable) reported in 14 health zones in North Kivu Province, as well as 3 health zones in Ituri Province. The case fatality rate has been 52% among confirmed cases. Since the start of vaccination on August 8th, there have been 34,091 persons vaccinated. Among the new cases that occurred in the Kalunguta health zone of North Kivu, a 6-day old neonate died of Ebola virus disease on November 4th. The infant's mother had developed symptoms of Ebola infection 5 days before delivering her son; neither had received the vaccine. Also among the newest reported cases have been 7 newborn babies and infants ages less than 2 years, 3 children aged 2-17 years, and 3 mothers who were pregnant or breastfeeding. Challenges in the control of this outbreak are similar to those existing during the 2013–2015 West African epidemic—families concealing persons with potential or probable infection, refusals to permit health-care providers to take patients to the Ebola treatment center (ETC) or to be quarantined, delays by persons in reaching the ETC after developing symptoms, refusal of treatment, or, in this present outbreak, vaccination, unsafe burials, weak infection prevention and control procedures in health facilities leading to disease transmission, and the occurrence of violent incidents against medical staff and care facilities. According to Dr. Erik Mukama, a DRC physician coordinating Ebola relief in Beni, a vehicle containing humanitarian staff members that was transporting an Ebola patient's dead body for burial was attacked and had passengers injured. As in previous Ebola outbreaks, a significant number (30) of health-care workers have become infected and 3 have died. Once again, pregnant and lactating women have been excluded from receiving the vaccine rVSV-ZEBOV—this despite a plea by public health experts to reverse this restriction. However, some pregnant women have inadvertantly been administered the Ebola vaccine, and they are being followed by health officials. This Ebola outbreak has also been especially dangerous to children, who have been reported to be dying at an unprecendented rate. According to data from the Ministry of Health, 30 of 120 confirmed cases of EVD in the epicenter of Beni are children under 10 years of age, and 27 of them have died largely as a result of unhygienic practices at clinics run by traditional healers.

Although this book focuses on the tragedy of the West African Ebola epidemic, and, more specifically, as it affected pregnant and nonpregnant women and their children, the experience and knowledge gained during that time has been life-saving during the recent two outbreaks in the DRC. The editors and authors of this book all hope for a speedy conclusion to this present outbreak, and a very long interval until the next one.

Augusta, GA, USA David A. Schwartz

Acknowledgments

The editors would like to acknowledge the enthusiasm of all the authors in this book. They were willing to share their experiences, challenges, expertise, and stories with one another and the editors, hoping that this book would be of value in helping to organize and implement the response to future outbreaks of Ebola virus disease and other hemorrhagic fevers. In addition, we would like to express our sincere gratitude to our Springer Editor, Janet Kim, MPH, who energetically supported this project from its inception and provided her editorial expertise which immeasurably helped bring this book to fruition. We would also like to thank our own families who patiently supported us during the preparation of this book.

Contents

Contributors

Sharon A. Abramowitz, Ph.D. Department of Anthropology, Rutgers University, New Brunswick, NJ, USA

Julienne Ngoundoung Anoko, Ph.D., M.S., M.S. University of Rene Descartes Paris V La Sorbonne, Paris, France

Janine Barden-O'Fallon, Ph.D. Carolina Population Center, University of North Carolina at Chapel Hill, Chapel Hill, NC, USA

Department of Maternal and Child Health, Gillings School of Global Public Health, University of North Carolina at Chapel Hill, Chapel Hill, NC, USA

Kathryn Barron Partners in Health, Boston, MA, USA

Mamadou Alimou Barry, Pharm.D., M.P.H., M.S.C. John Snow, Inc., Chapel Hill, NC, USA

Emily Bayne, M.B.B.S., M.R.C.E.M. National Health Service, Kendal, Cumbria, Great Britain

International Medical Corps, Bombali District, Sierra Leone

Lisa M. Bebell, M.D. Division of Infectious Diseases, Massachusetts General Hospital and Harvard Medical School, Boston, MA, USA

Margaret Bentley, M.A., Ph.D. Department of Nutrition, Gillings School of Global Public Health, University of North Carolina at Chapel Hill, Chapel Hill, NC, USA

Alessandro Bertoldo, Sc.D. Zerouno Procreazione, Centro di Medicina, Venezia Mestre (VE), Italy

Paul Henry Brodish, Ph.D., M.S.P.H. Carolina Population Center, University of North Carolina at Chapel Hill, Chapel Hill, NC, USA

Veronica Fynn Bruey, Ph.D., L.L.B., L.L.M., M.P.H. Seattle University School of Law, Seattle, WA, USA

Alexandria Buller Global Studies Department, Abilene Christian University, Abilene, TX, USA

Gillian Burkhardt, M.D. Médecins Sans Frontières (Operational Center Barcelona), Freetown, Sierra Leone

Department of Obstetrics and Gynecology, University of New Mexico, Albuquerque, NM, USA

Charles W. Callahan, D.O., M.A. University of Maryland, Baltimore, MD, USA

University of Maryland Medical Center, University of Maryland, Baltimore, MD, USA

Maforki Ebola Holding and Treatment Centre, Port Loko, Sierra Leone

Uniformed Services University of the Health Sciences, Bethesda, MD, USA

Severine Caluwaerts, M.D., M.P.H. Médecins Sans Frontières, Operations Center, Brussels, Belgium

Department of Clinical Sciences, Institute of Tropical Medicine, Antwerp, Belgium

Allison M. Connolly, M.P.H., M.A. Palladium (Data, Informatics and Analytical Solutions Practice Area), Washington, DC, USA

Palladium International Development, Raleigh, NC, USA

Janice L. Cooper, Ph.D., M.P.H. The Carter Center Mental Health Program, Monrovia, Liberia

Incident Management Team and National Ebola Survivor Network, Monrovia, Liberia

National Ebola Survivor Network, Monrovia, Liberia

Caroline Crystal Center for Infectious Disease Modeling and Analysis (CIDMA), Yale School of Public Health, New Haven, CT, USA

Alexandra Davis, M.P.H. Department of Global Health, Boston University School of Public Health, Boston, MA, USA

The ALS Association, Greater New York Chapter, New York, NY, USA

School of Public Health, Boston University, Boston, MA, USA

Kerry L. Dierberg, M.D., M.P.H. Partners in Health, Boston, MA, USA

Division of Infectious Diseases and Immunology, New York University, School of Medicine, New York, NY, USA

David P. Durham, Ph.D. Center for Infectious Disease Modeling and Analysis (CIDMA), Yale School of Public Health, New Haven, CT, USA

Michelle M. Dynes, Ph.D., M.P.H., M.S.N., R.N. Division of Global Health Protection (at time of Ebola epidemic in West Africa), Centers for Disease Control and Prevention, Atlanta, GA, USA

Division of Reproductive Health, Centers for Disease Control and Prevention, Atlanta, GA, USA

Elin Erland, R.N., R.M. Médecins Sans Frontières (Operational Center Barcelona), Freetown, Sierra Leone

Telemark Hospital, Skien, Norway

Mosoka P. Fallah, Ph.D., M.P.H. Center for Infectious Disease Modeling and Analysis (CIDMA), Yale School of Public Health, New Haven, CT, USA

National Institute of Allergy and Infectious Diseases, PREVAIL-III Study, Monrovia, Liberia

A.M. Dogliotti College of Medicine, University of Liberia, Monrovia, Liberia

National Public Health Institute of Liberia, Monrovia, Liberia

John Fankhauser, M.D. ELWA Hospital, Paynesville, Monrovia, Liberia

David Geffen School of Medicine at UCLA, Los Angeles, CA, USA

Paul E. Farmer, M.D., Ph.D. Partners in Health, Boston, MA, USA

Division of Global Health Equity, Brigham and Women's Hospital, Boston, MA, USA

Department of Global Health and Social Medicine, Harvard Medical School, Boston, MA, USA

Hilary Flumo National Institute of Allergy and Infectious Diseases, PREVAIL-III Study, Monrovia, Liberia

Alison P. Galvani, Ph.D. Center for Infectious Disease Modeling and Analysis (CIDMA), Yale School of Public Health, New Haven, CT, USA

A.M. Dogliotti College of Medicine, University of Liberia, Monrovia, Liberia

Francesco Di Gennaro, M.D. Clinic of Infectious Diseases, University of Bari, Bari, Italy

Operational Research Unit, Doctors with Africa Cuamm, Padua, Italy

Peter Matthew George, M.B., Ch.B. Sierra Leone Ministry of Health, Port Loko, Sierra Leone

University of Sierra Leone, Freetown, Sierra Leone

Meekie J. Glayweon, M.A. Incident Management Team and National Ebola Survivor Network, Monrovia, Liberia

PREVAIL-III Natural History Study, J.F.K. Hospital, Monrovia, Liberia

National Ebola Survivor Network, Monrovia, Liberia

Christine L. Godwin, M.S.P.H., Ph.D. Department of Maternal and Child Health, Gillings School of Global Public Health, University of North Carolina at Chapel Hill, Chapel Hill, NC, USA

Rebecca Henderson, Ph.D., M.D. University of Florida, Gainesville, FL, USA

Doug Henry, Ph.D. Department of Anthropology, University of North Texas, Denton, TX, USA

Theresa Jones, Clin.Psy.D. International Rescue Committee, Monrovia, Liberia

Dominique de Juriew, Ph.D. UNICEF and UNHCR, Montreal, QC, Canada

Patricia Kahn, Ph.D. Médecins Sans Frontières, New York, NY, USA

Mohamed Hajidu Kamara, M.B.Ch.B. (USL) Ministry of Health, Freetown, Sierra Leone

Ministry of Health, Magburaka, Sierra Leone

Katherine E. Kralievits, M.S. Partners in Health, Boston, MA, USA

Division of Global Health and Social Medicine, Harvard Medical School, Boston, MA, USA

Jonah Lipton, Ph.D. Firoz Lalji Centre for Africa, London School of Economics, London, UK

Regan H. Marsh, M.D., M.P.H. Partners in Health, Boston, MA, USA

Division of Global Health Equity, Brigham and Women's Hospital, Boston, MA, USA

Department of Global Health and Social Medicine, Harvard Medical School, Boston, MA, USA

Kristen McLean, M.P.H. Yale University, New Haven, CT, USA

Department of Anthropology, Yale University, New Haven, CT, USA

Laura Miller, M.P.H. IRC Sierra Leone (at time of Ebola epidemic in West Africa), Freetown, Sierra Leone

Health Programs, International Rescue Committee, Atlanta, GA, USA

Olive Melissa Minor, Ph.D., M.P.H. International Rescue Committee, New York, NY, USA

Humanitarian Department, Oxfam, Oxford, Great Britain

Emma Mulhern, L.L.M. Irish Aid, Dublin, Ireland

Brighton, UK

Tolbert Nyenswah, J.D., M.P.H. National Public Health Institute of Liberia, Monrovia, Liberia

Monica Adhiambo Onyango, Ph.D., M.S., M.P.H., R.N. Department of Global Health, Boston University School of Public Health, Boston, MA, USA

Baron Peter Piot, K.C.M.G., M.D., Ph.D., D.T.M. London School of Hygiene and Tropical Medicine and Handa Professor of Global Health, London, UK

Damiano Pizzol, M.D., Ph.D. Doctors with Africa CUAMM, Padova, Italy

Giovanni Putoto, M.D., D.T.M&H., M.A.H.M.P.P. Doctors with Africa CUAMM, Padova, Italy

Medici con l'Africa CUAMM (Doctors with Africa CUAMM), Padua, Italy

GianLuca Quaglio, M.D. Directorate-General for Parliamentary Research Services, European Parliament, Brussels, Belgium

John T. Redd, M.D., M.P.H., F.A.C.P. Center for Global Health, Division of Global Health Protection, Centers for Disease Control and Prevention, Atlanta, GA, USA

Kirsten Resnick, M.S. Boston University, Boston, MA, USA

Rick Sacra, M.D. University of Massachusetts Medical School, Worcester, MA, USA

ELWA Hospital, Paynesville, Monrovia, Liberia

Tamba Sam, M.P.H. IRC Sierra Leone (at time of Ebola epidemic in West Africa), Freetown, Sierra Leone

David A. Schwartz, M.D., M.S. Hyg., F.C.A.P. Department of Pathology, Medical College of Georgia, Augusta University, Augusta, GA, USA

Springer Series Editor, Global Maternal and Child Health: Medical, Anthropological and Public Health Perspectives, New York, NY, USA

Rupal Ramesh Shah, M.S., M.P.H. Partners in Health, Boston, MA, USA

Zanmi Lasante, Mirebalais, Haiti

Mohamed G. Sheku, M.B.Ch.B. Koidu Government Hospital, Koidu, Sierra Leone

Kavita Singh, Ph.D. Department of Maternal and Child Health, Gillings School of Global Public Health, University of North Carolina at Chapel Hill, Chapel Hill, NC, USA

Laura A. Skrip, Ph.D., M.P.H. Center for Infectious Disease Modeling and Analysis (CIDMA), Yale School of Public Health, New Haven, CT, USA

National Public Health Institute of Liberia, Monrovia, Liberia

Adrienne E. Strong, Ph.D., M.A. Department of Anthropology, University of Florida, Gainesville, FL, USA

Barbara Tomczyk, Dr.P.H., M.P.H., M.S., B.S.N. Center for Global Health, Centers for Disease Control and Prevention, Atlanta, GA, USA

Indi Trehan, M.D., M.P.H., D.T.M&H. Lao Friends Hospital for Children, Luang Prabang, Lao PDR

Washington University in St. Louis, St. Louis, MO, USA

Maforki Ebola Holding and Treatment Centre, Port Loko, Sierra Leone

Mohamad Alex Vandi, M.D. Kenema District Health Management Team, Sierra Leone Ministry of Health and Sanitation, Kenema, Sierra Leone

Sinead Walsh, Ph.D. Department of Foreign Affairs and Trade, Irish Aid, Dublin, Ireland

Gretchen Williams, M.S. Partners in Health, Boston, MA, USA

Division of Global Health and Social Medicine, Harvard Medical School, Boston, MA, USA

Alyssa J. Young, M.S.P.H. Clinton Health Access Initiative (Malaria Analytics and Surveillance), Boston, MA, USA

GOAL Global, Dublin, Ireland

About the Editors

David A. Schwartz, M.D., M.S. Hyg., F.C.A.P. has an educational background in Anthropology, Medicine, Emerging Infections, Maternal Health, and Medical Epidemiology and Public Health. He has professional and research interests in reproductive health, diseases of pregnancy, and maternal and infant morbidity and mortality in both resource-rich and resource-poor countries. In the field of Medicine, his subspecialties include Obstetric, Placental, and Perinatal Pathology as well as Emerging Infections. An experienced author, editor, investigator, and consultant, Dr. Schwartz has long experience investigating the anthropological, biomedical, and epidemiologic aspects of pregnancy and its complications as they affect society, in particular among indigenous populations and when they involve emerging infections. Dr. Schwartz has been a recipient of many grants, was a Pediatric AIDS Foundation Scholar, and has organized and directed projects involving maternal health, perinatal infectious diseases, and placental pathology for such agencies as the US Centers for Disease Control and Prevention, National

Institutes of Health, and the United States Agency for International Development, as well as for the governments of other nations. He has published two previous books on pregnancy-related morbidity and mortality, the first in 2015 entitled *Maternal Mortality: Risk Factors, Anthropological Perspectives, Prevalence in Developing Countries and Preventive Strategies for Pregnancy-Related Deaths* and more recently a book published in 2018 for Springer entitled *Maternal Death and Pregnancy-Related Morbidity Among Indigenous Women of Mexico and Central America: An Anthropological, Epidemiological and Biomedical Approach.* Dr. Schwartz is the editor of the Springer book series *Global Maternal and Child Health: Medical, Anthropological and Public Health Perspectives*, of which this book is a volume. Currently involved with maternal-fetal aspects of the Zika virus pandemic and Ebola virus infections, Dr. Schwartz serves on the Editorial Boards of four international journals and is Clinical Professor of Pathology at the Medical College of Georgia in Augusta, Georgia.

Julienne Ngoundoung Anoko, Ph.D., M.S., M.S. is a social anthropologist from the Sorbonne University in France. She completed her academic preparation in the area of Epidemiology and Public Health and Gender and Health with master's degrees from Universidad Rey Juan Carlos in Spain. For greater than 17 years, she has been demonstrating and putting in practice how the use of socio-anthropological evidence from research/action research can contribute to efficient programming and accountability of development and humanitarian interventions, in areas of communication for development, child protection, emergency outbreaks, and gender. Since 2005, Dr. Anoko has supported the World Health Organization (WHO) and UNICEF during emergency response outbreaks such as Ebola, Marburg, H1N1 influenza pandemic, Zika, and most recently the plague in both developed and developing countries. Between the 2014 and 2016 Ebola outbreak in West Africa, she worked for the WHO, the UN Mission for Ebola Emergency Response (UNMEER), and UNICEF in Guinea to coordinate, support, and leverage the Social Mobilization and Community Engagement pillar in order to implement interventions compatible with sociocultural contexts to gain community trust and participation into the overall response. Between 2017 and 2018, Dr. Anoko has supported the WHO in the plague response outbreak in Madagascar. Dr. Anoko has been appointed in 2017–2018 as member of Scientific Advisory Groups and Independent Advisory Boards of several health organizations and institutions in Europe and Africa. Dr. Anoko has published books and papers and contributed to developing several guidelines for United Nations agencies dealing with her areas of expertise. Dr. Anoko had been featured in articles and broadcasting programs from the National Public Radio (USA), National Geographic, *The Washington Post*, France Culture, World Health Organization, and others. She has been distinguished twice with "Certificate of Recognition" for her work by the governments of Niger (2013) and Madagascar (2018). She is recipient of both the Research and

Innovation 2015 Award for her engagement in the field during the West African Ebola epidemic, from the French Red Cross Humanitarian Fund, and the Marsh Award for Anthropology in the World from the Royal Anthropological Institute of the United Kingdom.

Sharon A. Abramowitz, Ph.D. is a sociocultural and medical anthropologist who specializes in the anthropology of humanitarian intervention, mental health, gender-based violence, health sector transitions, and post-conflict reconstruction. Since 2014, she has been a prominent anthropologist of the West African Ebola outbreak in Liberia, Guinea, and Sierra Leone and led the American Anthropological Association's Emergency Ebola Anthropology Initiative. Dr. Abramowitz is based in Boston, MA, and has training in sociology and epidemiology, as well as anthropology and African studies. She is the author of *Searching for Normal in the Wake of the Liberian War*, coeditor of the book *Medical Humanitarianism: Ethnographies of Practice*, and 20 peer-reviewed publications on the West African Ebola outbreak. During the West African Ebola response, she provided research and analysis to a range of organizations working in Liberia and Sierra Leone, including UNICEF, the World Health Organization, USAID, and Save the Children. Presently, she is consulting with UNICEF on the development of minimum standards and indicators for community engagement in development and humanitarian contexts. Dr. Abramowitz holds research affiliations with the Rutgers University Department of Anthropology and Yale University's MacMillan Center for International and Area Studies Program on Conflict, Resilience, and Health.

Part I

The West African Ebola Epidemic, Women, and Their Children

Ebola's Assault on Women, Children, and Family Reproduction: An Introduction to the Issues

1

Sharon A. Abramowitz

1.1 An Introduction

The subject of this book—pregnancy, women, and children during the West Africa Ebola epidemic of 2013–2015—is challenging to capture in an introduction. It is a field that has yet to be defined; an agenda that has yet to be populated with interests, advocates, and defined areas of research and action; and with data still emerging. When Dave Schwartz first proposed the idea of an edited volume on Ebola virus disease in pregnancy and infancy to Julienne Anoko and me in 2015, the unknowability of the subject seemed too heavy to countenance. Pregnancy and early childhood was known to be one of the least documented, least attended to areas of documented clinical, epidemiological, and socio-cultural research from the West Africa Ebola epidemic. How would it be possible to capture the entire span of obstetrics, pediatrics, health systems research, anthropology, epidemiology, clinical medicine, and memoir from this devastated West African region? For long intervals, clinics were so overwhelmed by Ebola infection that they didn't administer pregnancy tests, or even count confirmed Ebola cases by gender.

Our concerns were echoed by many of the contributors whom we approached to contribute to this book. How, some asked, could they write about something that they had so little knowledge and information about, even during the epidemic? Some worried that they didn't have adequate data and would be drawing on their own experiences. Others indicated that the data that they held were politically and institutionally sensitive and would need to go through internal review prior to publication. Still others initiated novel ethnographic research, or provided research in progress, in order to advance the goal of a collectively compiled work that would aggregate clinical, research, patient, and policy makers' experiences, insights, analyses, discoveries, and regrets.

What was certain is that pregnancy and childbirth, even during global health disasters and humanitarian emergencies, will always happen. This is both self-evident and undervalued in global health and humanitarian policy. Pregnancy and childbirth, whether in concentration camps, conflict zones, refugee camps, or quarantined communities, defies demands for prevention, behavior change, and no-touch policies. They are inexorably deterministic and cannot be stopped by the collapse of state

The findings and conclusions in this manuscript are those of the author and do not reflect the position of any affiliate institutions.

S. A. Abramowitz (✉)
Department of Anthropology, Rutgers University, New Brunswick, NJ, USA

D. A. Schwartz et al. (eds.), *Pregnant in the Time of Ebola*, Global Maternal and Child Health,
https://doi.org/10.1007/978-3-319-97637-2_1

healthcare systems. Women will continue to have children, no matter what natural disaster, infections, or human agency is wreaking destruction in their vicinity. It's self-evident: pregnancy continues to be high-risk in low-resource settings and relatively safe and predictable in high-resource settings.

The cultural importance of pregnancy as the intersection of life and death (for example, Julienne Anoko's now-famous intervention to ensure the safe burial of a pregnant woman who had died of Ebola (Epelboin et al. 2008)) impacted whole communities and undermined well-intentioned efforts to slow down Ebola transmission within households and communities. Thomas Eric Duncan, the Liberian man who died of Ebola infection while in Texas—and directly provided the impetus for the expansion of U.S. President Barack Obama's engagement of the U.S. government in the West Africa Ebola emergency—contracted the virus as he assisted a pregnant friend get a taxi so that she could go to a hospital (Botelho and Wilson 2014).

However, the problem of pregnancy and childhood during the West Africa Ebola epidemic posed specific challenges and unanswered opportunities that require a biosocial approach (Richardson et al. 2016), a respect for phenomenological experience (Matua and Wal 2015), and an unprecedented level of seriousness about sociocultural factors (Ravi and Gauldin 2014; Strong and Schwartz 2016). While journalists feverishly wrote that Ebola would tear apart the basic fabric of society and change culture, traditions, and norms forever (Maxmen and Muller 2015), the nonnegotiable realities of sisters, daughters, mothers, and friends having babies without the help of doctors or clinics, and the urgent necessity of orphans needing homes and care and education (Abramowitz et al. 2015), imposed counter-demands that communities and cultures hold together to support its most vulnerable. The view of the situation was murky. In the three affected countries, qualitative research documented that pregnant women were being turned away from hospitals, clinics (Sadaphal et al. 2018), and community care centers (Abramowitz et al. 2016), while clinicians and social mobilizers battled rumors circulating among women that discouraged them from coming to clinics and hospitals (Strong and Schwartz 2016). Where Ebola destroyed human connections, pregnancy, childbirth, and children tethered *everyone*—West Africans and international expatriates—to each other and to the recognition of their own humanity and, in many cases, faiths.

Pregnancy brought into focus inherent tensions in the architecture of humanitarian assistance and post-conflict health systems reconstruction missions. In the decade before Ebola's introduction into human populations in West Africa, significant gains had been achieved in maternal, infant, and childhood health and mortality in the countries most affected by Ebola tethered Liberia, Guinea, and Sierra Leone (Table 1.1). The Ebola epidemic erased these gains (Practicing Midwife 2014).

Despite considerable progress towards achieving the United Nations Sustainable Development Goal 3 targets of reducing maternal, infant, and early childhood mortality rates, care for women during pregnancy, labor, delivery, and postpartum will all be compromised in contexts in which the Sustainable Development Goals addressing the basic healthcare requirements for these conditions have not been established.

Table 1.1 Gains in maternal health being wiped out by the Ebola epidemic—United Nations Development Programme (UNDP) Sustainable Development Goal 3 indicators for countries affected by Ebola 2014

	SDG 3 target	Liberia		Guinea		Sierra Leone	
Year		**2005**	**2014**	**2001**	**2015**	**2000**	**2015**
Maternal mortality ratio (per 100,000)	70	1020	741	973	679	1800	1165
Neonatal mortality ratio (per 1000 live births)	12	34.5	24.7	96.7	60	170	92
Under-5 mortality ratio (per 1000 live births)	25	124.7	72.9	158.3	92	286	156

Source: World Bank Data Bank

Childhood in the context of the Ebola epidemic raised still more complexities. If local populations were supposed to absorb the unknown numbers of Ebola orphans left in communities after the death of their parents and families, how were they supposed to provide for their food, education, and healthcare in some of the poorest economies of the world? In a region that has been involved in international humanitarian actions for nearly three decades, the lack of child welfare, orphan support, and child protection measures for children within and outside ETUs and quarantine centers provided widespread and immediate evidence of the limited frameworks that have been used to imagine and instantiate norms around *child protection, child support,* and *child welfare* during Sierra Leone and Liberia's post-conflict periods (Risso-Gill and Finnegan 2015; Sierra Leone Ministry of Social Welfare et al. 2014).

During the West Africa outbreak, pregnant women posed a particularly serious problem for Ebola responders. Because of the overlapping similarities of Ebola virus disease symptoms with pregnancy complications that have nothing to do with Ebola infection with those of infection from the virus, it may never be accurately known how many women experienced or died from EVD-related complications either during or after pregnancy outside of clinical reports. We do not know how many children were lost to Ebola. We do not know what EVD's long-term impact is on women EVD survivors' future pregnancies, but emerging evidence suggests that miscarriage and stillbirth is likely to be a long-term reality for women survivors (Fallah et al. 2016).

We *do* know, now, that, for many pregnant women who become infected with EVD, it is not the certain death that many believed it to be. The tragedy of this insight is that clinical and testing protocols were based on faulty assumptions regarding maternal mortality and may have cost additional lives. We know that most infants born with EVD will die—but not all, as MSF's experience bringing Baby Nubia into the world demonstrated (Caluwaerts 2017). We still don't know how to balance the threat of infecting healthcare workers against the life of a mother experiencing a crisis during labor while she has a hemorrhagic fever; but much has been learned about task shifting (Milland and Bolkan 2015), patient care (Oduyebo et al. 2015; Henwood et al. 2017), and patient survival. As the chapters in this volume demonstrate, clinics and hospitals were confronted with grey zones of judgement around referring pregnant women and children to ETUs or hospitals, and there were no clear answers (Deaver and Cohen 2015).

As a mother and as a medical anthropologist, what keeps me awake at night is the invisible epidemiology of Ebola among pregnant women in their communities. There will always be women and children seeking care, and there will always be care-seeking pathways populated by more or less honest brokers of healthcare, traditional medicines, or informal support (McLean et al. 2018). Lacking recognition of these facts forced many women to pursue "choiceless choices" that women, families, children, and traditional birth attendants faced during the Ebola epidemic; and they put their faith in God and family to see them through. As one mother, speaking to a USAID evaluator, reported:

> [Mother] *Life was upside down. It was very a tough time we went through. No clinic or hospital was opened. I had my daughter; she was pregnant at that time, during the heat of the crisis, no clinic at all. The information hit me when pain cut her, at that time, I was in the garden. I was discouraged. Where do I carry her at that time? So as I previously said, it was just by God's grace, so I had no alternative, but I put the problem in God's hand. I said God! This is the problem for you and not for me. You know the tough time we were going through; you take control of the situation. And definitely God was on my side and everything was fine for us here*. (Sadaphal et al. 2018).

A midwife echoed the same themes of abandonment and impossible choices, often seen as "community resistance" by officials in the Ebola response:

> [Midwife] *It happened my sister daughter was pregnant, she was in Kakata,*[1] *she was in labour pain, they carried her to the hospital there, and they refused her and end up bringing her here, and that's our chairman's*

[1]Kakata is the capital city of Liberia's Margibi County, located in the Kakata District.

> *daughter. And the midwife here said that, they people say we must not touch anybody. The girl's mother said in God and work on this girl, if we leave her like that, either she dies or the child die. So the woman trust God and took care of the girl, and she delivered. So we were only depending on God, and we continue to depend on him.* (Sadaphal et al. 2018).

There is agreement that there was a sharp decline in the utilization of healthcare services by pregnant and postpartum women, caused by a wide range of factors that require further investigation. But where did pregnant women go for pregnancy care outside of the limits of clinics and hospitals? Data on private healthcare services, alternative and traditional healthcare provision, informal (e.g., "the backdoor of the clinic") services provided by trained healthcare providers, and family and social systems are mostly lacking; with the result that it cannot be integrated into wider analysis of healthcare access, community-based transmission, and social networks involved in reproductive health.

For this reason, we, the editors, feel that pregnancy, childbirth, and childhood need to be prioritized as a focal area for research, clinical practice, and social work interventions during infectious disease crises. The time has come to establish an agenda for recognizing the distinct needs of pregnant women, infants, and children during infectious disease crises. This includes setting forth the ethics, foundational principles and policy, and programs and priorities needed to address the special needs of pregnant women and children. Existing ethics have been found to be deeply problematic (Gomes et al. 2017), and this epidemic saw international pharmaceutical companies beginning to address local demands to respond to the compassionate use of unproven Ebola vaccines in pregnant women (Henao-Restrepo et al. 2015; Shields and Lyerly 2013; Carazo Perez et al. 2017; Largent 2016; Alirol et al. 2017) Better protocols are needed to support women who seek healthcare during epidemic outbreaks to ensure that they don't experience referrals to other facilities, or complex decisions around pregnancy and virus testing as rejection (Deaver and Cohen 2015; Nam and Blanchet 2014).

The contributors' chapters in this book that documented a wide range of experiences of maternal and fetal mortality, Ebola healthcare workers' experiences, childcare and child protection challenges, and societal burdens broke my heart and challenged my personal, activist, and research ethics, as I expect they will for the reader. The tragedy documented in these pages is total and complete. Pregnant women's bodies fell on the sharp knife's edge that serrated global demand for data about the epidemic and the disease and the local burden of care that fell on households, families, communities, and hospitals and clinics. Clinical providers were unable to anticipate what would happen to mothers and infants exposed to Ebola, but the presumption was an expectation of nearly full mortality.

This book sets forth the framework of a cross-cutting research agenda that is derived from the necessarily partial and incomplete presentations of evidence and experiences by the contributors that took the risk of sharing what they knew to inform this domain. Each chapter offers insight into one facet of the kaleidoscopic biology, epidemiology, clinical care, and human experience of pregnancy, infancy, and early childhood during the EVD epidemic that remain elusive as a whole. One contributor's epidemiology reveals that the assumptions underlying another contributor's clinical protocols may have been based on misinformation, leading to complex questions about how our epidemiological assumptions may indirectly shape clinical morbidity and mortality outcomes. Another contributor's memoir reveals the burden of care and time that was required to keep children alive, safe, and contained in EVD units under impossible circumstances caused by the limitations of PPE; and the craving that their healthcare providers had to be able to touch and give comfort to people who were suffering.

Tragically, much of what we did not know shaped the courses of action available to healthcare workers who staffed clinics and hospitals across the most infected countries of Guinea, Sierra Leone, and Liberia, and the policies and practices of EVD responders. Médecins Sans Frontières (MSF), which played a major role in providing clinical care during the EVD response, progressively developed protocols and guidelines over the course of the epidemic. Pregnancy-specific difficulties with triage and testing for EVD created clinical and research challenges, and common interventions for pregnancy were less feasible due to EVD-related complications. The continuing knowledge gaps that surround our understanding of the interaction between pregnancy, delivery, post-partum recovery, and early infancy and childhood in the context of EVD are cross-cutting across social, cultural, biomedical, and public health issues. Moreover, these kinds of knowledge gaps are duplicated for other infectious disease. A comprehensive, targeted research agenda that prioritizes women, children, and reproductive health during infectious disease outbreaks is needed and past due.

In the fog of the epidemic response, it was widely suspected that pregnant women were unable to access care. Rumors about pregnant women being turned away from healthcare facilities were reported in Guinea, Liberia, and Sierra Leone; and late-term pregnancy created difficulties with mobility that further prevented women from accessing care. Several of the contributors to this volume highlight the fact that providing medical care to patients was a third-order priority that ranked below isolating infected individuals with EVD from their communities and keeping EVD responders free of infection. As Baynes reported, this resulted in additional concerns about providing medical support and intervention to women who were delivering children, given the large quantities of bodily fluid and bleeding. Correspondingly, fear of nosocomial transmission of EVD, as well as fears of poisoning, murder, and organ theft rumors spreading through local networks, kept women and children away from clinics for months.

Such concerns had a profound impact on all aspects of EVD response policy that determined when foreign and national healthcare workers would intervene and shaped responders' perceptions about the availability of healthcare services to women and children. Caluwarts and Kahn noted that *"medical teams were severely hampered by the lack of knowledge and clinical guidance on managing pregnancy patients, which was seen as essentially providing palliative care for mother and fetus."*

Contemporary state-of-the-art indicators of maternal, newborn, and child (MNCH) health provide evidence of healthcare access, utilization, and outcomes. Such indicators—many of which have been moved into the center of global health evidence by the Millenium Development Goals process—include the number of women giving birth in health facilities [facility-based deliveries]; the number of women giving birth with a skilled birth attendant present (SBA); maternal mortality ratios, the contraceptive prevalence rate; under-five mortality rates; and overall and disease-specific immunization rates. These metrics are often correlated with other indicators of health system capacity, like health workforce density, the number of hospital beds, and the number of specialists per 100,000 in the population.

Such measures, while indicative of overall trends in MNCH capacities, cannot present a full picture of the situation on the ground in a public health emergency. In order to reconcile official figures with grassroots sociocultural processes, nuanced qualitative research, historical analysis, and clinical records-based reviews need to take place in order to explain what happened, why it happened, and to give insights into how to prevent the collapse of maternal and child healthcare in future outbreaks.

Research on the capacities of healthcare systems remains compromised by key challenges that prevent us from fully seeing the complete picture of what occurs to pregnant and post-partum women, infants, and young children during epidemic outbreaks in Sub-Saharan Africa. This leads to other questions. When emergency response infrastructures are established in response to epidemic outbreaks, what specific capacities are needed to onboard already-existing anthropological knowledge for the region about maternal and child health? What other capacities are underdeveloped in epidemic response?

One capacity that is insufficiently developed involves restrictions on the handling of biological material, including biological samples. Such restrictions have prevented such important procedures for understanding the mechanisms of maternal-fetal infection and transmission as infant autopsy and placental pathology studies (Schwartz 2013, 2017). Even more seriously, women's and children's experiences of pregnancy, labor, delivery, and early childhood during epidemic outbreaks are rendered invisible. As Barden and O'Fallon indicated, when the data collection used in private healthcare systems is not integrated into national surveillance and reporting systems, those data can be excluded from clinical and epidemiological reviews due to a lack of consistency with national and subnational official statistics. During healthcare emergencies, divisions of social class, wealth, urban and rural settlement, and ethnicity can mean the difference between access to "backdoor healthcare services" providing reliable drug and healthcare resources and an utter lack of resources during critical periods like labor and delivery, when the possibility for community and household-based transmission can be exceptionally high.

New developments, like the development of a rapid EVD test, make it possible to consider disease interactions between EVD and common co-occurring endemic diseases in EVD-affected regions, such as malaria. To quote former WHO Director-General Margaret Chan, "*What gets measured, gets done*." If we cannot "see" the holistic context of maternal, newborn, and child health during epidemic outbreaks like Ebola, we cannot act. We—the national governments, multilateral organizations, nongovernmental organizations, medical providers, scientific and social science researchers, community activists, and advocates—must understand that healthcare begins far beyond the clinic. The world beyond the clinic and the hospital *is* the majority of the ecosystem of maternal, newborn, and child healthcare. Therefore, the failure to see it, count it, and respond to the ethical, technical, and clinical imperatives it poses means that health emergencies will continue to disproportionately impact the part of ourselves that makes us the most human.

References

Abramowitz, S. A., McLean, K. E., McKune, S. L., Bardosh, K. L., Fallah, M., Monger, J., et al. (2015). Community-centered responses to Ebola in urban Liberia: The view from below. *PLOS Neglected Tropical Diseases, 9*(4), e0003706. https://doi.org/10.1371/journal.pntd.0003706. Retrieved March 31, 2018, from http://journals.plos.org/plosntds/article?id=10.1371/journal.pntd.0003706.

Abramowitz, S., Rogers, B., Aklilu, L., Lee, S., & Hipgrave, D. (2016). *Ebola community care centers: Lessons learned from UNICEF's 2014-2015 experience in Sierra Leone*. New York: UNICEF. Retrieved March 31, 2018, from https://www.unicef.org/health/files/CCCReport_FINAL_July2016.pdf.

Alirol, E., Kuesel, A. C., Guraiib, M. M., Fuente-Núñez, V., Saxena, A., & Gomes, M. F. (2017). Ethics review of studies during public health emergencies—The experience of the WHO ethics review committee during the Ebola virus disease epidemic. *BMC Medical Ethics, 18*, 43. https://doi.org/10.1186/s12910-017-0201-1. Retrieved March 31, 2018, from https://www.ncbi.nlm.nih.gov/pmc/articles/PMC5485606/.

Botelho, G., & Wilson, J. (2014). Thomas Eric Duncan: First Ebola death in U.S. *CNN*. Retrieved March 31, 2018, from https://www.cnn.com/2014/10/08/health/thomas-eric-duncan-ebola/index.html.

Caluwaerts, S. (2017). Nubia's mother: Being pregnant in the time of experimental vaccines and therapeutics for Ebola. *BMC Reproductive Health, 14*(Suppl 3), 157. Retrieved March 30, 2018, from https://reproductive-health-journal.biomedcentral.com/articles/10.1186/s12978-017-0429-8.

Carazo Perez, S., Folkesson, E., Anglaret, X., Beavogui, A.-H., Berbain, E., Camara, A.-M., et al. (2017). Challenges in preparing and implementing a clinical trial at field level in an Ebola emergency: A case study in Guinea, West Africa. *PLOS Neglected Tropical Diseases, 11*(6), e0005545. Retrieved March 31, 2018, from https://www.ncbi.nlm.nih.gov/pmc/articles/PMC5480829/.

Deaver, J. E., & Cohen, W. R. (2015). Ebola virus screening during pregnancy in West Africa: Unintended consequences. *Journal of Perinatal Medicine, 43*(6), 649–655. https://doi.org/10.1515/jpm-2015-0118.

Epelboin, A., Formenty, P., Anoko, J., & Allarangar, Y. Allarangar, Y. (2008). Humanisation and informed consent for people and populations during responses to VHF in Central Africa (2003-2008). In: *Humanitarian borders* (pp. 25–37). Retrieved March 30, 2018, from http://www.crcf.sn/wp-content/uploads/2014/08/Epelboin2008MSFHumanisationOutbreakResponse_En.pdf.

Fallah, M. P., Skrip, L. A., Dahn, B. T., Nyenswah, T. G., Flumo, H., Glayweon, M., et al. (2016). Pregnancy outcomes in Liberian women who conceived after recovery from Ebola virus disease. *The Lancet Global Health, 4*(10), e678–e679. https://doi.org/10.1016/S2214-109X(16)30147-4. Retrieved March 30, 2018, from http://www.thelancet.com/journals/langlo/article/PIIS2214-109X(16)30147-4/fulltext.

Gomes, M. F., Fuente-Núñez, V., Saxena, A., & Kuesel, A. C. (2017). Protected to death: Systematic exclusion of pregnant women from Ebola virus disease trials. *Reproductive Health, 14*(Suppl 3), 172. https://doi.org/10.1186/s12978-017-0430-2. Retrieved March 31, 2018, from https://www.ncbi.nlm.nih.gov/pmc/articles/PMC5751665/.

Henao-Restrepo, A. M., Longini, I. M., Egger, M., Dean, N. E., Edmunds, J., Camacho, A., et al. (2015). Efficacy and effectiveness of an rVSV-vectored vaccine expressing Ebola surface glycoprotein: Interim results from the Guinea ring vaccination cluster-randomised trial. *The Lancet, 386*(29), 857–866. https://doi.org/10.1016/S0140-6736(15)61117-5. Retrieved March 31, 2018, from http://www.thelancet.com/journals/lancet/article/PIIS0140-6736(15)61117-5/fulltext.

Henwood, P. C., Bebell, L. M., Roshania, R., Wolfman, V., Mallow, M., Kalyanpur, A., et al. (2017). Ebola virus disease and pregnancy: A retrospective cohort study of patients managed at 5 Ebola treatment units in West Africa. *Clinical Infectious Diseases, 65*(2), 292–299. https://doi.org/10.1093/cid/cix290. Retrieved March 31, 2018, from https://www.ncbi.nlm.nih.gov/pmc/articles/PMC5850452/.

Largent, E. A. (2016). Ebola and FDA: Reviewing the response to the 2014 outbreak, to find lessons for the future. *Journal of Law and the Biosciences, 3*(3), 489–537. https://doi.org/10.1093/jlb/lsw046. Retrieved March 31, 2018, from https://academic.oup.com/jlb/article/3/3/489/2548359.

Matua, G. A., & Wal, D. M. (2015). Living under the constant threat of Ebola: A phenomenological study of survivors and family caregivers during an Ebola outbreak. *Journal of Nursing Research, 23*(3), 217–224. https://doi.org/10.1097/jnr.0000000000000116.

Maxmen, A., & Muller, P. (2015). How the fight against Ebola tested a culture's traditions. *National Geographic*. Retrieved from https://news.nationalgeographic.com/2015/01/150130-ebola-virus-outbreak-epidemic-sierra-leone-funerals/.

McLean, K. E., Abramowitz, S. A., Ball, J. D., Monger, J., Tehoungue, K., McKune, S. L., et al. (2018). Community-based reports of morbidity, mortality, and health-seeking behaviours in four Monrovia communities during the West African Ebola epidemic. *Global Public Health, 13*(5), 528–544. https://doi.org/10.1080/17441692.2016.1208262.

Milland, M., & Bolkan, H. A. (2015). Enhancing access to emergency obstetric care through surgical task shifting in Sierra Leone: Confrontation with Ebola during recovery from civil war. *Acta Obstetricia et Gynecologica Scandinavica, 94*(1), 5–7. https://doi.org/10.1111/aogs.12540.

Nam, S. L., & Blanchet, K. (2014). We mustn't forget other essential health services during the Ebola crisis. *British Medical Journal, 349*, g6837. https://doi.org/10.1136/bmj.g6837. Retrieved April 1, 2018, from https://www.bmj.com/content/349/bmj.g6837.full.

Oduyebo, T., Pineda, D., Lamin, M., Leung, A., Corbett, C., & Jamieson, D. J. (2015). A pregnant patient with Ebola virus disease. *Obstetrics and Gynecology, 126*(6), 1273–1275. https://doi.org/10.1097/AOG.0000000000001092.

Practicing Midwife. (2014). Gains in maternal health being wiped out by Ebola. *Practicing Midwife, 17*(11), 6.

Ravi, S. J., & Gauldin, E. M. (2014). Sociocultural dimensions of the Ebola virus disease outbreak in Liberia. *Biosecurity and Bioterrorism, 12*(6), 301–305. https://doi.org/10.1089/bsp.2014.1002.

Richardson, E. T., Barrie, M. B., Kelly, J. D., Dibba, Y., Koedoyoma, S., Farmer, P. E., et al. (2016). Biosocial approaches to the 2013-2016 Ebola pandemic. *Health and Human Rights, 18*(1), 115–128.

Risso-Gill, I., & Finnegan, F. (2015). Children's Ebola recovery assessment: Sierra Leone. Save the Children Fund, World Vision International, Plan International, UNICEF.

Sadaphal, S., Leigh, J., Toole, M., Hansch, S., & Cook, G. (2018). Evaluation of the USAID/OFDA Ebola virus disease outbreak response in West Africa 2014–2016. Objective 1: Effectiveness of the response. Evaluation report to USAID/OFDA. Vienna, VA

Schwartz, D. A. (2013). Challenges in improvement of perinatal health in developing nations—Role of perinatal pathology. *Archives of Pathology & Laboratory Medicine, 137*(6), 742–746. https://doi.org/10.5858/arpa.2012-0089-ED. Retrieved March 31, 2018, from http://www.archivesofpathology.org/doi/pdf/10.5858/arpa.2012-0089-ED.

Schwartz, D. A. (2017). Viral infection, proliferation and hyperplasia of Hofbauer cells and absence of inflammation characterize the placental pathology of fetuses with congenital Zika virus infection. *Archives of Gynecology and Obstetrics, 295*(6), 1361–1368. https://doi.org/10.1007/s00404-017-4361-5. Retrieved March 30, 2018, from http://link.springer.com/article/10.1007/s00404-017-4361-5/fulltext.html.

Shields, K. E., & Lyerly, A. D. (2013). Exclusion of pregnant women from industry-sponsored clinical trials. *Obstetrics & Gynecology, 122*(5), 1077–1081. https://doi.org/10.1097/AOG.0b013e3182a9ca67.

Sierra Leone Ministry of Social Welfare, Gender and Children's Affairs, UN Women Sierra Leone, OXFAM Sierra Leone, and Statistics Sierra Leone. (2014). Report of the multisector impact assessment of gender dimensions of the Ebola virus disease (EVD) in Sierra Leone.

Strong, A., & Schwartz, D. A. (2016). Sociocultural aspects of risk to pregnant women during the 2013–2015 multinational Ebola virus outbreak in West Africa. *Health Care for Women International, 37*(8), 922–942. https://doi.org/10.1080/07399332.2016.1167896.

Effects of the West African Ebola Epidemic on Health Care of Pregnant Women: Stigmatization With and Without Infection

2

Adrienne E. Strong and David A. Schwartz

2.1 Introduction: Stigma, Ebola Virus, and Pregnant Women

Most scholarship on stigma begins with an explanation of Erving Goffman's *Stigma: Notes on the Management of Spoiled Identity* (Goffman 1963). However, since Goffman's now-classic work on the subject, there have been many linguistic shifts, conceptual slippages, and new contributions to the theorization and understanding of stigma. More recent interpretations of stigma emphasize the psychological and social aspects of stigma, as well as highlight the critical role of power and systemic inequalities which can work to produce or reinforce existing stigmas (Castro and Farmer 2005; Parker and Aggleton 2003). Link and Phelan (2001) stipulate that, in order for stigma to develop, there must be an element of power involved; they do not specifically state who it is that is exercising the power they mention, but surely the differential exists in favor of those doing the stigmatizing. Similarly, Castro and Farmer (2005) argue that for a more nuanced understanding of stigma and its effects in an infectious disease and treatment context, the power structures and inequities that produce poor access to care and treatment should be an integral component of the analysis of stigma production. Parker and Aggleton (2003:13) also argue that stigma "*feeds upon, strengthens and reproduces existing inequalities of class, race, gender, and sexuality.*"

There are a number of possible definitions of stigma. Parker and Aggleton (2003) suggest that many times, stemming from Goffman (though, they say, perhaps mistakenly so), authors treat stigma as an individual process and as a static, fixed item that exists. However, they argue that stigma also "*plays a key role in producing and reproducing power relations and control…To properly understand issues of stigmatization and discrimination…requires us to think more broadly about how some individuals and groups come to be socially excluded, and about the forces that create and reinforce exclusion in different settings,*" advocating for analysis of stigma that takes into account its fluid and generative capabilities (Parker and Aggleton 2003:16). Parker and Aggleton (2003) further argue that conceptualizing

A. E. Strong, Ph.D. (✉)
Department of Anthropology, University of Florida, Gainesville, FL, USA
e-mail: adrienne.strong@ufl.edu

D. A. Schwartz
Department of Pathology, Medical College of Georgia, Augusta University, Augusta, GA, USA

D. A. Schwartz et al. (eds.), *Pregnant in the Time of Ebola*, Global Maternal and Child Health,
https://doi.org/10.1007/978-3-319-97637-2_2

stigma as an individual process may work in cultures that are more individualistic but, in more communal cultures, stigma clearly works at the level of entire groups of people, affecting them through the connections of kin, community, or village. Barrett and Brown (2008:S34) explain that since Goffman's time, stigma has come to mean a process of negative discrimination "*against people with certain physical, behavioral, or social attributes*," going on to say that stigma is "*an illness in itself, comorbid with respect to its marked physical conditions*" (Barrett and Brown 2008:S34).

There are several similarities between Ebola virus disease (EVD) and other stigma-inducing conditions, most particularly HIV/AIDS (see also Davtyan et al. 2014). Maman et al. (2009) found in their multinational study of HIV stigma and attendant discrimination that the factors contributing to HIV stigma and discrimination were: "*fear of transmission, fear of suffering and death, and the burden of caring*" for someone living with HIV/AIDS (Maman et al. 2009). Unlike in the case of HIV/AIDS, in which a patient might now expect, with access to proper health care and antiretroviral drugs, to live many years with few outward signs and symptoms of the disease, Ebola virus disease strikes relatively quickly, with an incubation period of between 2 and 21 days (WHO 2017), and can rapidly ravage the bodies of victims. In the most severe cases, the sufferer's body hemorrhages in a gruesome display. Though the timeline of EVD is accelerated in comparison to that of HIV, Maman et al.'s (2009) criteria for stigma and discrimination are most certainly all present. In other settings, the fear of infection appears to be a primary driver of stigma, as with tuberculosis stigma in Ghana (Dodor et al. 2008). On a group level, stigma might be collectively produced or reproduced through fear of a condition or state of being, thereby acting as a protective mechanism when individuals or a society feel they are at risk due to the condition (c.f. Dodor et al. 2008).

Recently, scholars have been interested in stigma facing people with HIV/AIDS, but stigma has long been attached to other behaviors and medical conditions. For example, many scholars have written of stigma in the context of people living with tuberculosis and leprosy (for a small sampling see: Adhikari et al. 2014; Amo-Adjei 2016; Barrett 2005; Chinouya and Adeyangu 2017; Coreil et al. 2010; Courtwright and Turner 2010; Garbin et al. 2015). In the case of leprosy, the infected person, without proper access to early and effective treatment, can often exhibit physical deformities, including the loss of digits or open ulcers. Adhikari et al. (2014) found these outwardly visible manifestations of the disease to be correlated with higher degrees of stigmatization for leprosy patients. In such instances, it is difficult for the sufferer to conceal their condition, opening them to increased stigma. Both Luka in the Sudan (2010) and Heijnders in Nepal (2004) found that the stigmatization of people with leprosy occurred in two stages. The first stage, the cognitive dimension, describes the degree of influence that leprosy has on the person's life. The patients "*pass through the concealability course, disruptive, aesthetic, origin and peril dimensions*" (Bainson and van den Borne 1998). In the second stage, or the affective stage, the person with leprosy undergoes social devaluation. Heijnders also found that stigma reinforced those preexisting inequalities that already existed in the community with regard to social class, gender, and age (Heijnders 2004).

Unlike leprosy, with its mentions in ancient religious texts, for example (Jopling 1991; Rao 2010), or tuberculosis (Dodor et al. 2008) which has been found in Egyptian mummies greater than 5000 years old (Daniel 2006), Ebola is a recently recognized infectious disease, termed an emerging infection. A member of the Filovirus (*Filoviridae*) family of viruses, the Ebola virus was first identified in 1976, at which time it was initially believed to represent a new strain of another filoviral agent, the Marburg virus. Western investigators' attention was first drawn to a human outbreak of the virus that occurred in late August, 1976, when people with a hemorrhagic fever syndrome began to arrive at the rural Yambuku Mission Hospital in northwestern Zaire (now the Democratic Republic of Congo).[1]

[1]The initial patient specimens were evaluated at the Institute of Tropical Medicine, Antwerp, Microbiological Research Establishment, Porton Down, United Kingdom, and the Institute Pasteur, Paris, France. Eventually, the Special Pathogens Branch of the Centers for Disease Control and Prevention (CDC) in Atlanta, USA, isolated the virus and

Stigma, the subject of this chapter, constituted one of the major deciding factors that determined the name for the new virus in 1976. Previous viral agents had been named after the villages where they first occurred.[2] In the case of the Ebola virus, to avoid permanently stigmatizing the village from which it was first recognized, it was decided instead to assign it a different name than that of the village of Yambuku where the index epidemic had occurred (Gholipour 2014).[3]

Despite its recent identification, it is generally believed that the Ebola virus has circulated for a long time among humans as well as nonhuman primates living in the endemic areas of Central Africa (Leroy et al. 2004).[4] Thus, it seems highly probable that there were at least sporadic human infections, if not small outbreaks, of Ebola virus disease occurring for at least several hundred years prior to its initial recognition by Western investigators in 1976 (Duerkson 2014).

Patients infected with these three diseases—leprosy, tuberculosis, and EVD—have, either through necessity (i.e., to prevent transmission) or for other reasons, been subjected to quarantines in which they were removed from normal (uninfected) society—think of leper colonies[5] and sanatoriums for TB patients.[6] In the case of EVD, persons with the infection, or even suspected of having infection, were removed from their home settings and quarantined in sites specifically set up for the purpose. This act of quarantining people created ruptures in the preexisting social structures and reinforced the social distance of EVD patients from their social networks through physical removal to other, specially designated spaces. Their removal to sites often constructed by outsiders (humanitarian organizations, for example) could have served to reinforce the patient's strangeness, abnormality, and their threat to other, uninfected individuals. From there, it was only a matter of time until the mere suspicion of Ebola virus infection could start the work of this separation, contributing to stigma production. Dodor et al. (2008) also suggest, in Ghana, that the ways in which health care workers treated TB patients, separating them from family members and restricting their access to visitors, donning extra personal protective equipment (PPE) when attending the patients, and barring families from returning corpses of deceased TB patients to their homes, all contributed to community perceptions of TB as a shameful disease, increasing community fear of the condition. Many of these same aspects of infection prevention and control were present in Ebola-affected areas during the 2013–2015 outbreak but on a much larger scale, and were, out of necessity, undertaken with greater rapidity. These behaviors or activities borne of infection prevention protocols would fall into what Brown et al. (2015:3) term "*reorganizations of spatial and material worlds*" which "*are among the most striking interventions to manage Ebola*" and could have

discovered that it represented a newly recognized filovirus.

[2]When the Lassa fever virus, which initially emerged in the town of Lassa, in Borno State, Nigeria in 1969, it was named after the town in which it occurred. This had the unfortunate result that it forever stigmatized the town of Lassa as being associated with the hemorrhagic fever epidemic.

[3]How the investigators reached this decision to avoid stigmatizing the town is interesting. While the investigators discussed this issue one evening while drinking Kentucky bourbon, it was initially suggested to name the virus after the town of Yambuku, but Dr. Joel Bremen of the CDC suggested that it be assigned a different name to avoid undue stigma. Another CDC scientist, Dr. Karl Johnson, suggested they name it after a river, and looking at a small map pinned to the wall. They saw that the Ebola river (meaning Black river in Lingala, the local language) ran near Yambuku. Of their choice, Dr. Piot stated "*It seemed suitably ominous.*" It was only after the virus was named that they discovered the Ebola River was not the closest river to the town of Yambuku but, as Dr. Piot relates, "*In our entirely fatigued state, that's what we ended up calling the virus: Ebola*" (Gholipour 2014; Piot 2012).

[4]Genetic studies of the relationship between Ebola virus and the related Marburg virus indicate that the two viruses evolved from a common filovirus ancestor approximately 700–850 years ago, a time when larger Bantu-speaking societies were emerging in the same regions of Central Africa (Carroll et al. 2013).

[5]The second author of this chapter worked at the Tala Leprosarium, also known as the Central Luzon Sanitarium, in the Philippines, which has a bed capacity for 2000 and provides care for over 1000 persons with Hansen's Disease (leprosy).

[6]The well-known leprosarium Kalaupapa on Molokai, Hawaii, housed up to 1200 persons at the height of its functioning.

multivalent meanings for social relations within communities that would, ultimately, disrupt containment efforts. The ways in which local and international organizations enacted many of these containment protocols both engendered and exacerbated community resentment, mistrust, and, at times, hostility that transformed into outright violence directed at the people working to implement public health measures (Calain and Poncin 2015).

2.2 Rethinking Stigma in the Context of an Acute Epidemic

When faced with rapidly deteriorating health infrastructure, there is some evidence to suggest, as do Blair et al. (2017) from their work in Liberia, that people with low confidence in formal state functions, institutions, and structures were more likely to ignore recommendations meant to control the spread of EVD. This could, in turn, suggest that countries in which citizens have low confidence in state structures might experience greater levels of EVD-related stigma; as a population at risk, with little other recourse (in the absence of confidence in the state to be able to protect them), began ostracizing the members of society they considered to be the greatest threat to them. Pregnant women, having unpredictable bodily fluids and, upon delivery, Ebola-infected amniotic fluid, hemorrhage, and placental tissue, would certainly fall into that category. The health care workers involved in the care of EVD patients also fell into this category, as well as those people thought to be even remotely affiliated with health facilities or locations in which EVD patients had been.

In their discussion of HIV/AIDS stigma, Castro and Farmer (2005) argue that a driving force behind the stigma is a lack of access to treatment. Stigma is not the determining factor in treatment seeking but, instead, economic and logistical barriers decide who will be able to avail themselves of care (Castro and Farmer 2005). In the case of EVD, poor infrastructure and weak public health systems were undeniable contributors to the spread of the disease. Additionally, as Castro and Farmer (2005) indicate, EVD differentially affected people based upon economic resources. In one of the stories we analyze later in this chapter, the family was able to pay an exorbitant sum for the pregnant wife to access care, but many others did not have the economic means to pursue even regular obstetric care at any cost. These infrastructure and economic constraints, when combined with a high case fatality rate and rapid progression of clinical EVD, dramatically set Ebola apart from HIV/AIDS. Taken together with the criteria of Maman et al. (2009) for stigma, particularly as related to the fears of suffering and death and the fear of transmission, the weak state in many of the affected countries provided the perfect storm for the development of strong and pervasive Ebola stigma. We argue here that it is likely that health care workers and community members fell back on stigmatization as, what they saw as, one of only a few possible ways of protecting themselves. Pregnant women, even those unaffected by the virus, suffered the consequences of people's fears, which took the form of increased stigma that caused many women to be refused care at health facilities. The health care system was undeniably overwhelmed by the number of Ebola patients and the particular needs for quarantine and supplies these patients presented. The fact that an Ebola epidemic was able to effectively shut down any and all routine health care services, particularly vital functions related to obstetric care, is a testament to the long-standing weaknesses of these systems (Chothia 2014).

While others have begun to write about the lasting stigmatization that Ebola survivors (Karafillakis et al. 2016; van Bortel et al. 2016) or Ebola orphans (Denis-Ramirez et al. 2017) have been experiencing since the end of the outbreak, we are most interested here in examining stigma that occurred *during* the active phase of the outbreak. Likewise, we have chosen to limit our discussion to the experiences of women who were pregnant at the time of Ebola. In this chapter, we argue for a reexamination of stigma in the context of transient, acute crises. Without first understanding more about the ways in which people use or suffer from the effects of stigmatization under these circumstances, it will be impossible to mitigate the effects of stigma in similar situations. We use a meaning-centered approach to stigma, thinking through what is at stake for both the stigmatized and stigmatizing groups. Next, we put forth examples of three ways in which pregnant women were particularly affected by stigma in the health care setting during the 2014–2015 Ebola virus outbreak in West Africa: (1) Women/community stigmatization and avoidance of health care facilities; (2) Stigmatization of health care workers due to association with Ebola; and (3) Stigmatization directed toward pregnant women (or perceived stigma), which prevented women from receiving care when pregnant or during delivery and in the postpartum period. We then explore the consequences of this stigmatization, seeking to answer questions such as: how did women and/or their family members seek to combat stigmatization? How did pregnant women's stigmatization of other parties (namely health care workers) affect their access to obstetric care? What were the intended effects of stigmatizing health workers? What were the protective effects of stigmatizing pregnant women and for whom? How did reluctance to care for pregnant women result in compounding adverse effects for a segment of the population that was already at far greater risk for death—both from pregnancy-related causes, and Ebola virus itself? The residual effects of EVD, even in people who have survived and recovered, are bound to increase the chances that any similar future outbreak of EVD could generate even more severe stigma because people will have the collective memory of the previous outbreak and its devastating effects.

2.3 Gendering Outbreaks

In outbreaks of infectious diseases, sex, gender, and age all play important roles. Women typically have different gender-assigned roles than do men—even when they are pregnant. They prepare meals, are caregivers for the sick, and often help to prepare the dead, in addition to the daily work of childcare, and, in many areas of Africa, fetching water and firewood, washing clothes, and ensuring the availability of daily provisions for the family, as well as engaging in subsistence farming (Avotri and Walters 1999; WHO 2007). Broader structural factors and power relations influence women's ability to access health care services even under normal conditions.

Women who are pregnant are at an especially high risk for becoming infected with EVD, and then having poor clinical outcomes, for several reasons. Depending on the community norms of their particular ethnic group, women, and pregnant women perhaps even more so, typically have less mobility than do other members of the society, are often dependent upon men or more senior women for decisions related to seeking health care, are poorer than men, and have less access and control over economic resources (Coulter 2009). There are frequently concomitant biological factors which may make pregnant women even more susceptible such as anemia, altered immunity due to pregnancy or reduced health over their life course, low body mass index (BMI), malnutrition and poor nutritional quality of the diet, and other complications of pregnancy including diabetes and hypertensive diseases. Pregnant women may spend more time in health care settings, such as clinics and hospitals, due to the pregnancy itself or due to monthly visits with their young children, or as they accompany others seeking care, and thus have greater exposure to nosocomial transmission of all diseases, not simply Ebola. Pregnant women may also have fewer treatment options or choice of medications because of potential teratogenic effects on the fetus and, if postpartum, due to breastfeeding. As was seen in previous Ebola outbreaks in Uganda and Democratic Republic of the Congo, women are often placed at greater risk for infection due to gender specific roles and activities, particularly those related to funerary practices, which have been well-studied in these previous outbreaks and have taken on a prominent role in gender-specific risk (Anoko 2014; Barbato 2014; Hewlett and Amola 2003; Segers 2014; World Health Organization 2007).

Infants and children are another vulnerable part of society in these countries. In previous outbreaks of the Ebola virus, there have been no documented neonatal survivors from pregnant women infected with the virus (CDC 2016a). The most recent data from the West African epidemic indicate that there are very few infants born to Ebola-infected mothers who survived as neonates for even a few days (Howard 2005; Jamieson et al. 2014).[7] Those newborns who have survived following their mother's Ebola virus infection have been highly profiled in the media (Awford 2015; Chicago Tribune 2015; Martel 2014; Médecins Sans Frontières 2015).[8] Women and children are also vulnerable to the continued effects of destabilization within their countries after the official termination of the outbreak. We are now seeing the beginning signs of lasting physiological and psychological effects of Ebola even on those who made a full recovery after initial infection (Clark et al. 2015; Maron 2015). In these instances, it is clear that Ebola survivors will continue to face the continuing impact of their infection for years to come in the form of the physiological effects mentioned above, but also in the form of social effects. Early findings suggest survivors and children left orphaned due to the outbreak may suffer from on-going stigmatization (Denis-Ramirez et al. 2017). It is also expected that comorbid diseases, especially those that are vaccine-preventable such as measles and meningitis, may now have a resurgence as a result of interruptions of routine immunization practices caused by the Ebola epidemic, which further weakened already strained health systems (Takahashi et al. 2015).

2.4 Ebola and Women

During the 2013–2015 outbreak, women in Liberia, Sierra Leone, and Guinea faced the unwelcome prospect of a triple burden of death—they could die from Ebola virus, during pregnancy, or during childbirth. Early during the development of this outbreak, a preponderance of women were affected, with infected women outnumbering infected men by a 3 to 1 ratio in Liberia, and 2 to 1 in both Guinea and Sierra Leone (Barbato 2014; Bofu-Tawamba 2014; Hogan 2014; Life for African Mothers 2014). This was the result of customs which have been well-described—women had the major roles in providing care for the ill and preparing the dead for burial (Menéndez et al. 2015). It was not until much later in the epidemic that the ratio of new cases normalized to equal numbers of both sexes being infected by early 2015 (Saul 2015; Thomas 2014; WHO 2015). While it has not been systematically confirmed, this equalization perhaps represented effective education concerning modes of transmission, risk behaviors, and an increase in resources coming from governments, volunteers, and nongovernmental organizations (NGOs), and their assistance in the removal and burial of the dead.

With the onset of the Ebola outbreak, the progress which Liberia, Sierra Leone, and Guinea had made in the improvement of maternal health prior to the epidemic had all but disappeared. The few hospitals and clinics that were in existence in these countries were converted largely, or in some cases exclusively, into Ebola treatment centers (ETCs). This had dire consequences even for women who were not infected with the virus—during the height of the epidemic from October 2014 to October 2015, it was estimated that 800,000 women would give birth in Liberia, Sierra Leone, and Guinea, and

[7] During this outbreak, young age was a predictor of mortality due to EVD—children tended to rapidly progress to severe disease and deteriorate, with a median time of admission to an Ebola treatment center (ETC) to death of only 3 days (Fitzgerald et al. 2016; Rojek et al. 2017; WHO Ebola Response Team et al. 2015).

[8] The infant who is perhaps the most well-known survivor of Ebola infection is Nubia, delivered to an Ebola-infected mother in Guinea on October 27th, 2015. Although her mother died of EBV, Baby Nubia was given two experimental medicines—monoclonal antibodies (ZMapp) and the broad-spectrum antiviral GS-5734—together with a buffy coat transfusion from an Ebola survivor by her Médecins Sans Frontières treatment team. Nubia became the first neonate to survive congenital Ebola virus infection, and she remains alive today (Dörnemann et al. 2017; Médecins Sans Frontières 2015).

that up to 120,000 mothers could die if denied access to emergency obstetrical care; this is equivalent to almost 330 women dying from pregnancy each day (Boseley 2014; Hayden 2014; UNFPA 2014c, d). At that time, the United Nations Population Fund (UNFPA) had estimated that, unless sufficient emergency obstetrical care was provided, the maternal death rates in these three countries could effectively double to levels seen in the 1990s during times of political unrest, civil war, and violence. This increase would represent a maternal mortality ratio (MMR)[9] of 1000 in Guinea and Liberia, and over 2000 in Sierra Leone (UNFPA 2015).

There is no evidence to demonstrate that pregnant women are biologically more susceptible to acquiring Ebola virus infection following exposure; however, there are studies which suggest that the Ebola virus is much more deadly to pregnant women and their infants than it is to the general population (Mupapa et al. 1999; Doucleff 2014; Jamieson et al. 2014; Sieff 2015). The Ebola virus epidemic in West Arica had a case fatality rate (CFR) among people of all ages and sexes of from 60 to 70%, depending upon the specifics of the group evaluated. Data from the Centers for Disease Control and Prevention reported a total of 28,616 confirmed cases of EVD occurring in Liberia, Sierra Leone, and Guinea, with a total number of 11,310 deaths—a mortality rate of 74% (CDC 2016b). Among pregnant women and their infants, however, there is evidence to suggest that the mortality rates are higher. In an analysis of 12 published studies of 108 pregnant women and 110 fetal outcomes following Ebola virus infection in Guinea, Liberia, and Sierra Leone, there were 91 maternal deaths, a case fatality rate of 84.3%, and only one surviving fetus (Garba et al. 2017). In a study of 111 reported cases of EVD among pregnant women in all three West African countries, Bebell et al. (2017) found an aggregate maternal death rate of 86% (Bebell et al. 2017).

The fetus from a pregnant woman with Ebola virus disease (EVD) will almost certainly become infected, likely via the virus passing through the placenta, and be stillborn or die shortly after birth. Dr. Denise Jamieson, an obstetrician with the Division of Reproductive Health of the Centers for Disease Control and Prevention (CDC) in Atlanta, stated, "There have been no neonatal survivors" (Sieff 2015). During the Kikwit, Zaire (currently Democratic Republic of Congo), outbreak of Ebola virus infection in 1995, the case fatality rate (CFR) among infected pregnant women was 95.5% (as compared with a case fatality rate of 70% for nonpregnant women). In this Kikwit outbreak, the infection was uniformly fatal to all unborn infants (CFR = 100%) (Mupapa et al. 1999).

2.5 Stigmatization of Health Facilities and Health Care Workers: Avoidance of Hospitals and Birthing Centers

Liberia, Sierra Leone, and Guinea all had a severe shortage of health workers before the outbreak began in 2014. In Liberia, a nation with 4.3 million people, there were only 51 physicians, 978 nurses and midwives, and 269 pharmacists in the country. Sierra Leone, a more populous country with six million inhabitants, had 136 doctors, 1017 nurses and midwives, and 114 pharmacists (Chothia 2014). Health care resources in the most affected countries were already stretched thin, with Guinea having 10 physicians per 100,000 people, Sierra Leone having 2 physicians per 100,000 people, and Liberia having only 1.4 physicians per 100,000 people (in contrast, Sweden has 380 physicians per 100,000 people) to assist in the treatment of infected persons, as well as the treatment of endemic illnesses and chronic conditions (CIA 2015). And among these limited contingents of physicians present at the

[9]The maternal mortality ratio, or MMR, is a standard statistic utilized to evaluate the level of maternal death in a population. It is derived from the ratio of the number of maternal deaths during a given time period from any cause related to or aggravated by pregnancy or its management (excluding accidental or incidental causes) per 100,000 live births during the same time-period. It is the most widely used statistic to compare maternal deaths between countries.

beginning of the epidemic, only a very few were obstetricians, or had even received any additional specialized training in Obstetrics. It has been estimated that there were only six obstetricians in all of Liberia, some of whom also had administrative roles, and many health facilities lacked a single midwife (Sepkowitz and Haglage 2014).

To make matters worse, as the epidemic continued in these three countries, the small contingent of physicians became even smaller as a result of exposure to infected patients. In just 10 months, the total number of physicians in Sierra Leone was reduced by 8.2%—a total of 11 doctors in that nation had died from Ebola virus infection as of December 2014 (Frankel 2014). As of November 4, 2015, there were 378 cases of Ebola infection among all types of health care workers in Liberia, resulting in 192 deaths; 196 cases of Ebola infection in Guinea among health care workers resulting in 100 deaths; and, in Sierra Leone, 307 infections among health care workers causing 221 deaths (Statista 2015) —a total of 831 infected health care workers and 512 deaths as of January 14, 2016 (The Economist 2016). According to the CDC, the confirmed incidence of Ebola virus infection was 103-fold greater in health care workers than in the general population of Sierra Leone (Kilmarx et al. 2014).

Because of these grim statistics, health care workers became associated with Ebola, death, and the spread of the disease. After the end of the outbreak, many health care workers continued to experience the social repercussions of carrying out their professional duties—they have been ostracized and stigmatized in their communities (McMahon et al. 2016). The health care workers themselves lamented the drastic changes in patient-provider interactions that the Ebola outbreak necessitated; gone were many of the small gestures that communicate closeness and caring in these contexts, such as sitting in close proximity, touching a patient's arm or hand, or helping a breastfeeding mother with her baby's latch (McMahon et al. 2016). Quarantines imposed on health facilities that experience compromised contact with people with EVD "*induced panic within facilities and the broader community*" (McMahon et al. 2016:1235) and health care providers in such communities reported that many people delayed seeking treatment because they associated facilities with points of Ebola transmission (McMahon et al. 2016). McMahon et al. (2016:1235) additionally state that one provider told them, "'*If we have not done triage, we are not going to take care of that person even if she is in labor...we have to abandon her on the street.*'" McMahon et al.'s (2016) findings clearly demonstrate the ways in which all three forms of stigma were at play in these communities—people came to associate facilities with transmission points for Ebola; health care workers were stigmatized due to their work caring for patients and for changed behavior necessitated by the outbreak; and even when pregnant women did elect to seek health care services, they might not be treated but left on the street to give birth alone, no one willing to risk assisting her.

Pregnant women avoided going to clinics and hospitals where Ebola patients were concentrated for fear of becoming infected with Ebola, despite the need for antenatal and intrapartum care. This behavior presents one way in which Ebola stigma affected maternal health—by not seeking care for fear of infection or contact with possible Ebola patients, pregnant mothers put themselves at risk for complications left undetected or untreated. There was also fear of physicians, nurses, and other health workers among some people, fueled by rumors and conspiracy theories including that health personnel and aid organizations were responsible for spreading the virus. In one occurrence in Guinea, these fears resulted in the deaths of eight health workers who were killed by members of the community during a health education campaign (BBC 2014). The murder of these health care workers was, arguably, an extreme example of enacted stigma as community members killed those thought responsible for the disease and its effects. The deaths of health care workers in countries that already suffered from a severe health worker deficit further limited the number of those who would have been able to serve pregnant women, when pregnant women overcame their fears of health facilities and providers.

As a result of the murders, governments and agencies organized attempts to educate the populace about the need to seek medical attention, especially targeting pregnant women. For example, the

United Nations Population Fund (UNFPA) distributed the following radio message—"*Pregnant women: health workers are there for you, to give you all the care and advice you need for a safe pregnancy and delivery... they ensure your safety and the safety of your child during the Ebola outbreak*" (UNFPA 2014b). In the face of strong Ebola stigma and fears of infection, the effects of such an announcement were debatable, particularly when paired with low levels of local confidence in state structures or outsiders (in the form of aid and humanitarian organizations).

Prior to the outbreak, the maternity ward at the West Point[10] clinic, located in an impoverished residential area outside of Monrovia, would have 10–15 births each week—during the height of the outbreak, the clinic saw one or two pregnant women per week, and not all stayed to deliver. "*People won't go to the hospital, the clinic,*" according to a 10-year staff member there, Comfort Tapeh, "*They say that when you go to the hospital, the nurses kill you. I tell them, 'So who's killing the nurses?'*" (Moore 2014). At the same maternity ward, a pregnant women developed a fever, but the woman resisted attempts by the staff to medicate her. She yelled to the staff, "*My husband told me I shouldn't take any injections here!*" When the staff spoke with her husband, he refused to permit them to treat his pregnant wife and brought her home. Jemimah Kargbo was on duty at the maternity ward when the woman left. "*She will die. With a fever like that, she will definitely die. Maybe the baby too,*" she said (Moore 2014). These quotes raise a number of other issues related to the differential effect of the Ebola virus outbreak on pregnant women, as well as the effects of stigma associated with EVD itself, and the stigma those people in close contact with Ebola victims attracted in the course of caring for them—be they biomedical health workers, Red Cross volunteers, or family members caring for relatives in their homes. The stigma associated with working in proximity to Ebola patients, or associated with the perception that one was working with Ebola patients, effectively led some pregnant women, such as Comfort Tapeh above, to reject assistance from health care workers. Comfort's case demonstrates, too, the ways in which stigma contributed to Comfort's desire to risk possible death due to infection instead of, what she perceived to be, the worse fate of risking infection from the Ebola virus while in the hospital.

Even in the best of circumstances, in communities with resource-poor health facilities there is often a deep mistrust and suspicion of health care providers (Strong 2017); throughout many resource-poor communities, people speculate that there are no medications or even gloves, because the nurses were selling these supplies in their own shops. In those instances, people became reluctant to seek care in the frontline health facilities, preferring to go further afield or stay at home with local birth attendants for economic, logistic, or social reasons. If the situation were to deteriorate further, as it did during the Ebola outbreak, this suspicion and mistrust would easily multiply.

Data from the CDC in Atlanta have confirmed that fewer pregnant women were seeking care in clinics and hospitals—Dynes et al. (2015) and Luginaah et al. (2016) state the evidence suggests the outbreak most likely eroded recent gains in utilization of prenatal and delivery care in Libera. In the Kenema District of Sierra Leone, a high prevalence area for Ebola virus infections, during the period May to July 2014, there were 29% fewer antenatal care visits and 21% fewer postnatal care visits than occurred prior to the outbreak. The CDC researchers conducted a focus group among pregnant and lactating women and they found that these declines in care-seeking behavior were motivated by fear of becoming infected with Ebola at the facilities and mistrust of the physicians (Dynes et al. 2015). Here, we see both the first and second forms of stigma we proposed above—women and their relatives were stigmatizing health care workers and the facilities in which they worked. This stigmatization or discrimination was motivated by fear and a lack of trust. Returning to Parker and Aggleton's (2003)

[10] West Point is a township that is one of Monrovia's most densely populated slums. With approximately 75,000 impoverished persons concentrated on a peninsula which projects out into the Atlantic Ocean, the living conditions are squalid, there are insufficient public toilets and sanitation, and infectious diseases such as tuberculosis are rampant.

assertions that stigma reveals power relations and social processes, we can read women's stigmatization of health care facilities and workers as a different form of power relations. The stigmatization of these health care providers during the Ebola outbreak had roots in a much longer history of distrust of state institutions and health care workers who were unable to provide consistently high quality care due to a persistently weak health care system in many of these countries (Blair et al. 2017). While the pregnant women themselves may not have had any real form of power over health care workers, we suggest that their stigmatization of health providers and facilities, leading to avoidance of services, was a result of systemic and on-going disempowerment of women and community members in the biomedical system, produced by weak state institutions. It is also possible that pregnant women were acting out of perceived, or internalized, stigma (Luginaah et al. 2016), a fear of being stigmatized when they reported to health facilities, as we discuss further below. According to the chief executive of ActionAid, Justin Forsyth, "*Ebola is having a huge impact on wider health issues like maternal healthcare. No children have gone to school since March and pregnant mums are avoiding health clinics and hospitals. One clinic I went to said the admissions had plummeted from 80 a day to 20–a worrying stat when the UN estimates that 800,000 mums will give birth in the coming year across the region*" (Boseley 2014).

The clinics and hospitals were so overwhelmed with caring for victims of the Ebola outbreak that providing obstetrical services for noninfected women became less of a priority. An important result of this was decreased prenatal care for pregnant women, placing them and their unborn children at higher risk for undetected complications such as other infectious diseases including malaria, HIV and tuberculosis, hypertensive diseases of pregnancy, anemia, and obstructed labor. "*The lack of access by women, especially pregnant women, to reproductive health services is a major health disaster in waiting*," said John K. Mulbah, Chairman of the Obstetrics and Gynaecology Department at the University of Liberia (UNFPA 2014a). "*As a result of the outbreak, there has been an increase in pregnant women dying from preventable causes, including antepartum and post-partum hemorrhage, ruptured uterus, as well as hypertensive disease*," said Dr. Mulbah. Pregnant women had to give birth without medical supervision, often without a midwife, and occasionally even alone. In Monrovia, 36 year-old Comfort Fayiah went into labor and was taken by her husband to four different clinics and hospitals to deliver her twins, but she was refused admittance to all four. Her husband, Victor Fayiah, said, "*I begged them, I cried, but they bluntly refused*." On the final day of hospitalization, an up-front payment of $450 for a surgical delivery was demanded. "*The hospital administration requested a cash down payment of $450 before my wife would be touched*," said her husband. "*Upon realizing that we did not have the money, and for fear that my wife could pass away in their premises, a man acting on the order of the hospital physically pushed my wife out. He said, 'Get outside! Do you think this is a free hospital?'*" After leaving in the rain, they hadn't walked more than 10 m before Mrs. Fayiah fell onto the street and began to deliver her twins. She received no care from the hospital staff, but instead was assisted by people passing by and onlookers. A nurse assistant on a motorbike stopped to help deliver the twins (UNFPA 2014d). Hayden (2014) relates another situation in which a laboring woman was left to give birth alone in the back of an ambulance in Liberia. When the mother arrived at an MSF Ebola management center, the obstetrician found that the woman's placenta and now-dead baby were still between her legs; tests later determined neither the mother nor baby had had Ebola (Hayden 2014). Despite the fact that the woman was not infected with the Ebola virus, health care workers had been too afraid of contamination from the copious bodily fluids involved in birth to chance assisting her without full-body protective equipment.

Other obstetric causes of hemorrhage could simulate the effects of the Ebola virus infection during pregnancy. During the Ebola outbreak in Liberia, Mamie Tarr arrived at a clinic in her fifth month of pregnancy while hemorrhaging, holding her abdomen and complaining of severe pain. Her husband Edwin was ill and previously had been brought to an Ebola Treatment Center for care. "*My wife saw me being carried to the Ebola treatment center as a suspected Ebola case, and she thought I was going to die,*" he stated. But Mamie did not want to raise the infant without her husband. So, without informing her sick husband, the 30-year-old mother sought out an illegal and unsafe abortion from a "backstreet abortionist" who performed the procedure using a combination of herbs, a rusty syringe, and chalk, known in Liberia as "spoiling the belly." Her husband said "*She got scared about raising the child without me, so she spoiled her belly. Then she just closed her mouth, she said nothing, until she started bleeding*." Mamie died as a result of the unsafe abortion; her husband Edwin was released from the ETC without ever having had the Ebola virus infection (Thomas 2017).

The Ebola crisis became more severe and, as it spread during 2014 and into 2015, country officials diverted critical resources away from pregnant women, who already faced limited access to adequate health care in these nations. The majority of Ebola units in these countries did not have staff trained in obstetrics or midwifery. For example, in Bong, one of the most populous counties in Liberia, the ambulance that was used for obstetric emergencies was instead being used for the Ebola response, diverting this already scarce and much-needed resource from its original purpose, which had not disappeared, only been overwhelmed by a louder, more urgent disaster. And the surgical and emergency departments at JFK Hospital in Monrovia, one of the country's major referral hospitals, were closed during part of the outbreak (Sepkowitz and Haglage 2014). Giving birth at home is dangerous—there is a risk of hepatitis or HIV infections being transmitted because instruments are not sterilized, as well as life-threatening hemorrhage from a placental abruption, uterine rupture, placenta previa, velamentous cord, retained placenta, or coagulopathy (Schwartz 2015). Giving birth at home also generally takes place with the assistance of untrained relatives or perhaps local midwives who may or may not have had any experience or training related to pregnancy or delivery complications, and may not be able to recognize and deal with an emergency should one arise. This further endangers the lives of both the mother and her baby; a woman can live with hepatitis or HIV, but without proper emergency treatment, will most likely not recover from hemorrhage leading to hypovolemic shock. Should an obstructed delivery occur, which is one of the most frequent causes of maternal death in sub-Saharan Africa, there is no provision for performance of a life-saving Cesarean section at home (Schwartz 2015).

2.6 Stigmatization of Pregnant Women Due to Ebola Infection

Social stigmatization can result from the fear of disease, as well as the fear of people who are understood to be different, often due to perceived connections with the contagion. During an infectious disease outbreak, these two fears can occur simultaneously, resulting in the stigmatization of strangers with or without the disease. This stigmatization creates numerous problems for both care givers and care seekers, increasing the suffering of infected people and interfering with public health measures to control the outbreak, as well as the provision of health care (Des Jarlais et al. 2006; Goffman 1963). From a contemporary historical perspective, stigmatization has been an important sociological aspect of many infectious diseases including Hansen's disease (leprosy), AIDS, tuberculosis, and SARS (severe acute respiratory syndrome) to name but a few (Obilade 2015). During the 2013–2015 Ebola epidemic, besides the problem of pregnant women intentionally avoiding health care facilities, many others were stigmatized when seeking treatment at a clinic or hospital due to the widespread general fear regarding the Ebola outbreak. As a result, when pregnant women sought health care, in some cases they were turned away for fear that they were infected with Ebola virus. For example, in August 2014, Sierra Leonean filmmaker Arthur Pratt was working in Freetown to educate people about the Ebola outbreak. Because all hospitals there were closed to non-Ebola-infected cases, he was forced to drive his 8-month pregnant wife to her mother's village, many hours distant, in order to give birth. *"A lot of women are having to give birth in the house now,"* he said during an interview, *"All of the hospitals are closed. They have nowhere to go"* (Sepkowitz and Haglage 2014). In the case of Aminata, a young pregnant woman who presented to a health care facility in Sierra Leone with a life-threatening obstructed labor, medical staff at the health care facility were afraid to treat her because she might have Ebola virus disease. While waiting for the results of her blood test for the infection, she was transferred to an Ebola holding center so that she could be rushed for a cesarean section if her test was negative. Aminata and her unborn fetus died from complications of the obstructed labor while waiting for the test results—they were found to be negative for Ebola virus hours later (Farmer and Koroma 2015).

Childbirth is a messy process in the safest of circumstances—hemorrhage, amniotic fluid, secretions, sweat, and expulsion of a bloody placenta are all normal processes in childbirth, except that all of these fluids and tissues may contain the virus in infected women. Pregnant women infected with the Ebola virus have been found to have high levels of the viral nucleic acid persisting in amniotic fluid even following the clearance of the virus from maternal blood (Baggi et al. 2014).

Even in the first author's field site in southwestern Tanzania, there was great fear in 2014 that Ebola cases from Democratic Republic of Congo (DRC) or Uganda would jump borders and the virus could find its way into Tanzania. The maternity ward nursing staff repeatedly discussed a deficiency of personal protective equipment (PPE) available at the hospital under normal circumstances and wanted to know why the maternity ward staff had not been the first to be trained in proper techniques for isolating Ebola patients and recognizing the virus's early symptoms. The nurses were of the opinion that the maternity ward staff, with their intimate and nearly continual contact with vast quantities of bodily fluids, would be the first to be infected and die from Ebola in the event that the virus became an issue in their region. Under normal conditions, these midwives expressed a concern with their greater than average exposure to bodily fluids and a lack of vaccines for diseases such as Hepatitis B (Strong 2018). This fear and concern most certainly was multiplied exponentially for midwives working in the areas most heavily affected by the EVD epidemic. Surely, any amount of money to deal with bodily fluids that may or may not be infected was poor incentive for providers to expose themselves to infection and a terrifying death as they watched or heard of their colleagues dying around them.

Compounding the stigmatization of pregnant women due to their potential Ebola infection was a much older, geographically widespread stigma that ties women's bodily fluids to concepts of pollution, dirt, and contamination. In some places, only women from groups which are already constructed as polluted or polluting are considered suitable birth attendants in order to contain the stigma associated with contamination from the bodily by-products of birth. In India, for example, the local birth attendants, *dais*, are generally from the lowest caste (Rai 2007:177; van Hollen 2003). Even in Western biomedical settings, health care providers and a woman's relatives act to mitigate their contamination from women's bodily fluids due to concerns about germs or infection, but Callaghan (2007) argues, also due to much older, ingrained ideas about the inherently polluting nature of women's bodies that continue to float below the surface of action and thought. In West Africa, a history of menstrual taboos or seclusion practices and other taboos related to sex or reproduction may hint at beliefs regarding the underlying dangers of exposing others to the products of women's bodies (Sterner and David 1991; Strassmann 1992), even if these practices are not in place any longer. It should be noted that restrictions related to menstruation, for example, are not universally damaging to women (Buckley and Gottlieb, 1988). With the spread of biomedicine and its conceptions of microscopic agents of infection (viruses and bacteria) has come a newer understanding of what might be hiding in the substances produced by bodies, particularly women's bodies—in this case, infectious, potentially deadly pathogens such as HIV, Ebola, hepatitis, and others.

Exacerbating the potential stigmatization of caring for pregnant women during an Ebola outbreak were data indicating that some women who were pregnant and infected with Ebola virus may not have had symptoms of the infection at the time of labor. A 31-year-old woman was presented to a Monrovia hospital in the late stage of pregnancy with suspected premature rupture of membranes (PROM), abdominal pain, and mild uterine contractions. She was afebrile, reported no contact with persons having Ebola virus disease, and did not meet the existing case criteria for having Ebola infection. Subsequently, her blood was found to contain a high viral titer of Ebola virus. Three days after admission, she became febrile, and both she and her unborn fetus died of Ebola virus disease on the seventh day (Akerlund et al. 2015).

Those women with Ebola infection who are pregnant can have copious hemorrhaging. Compounding the hemorrhaging that accompanied delivery in women with Ebola infections was the occurrence of obstetric emergencies that could occur in any cohort of women during pregnancy or in labor. These included such hemorrhagic events as placental abruptions (or abruptio placentae; the premature separation of the placenta before or during labor), placenta previa, retained products of conception, rupture of the uterus, cervical lacerations, ruptured ectopic pregnancy, incomplete abortions, and others (Bolkan 2017). There are no statistics available detailing the prevalence of the problem of obstetrical hemorrhage with Ebola infection, but there are published reports. The chief of ActionAid in Liberia, Korto Williams, has stated, "*But we know that many women have had to give birth on their own because people were afraid they had Ebola and women have died because of a lack of care*" (Boseley 2014). Korto Williams also said that there are online videos showing women giving birth in the streets alone and without assistance because bystanders were scared they were infected with Ebola (Feminist Newswire 2014). Some women were unable to leave quarantine to give birth (Boseley 2014). The inexperienced staff members found themselves having to risk contamination by caring for a pregnant woman who they believed would die regardless of treatment, thus also wasting valuable supplies and medicines (Lang 2014). The conditions in many clinics and hospitals, overwhelmed with providing care and housing for patients with Ebola, resulted in providers discouraging pregnant women from receiving antenatal care or giving birth within their walls as a result of fear of bodily fluids (Feminist Newswire 2014). There were reports of some clinics and hospitals refusing to admit Ebola-infected pregnant women into their Ebola wards with other patients. At the West Point maternity ward in Monrovia, the staff turned away pregnant women with red eyes and weakness, symptoms of Ebola infection, who were seeking maternity care. A staff nurse there, Tarpeh, said "*We are afraid-o, sister. If I had money to sustain my family for six months, I would go home. I'm not lying to you*" (Moore 2014). Based upon interviews with nurses caring for Ebola patients in Sierra Leone, the magazine *New Yorker* reported that an unofficial protocol developed among health workers in that nation: deny infected pregnant women access to Ebola wards, or if entry was permitted, triage them last (Lang 2014). It was reported from Liberia that rumors were common that most health workers who became infected with Ebola virus contracted the disease as a result of caring for pregnant women—it is well-known that this is how the American physician, Dr. Rick Sacra, who was equipped with full personal protective equipment, became infected in August 2014 (Dantzer 2015).

Speaking on the condition of anonymity, a Sierra Leonean nurse told Joshua Lang, a medical student at the University of California San Francisco, "*The hospitals are neglecting [pregnant women]—they won't even allow them in.*" Gabriel Warren, who runs West African Medical Missions, a nonprofit organization in Sierra Leone, saw the effect of this exclusion at a variety of treatment centers. He told Joshua Lang, "*They aren't given preferential treatment,*" he said, "*They aren't even given beds. They get put in an area where they get no interventions. They are assumed to die*" (Lang 2014). The fear of infection compounded with previously existing poor quality maternity care, which was evidenced by the outstandingly high MMR even before the outbreak had begun. Any tendency towards neglect or abuse (Bohren et al. 2015) could then be amplified by providers' fears of infection and contamination, which were taken out on the women for whom they cared.

The equipment in many of the maternity wards and clinics was insufficient to properly protect the staff, especially in the earlier stages of the outbreak, similar to previous outbreaks in Central African countries. Lucy Barh, the president of the Liberia Midwifery Association, said that feelings about inadequate protection were common to the midwives she led, "*That's why some of the clinics are closed—fear, fear, fear. There's no protective equipment, and the lives of those health care workers are threatened*" (Moore 2014). In addition to the lack of personal protective equipment, nurses often faced stigma within their own families or communities due to their exposure or potential exposure to the deadly virus (Hewlett and Hewlett 2005; McMahon et al. 2016).

In Monrovia, Fatuma Fofana was a 34-year-old mother of five who was pregnant with her sixth child when she developed severe abdominal pain. Her brother Sheriff took her to a nearby clinic, but they refused to see her—the staff was terrified that she might have Ebola infection. Her brother took her to two other clinics, where she was also refused treatment. Finally, a fourth clinic let them in, but by that time her baby had died—and she did not have Ebola. The staff member did not have the supplies to further treat her and the retained dead infant, so Sheriff called numerous doctors, prepared to offer them a large sum of money ($400 USD). But no doctor answered his calls, and his sister died, adding to the maternal mortality of the epidemic (Moore 2014). Stories abound of uninfected pregnant women who died as a result of pregnancy complications or miscarriages because they could not get health care. Fatuma's story, similar to Comfort Fayiah's above, also demonstrates an additional burden on poor women; even though her brother was prepared to offer a very large sum of money for her care, it was still impossible to get appropriate medical attention. For many other women and their families, such sums of money would be absolutely out of their reach. Under more normal circumstances (i.e., without the added stressors of a highly fatal epidemic), a family might fundraise from relatives and/or neighbors and friends to acquire enough money for treatment. However, with so many people infected, and rapidly deteriorating infrastructure making daily life more difficult than normal, this network might, logically, rapidly start to fray and become less reliable. It must be considered that women coming from wealthier families would not only have more cultural capital to exert when trying to access care during pregnancy or while giving birth, but would have more physical capital to put on the line in the search for care, as well. Women and families with more funds would be able to pay higher costs that might be charged for scarce medical supplies or that might be given as incentives for transportation or health care services. This differential access to capital would further enlarge the already present gap between those women living at a subsistence level and those who might be considered more middle or upper middle class in these countries. Being able to put resources on the line in the effort to attain care could possibly have helped to mitigate the stigma associated with pregnant women and treating them. However, the story of Fatuma suggests that pregnant women were so highly feared and stigmatized that even large sums of money could not induce health care workers to overcome this stigma.

In some cases of pregnant women who arrived at a clinic or hospital with an urgent pregnancy complication which included bleeding, a common pregnancy-related occurrence with or without Ebola infection, there was a delay in their initial evaluation and treatment. The staff were hesitant to expose themselves to potentially infectious bodily fluids, and there were delays while waiting for the results of testing for the Ebola virus. Dr. Benjamin Black, an obstetrician, has said that poor infrastructure and limited access to laboratory services meant that test results for suspected Ebola patients could take more than 24 h to arrive, during which time a woman and her fetus could die (Guilbert 2015). Poor infrastructure was one of the significant factors in these countries that caused them to be ripe for the spread of this epidemic. While many knew this weak infrastructure was already a barrier to significant improvement in a diverse array of health outcomes in these countries, the Ebola outbreak threw this into stark relief, drastically highlighting the shortcomings and inefficiencies of the preexisting weak health care sector. The existing social inequalities and health system weaknesses helped contribute to the forms of stigma affecting pregnant women, which we have discussed here.

2.7 Conclusions

Despite many of the representations, if we look deeper, pregnant women were not only victims of stigma, but perpetrators, as well. They avoided health facilities and health care workers due to their associations with the disease and its spread. Avoidance of facilities and distrust of health care workers thought to have almost certainly been exposed to Ebola led to the almost complete erasure of previous gains in maternal health outcomes. In a mutually reinforcing cycle, health care workers feared and shunned pregnant women and pregnant women avoided providers and facilities. These forms of stigmatization had fertile ground in which to prosper due to the preexisting health system weaknesses in Guinea, Liberia, and Sierra Leone, compounded by the rapid spread of the epidemic. Gender inequalities meant women had, at least during the first part of the epidemic, a greater risk of contracting Ebola and worse outcomes when they became infected. Pregnant women, both in 2013–2015 and moving forward in time, perhaps suffered the greatest burden through near 100% fatality when infected, but also through the reduction in obstetric services of any kind. It will doubtlessly take these countries several years to recover from the impact of the epidemic and pregnant women will continue to suffer the consequences via a decimated health care profession and weakened networks of care and services. In what was perhaps the most shocking example of the stigma related to pregnant women, health care workers, and EBV infection, one of *TIME Magazine's* "People of the Year" from 2014, an "Ebola Fighter," Salome Karwah, who was a nursing assistant and herself an Ebola survivor, died on February 21, 2017, due to complications following the Cesarean delivery of her third child (Baker 2017). Karwah used the lasting immunity from her previous Ebola infection to care for patients during the epidemic whom no one else would touch. The outbreak also claimed many of her immediate relatives, including her father who was also a physician (Baker 2017). Of Salome's death following her son's birth, Baker (2017) writes, *"Within hours of coming home, Karwah lapsed into convulsions. Her husband and her sister rushed her back to the hospital, but no one would touch her. Her foaming mouth and violent seizures panicked the staff. 'They said she was an Ebola survivor,' says her sister by telephone. 'They didn't want contact with her fluids. They all gave her distance. No one would give her an injection.'"* Salome's sister directly pointed to stigma associated with Ebola as a cause of Salome's death. An MSF health promoter, Ella Watson-Stryker, who worked with Salome during the outbreak said, "To survive Ebola and then die in the larger yet silent epidemic of health system failure… I have no words" (Baker 2017). News of Salome's death spread around the world due to her appearance on the cover of TIME, but she holds a place for the unknown number of pregnant women who died during, and after, the 2013–2015 epidemic due to the effects of stigma and fear.

References

Adhikari, B., Kaehler, N., Chapman, R. S., Raut, S., & Roche, P. (2014). Factors affecting perceived stigma in leprosy affected persons in Western Nepal. *PLOS Neglected Tropical Diseases, 8*(6), e2940–e2948. Retrieved January 16, 2018, from http://journals.plos.org/plosntds/article?id=10.1371/journal.pntd.0002940.

Akerlund, E., Prescott, J., & Tampellini, L. (2015). Shedding of Ebola virus in an asymptomatic pregnant woman. *New England Journal of Medicine, 372*(25), 2467–2469. Retrieved January 16, 2018, from http://www.nejm.org/doi/full/10.1056/NEJMc1503275.

Amo-Adjei, J. (2016). Individual, household, and community level factors associated with keeping tuberculosis status secret in Ghana. *BMC Public Health, 16*, 1196. Retrieved January 16, 2018, from https://bmcpublichealth.biomedcentral.com/articles/10.1186/s12889-016-3842-y.

Anoko, J. N. (2014). Communication with rebellious communities during an outbreak of Ebola virus disease in Guinea: An anthropological approach. *Ebola Response Anthropology Platform*. Retrieved September 18, 2017, from http://www.ebola-anthropology.net/case_studies/communication-with-rebellious-communities-during-an-outbreak-of-ebola-virus-disease-in-guinea-an-anthropological-approach/.

Avotri, J. Y., & Walters, V. (1999). "You just look at our work and see if you have any freedom on earth:" Ghanaian women's accounts of their work and their health. *Social Science & Medicine, 48*(9), 1123–1133. https://doi.org/10.1016/S0277-9536(98)00422-5.

Awford, J. (2015). Woman who became the first to survive Ebola in Sierra Leone but lost 21 relatives to the disease gives birth to a baby boy. *Daily Mail*. Retrieved January 16, 2018, from http://www.dailymail.co.uk/news/article-3214664/Sierra-Leone-s-Ebola-survivor-gives-birth- baby-boy.html.

Baggi, F. M., Taybi, A., Kurth, A., Van Herp, M., Di Caro, A., Wölfel, R., et al. (2014). Management of pregnant women infected with Ebola virus in a treatment centre in Guinea, June 2014. *Eurosurveillance, 19*(49), 20983. Retrieved November 18, 2017, from http://eurosurveillance.org/content/10.2807/1560-7917.ES2014.19.49.20983.

Bainson, K., & van den Borne, B. (1998). Dimensions and process of stigmatization in leprosy. *Leprosy Review, 69*, 341–350.

Baker, A. (2017). Liberian Ebola fighter, a TIME Person of the Year, dies in Childbirth. *Time*. Retrieved January 16, 2018, from http://time.com/4683873/ebola-fighter-time-person-of-the-year-salome-karwah/.

Barbato, L. (2014). Most Ebola victims are women, for one big reason. *Bustle*. Retrieved January 16, 2018, from https://www.bustle.com/articles/36718-most-ebola-victims-are-women-for-one-big-reason.

Barrett, R. (2005). Self-mortification and the stigma of leprosy in Northern India. *Medical Anthropology Quarterly, 19*(2), 216–230.

Barrett, R. & Brown, P. J. (2008). Stigma in the time of influenza: Social and institutional responses to emergencies. *The Journal of Infectious Diseases, 197*(S1), s34–s37.

BBC. (2014). Ebola outbreak: Guinea health team killed. *BBC News*. Retrieved January 16, 2018, from http://www.bbc.com/news/world-africa-29256443.

Bebell, L. M., Oduyebo, T., & Riley, L. E. (2017). Ebola virus disease and pregnancy: A review of the current knowledge of Ebola virus pathogenesis, maternal, and neonatal outcomes. *Birth Defects Research, 109*(5), 353–362.

Blair, R. A., Morse, B. S., & Tsai, L. L. (2017). Public health and public trust: Survey evidence from the Ebola virus disease epidemic in Liberia. *Social Science & Medicine, 172*, 89–97. https://doi.org/10.1015/j.socscimed.2016.11.016.

Bofu-Tawamba, N. (2014). African women face Ebola triple jeopardy. *Al Jazeera America*. Retrieved January 16, 2018, from http://america.aljazeera.com/opinions/2014/11/ebola-response-africanfemalehealthworkersculturecaregivingroles.html.

Bohren, M. A., Vogel, J. P., Hunter, E. C., Lutsiv, O., Makh, S. K., Souza, J. P., et al. (2015). The mistreatment of women during childbirth in health facilities globally: A mixed-methods systematic review. *PLoS Med, 12*(6), e1001847. https://doi.org/10.1371/journal.pmed.1001847.

Bolkan, H. A. (2017). How Ebola affected a clinical officer training program in Sierra Leona and the decline of surgical care. In S. M. Wren & A. L. Kushner (Eds.), *Operation Ebola: Surgical care during the West African outbreak* (pp. 49–59). Baltimore: Johns Hopkins University Press.

Boseley, S. (2014). One in seven pregnant women could die in Ebola-hit countries, say charities. *The Guardian*. Retrieved January 16, 2018, from http://www.theguardian.com/world/2014/nov/10/ebola-one-in-seven-pregnant-women-could-die.

Brown, H., Kelly, A. H., Saez, A. M., Fichet-Calvet, E., Ansumana, R., Bonwitt, J., et al. (2015). Extending the "social": Anthropological contributions to the study of viral haemorrhagic fevers. *PLOS Neglected Tropical Diseases, 9*(4), e0003651. https://doi.org/10.1371/journal.pntd.0003651. Retrieved January 16, 2018, from http://journals.plos.org/plosntds/article?id=10.1371/journal.pntd.0003651.

Buckley, T., & Gottlieb, A. (1988). *Blood magic: The anthropology of menstruation*. Berkeley: University of California Press.

Callaghan, H. (2007). Birth dirt. In M. Kirkham (Ed.), *Exploring the dirty side of women's health* (pp. 8–25). New York: Routledge.

Calain, P. & Poncin, M. (2015). Reaching out to Ebola victims: Coercion, persuasion or an appeal for self-sacrifice? Social Science & Medicine, 147,126–133.

Carroll, S. A., Towner, J. S., Sealy, T. K., McMullan, L. K., Khristova, M. L., Burt, F. J., et al. (2013). Molecular evolution of viruses of the family Filoviridae based on 97 whole-genome sequences. *Journal of Virology, 87*(2013), 2608–2616.

Castro, A., & Farmer, P. (2005). Understanding and addressing AIDS-related stigma: From anthropological theory to clinical practice in Haiti. *American Journal of Public Health, 95*(1), 53–59. Retrieved January 16, 2018, from https://www.ncbi.nlm.nih.gov/pmc/articles/PMC1449851/.

CDC. (2016a). *Guidance for screening and caring for pregnant women with Ebola virus disease for healthcare providers in U.S. hospitals*. Retrieved October 12, 2017, from https://www.cdc.gov/vhf/ebola/healthcare-us/hospitals/pregnant-women.html#R1.

CDC. (2016b). *2014 Ebola outbreak in West Africa—Case counts*. Retrieved October 21, 2017, from https://www.cdc.gov/vhf/ebola/outbreaks/2014-west-africa/case-counts.html.

Chicago Tribune. (2015). *More heartbreak for Ebola survivor after baby boy dies*. Retrieved November 5, 2017, from http://www.chicagotribune.com/news/nationworld/ct-ebola-survivor-baby-dies-20150904-story.html.

Chinouya, M., & Adeyanju, O. (2017). A disease called stigma: The experience of stigma among African men with TB diagnosis in London. *Public Health, 145*, 45–50. https://doi.org/10.1016/j.puhe.2016.12.017.

Chothia, F. (2014). Ebola drains already weak West African health systems. *BBC News*. Retrieved January 16, 2018, from http://www.bbc.com/news/world-africa-29324595.

CIA (Central Intelligence Agency). (2015). *The World Factbook*. Retrieved from https://www.cia.gov/library/publications/the-world-factbook/.

Clark, D. V., Kibuuka, H., Millard, M., Wakabi, S., Lukwago, L., Taylor, A., et al. (2015). Long-term sequelae after Ebola virus disease in Bundibugyo, Uganda: A retrospective cohort study. *The Lancet Infectious Diseases, 15*(8), 905–912. Retrieved January 16, 2018, from http://www.thelancet.com/journals/laninf/article/PIIS1473-3099(15)70152-0/abstract.

Coreil, J., Mayard, G., Simpson, K., Lauzardo, M., Zhu, Y., & Weiss, M. (2010). Structural forces and the production of TB-related stigma among Haitians in two contexts. *Social Science & Medicine, 71*(8), 1409–1417. Retrieved January 16, 2018, from https://www.ncbi.nlm.nih.gov/pmc/articles/PMC3430377/.

Coulter, C. (2009). *Bush wives and girl soldiers: Women's lives through war and peace in Sierra Leone*. Ithaca: Cornell University Press.

Courtwright, A., & Turner, A. (2010). Tuberculosis and stigmatization: Pathways and interventions. *Public Health Reports, Social Determinant of Health, 4*(Suppl), 34–42. Retrieved January 16, 2018, from https://www.ncbi.nlm.nih.gov/pmc/articles/PMC2882973/.

Daniel, T. M. (2006). The history of tuberculosis. *Respiratory Medicine, 100*(11), 1862–1870.

Dantzer, E. (2015). *Reproductive health: The biggest casualty of the Ebola epidemic?* Retrieved January 16, 2018, from http://thepump.jsi.com/reproductive-health-the-biggest-casualty-of-the-ebola-epidemic/.

Davtyan, M., Brown, B., & Folayan, M. O. (2014). Addressing Ebola-related stigma: Lessons learned from HIV/AIDS. Global Health Action, 7(1), 258–260.

Denis-Ramirez, E., Sorensen, K. H., & Skovdal, M. (2017). In the midst of a 'perfect storm': Unpacking the causes and consequences of Ebola-related stigma for children orphaned by Ebola in Sierra Leone. *Children and Youth Services Review, 73*, 445–453. https://doi.org/10.1016/j.childyouth.2016.11.025.

Des Jarlais, D. C., Galea, S., Tracy, M., Tross, S., & Vlahov, D. (2006). Stigmatization of newly emerging infectious diseases: AIDS and SARS. *American Journal of Public Health, 96*(3), 561–567. Retrieved January 13, 2018, from http://www.ncbi.nlm.nih.gov/pmc/articles/PMC1470501/.

Dodor, E. A., Neal, K., & Kelly, S. (2008). An exploration of the causes of tuberculosis stigma in an urban district in Ghana. *International Journal of Lung Disease, 12*(9), 1048–1054.

Dörnemann, J., Burzio, C., Ronsse, A., Sprecher, A., De Clerck, H., Van Herp, M., et al. (2017). First newborn baby to receive experimental therapies survives Ebola virus disease. *Journal of Infectious Diseases, 215*(2), 171–174. Retrieved January 13, 2018, from https://www.ncbi.nlm.nih.gov/pmc/articles/PMC5583641/.

Doucleff, M. (2014). Dangerous deliveries: Ebola leaves moms and babies without care. *National Public Radio*. Retrieved January 13, 2018, from https://www.npr.org/sections/goatsandsoda/2014/11/18/364179795/dangerous-deliveries-ebola-devastates-womens-health-in-liberia.

Duerkson, M. (2014). Arcade Africa. *Ebola's history. 3: pre-1976*. Retrieved January 10, 2018, from https://markduerksen.com/2014/08/04/ebolas-history-3-pre-1976/#_ftn2.

Dynes, M. M., Miller, L., Sam, T., Vandi, M. A., & Tomczyk, B. (2015). Perceptions of the risk for Ebola and health facility use among health workers and pregnant and lactating women—Kenema District, Sierra Leone, September 2014. *MMWR Morbidity and Mortality Weekly Report, 63*(51), 1226–1227. Retrieved January 12, 2018, from http://www.cdc.gov/mmwr/preview/mmwrhtml/mm6351a3.htm.

Farmer, P., & Koroma, A. (2015, April 8). *PIH, Sierra Leone address needs of pregnant women amid Ebola*. Retrieved September 2, 2015, from http://www.pih.org/blog/pih-sierra-leone-address-needs-of-pregnant-women-amid-ebola.

Feminist Newswire. (2014). *Maternal death rate in West Africa expected to rise due to Ebola*. Retrieved January 13, 2018, from https://feminist.org/blog/index.php/2014/11/13/maternal-death-rate-in-west-africa-expected-to-rise-due-to-ebola/.

Fitzgerald, F., Naveed, A., Wing, K., Gbessay, M., Ross, J. C., Checchi, F., et al. (2016). Ebola virus disease in children, Sierra Leone, 2014–2015. *Emerging Infectious Diseases, 22*, 1769–1777. Retrieved January 9, 2018, from https://www.ncbi.nlm.nih.gov/pmc/articles/PMC5038433/.

Frankel, T. C. (2014, December 22). Imagine a disease wiping out 64,000 U.S. doctors. Now, you understand Ebola in Sierra Leone. The Washington Post. Retrieved from http://www.washingtonpost.com/news/storyline/wp/2014/12/22/imagine-a-disease-wiping-out-63000-u-s-doctors-now-you-understand-ebola-in-sierra-leone/.

Garba, I., Dattijo, L. M., & Habib, A. G. (2017). Ebola virus disease and pregnancy outcome: A review of the literature. *Tropical Journal of Obstetrics and Gynaecology, 34*(1), 6–10. Retrieved November 15, 2017, from http://www.tjogonline.com/article.asp?issn=0189-5117;year=2017;volume=34;issue=1;spage=6;epage=10;aulast=Garba.

Garbin, C., Garbin, A., Carloni, M., Rovida, T., & Martins, R. (2015). The stigma and prejudice of leprosy: Influence on the human condition. *Revista de Sociedade Brasileira de Medicina Tropical, 48*(2), 194–201. https://doi.org/10.1590/0037-8682-0004-2015.

Gholipour, B. (2014). *How Ebola got its name*. Retrieved October 12, 2017, from https://www.livescience.com/48234-how-ebola-got-its-name.html.

Goffman, E. (1963). *Stigma: Notes on the management of a spoiled identity*. New York: Simon & Schuster.

Guilbert, K. (2015). Ebola health workers face life or death decision on pregnant women—experts. *Reuters*. Retrieved January 12, 2018, from http://www.reuters.com/article/2015/01/14/us-ebola-africa-pregnancy-idUSKBN0KN1T620150114.

Hayden, E. C. (2014). Ebola's lost ward: A hospital in Sierra Leone has struggled to continue its research amid the worst Ebola outbreak in history. *Nature, 513*, 474–477. Retrieved January 13, 2018, from http://ebola-honors210g.wikispaces.umb.edu/file/view/Ebola%27s%20lost%20ward%202014.pdf/560012601/Ebola%27s%20lost%20ward%20 2014.pdf.

Heijnders, M. (2004). The dynamics of stigma in leprosy. *International Journal of Leprosy and Other Mycobacterial Diseases, 2004*(72), 437–447.

Hewlett, B. S., & Amola, R. P. (2003). Cultural contexts of Ebola in northern Uganda. *Emerging Infectious Diseases, 9*(10), 1242–1248. Retrieved January 10, 2018, from https://wwwnc.cdc.gov/eid/article/9/10/02-0493_article.

Hewlett, B. L., & Hewlett, B. S. (2005). Providing care and facing death: Nursing during Ebola outbreaks in Central Africa. *Journal of Transcultural Nursing, 16*(4), 289–297. https://doi.org/10.1177/1043659605278935.

Hogan, C. (2014). Ebola striking women more frequently than men. *The Washington Post*. Retrieved January 16, 2018, from https://www.washingtonpost.com/national/health-science/2014/08/14/3e08d0c8-2312-11e4-8593-da634b334390_story.html?utm_term=.a0940f34a8ff.

Howard, C. R. (2005). *Viral haemorrhagic fevers*. Amsterdam: Elsevier.

Jamieson, D. J., Uyeki, T. M., Callaghan, W. M., Meaney-Delman, D., & Rasmussen, S. A. (2014). What obstetrician-gynecologists should know about Ebola: A perspective from the Centers for Disease Control and Prevention. *Obstetrics and Gynecology, 124*(5), 1005–1010. Retrieved January 15, 2018, from http://ldh.louisiana.gov/assets/oph/Center-PHCH/Center-CH/infectious-epi/EpiManual/ObstetricsEbola.pdf.

Jopling, W. (1991). Leprosy stigma. *Leprosy Review, 62*, 1–12.

Karafillakis, E., Nuriddin, A., Larson, H. J., Whitworth, J., Lees, S., et al. (2016). 'Once there is life, there is hope' Ebola survivors' experiences, behaviours and attitudes in Sierra Leone, 2015. *BMJ Global Health, 1*(3), e000108. https://doi.org/10.1136/bmjgh-2016-000108. Retrieved January 13, 2018, from http://gh.bmj.com/content/1/3/e000108.

Kilmarx, P. H., Clarke, K. R., Dietz, P. M., Hamel, M. J., Husain, F., McFadden, J. D., et al. (2014). Ebola virus disease in health care workers—Sierra Leone, 2014. *Morbidity and Mortality Weekly Report (MMWR), 63*(49), 1168–1171.

Lang, J. (2014). Ebola in the maternity ward. *The New Yorker*. Retrieved January 9, 2018 from https://www.newyorker.com/tech/elements/ebola-maternity-ward.

Leroy, E. M., Telfer, P., Kumulungui, B., Yaba, P., Rouquet, P., Roques, P., et al. (2004). A serological survey of Ebola virus infection in Central African nonhuman primates. *Journal of Infectious Diseases, 190*(2004), 1895–1899. https://doi.org/10.1086/425421.

Life for African Mothers. (2014). Most victims of Ebola virus are women. *Life For African Mothers*. Retrieved January 10, 2018, from http://lifeforafricanmothers.org/most-victims-of-ebola-virus-are-women/.

Link, B. C., & Phelan, J. C. (2001). Conceptualizing stigma. *Annual Review of Sociology, 27*, 363–385.

Luginaah, I. N., Kangmennaang, J., Fallah, M., Dahn, B., Kateh, F., & Nyenswah, T. (2016). Timing and utilization of antenatal care services in Liberia: Understanding the pre-Ebola epidemic context. *Social Science & Medicine, 160*, 75–86. https://doi.org/10.1016/j.socscimed.2016.05.019.

Luka, E. E. (2010). Understanding the stigma of leprosy. *South Sudan Medical Journal*, 3(3). Retrieved October 1, 2017, from http://www.southsudanmedicaljournal.com/archive/august-2010/understanding-the-stigma-of-leprosy.html.

Maman, S., Abler, L., Parker, L., Lane, T., Chirowodza, A., Ntogwisangu, J., et al. (2009). A comparison of HIV stigma and discrimination in five international sites: The influence of care and treatment resources in high prevalence settings. *Social Science & Medicine, 68*, 2271–2278. https://doi.org/10.1016/j.socscimed.2009.04.002. Retrieved January 13, 2018, from https://www.ncbi.nlm.nih.gov/pmc/articles/PMC2696587/.

Maron, D. F. (2015). Thousands of Ebola survivors face persistent joint pain and other problems. *Scientific American*. Retrieved January 8, 2018, from https://www.scientificamerican.com/article/thousands-of-ebola-survivors-face-persistent-joint-pain-and-other-problems/.

Martel, F. (2014). Two-week-old baby girl survives Ebola in Sierra Leone. *Breitbart*. Retrieved September 18, 2017, from http://www.breitbart.com/national-security/2014/10/31/two-week-old-baby-girl-survives-ebola-in-sierra-leone/.

McMahon, S., Ho, L., Brown, H., Miller, L., Ansumana, R., & Kennedy, C. (2016). Healthcare providers on the frontlines: A qualitative investigation of the social and emotional impact of delivering health services during Sierra Leone's Ebola epidemic. *Health Policy and Planning, 31*, 1232–1239. https://doi.org/10.1093/heapol/czw055. Retrieved January 10, 2018, from https://academic.oup.com/heapol/article/31/9/1232/2452988.

Médecins Sans Frontières. (2015). *Nubia. First newborn to survive Ebola*. Retrieved January 15, 2018, from https://msf.exposure.co/nubia.

Menéndez, C., Lucas, A., Munguambe, K., & Langer, A. (2015). Ebola crisis: The unequal impact on women and children's health. *The Lancet Global Health, 3*(3), e130. Retrieved January 15, 2018, from http://www.thelancet.com/journals/langlo/article/PIIS2214-109X(15)70009-4/fulltext.

Moore, J. (2014). How Ebola can kill you—Even when you don't have Ebola. *BuzzFeed News*. Retrieved January 15, 2018, from https://www.buzzfeed.com/jinamoore/how-ebola-can-kill-you-even-when-you-dont-have-ebola?utm_term=.gmAO8kgqX1#.cxvxe9Xq2d.

Mupapa, K., Mukundu, W., Bwaka, M. A., Kipasa, M., De Roo, A., Kuvula, K., et al. (1999). Ebola hemorrhagic fever and pregnancy. *Journal of Infectious Diseases, 179*(Suppl 1), S11–S12. Retrieved September 18, 2017, from https://academic.oup.com/jid/article/179/Supplement_1/S11/879962.

Obilade, T. T. (2015). Ebola virus disease stigmatization; the role of societal attributes. *International Archives of Medicine, 8*(14), 1–19. Retrieved January 12, 2018, from http://imed.pub/ojs/index.php/iam/article/view/1007.

Parker, R., & Aggleton, P. (2003). HIV and AIDS-related stigma and discrimination: A conceptual framework and implications for action. *Social Science & Medicine, 57*, 13–24.

Piot, P. (2012). *No time to lose: A life in pursuit of deadly viruses*. New York: W. W. Norton.

Rai, S. (2007). Listening to *dais* speak about their work in Gujarat, India. In M. Kirkham (Ed.), *Exploring the dirty side of women's health*. New York: Routledge.

Rao, P. (2010). Study on differences and similarities in the concept and origin of leprosy stigma in relation to other health-related stigma. *Indian Journal of Leprosy, 82*, 117–121.

Rojek, A., Horby, P., & Dunning, J. (2017). Insights from clinical research completed during the West Africa Ebola virus disease epidemic. *Lancet Infectious Diseases, 17*(9), e280–e292. Retrieved November 10, 2017, from http://www.thelancet.com/journals/laninf/article/PIIS1473-3099(17)30234-7/fulltext.

Saul, H. (2015). Ebola crisis: This is why '75%' of victims are women. *The Independent*. Retrieved January 13, 2018, from http://www.independent.co.uk/news/world/africa/ebola-virus-outbreak-this-is-why-75-of-victims-are-women-9681442.html.

Schwartz, D. A. (2015). Pathology of maternal death—The importance of accurate autopsy diagnosis for epidemiologic surveillance and prevention of maternal mortality in developing countries. In D. A. Schwartz (Ed.), *Maternal mortality: Risk factors, anthropological perspectives, prevalence in developing countries and preventative strategies for pregnancy-related death* (pp. 215–253). New York: Nova Scientific Publishing.

Segers, N. (2014). The underestimated gender(ed) politics of Ebola. Why a preventive and remedial focus on women is critical to save lives. *The New Federalist*. Retrieved January 15, 2018, from http://www.thenewfederalist.eu/7035.

Sepkowitz, K., Haglage, A. (2014). The only thing more terrifying than Ebola is being pregnant with Ebola. *The Daily Beast*. Retrieved January 14, 2018, from https://www.thedailybeast.com/the-only-thing-more-terrifying-than-ebola-is-being-pregnant-with-ebola.

Sieff, K. (2015). Could a pregnant woman change the way we think about Ebola? *The Washington Post*. Retrieved January 15, 2018, from https://www.washingtonpost.com/world/could-a-pregnant-woman-change-the-way-we-think-about-ebola/2015/01/04/a5ed5a7f-73a6-427b-b515-6f8f7e77801e_story.html?utm_term=.c40e7273d91a.

Statista. (2015). *Ebola cases and deaths among health care workers due to the outbreaks in African countries as of November 4, 2015*. Retrieved January 14, 2018, from https://www.statista.com/statistics/325347/west-africa-ebola-cases-and-deaths-among-health-care-workers/.

Sterner, J., & David, N. (1991). Gender and caste in the Mandara Highlands: Northeastern Nigeria and Northern Cameroon. *Ethnology, 30*(4), 355–369.

Strassmann, B. I. (1992). The function of menstrual taboos among the Dogon: Defense against cuckoldry? *Human Nature, 3*(2), 89–131.

Strong, A. E. (2017). Working in scarcity: Effects on social interactions and biomedical care in a Tanzanian hospital. *Social Science & Medicine, 187*, 217–224.

Strong, A. E. (2018). Causes and effects of occupational risk for healthcare workers on the maternity ward of a Tanzanian hospital. *Human Organization, 77*(3), 273–286.

Takahashi, S., Metcalf, C. J., Ferrari, M. J., Moss, W. J., Truelove, S. A., Tatem, A. J., et al. (2015). Reduced vaccination and the risk of measles and other childhood infections post-Ebola. *Science, 347*(6227), 1240–1242. Retrieved January 13, 2018, from https://www.ncbi.nlm.nih.gov/pmc/articles/PMC4691345/.

Thomas, K. (2014). Ebola's gender bias: The triple threat facing women. *Ebola Deeply*. Retrieved January 15, 2018, from http://archive.eboladeeply.org/articles/2014/12/6897/ebolas-gender-bias-%20triple-threat-facing-women-affected-countries/.

Thomas, K. (2017). *'Spoiling the belly': The dangers of backstreet abortions in Liberia*. Retrieved November 22, 2017, from https://www.newsdeeply.com/womenandgirls/articles/2017/04/21/spoiling-the-belly-the-dangers-of-backstreet-abortions-in-liberia.

United Nations Population Fund (UNFPA). (2014a). *Liberia's Ebola outbreak leaves pregnant women stranded*. Retrieved January 15, 2018, from http://www.unfpa.org/news/liberias-ebola-outbreak-leaves-pregnant-women-stranded.

United Nations Population Fund (UNFPA). (2014b). *Fear of health workers fuels Ebola crisis in Guinea*. Retrieved January 15, 2018, from http://www.unfpa.org/news/fear-health-workers-fuels-ebola-crisis-guinea.

United Nations Population Fund (UNFPA). (2014c). *Ebola wiping out gains in safe motherhood*. Retrieved January 13, 2018, from http://esaro.unfpa.org/news/ebola-wiping-out-gains-safe-motherhood.

United Nations Population Fund (UNFPA). (2014d). *Pregnant in the shadow of Ebola: Deteriorating health systems endanger women*. Retrieved January 10, 2018, from http://www.unfpa.org/news/pregnant-shadow-ebola-deteriorating-health-systems-endanger-women.

United Nations Population Fund (UNFPA). (2015). *$56 Million needed to provide services in Ebola-affected countries to avoid maternal death toll of civil wars years*. Retrieved January 11, 2018, from http://www.unfpa.org/press/56-million-needed-provide-services-ebola-affected-countries-avoid-maternal-death-toll-civil.

Van Bortel, T., Basnayake, A., Wurie, F., Jambai, M., Koroma, A. S., Muana, A. T., et al. (2016). Psychosocial effects of an Ebola outbreak at individual, community and international levels. *Bulletin of the World Health Organization, 94*(3), 210–214. Retrieved January 13, 2018, from http://www.who.int/bulletin/volumes/94/3/15-158543/en/.

Van Hollen, C. (2003). *Birth on the threshold: Childbirth and modernity in South India*. Berkeley: University of California Press.

WHO Ebola Response Team, Agua-Agum, J., Ariyarajah, A., Blake, I. M., Cori, A., Donnelly, C. A., et al. (2015). Ebola virus disease among children in West Africa. *New England Journal of Medicine, 372*, 1274–1277. Retrieved November 19, 2017, from http://www.nejm.org/doi/10.1056/NEJMc1415318.

World Health Organization (WHO). (2007). *Addressing sex and gender in epidemic-prone infectious diseases*. Retrieved January 15, 2018, from http://www.who.int/csr/resources/publications/sexandgenderinfectiousdiseases/en/.

World Health Organization (WHO). (2015). *Ebola situation report—4 March 2015*. Retrieved from http://apps.who.int/ebola/current-situation/ebola-situation-report-4-march-2015.

World Health Organization (WHO). (2017). *Ebola virus disease*. Retrieved October 15, 2017, from http://www.who.int/mediacentre/factsheets/fs103/en/.

3 The Challenges of Pregnancy and Childbirth Among Women Who Were Not Infected with Ebola Virus During the 2013–2015 West African Epidemic

Regan H. Marsh, Katherine E. Kralievits, Gretchen Williams, Mohamed G. Sheku, Kerry L. Dierberg, Kathryn Barron, and Paul E. Farmer

3.1 Introduction: The Story of Aminata

Aminata's story is the story of thousands of women who lost their lives during the West African Ebola epidemic that began in 2013. It brings into relief the impact of the Ebola epidemic not on just one woman and her family, but on millions of women and theirs.

In late December 2014, Aminata became one of the many counted (and uncounted) maternal deaths that occurred during the epidemic. She was a 37-year-old mother of two and pregnant for a third time. She had not received antenatal care during this pregnancy, as health services were limited during the peak of the outbreak in Sierra Leone. She went into labor at full-term at home, but recognized, as did

R. H. Marsh · P. E. Farmer
Partners In Health, Boston, MA, USA

Division of Global Health Equity, Brigham and Women's Hospital, Boston, MA, USA

Department of Global Health and Social Medicine, Harvard Medical School, Boston, MA, USA

K. E. Kralievits (✉) · G. Williams
Partners In Health, Boston, MA, USA

Department of Global Health and Social Medicine, Harvard Medical School, Boston, MA, USA
e-mail: katherine_kralievits@hms.harvard.edu

M. G. Sheku
Koidu Government Hospital, Koidu, Sierra Leone

K. L. Dierberg
Partners In Health, Boston, MA, USA

Division of Infectious Diseases and Immunology, New York University, School of Medicine, New York, NY, USA

K. Barron
Partners In Health, Boston, MA, USA

D. A. Schwartz et al. (eds.), *Pregnant in the Time of Ebola*, Global Maternal and Child Health,
https://doi.org/10.1007/978-3-319-97637-2_3

those attending her, that there was a problem a day later. Her delivery was not progressing in the same way as the previous two pregnancies had.

Her family took her to the local district hospital for care only to find it closed due to active Ebola virus transmission occurring within it. It had been shuttered a week earlier on the advice of the U.S. Centers for Disease Control and Prevention (CDC) and by order of the Sierra Leone Ministry of Health and Sanitation. A local nurse saw her privately. Following this visit, Aminata's family raised money for her to take an hour-long taxi ride to the district's other public hospital, which she reached in the early afternoon with a note from the nurse in hand, documenting a likely obstructed labor.

On arrival, she was met only by a volunteer nurse and cleaner. Three of the hospital's four midwives on staff had been reassigned to work in Ebola treatment units (ETUs), and the fourth was not available in the hospital that afternoon. The hospital's two doctors were also actively engaged in the Ebola response and unavailable. Aminata was brought into the delivery room and laid on the delivery bed; there were no mattresses on the labor ward beds, as they had been burned as an infection-control precaution.

About 10 minutes later, before a midwife could be summoned, Aminata had a seizure, likely due to eclampsia (diagnosis of the syndrome requires evidence of elevated blood pressure and proteinuria, which were not obtained). Aside from Aminata's brother, who had traveled with her, the maternity ward was empty. There was no skilled provider available; no intravenous lines to place; no diazepam or magnesium available to stop her seizure; no oxygen or suction (and in fact, no power at the hospital); no resuscitation equipment; and no surgeon to try to save her or the baby. Still unconscious, Aminata then vomited, aspirated, and died—unattended—of one of the most common causes of maternal mortality.

Aminata did not have Ebola.

With access to routine care, early signs of eclampsia (hypertension and protein in the urine) might have been detected; even if they had not, standard maternity care, basic intravenous medications, oxygen, and access to safe, urgent treatment and delivery might have saved her life. And while eclampsia was the immediate and proximate cause of death—the cause noted in the records and registers—the true root etiology of Aminata's death was that of a dysfunctional and under-resourced health system, stripped even further of resources by the Ebola epidemic.

3.2 Partners In Health

In September 2014, as Ebola virus disease spread rapidly across West Africa, Partners In Health—a nonprofit dedicated to providing quality healthcare to those living in poverty—made a commitment to respond and began working in Liberia and Sierra Leone (Partners In Health 2016a). Founded in 1987, PIH was initially established as a community health project based in a rural squatter settlement in central Haiti. Today, Partners In Health (PIH) works in ten countries: Haiti, Rwanda, Malawi, Lesotho, Mexico, Russia, Peru, the Navajo Nation, Liberia, and Sierra Leone. At the heart of PIH's work is its mission to provide a preferential option for the poor in healthcare by forging long-term relationships with Ministries of Health and local sister organizations that work in solidarity with poor communities. This work is built on a foundation of integrated service delivery, training, and research.

In Haiti, PIH pioneered the use of *accompagnateurs*—community health workers that offer direct medical, social, and economic support to patients—to treat the poorest and hardest to reach patients; this model continues to guide PIH's work at each of its sites. Critically, PIH works within the public sector, working with governments and other local organizations to support and strengthen health systems, making long-term commitments to the communities served. This model has proven that it is possible to prevent and treat complex and difficult illnesses, including cancer, AIDS, and multidrug-resistant tuberculosis, among impoverished populations. It has also sought to prevent unplanned or undesired pregnancies and to deliver facility-based obstetrics and surgical care. Now with 18,000 staff members (most of them community health workers), PIH programs reflect local needs and seek to strengthen health systems in order to increase their ability to respond effectively to unforeseen problems.

In 2010, PIH was among the organizations that responded to the earthquake in Haiti, which devastated much of the country's medical and other infrastructure. Although not explicitly an emergency-response organization, PIH felt a moral imperative to provide relief in the immediate aftermath, which ended or indelibly impacted the lives of many long-time PIH colleagues and their families, and soon became a leader in the fight against cholera, which began explosively seven months after the quake. PIH operated 11 cholera treatment centers, recruited and trained community health workers, and later provided cholera vaccinations in partnership with the Ministry of Public Health and Population; this work continues today as cholera remains a threat in the wake of Hurricane Matthew and other storms. After the earthquake, the government of Haiti asked PIH to aid in post-disaster reconstruction by building an academic teaching hospital outside of Port-au-Prince. In 2013, PIH opened the 300-bed teaching hospital, Hôpital Universitaire de Mirebalais, in Haiti's Central Plateau. In partnership with the Government of Rwanda, PIH has also undertaken significant projects in that country, including the opening of Butaro Hospital in 2011, the Butaro Cancer Center of Excellence (the first of its kind in rural Africa) in 2013, and most recently and at the same site, the University of Global Health Equity in 2015. Amid this work in Haiti, Rwanda, and elsewhere—and with equity in mind—PIH and its partners undertook similar, long-term endeavors in one of the great clinical deserts of the twenty-first century, Upper West Africa.

3.3 The Ebola Epidemic and the Partners In Health Response

Eight months after Ebola virus disease was retrospectively alleged to have claimed the life of a toddler in eastern Guinea—"Patient Zero" of the 2013–2015 outbreak—the World Health Organization (WHO) declared the Ebola outbreak a "public health emergency of international concern" (World Health Organization 2014). The already beleaguered—and, in many cases, nearly nonexistent—health systems in these countries were no match for the disease, which targeted caregivers (healthcare professionals, family members, and traditional healers) and required diagnostic, preventive, and supportive therapeutic capabilities not readily available in these settings.

Earlier, in March 2014, the WHO announced that the previously unidentified febrile illness that had been causing deaths in Guinea—and across its borders—was in fact Ebola virus disease (EVD). Within a week and a half of this diagnosis, Médecins Sans Frontières (MSF) opened and staffed Ebola isolation wards in three locations in Guinea, where it had already been working. The organization expressed serious concerns about the continued escalation of the epidemic, which was spreading undiagnosed in multiple regions, including within Sierra Leone and Liberia. On June 21st, MSF announced the epidemic to be "out of control," which stood in contrast to earlier assessments by other responders and global institutions that the epidemic was waning (Sack et al. 2014).

By August 6th, when the WHO did declare the epidemic as a public-health emergency, the disease had already severely disrupted healthcare delivery, caused widespread panic across West Africa, and taken 936 lives, although this likely underrepresented the true burden of disease due to underreporting (Garrett 2015; WHO 2014). In September, the region experienced the epidemic's peak incidence, with significant variations between the three countries (WHO Ebola Response Team 2015). That same month, as the United Nations Security Council and the CDC issued dire warnings regarding the scope and potential trajectory of the disease, members of the United States and British militaries were deployed to West Africa (Garrett 2015). The UN Mission for Ebola Emergency Response was also established, with the aim of coordinating efforts across the region (United Nations 2016).

Among the quickest to respond to the outbreak—beyond the West African healthcare providers that were on the frontlines from the beginning—were large nongovernmental organizations (NGOs), which primarily relied on the deployment of expatriate clinicians and other responders. Such organizations included MSF, the International Federation of the Red Cross and Red Crescent Societies, International Medical Corps, and GOAL. These organizations faced some level of distrust in the local communities, which were skeptical of the motivations of the governments and outside responders and as an epidemic of fear gripped the region and others largely untouched by the Ebola outbreak itself (Sack et al. 2014). Those working to stem the epidemic faced immense challenges amid weak existing health systems, poor coordination among responding bodies, inadequate and slow responses from the global community, and insufficient resources to address the spread of the disease and save lives. MSF reports caring for 35% of all confirmed Ebola cases in the region; their six-pronged approach included caring for the sick, providing training for and conducting safe burials, raising awareness about how Ebola is spread, disease surveillance, contact tracing, and providing some healthcare services separate from efforts to curb Ebola (Médecins Sans Frontières 2017).

In September 2014, as the disease incidence mounted in Sierra Leone and plateaued in Liberia, PIH began working with these countries' Ministries of Health and local partners (most notably, Wellbody Alliance in Sierra Leone and Last Mile Health in Liberia). By then, most recognized the profound impact of weak health systems on disease surveillance and the quality of clinical services for those with Ebola. Patients needing routine care were, however, often forgotten amid the crisis.

For those working in Haiti, Rwanda, and many other settings, the framing of this epidemic as an "acute-on-chronic" process that overwhelmed already weak health systems was a familiar dynamic. PIH's early response focused on stemming the acute epidemic by treating cases, employing active case finding, and operating Ebola treatment units in partnership with district and national systems, while also delivering healthcare services for illnesses and conditions other than Ebola. PIH deployed nearly 200 expatriate clinicians, logisticians, and support staff to operate these units, while also recruiting and training 2,000 local people as community health workers, drivers, data collectors, clinical aides, and sundry other jobs (Partners In Health 2016a). Working closely with the Ministry of Health and Social Welfare in Liberia, PIH targeted its response in the southeast, including Maryland County, a 20-hour drive from the capital city of Monrovia. There, PIH supported care in two Ebola treatment units (ETUs) and three community care centers—clinics designated for suspected cases of Ebola with confirmed cases transferred to ETUs, when possible. Even before the onset of the epidemic, the region's foremost problem was already the collapse of primary care and delivery of care to the critically ill and injured, impacting the ability to respond to this crisis.

At the request of the Ministry of Health and Sanitation in Sierra Leone, PIH focused its efforts in three districts: Freetown (the capital city), Port Loko (about two hours outside the capital), and Kono (in the eastern part of the country). In Freetown, we worked at the Princess Christian Maternity Hospital (PCMH), the country's only referral hospital for pregnant women facing serious complications. In collaboration with PCMH, our team established formal Ebola screening for all patients and visitors entering the hospital, and an Ebola triage and isolation unit, where women with symptoms of Ebola could receive diagnostic testing and care. From December 2014 to December 2015, over 42,000 women were screened for symptoms of Ebola; 610 met case definitions for isolation and 29 tested positive for Ebola (Garde et al. 2016).

In Port Loko, PIH helped to operate and staff the Port Loko Government Hospital (PLGH), Maforki Ebola Treatment Unit, and seven other community care centers (CCCs). In many cases, responders relied on temporary, makeshift isolation centers to manage the overwhelming caseload. PLGH itself was filled with Ebola patients (and corpses) upon our arrival. Given the overwhelming need and to support government efforts, PIH partnered with the District Health Management Team, which was utilizing an abandoned vocational school as an ETU at Maforki; construction of a new purpose-built unit would have taken too long to complete. Upon initiation of our clinical and logistical support, the Maforki ETU was the only Ebola care unit operational in the district—an Ebola hotspot which would go on to manage 1,884 admitted patients in the ETU and have 978 cases in the district—and was operating at 100% capacity with extremely limited government resources. In Kono District, PIH supported operations and clinical care at Koidu Government Hospital (KGH), Wellbody Clinic, and four CCCs. At both district hospitals (PLGH and KGH), our teams worked to establish screening and triage procedures, as well as to provide direct clinical services to care for the thousands of patients needing usual services and to restore community trust in the public hospitals. By January 2015, PIH supported a total of 21 health facilities across Liberia and Sierra Leone (Partners In Health 2016a). Community health workers, many of whom were Ebola survivors, were instrumental to these efforts, as they helped to identify cases and connect the sick to healthcare while dispelling stigma surrounding Ebola survivorship. Responders to the epidemic, including local healthcare workers, faced a paucity of the staff, stuff, space, and systems essential to strong health systems. Compounded by the Ebola emergency, which demanded and diverted the scant resources available, pregnant women had limited access to essential care, causing an escalation of otherwise preventable maternal deaths.

In January 2016, the WHO declared that all known chains of transmission of Ebola in West Africa had been halted, although sporadic cases continued in Guinea and Liberia until April 2016. Finally, the WHO announced the end of transmission in those locations in June 2016 (U.S. Centers for Disease Control and Prevention 2016; WHO 2016a). There were more cases and deaths during this epidemic than in all previous Ebola virus outbreaks combined (WHO 2016b). While likely an underestimate due to weaknesses in surveillance and reporting—and the presence of minimally symptomatic cases of infection—there had been 28,616 recorded cases of and 11,310 deaths from Ebola reported across Guinea, Sierra Leone, and Liberia. There were a further 36 locally acquired or imported cases and 15 deaths in seven countries: Nigeria, Senegal, Spain, United States, Mali, United Kingdom, and Italy (WHO 2016b).

Multiple studies conducted during the course of the epidemic indicated that, despite the complex challenges presented by the spread of the disease in urban areas, Ebola epidemics are preventable, provided there is an immediate response aimed at interrupting transmission and limiting spread, including active case finding and rapid diagnosis, as well as prompt hospitalization with adequate care (WHO Ebola Response Team 2015). However, while international emergency responses to infectious or natural disasters typically prioritize rapid mobilization and relief, they are rarely linked to long-term strategies for disease surveillance and control, which are essential in strengthening health systems and preventing future outbreaks. They are even more rarely linked to the provision of quality clinical care, further compounding distrust of health services and of disease-control efforts.

This epidemic was characterized by the collision of a highly contagious virus with weak health systems—overwhelmed and unable to adequately respond—and a sluggish international response. The unprecedented spread of Ebola in Upper West Africa was itself a symptom of such dysfunction. This epidemic also highlighted the many ways that Ebola differentially affects women, given their roles as caregivers in the health system and at home; indeed, women accounted for 75% of the deaths early in the epidemic prior to implementation of public-health measures, and 51% of laboratory-confirmed cases overall (Luginaah et al. 2016, WHO 2016c). This gender divide has been mirrored in previous outbreaks of the disease in Uganda and Sudan (Hogan 2014).

However, the second story—often less reported—is the impact of the epidemic on women who did not have Ebola; women like Aminata, who needed access to both routine and emergency care, but could not access these services because of the devastation to an already weak health sector as a secondary result of the epidemic.

3.4 The Health System in West Africa Prior to Ebola

The healthcare systems in Sierra Leone, Liberia, and Guinea were weak, allowing Ebola to explode rapidly across the region in 2014. The social, economic, and political history of West Africa is complex (and outside the scope of this chapter), including centuries of exploitation of resources (both human and physical), colonialism, and more recently civil war. The civil wars in Liberia (1989–2003) and Sierra Leone (1991–2002) were characterized by brutal violence, weakening of the government, and destruction of infrastructure and systems. While relatively more stable, Guinea has faced coups and intermittent military rule. Left in the wake of these disruptions were some of the weakest healthcare systems in the world. During the Ebola emergency, these underfunded and understaffed health systems were challenged—and often unable—to maintain routine health services, especially for children and pregnant women, even those not infected with the disease.

Maternal mortality is often used as a metric to evaluate the overall state of a health system. Sierra Leone had one of the highest maternal mortality ratios (MMRs) in the world prior to the Ebola epidemic, with an adjusted maternal mortality ratio of 1,100 maternal deaths per 100,000 live births and lifetime risk of maternal death for all women of childbearing age of 1 in 21 (WHO 2016b). Guinea had shown a great improvement in maternal mortality, reducing their maternal mortality from 1,040 maternal deaths per 100,000 live births in 1990 to 679 maternal deaths per 100,000 live births in 2015, but still representing stark excess mortality. For reference, the maternal mortality ratio in the United States was 23.8 maternal deaths per 100,000 live births in 2013, which is more than triple the rate of Canada—sparking a recent outcry in the public-health news (Tavernise 2016). As evidenced, the majority of maternal deaths worldwide occur in low-resource settings, like Sierra Leone, Liberia, and Guinea, and most of them can be readily prevented.

In an effort to reverse the trend of high maternal mortality, in 2010, the government of Sierra Leone launched the ambitious "Free Healthcare Initiative," which offers free health services in public facilities to pregnant women, lactating mothers, and children under five years old, as well to those needing HIV/AIDS and tuberculosis care. In theoretically eliminating the cost barrier for services, the government of Sierra Leone saw a doubling in utilization of services upon implementation of this program. However, there still remained several barriers to the program's objectives, including an insufficient healthcare workforce and infrastructure in place for women and children to receive care, and patients still having to pay despite the promise of "free" care (Pieterse and Lodge 2015). At the same time, other barriers remained unaddressed, including distance to facilities and limited options for transport (Sharkey et al. 2016).

As in Sierra Leone, Liberia had made improving maternal health a national priority after the civil war, before Ebola reached the country. In 2007, the Liberian government launched the "Emergency Human Resources Plan" to rebuild the healthcare workforce after its civil war ended. As a result, the number of working nurses doubled, and the country saw a substantial increase in facility-based deliveries, from 38% to 56% between 2007 and 2013. And, although still one of the highest in the world, maternal mortality had improved from 1,200 maternal deaths per 100,000 live births to 640 maternal deaths per 100,000 live births from 1990 to 2013 (Ly et al. 2016).

Even further, in 2013, and just before the Ebola outbreak, Liberia initiated the "Accelerated Action Plan to Reduce Maternal and Newborn Mortality." This plan aimed to further invest in a robust health workforce by training skilled birth attendants (SBAs); increasing coverage of and access to emergency obstetric care (EmOC) and essential maternal and newborn healthcare; increasing access to and utilization of family planning services; expanding and strengthening outreach and community-based services; and improving management of maternal and newborn health services. The country saw progress with this plan—in 2013, the Liberia Demographic and Health Survey reported that 96% of women received at least one antenatal care visit, and 61% gave birth with a skilled provider, with almost all of these encounters taking place in health facilities (Liberia Institute of Statistics and Geo-Information Services (LISGIS) et al. 2014). However, across Liberia, an overall disparity remained in the availability of care for those outside of Monrovia: 84% of poor women and 53% of rural residents reported little access to maternal health services (Luginaah et al. 2016). Similar trends were seen in Sierra Leone and Guinea.

Despite these ambitious national health initiatives, a variety of factors contributed to the frailty of the health system. The number of healthcare workers—defined as doctors, nurses, and midwives—available per population is a valuable metric to assess the strength (or weakness) of a health system. Prior to the Ebola epidemic, Guinea had only one health worker per 1,597 people, Liberia had one for every 3,472 people, and Sierra Leone had one per 5,319 people. This pales in comparison to the WHO's standard of one healthcare worker per 439 people (WHO 2016d). The statistics were even worse for specialist providers—in 2008, there were only 10 surgeons operating in government-run hospitals in Sierra Leone, and only seven obstetrician-gynecologists to serve the entire nation of six million people; in 2015, there were only three anesthesiologists nationwide (Ribacke et al. 2016). Data from Liberia, compiled just before the outbreak, revealed that there were only 51 doctors working in the entire country (WHO 2016d). Moreover, in all three nations, the healthcare workforce was unevenly distributed across the country, with the majority working in urban centers, inaccessible to the rural majority.

Of critical importance to pregnant women (who require Caesarean section in up to 10% of deliveries) and therefore maternal mortality, access to surgical care is another metric to evaluate the strength of a healthcare system. In a paper published prior to the Ebola epidemic, a Sierra Leonean surgeon, Dr. T.B. Kamara, revealed: "*Government hospitals in present day Sierra Leone lack the infrastructure, personnel, supplies, and equipment to adequately provide emergency and essential surgical care. In a comparison of present day Sierra Leonean and US Civil War hospitals, the US Civil War facilities are equivalent and in many ways superior*" (Crompton et al. 2010). This comparison illustrates the overall destitution of healthcare facilities in the country and the impact on its ability to deliver adequate health services. Beyond the limited number of trained surgeons to perform routine and emergency procedures, stockouts of essential surgical supplies, including sterile tools, suture kits, and oxygen, were commonly reported (Barden-O'Fallon et al. 2015). As expected, the Caesarean section rate in this region is low—in Sierra Leone, between 2.3 and 4.5% of all births are performed by Caesarean section. This is far below the WHO standard that a minimum of 10% of deliveries will require a Caesarean section (WHO 2017a). Further, with only 50% of deliveries occurring in a health facility, many women rely on traditional birth attendants (TBAs) or family members to assist in delivery, without any access to emergency obstetric and neonatal care (EmONC) (Ribacke et al. 2016). Even without an outbreak of epidemic disease, safe surgery, along with basic medical care, simply cannot be delivered where there is limited health infrastructure or a weak supply chain to procure and deliver medications, supplies, and necessary equipment.

The previously mentioned "symptoms" of a weak health system are in part due to a lack of adequate government funding for health services. The International Monetary Fund (IMF) recommends that governments spend a minimum $86 USD per capita per year for basic health services. Prior to the Ebola epidemic, the three most affected countries spent far less on healthcare: $9 USD in Guinea, $20 USD in Liberia, and $16 USD in Sierra Leone—shockingly representing a drastic increase from a decade earlier. Decades-old structural adjustment policies from the World Bank and IMF contributed to underinvestment in health and education. Through privatization of public goods, tax exemptions, and incentives given to private companies, NGOs, and embassies, it has been estimated that in Sierra Leone alone, the country lost approximately $200 million USD annually in revenue from 2010 to 2012 (Save the Children 2015). In the same period, the country invested only $20 million USD in the health sector annually (O'Hare 2015). This lack of capital and steady monetary support to establish and develop a functioning health system in these three countries contributed to the environment that allowed Ebola to explode rapidly across the region.

3.5 The Impact of Ebola on Health Systems

In any setting, the building blocks of a health system consist of the same essential components: staff, stuff, space, and systems. Each of these is essential to the provision of quality care, and without any one component, the health system—and the patients it serves—is vulnerable. In Upper West Africa, as described above, the three countries of Sierra Leone, Guinea, and Liberia had some of the weakest health systems in the world. And, in the words of the Medical Director of Koidu Government Hospital located in Kono District, "*the Ebola epidemic brought the health system in Sierra Leone to its knees*" (Marsh 2015).

In devastating fashion, the impact of the Ebola epidemic on the health system had its most significant impacts on those members of society who were the most vulnerable: women and children, people living with HIV and tuberculosis, and people with mental and physical disabilities. The impact on pregnant women—even those who did not have Ebola, like Aminata—was particularly profound (Strong and Schwartz 2016). In sub-Saharan Africa, nearly three quarters of maternal mortality is due to direct obstetric causes, such as prolonged and obstructed labor, hemorrhage, eclampsia and other hypertensive diseases, complications of abortion, and sepsis—conditions that can be averted with proper provision of emergency obstetric and neonatal care (O'Hare 2015).

Basic emergency obstetric and neonatal care is defined as seven essential "signal functions" that treat these major causes of maternal mortality: antibiotics to prevent puerperal infection; anticonvulsants for treatment of preeclampsia and eclampsia; oxytoxics to reduce postpartum hemorrhage; manual removal of placenta; assisted or instrumented vaginal delivery; removal of retained products of conception; and neonatal resuscitation (UNFPA 2014). Basic EmONC should be provided at the health center level. Comprehensive EmONC is the addition of surgical delivery by Caesarean section, blood transfusion, and advanced neonatal resuscitation—usually available at the hospital level. To reduce maternal mortality, health centers and hospitals should be staffed, equipped, and organized to provide this care.

3.5.1 Staff

The Ebola epidemic had a devastating impact on the healthcare workforce across the region. The first of the three losses of staff that directly impacted maternal healthcare services was death of health workers. Ebola is a caregiver's disease. While the majority of deaths occurred among families who cared for their sick loved ones and buried their dead, physicians and nurses—professional caregivers—were gravely affected. At the peak of the epidemic in Sierra Leone in 2014, the incidence of Ebola among healthcare workers was 103-fold higher than that of the general population, peaking in August 2014 (Kilmarx et al. 2014). Overall, healthcare workers were 21–32 times more likely to be infected than the general population across the three countries (Elston et al. 2017). Over the course of the Ebola epidemic, 881 confirmed infections and 513 deaths were reported among healthcare workers in total in the three countries—a particularly striking number when compared against the paucity of healthcare workers at the start of epidemic (WHO 2015a). As of late May 2015, Guinea, Liberia, and Sierra Leone, respectively, had lost 78, 83, and 79 doctors, nurses, and midwives to Ebola, which translates to 1.5%, 8.1%, and 6.9%, respectively, of the healthcare workforce that died (Evans et al. 2015). Tragically, prior to the outbreak, Liberia had only 12.8 physicians, midwives, and nurses per 10,000 people (far below the WHO benchmark of 23 per 10,000) and reported 175 healthcare worker deaths by the end of 2015 (Iyengar et al. 2015). To reach the minimum 80% health coverage targeted by the Millennium Development Goals (MDGs), 43,565 doctors, nurses, and midwives would need to be hired across the three countries—a likely impossible feat, but one that demonstrates the sheer magnitude of the fragility of the health system.

Another significant loss to the healthcare workforce was as a result of diversion into the Ebola response. Physicians, nurses, and midwives who would normally provide maternal health services were recruited to oversee operations and provide clinical service delivery in response to the outbreak. This happened as part of the national Ministries of Health strategy, as the health leadership of each country redirected its public employees to staff its public Ebola centers, but also occurred in an unplanned fashion, as NGOs responding to the epidemic hired trained workers to staff their facilities (Elston et al. 2017). While many NGOs tried to minimize hiring staff who were already employed in the public sector, given the profound lack of human resources, it was often unavoidable and ability to verify other employment was difficult. This meant that the few healthcare staff available were drawn out of public facilities into (higher-paying) jobs as part of the Ebola response. As a result, routine antenatal, delivery, and emergency obstetric care were reduced, as fewer doctors, surgeons, nurses, and trained birth attendants were available to provide care. For Aminata, three of the four midwives at the government hospital were engaged in the Ebola response, leaving her unattended when she presented for care.

Finally, though not as prominent as the previous two, another loss to the healthcare workforce was from fear of infection and subsequent "flight" from the health system. Healthcare workers reportedly abandoned posts for fear of infection or the stigma attached to caring for infected patients. This effect was particularly severe in maternity wards—as delivery is notable for significant loss of blood and other potentially Ebola-infected bodily fluids—exacerbating both risk and fear. This loss led to the abandonment of patients on wards by both family members and healthcare providers. Similarly, there were reports of staff refusing to care for patients, which unfortunately and disproportionately affected laboring women, for whom the risk to staff was perceived to be higher (Elston et al. 2017). Even where deliveries were still being performed, "*The few staff were demoralized and people lost faith in the healthcare system*" (Sheku 2018).

Sadly, as noted, there was a profoundly negative outcome of these staffing losses: "*In the context of the Ebola epidemic, the absence of healthcare providers offering relevant services, the inability to differentiate between Ebola and other febrile diseases at onset, and the fear of contracting Ebola at a health facility can also prevent pregnant women from seeking reproductive health services*" (Davtyan et al. 2014; Menendez et al. 2015; Walker et al. 2015).

3.5.2 Stuff

An efficient health system simply cannot function without adequate supplies—or "stuff"—including medications, consumable items, biomedical equipment, and laboratory necessities. Similar to the human resources, the supply chain was diverted into the Ebola response. This shift meant stockouts of essential drugs, such as antimalarial medications and antiretroviral treatment, and of supplies, including gloves, gowns, soap, and chlorine, as well as vehicles and the fuel required for these vehicles and generators. This diversion of supplies into the response resulted in an inability to continue routine services, which were already limited and insufficient prior to the beginning of the Ebola epidemic.

In Sierra Leone, public hospitals are provided quarterly distributions of medications and supplies from a national centralized supply. Many facilities reported baseline stockouts of essential medications prior to the epidemic (Barden-O'Fallon et al. 2015). While there is a possibility to purchase supplies from the private sector, hospitals have limited budgets with which to do so and, therefore, limited ability to manage stockouts locally. Additionally, inventory systems are both limited technologically and understaffed. When emergency supplies were delivered to hospitals, they were often limited in their ability to receive, inventory, and then dispense these items.

One of our colleagues in Liberia offered the following example during the peak of the epidemic: "*In one health center in Liberia, nurses complained of shortages of all drugs, except oral rehydration salts, for many months preceding the Ebola crisis.*" These stockouts persisted during the epidemic. In neighboring Sierra Leone, prior to the Ebola outbreak, a 170-bed capacity hospital, serving a catchment area of over 500,000 people, never had a connected X-ray machine and had just two doctors on staff.

Also with respect to access to EmONC, there remained a notable lack of oxygen and blood. Essential for maternal and neonatal resuscitation, oxygen can be delivered via oxygen cylinders or bedside concentrators. Oxygen cylinders must be filled at a central source; in Sierra Leone, there was only one oxygen plant in the public sector, at the teaching hospital in Freetown, which was unable to supply the country. Smaller bedside oxygen concentrators—while more available—require electricity, a variable resource from expensive fuel-dependent generators and limited national grids.

Access to safe transfusion is essential for EmONC, as hemorrhage is one of the leading causes of maternal mortality. While still extremely limited, all three countries had made significant progress in improving their blood supplies prior to the Ebola epidemic. However, this access plummeted during the epidemic. In Sierra Leone, voluntary donation was suspended, limiting the supply. In Guinea, the Director of the National Blood Service noted a drop in voluntary donations, citing fear of Ebola dissuading potential donors; and in Liberia, the Assistant Minister of Health stated that "*the blood service completely ceased up* [sic]" during the epidemic. Regulatory oversight dropped in all three countries (WHO 2017b). In many cases, family members were asked to donate for emergency cases or recruit other donors, increasing the risk for infection transmission and coercion (Raykar et al. 2015). Overall, blood became less available—or at least less readily available—in emergency cases, reducing EmONC and increasing risk of death for women with obstetric complications.

3.5.3 Space

Decades of poverty, structural adjustment policies that disincentivized investment in public infrastructure, followed by a decade of civil war in Sierra Leone and Liberia left the health infrastructure among the worst in the world. Road networks, telecommunications, hospitals, and laboratory systems were weak to nonexistent, particularly in rural areas. These gaps contributed to delays in recognition of the epidemic initially and then to delays in management and treatment (WHO 2015b).

At the request of the respective ministries of health, PIH worked to support public hospitals in two rural districts in Sierra Leone and in southeastern Liberia. At each facility, our experience was similar: power was available only intermittently; piped water was rare and none was potable; and general infrastructure (including roofs, windows, doors) had not been updated significantly in decades. No hospital had a functional X-ray machine, and laboratory capacity was minimal. Compounded by concern about nosocomial Ebola transmission, communities generally feared the hospitals. Unsurprisingly, these facilities were minimally prepared to perform the essentials of disease surveillance or control the spread of the epidemic.

Worsening the baseline limitations in infrastructure, hospital and health center closures due to hospital-based transmission of Ebola, and infectious risk further undermined access for women to preventative and therapeutic treatment, including family planning, antenatal, and delivery needs (Ly et al. 2016). Due to essentially nonexistent infrastructure for isolation at the start of the outbreak, even those hospitals that remained open were forced to convert functional wards into quarantine spaces—limiting access to care and heightening concern within the community. The situation was likely the worst in Liberia, where the WHO reported that two thirds of health services ceased functioning by August 2014 and that 62% of health facilities closed by September 2014. Additionally, the three largest hospitals in Monrovia—the site for the majority of clinical teaching and referral care—were all closed (Iyengar et al. 2015).

3.5.4 Systems

Delivery of quality healthcare relies on innumerable systems, including communications, WASH (water, sanitation, and hygiene), supply chain, triage, referral, financing, and many others. With the dramatic diversion of human, financial, and supply resources into the Ebola response, those (already weak) systems that were designed to care for pregnant women and their children were undermined by this further loss of inputs.

3.5.5 Referral Systems and Ambulances

Even before the epidemic, referral systems and ambulances for pregnant women were limited. In the hypothetical design of the health system, pregnant women would present to health centers for delivery; in cases of emergency, they would be transported to district hospitals for blood transfusion or surgical delivery—the two defining features of comprehensive EmONC. With the onset of the Ebola outbreak, referral became even more difficult. The few ambulances and other vehicles in the ministries of health's fleets were diverted into the response. Women with hemorrhage, eclampsia, or obstructed labor (like Aminata) had few options to access the care they needed. Families were often left to raise funds to pay for private transportation.

3.5.6 Travel and Movement Restrictions

Further challenging access to EmONC were Ebola-related travel restrictions. Travel between districts on major roads was often limited to daytime only and for those carrying special passes allowing travel. Between villages, movement was also often restricted, and community members were encouraged to report any unknown people for fear of disease. These restrictions—while important for EVD control—impacted access to non-Ebola care, particularly for women with complications of delivery, who needed care at Basic EmONC-capacitated health centers or at hospitals (for Caesarean section and blood transfusion).

3.5.7 Communication and Messaging

The nature of the Ebola epidemic demanded widespread communications systems to ensure public safety; this was especially the case due to the infection's rapid transmission across national borders involving multiple countries, as well as the occurrence of widespread infections not only in rural areas but also, for the first time, in large cities. However, it was often difficult to ensure messages were reaching the communities most in need. While ubiquitous, radio messaging is often expensive. Additionally, phone and the emergency call numbers were limited by the existing communications infrastructure in Upper West Africa. Initial communication around the Ebola outbreak further undermined communities' trust in government and public facilities. While early posters and billboards emphasized "ABC" (Avoid Bodily Contact) messaging, many focused on less common forms of EVD transmission (i.e., by the consumption of bushmeat and wild animals). With emphasis there (and not on infection prevention), these messages missed the opportunity to prevent infections among caregivers, and therefore, may have contributed to the early facility-based transmission of Ebola virus disease—ultimately resulting in significant fear of hospitals, deterring necessary women's healthcare. Further, much of the messaging around caring for people at home who were sick, as well as methods for burying the dead, were not matched with the enabling systems to do so. Families were left to manage the ill and deceased through usual means, further increasing risk.

3.5.8 Cost

Prior to 2013, in an effort to reduce their high maternal and neonatal mortalities, Sierra Leone and Liberia both implemented programs to eliminate user-fees and costs for pregnant women. The 2010 Free Healthcare Initiative (FHCI) in Sierra Leone made it illegal to charge fees for services for pregnant and lactating women. There were many successes of the FHCI, including significantly increased facility-based deliveries. However, even before the Ebola-related shortages, many facilities were routinely stocked out of essential free supplies provided by the government, and then charged women to purchase medications and necessary consumables (Pieterse and Lodge 2015). In a country where more than 60% of the population lives in absolute poverty (<$1.25/day), these costs created a real barrier to care (UNDP 2017). At PIH, we heard these concerns from patients. One woman said: "*I might be given a receipt to go the drug store, but I cannot afford these medicines. We are told medicines should be available, but they are not*" (personal communications with community members in Sierra Leone and Liberia, 2015).

3.5.9 Ebola Screening

The system for Ebola screening itself—while essential to the epidemic response—undermined critical maternity care for pregnant women who did not have Ebola. Ebola virus disease during pregnancy is associated with extremely high maternal mortality and nearly uniform fetal demise. Miscarriage and hemorrhage are common. Particularly early in the epidemic, treatment of Ebola in pregnancy was largely considered futile and high risk (Médecins Sans Frontières 2017). The WHO/CDC case definition (and therefore screening criteria) for Ebola virus disease included both (1) fever with any bleeding and (2) fever associated with any three other symptoms, including vomiting, diarrhea, abdominal pain, malaise, fatigue, anorexia, headache, muscle and joint pain, and respiratory difficulty. Many women with known and routine complications of pregnancy such as ante- or postpartum hemorrhage, miscarriage, eclampsia, or chorioamnionitis met these screening criteria, but did not have Ebola infections. When arriving at health facilities, they were therefore isolated for EVD polymerase chain reaction (PCR) confirmatory testing. Patients were required to have symptoms for at least 72 hours to ensure necessary sensitivity, and laboratory testing itself took a minimum of a day to receive results (and longer at the beginning of epidemic).

As such, many women with treatable obstetric conditions were held for several days in Ebola isolation units, where minimal care was available for the true etiology of their symptoms. Access to blood transfusion was rare and surgical delivery nonexistent. One study estimated that even at the peak of the epidemic, as many as 98.5% of women admitted to isolation would likely test negative for EVD, based on prevalence of EVD relative to the prevalence of other obstetric conditions, which would also meet this case definition. In this setting, in which treatment was largely withheld due to perceptions of futility and risk, these non-EVD-infected women—in a region with some of the highest maternal mortality rates in the world—were offered limited care for the true cause of their (often very treatable) presentations. While the policy of delaying obstetric inventions until Ebola was excluded was well-intentioned, it likely led to the death of many women without EVD (Deaver and Cohen 2015).

Additionally, in many hospitals, it was often the (formal or informal) policy that women who needed surgical delivery—even those who were well and did not meet the screening criteria for Ebola case definition—were required to have a negative EVD PCR test prior to Caesarean section, leading to delays in care for mother and neonate.

During the epidemic, these various systems—finance, communication, referral/transportation, and Ebola screening (among others)—left gaps in the ability of health systems to provide care for pregnant women, likely contributing to rise in maternal mortality and decline in routine services delivered during the epidemic (VSO 2015).

3.6 Impact of the Ebola Epidemic on Women Without Ebola Infection

3.6.1 The Story of Fatmata

Fatmata was fortunate. Born in Sierra Leone, she had a university education and a steady job at a bank in Freetown. Her husband worked as the executive director of a Sierra Leonean healthcare nonprofit and earned a respectable salary. It was early 2014; Fatmata was 22 years old and expecting her first child.

Concurrently, the Ebola epidemic was starting its spread across West Africa. As Fatmata's pregnancy progressed, the private clinics in Freetown where she was planning to seek care started to close—she was unable to receive any antenatal care. She and her husband became increasingly concerned about where she would go for a safe delivery once she went into labor.

As she reached full term, Fatmata's husband made increasingly worried calls to his various clinical colleagues, who were eventually able to secure her access to a free clinic in Freetown—one that usually provided care only to the poorest of women. Although her and her husband's economic status essentially disqualified them from receiving care there, she was promised admission once she went into labor.

However, in early August 2014, when she went into labor, Fatmata and her husband went to the clinic only to find "almost no staff there." They learned that there had been an Ebola case in the clinic, and fear of infection spread rapidly among the staff: many doctors and midwives were too afraid to come to the clinic to provide care for the patients. With persistence, eventually Fatmata was screened for Ebola and, since she did not meet case definition, she was allowed access into the clinic. Her husband and their family were asked—or forced—to remain outside and waited across the street. Her husband continually called his colleagues, who called the staff at the clinic for updates on Fatmata's progress. No one knew if a midwife or physician came to take care of her, while she labored inside. For a first pregnancy, it was a difficult and long delivery, and eventually she was delivered by one of the few nurses working at the time.

At 5 a.m., Fatmata was finally able to call her anxious family who had been waiting 24 hours to hear from her. After a long night, and without an intravenous line or any medications, she had successfully delivered a healthy baby girl. The nurse then spoke to her husband and asked him to come inside to take her home, as there were "no staff to take care of her." Inside the clinic, other women were laboring—and delivering—largely unattended. Having fortunately had an uncomplicated labor without the need for emergency obstetric care, and having the significant personal resources necessary to find even this basic care, Fatmata, her husband, and their new baby girl went home together, leaving the other women behind.

3.7 Impact on Maternal Health Services

A side effect of the Ebola epidemic itself, these profound disruptions to healthcare provision—including the staff, stuff, space, and systems—impacted women without Ebola (like Aminata and Fatmata) and their ability to access both routine preventative care and emergency treatment. This was exacerbated by fear and stigma, which further undermined relationships between communities and health facilities. As a result, there was a significant drop in attendance at facilities, and it is widely presumed that these women went unattended (Elston et al. 2017).

3.7.1 Maternal Mortality

Estimates suggest that during the peak of the epidemic, maternal mortality increased by as much as 38%, 74%, and 111% in Guinea, Sierra Leone, and Liberia, respectively. With a maternal mortality rate (MMR) of 1,100 deaths per 100,000 live births prior to the epidemic, Sierra Leone's maternal mortality reached 1,916 deaths per 100,000 live births, levels not seen since the civil war and erasing the progress that had been made since that time (Evans et al. 2015). Over the entire duration of the epidemic, it is estimated the Sierra Leone's maternal and neonatal mortality rates increased by 30% and 24%, respectively (VSO 2015).

3.7.2 Facility-Based Deliveries

Facility-based deliveries declined by 20–23% in Sierra Leone during the epidemic; with reductions of 40–50% reported in the most heavily affected regions during the peak of the epidemic (Ribacke et al. 2016).

Evidence is similar in Liberia. In one study of two Liberian counties from March 2014 (first case reported in Liberia) to December 2014 (worst month), total health facility deliveries dropped to less than 33% compared to March 2014 (Iyengar et al. 2015). In a different study in Liberia, Lori and colleagues also found that facility-based deliveries and use of maternity waiting homes dropped by 77% (from 500 to 113 deliveries per month) during the peak of the Ebola epidemic in a high-burden county (Lori et al. 2015).

Even in an area of rural southeast Liberia, where there were very few Ebola cases detected, there was a 30% decrease in odds of a facility-based delivery. Likelihood of facility-based delivery was associated with belief in the safety of health facilities: the odds of facility-based delivery were 41% lower in women who believed that facilities were a site of Ebola virus disease transmission versus those who did not share this belief (Ly et al. 2016).

There was a similar impact in Guinea, where prior to the epidemic, only 41% of deliveries were facility-based (Barden-O'Fallon et al. 2015). At the Matam maternity hospital in Conakry, there was a dramatic drop in utilization from 904 to 123 patients per quarter from July to September 2014 (Delamou et al. 2014). In Conakry and N'Zerekore overall, the number of women giving birth in a facility with a skilled birth attendant fell by 87% (Barden-O'Fallon et al. 2015).

Across all three countries, changes in facility-based deliveries were more notable at the hospital level when compared to health centers, suggesting that hospital services suffered more from Ebola-related stigma than did services at health centers (Barden-O'Fallon et al. 2015).

3.7.3 Surgical Delivery

Maternal and neonatal mortality increase when Caesarean sections rates fall below 10% (WHO 2017a). It has been noted that the rate of Caesarean sections remained approximately the same before and after the Ebola outbreak; however, the number of noninfected women seeking care at a health facility was dramatically lower. A similar reduction was seen in the private sector; Médecins Sans

Frontières performed fewer Caesarean section procedures during the Ebola outbreak and closed their large referral center in the southern province of Sierra Leone.

As mentioned above, the Caesarean section rate in Sierra Leone had increased steadily prior to 2013, likely due to the impact of the national Free Healthcare Initiative, but remained far below WHO recommendations. An analysis of all 61 public and private facilities that offered surgical care in Sierra Leone found a 20% decrease in Caesarean sections during the epidemic, reversing the gains of the previous decade (Ribacke et al. 2016). In 2014, only 5,025 Caesarean sections were reported in Sierra Leone, leaving as many as 17,000 women with an unmet need for surgical delivery (Elston et al. 2017).

In Guinea, national Caesarean section rates fell by 16% during the epidemic when compared to the previous year; in EVD-affected prefectures, the rates fell by as much as 90% during the peak of the epidemic (Elston et al. 2017). Ebola virus disease-affected districts in Sierra Leone also experienced similarly striking reductions in surgical delivery capacity. The Kenema Government Hospital had no one to provide surgical delivery care in summer 2014 and no viable place to transfer women with emergencies. At the Koidu Government Hospital, the operating theaters closed for three months without any documented surgical care.

3.7.4 Antenatal Care

As for Aminata and Fatmata, access to and utilization of preventive services, including antenatal care (ANC), also declined during the epidemic. In September 2014, in two counties in Liberia, ANC uptake fell to less than 9% of previous peak utilization and to 4% for intermittent preventive treatment for malaria (Iyengar et al. 2015). Overall in Sierra Leone, ANC fell by 18% and postnatal care by 22% (Jones et al. 2016). In Guinea, there was a 51% reduction in HIV testing at ANC during the six peak months of the epidemic. Sierra Leone experienced a 23% decline in HIV testing at ANC overall (Elston et al. 2017).

3.7.5 Economic Impact on Women

Women were disproportionately affected by the economic impact of the Ebola epidemic. The World Bank estimated that across the three economies of Sierra Leone, Liberia, and Guinea, $1.6 billion USD was forgone in 2015 due to EVD. Households experienced increased prices for necessary goods, as well as decreased employment. Job losses predominantly affected the informal and agricultural sectors in which more women are employed. Food insecurity also increased, with more significant direct impacts on pregnant women, lactating mothers, and young children (Elston et al. 2017).

3.7.6 Stigma and Mistrust

In addition to the direct impacts of the Ebola epidemic on maternal mortality, access to Caesarean sections, and utilization of antenatal care, there is concern that the outbreak indirectly contributed to increased stigmatization of pregnancy and delivery, even for women without Ebola (Strong and Schwartz 2016, 2019). In addition to the many barriers to supervised deliveries in West Africa, contributing to poor maternal and neonatal outcomes even prior to the Ebola epidemic, there was evidence that women "feared" seeking care at facilities due to the perceived risk of surgical intervention and the belief that "normal" deliveries occur in the village (Treacy and Sagbakken 2015). Given the high-risk nature of pregnancy during Ebola, the epidemic worsened stigmatization of pregnant women and delivery as a driver of the epidemic, and a service that was particularly risky for healthcare providers. This stigma likely contributed to reduced access to care for women. In addition to stigma imposed externally, Luginaah and colleagues argue that the underutilization of healthcare services may have been due to internal stigma, as pregnant women felt associated with Ebola, and therefore, missed necessary care to avoid feeling labeled (Luginaah et al. 2016).

Given the lack of human resources, reliable infrastructure or supplies, and robust systems for care, coupled with frequent out-of-pocket costs, there was often a profound lack of faith and mistrust in the health systems in Upper West Africa. These gaps were present well before the recent Ebola epidemic and widened during the epidemic, further undermining community trust. As a result, people sought care elsewhere—in private clinics, at pharmacies, and with traditional healers (Lori et al. 2015). High levels of transmission of Ebola virus disease occurred at the community level. Household members frequently took care of their sick loved ones, cleaning vomit, disposing of diarrhea, and sleeping in the same rooms with infected family members. By the time patients sought treatment in the formal healthcare sector, their condition was often very advanced. These fears were exacerbated by actual nosocomial transmission of Ebola between infected patients to staff, caregivers, and other patients (Dunn et al. 2016). In the Bombali district of Sierra Leone, nosocomial transmission of EVD occurred in a hospital maternity ward in which several mothers and their newborn infants became infected (Connolly et al. 2017; Connolly and Young 2019).

In the Kenema district in Sierra Leone, a qualitative study of healthcare workers found that there had been a perceived decrease in women presenting for antenatal care, preventative services, and delivery. The healthcare workers perceived that women were afraid of contracting EVD in hospital or outpatient settings, feared the presence of foreigners (who were perceived by some as having brought Ebola to the country), and feared that hospital staff were experimenting on the community (Dynes et al. 2015).

3.8 The Way Forward

As cases of Ebola declined across West Africa, PIH transitioned our response in Sierra Leone and Liberia to match the changing needs of the local communities and to build the staff, stuff, space, and systems needed to strengthen the health systems and promote resilience in the health sector. Building robust local health systems will help to prevent further Ebola infections, while also promoting primary care, maternal and child health, and the effective diagnosis and treatment of other infectious and chronic diseases—all of which had been largely neglected during the epidemic as Ebola demanded the health systems' already scant resources. In both countries, community health workers—many of whom are Ebola survivors—now serve as a crucial link between community members and new health, social, and economic resources. PIH has prioritized the remodeling of health facilities to ensure that they are equipped to deliver quality care; this has included installation of generators to prevent disruptions in electricity, fixing plumbing, stocking pharmacies, and updating patient wards.

In Liberia, PIH is now working in Grand Gedeh and Maryland counties. In the latter, we have refurbished and operate a hospital and a clinic, improving the space for healthcare delivery. At the PIH-supported J.J. Dossen Hospital, the emergency room, operating theaters, and maternity ward were renovated, along with the updating of a laboratory and blood bank. The electrical system has also been overhauled. Remarkably, in December 2016, triplets were successfully delivered there via Caesarean section (Partners In Health 2017a). PIH also rebuilt Pleebo Health Center in southeastern Liberia, which had fallen into severe disrepair. In the reconstruction of this center, the maternity ward was made a priority and the center now delivers approximately 100 babies per month, representing nearly 85% of those expected in its catchment area (Partners In Health 2017b). PIH is also working in neighboring districts to repair clinics and has begun planning in Maryland County to build a new national referral hospital for the underserved southeast region of the country.

It has also been critical to ensure that the necessary "stuff" is available to healthcare practitioners. In PIH-supported facilities, there are now fewer stockouts of essential medications and supplies; there is also new laboratory, sterilization, and biomedical equipment, as well as a new oxygen generator. At the J.J. Dossen Hospital, a new digital X-ray machine arrived in September 2016—the first X-ray machine at the hospital in 35 years. Beyond improving patient diagnosis and care, the machine provides a valuable teaching tool for student nurses and other clinicians; previously, the nearest X-ray machine was eight hours by road (Partners In Health 2016b).

To help build local human resources capacity (the "staff"), PIH is working alongside Harvard faculty members to provide mentorship at Tubman University, which houses Liberia's only public nursing program. In Monrovia, PIH has helped to renovate and staff a tuberculosis hospital. We are also working closely with the Ministry of Health and Social Welfare to provide training and staffing support (including Liberian nationals and expatriates), with priority placed on maternal and child health (e.g., hiring midwives and nurse anesthetists) (Partners In Health 2017c). Further, PIH has worked with our partners to develop a variety of systems, including supply chains, ambulance coordination, and health facility transfers, among others.

In Sierra Leone, PIH has been similarly engaged in efforts to bring the necessary staff, stuff, space, and systems to the communities in which we work. PIH is promoting access to healthcare largely neglected during the epidemic (e.g., treating malaria, malnutrition, HIV, and tuberculosis, and encouraging safe births) and continuing to support health facilities in Port Loko and Kono. At Port Loko Government Hospital (PLGH) and Koidu Government Hospital (KGH, in Kono District), PIH has rebuilt facility infrastructure: painting wards, installing new roofs, providing generators to prevent power disruption, fitting new plumbing for potable water, and improving driveways and walkways for patient and staff safety. In both facilities, PIH updated the operating rooms, including improved lighting and new operating tables. At the Wellbody Clinic in Kono, PIH helped to improve clinic infrastructure and has opened a maternity waiting home on the clinic campus.

In these PIH-supported facilities, there have been improvements made to the laboratories and laboratory equipment for faster and more accurate diagnostic testing and to ensure constant availability of blood. As they did in Liberia, PIH has installed the first X-ray machines in both Port Loko and Kono Districts. At KGH, PIH has installed the first functioning anesthesia machine since the war, which helps ensure safe Caesarean sections, among other surgical procedures. PIH has also purchased and installed biomedical equipment for its supported facilities, including pulse oximeters, ultrasound machines, electrocardiography machines, and a large oxygen concentrator, which can provide reliable oxygen both for the hospital and health centers in the district. Working closely with the Ministry of Health and Sanitation and the Central Medical Supply, PIH has ensured that the districts have an improved supply chain, particularly for essential medicines, including those used for antiretroviral therapy and treating tuberculosis.

PIH has supported public sector human resources for health by hiring key staff to fill essential gaps that are crucial to improving maternal and child mortality, such as midwives and nurse anesthetists. Nurse educators from PIH support the hospital nursing directors and clinical staff, work collaboratively with nurses on the wards, and teach and support emergency obstetric care and neonatal resuscitation. Physicians and mid-level clinical officers have been hired to care for admitted patients and perform essential surgery, including Caesarean sections. A laboratory manager was recruited to support service delivery and teaching in the laboratory and blood bank. Additionally, PIH has supported regular teaching for all cadres at the hospitals in which it works. In an effort to build durable systems to guide and buttress this work, PIH has worked with the Ministry of Health and Sanitation to support

District Health Management Team activities, held trainings on HIV/AIDS and tuberculosis in the districts, and led training on emergency obstetric care at the health center level. Each of these activities contributes to strengthening the health system in order to promote improved maternal and neonatal outcomes, while also building capacity to quickly identify Ebola or other infectious diseases so that a rapid, effective response may be initiated.

We have also continued to support Ebola survivors. In the wake of the epidemic, Ebola survivors have experienced a range of clinical sequelae—from joint pain to depression—associated with the disease that may require ongoing medical care following recovery from the acute illness. Among such complications, about 18% of survivors have developed uveitis, an inflammation of the eye that can cause blindness. The National Survivor Eye Care Program, which emerged from PIH's work in Port Loko, has screened over 4,000 Ebola survivors in Sierra Leone for uveitis and other related complications. To achieve comprehensive screening of survivors, PIH partnered with the Ministry of Health and Sanitation, local organizations, NGOs, clinics, mobile health clinics, and community health workers. By March 2016, 3060 survivors had been screened, with 379 people treated for uveitis (Partners In Health 2017d). PIH also continues to support routine care for Ebola survivors in Kono and Port Loko.

3.9 Conclusions

In the aftermath of the 2013–2015 Ebola epidemic that swept across Sierra Leone, Liberia, and Guinea, many often ask "Why these three countries and not others?" While Ebola made its way into several countries—including the United States and other "high-income" countries in Europe—the magnitude of the outbreak in Upper West Africa was uniquely profound, and the international response was highly focused on containment of the disease instead of comprehensive care of the Ebola-stricken. Following a decade of civil conflict in the region, and for a variety of reasons, little investment was made by the governments and international organizations to rebuild—or rather, build for the first time—functioning health systems. The lack of staff, stuff, space, and systems in these three countries allowed a preventable and treatable disease like Ebola to take so many lives of the poor and vulnerable.

Among the most vulnerable in such low-resource settings are pregnant women, as they are unable to access or receive the proper care they need to have a healthy pregnancy and safe delivery. The stories of Aminata and Fatmata have illustrated the adverse outcomes that can result across circumstances and social strata when a health system lacks the fundamental components to deliver adequate care for women and their newborns. Whenever the next epidemic or emergency happens—whether Ebola or Zika or a hurricane or earthquake—responders must consider and make long-term commitments to addressing the root causes of such losses. As Elston and colleagues note, "*the greatest benefit to populations is likely to come from slow but sustained health systems strengthening and public health development, both to prevent epidemics from happening in the first place, and to limit the deeper consequences of them afterwards*" (Elston et al. 2017). In such settings with the highest maternal mortality rates in the world, efforts to build a health system that can withstand turmoil—from epidemic disease to war to natural disasters—must be a local, national, and international priority.

References

Barden-O'Fallon, J., Barry, M. A., Brodish, P., & Hazerjian, J. (2015). Rapid assessment of Ebola-related implications for reproductive, maternal, newborn and child health service delivery and utilization in Guinea. *PLoS Currents, 4*, 7. Retrieved December 22, 2017, from https://www.ncbi.nlm.nih.gov/pmc/articles/PMC4542265/.

Connolly, A., Bayor, F. A., Edgerley, S., Sessay, T., & Jamieson, D. J. (2017). *An Ebola virus cluster linked to a hospital maternity ward in Bombali District, Sierra Leone: Implications for mother, infants and the community*. Presented at the annual meeting of the American Anthropological Society, Washington, DC.

Connolly, A.M., & Young, A.J. (2019). Ebola virus disease surveillance in two high-transmission districts of Sierra Leone during the 2013–2015 outbreak: Surveillance methods, implications for maternal and child health, and recommendations. In: D.A. Schwartz, J.N. Anoko, & S.A. Abramowitz (Eds.), *Pregnant in the time of Ebola: Women and their children in the 2013-2015 West African epidemic*. New York: Springer. ISBN 978-3-319-97636-5.

Crompton, J., Kingham, T. P., Kamara, T. B., Brennan, M. F., & Kushner, A. L. (2010). Comparison of surgical care deficiencies between US civil war hospitals and present-day hospitals in Sierra Leone. *World Journal of Surgery, 34*(8), 1743–1747.

Davtyan, M., Brown, B., & Folayon, M. O. (2014). Addressing Ebola-related stigma: Lessons from HIV/AIDS. *Global Health Action*, 7. https://doi.org/10.3402/gha.v7.26058. Retrieved December 20, 2017, from https://www.ncbi.nlm.nih.gov/pmc/articles/PMC4225220/.

Deaver, J. E., & Cohen, W. R. (2015). Ebola virus screening during pregnancy in West Africa: Unintended consequences. *Journal of Perinatal Medicine, 43*(6), 649–655.

Delamou, A., Hammonds, R. M., Caluwaerts, S., Utz, B., & Delveaux, T. (2014). Ebola in Africa: Beyond epidemics, reproductive health in crisis. *Lancet, 384*(9960), 2105. Retrieved December 20, 2017, from http://www.thelancet.com/journals/lancet/article/PIIS0140-6736(14)62364-3/fulltext.

Dunn, A. C., Walker, T. A., Redd, J., Sugerman, D., McFadden, J., Singh, T., et al. (2016). Nosocomial transmission of Ebola virus disease on pediatric and maternity wards: Bombali and Tonkolili, Sierra Leone, 2014. *American Journal of Infection Control, 44*(3), 269–272.

Dynes, M. M., Miller, L., Sam, T., Vandi, M. A., Tomcyzk, B., & Centers for Disease Control and Prevention (CDC). (2015). Perceptions of the risk for Ebola and health facility use among health workers and pregnant and lactating women—Kenema District, Sierra Leone, September 2014. *MMWR. Morbidity and Mortality Weekly Report, 63*(51), 1226–1227. Retrieved December 22, 2017, from https://www.cdc.gov/mmwr/preview/mmwrhtml/mm6351a3.htm.

Elston, J. W., Cartwright, C., Ndumbi, P., & Wright, J. (2017). The health impact of the 2014-15 Ebola outbreak. *Public Health, 143*, 60–70.

Evans, D. K., Goldstein, M., & Popova, A. (2015). Health-care worker mortality and the legacy of the Ebola epidemic. *The Lancet Global Health, 3*(8), e439–e440.

Garde, D. L., Hall, A. M. R., Marsh, R. H., Barron, K. P., Dierberg, K. P., & Koroma, A. P. (2016). Implementation of the first dedicated Ebola screening and isolation for maternity patients in Sierra Leone. *Annals of Global Health, 82*(3), 418. Retrieved December 19, 2017, from http://www.annalsofglobalhealth.org/article/S2214-9996(16)30191-6/fulltext#sec3.

Garrett, L. (2015). Ebola's lessons: How the W.H.O. mishandled the crisis. *Foreign Affairs*. Retrieved December 22, 2017, from https://www.foreignaffairs.com/articles/west-africa/2015-08-18/ebolas-lessons.

Hogan, C. (2014). Ebola striking women more frequently than men. *Washington Post*. Retrieved November 17, 2017, from https://www.washingtonpost.com/national/health-science/2014/08/14/3e08d0c8-2312-11e4-8593-da634b334390_story.html?utm_term=.660dc6b7fb83.

Iyengar, P., Kerber, K., Howe, C. J., & Dahn, B. (2015). Services for mothers and newborns during the Ebola outbreak in Liberia: The need for improvement in emergencies. *PLoS Currents, 16*, 7. Retrieved December 22, 2017, from http://currents.plos.org/outbreaks/article/services-for-mothers-and-newborns-during-the-ebola-outbreak-in-liberia-the-need-for-improvement-in-emergencies/.

Jones, S. A., Gopalakrishnan, S., Ameh, C. A., White, S., & van den Broek, N. R. (2016). 'Women and babies are dying but not of Ebola': The effect of the Ebola virus epidemic on the availability, uptake and outcomes of maternal and newborn health services in Sierra Leone. *BMJ Global Health, 1*(3), e000065. Retrieved December 20, 2017, from https://www.ncbi.nlm.nih.gov/pmc/articles/PMC5321347/.

Kilmarx, P. H., Clarke, K. R., Dietz, P. M., Hamel, M. J., Husain, F., & McFadden, J. D. (2014). Ebola virus disease in healthcare workers—Sierra Leone, 2014. *MMWR. Morbidity and Mortality Weekly Report, 63*(49), 1168–1171. Retrieved December 19, 2017, from https://www.cdc.gov/mmwr/preview/mmwrhtml/mm6349a6.htm.

Liberia Institute of Statistics and Geo-Information Services (LISGIS), Ministry of Health and Social Welfare [Liberia], National AIDS Control Program [Liberia], & ICF International. (2014). *Liberia demographic and health survey 2013*. Monrovia: Liberia Institute of Statistics and Geo-Information Services (LISGIS) and ICF International.

Lori, J. R., Rominski, S. D., Perosky, J. E., Munro, M. L., Williams, G., Bell, S. A., et al. (2015). A case series study on the effect of Ebola on facility based deliveries in rural Liberia. *BMC Pregnancy and Childbirth, 15*, 254. Retrieved December 20, 2017, from https://bmcpregnancychildbirth.biomedcentral.com/articles/10.1186/s12884-015-0694-x.

Luginaah, I. N., Kangmennaang, J., Fallah, M., Dahn, B., Kateh, F., & Nyenswah, T. (2016). Timing and utilization of antenatal care services in Liberia: Understanding the pre-Ebola epidemic context. *Social Science and Medicine, 160*, 75–86.

Ly, J., Sathananthan, V., Griffiths, T., Kanjee, Z., Kenny, A., Gordon, N., et al. (2016). Facility-based delivery during the Ebola virus disease epidemic in rural Liberia: Analysis from a cross-sectional, population-based household survey. *PLoS Medicine, 13*(8), e1002096. Retrieved December 22, 2017, from http://journals.plos.org/plosmedicine/article?id=10.1371/journal.pmed.1002096.

Marsh, R. (2015). Partners In Health. Personal communication.

Médecins Sans Frontières. (2017). *Ebola*. Retrieved December 22, 2017, from http://www.doctorswithoutborders.org/our-work/medical-issues/ebola.

Menendez, C., Lucas, A., Munguambe, K., & Langer, A. (2015). Ebola crisis: The unequal impact on women and children's health. *Lancet Global Health, 3*(3), e130. Retrieved December 19, 2017, from http://www.thelancet.com/journals/langlo/article/PIIS2214-109X(15)70009-4/fulltext.

O'Hare, B. (2015). Weak health systems and Ebola. *The Lancet Global Health, 3*(2), e71–e72. Retrieved December 20, 2017, from http://www.thelancet.com/journals/langlo/article/PIIS2214-109X(14)70369-9/fulltext.

Partners In Health. (2016a). *Ebola*. Retrieved January, 2017, from http://www.pih.org/blog/partners-in-health-ebola-response.

Partners In Health. (2016b). *An X-ray machine that inspires devotion*. Retrieved November 17, 2017, from http://www.pih.org/blog/x-rays-help-inspire.

Partners In Health. (2017a). *Three boys, three joys: Triplets born in Liberia*. Retrieved November 17, 2017, from http://www.pih.org/blog/triplets-born-at-j.j.-dossen-hospital-in-liberia.

Partners In Health. (2017b). *Pleebo*. Retrieved November 17, 2017, from http://www.pih.org/pages/pleebo.

Partners In Health. (2017c). *Liberia*. Retrieved November 17, 2017, from http://www.pih.org/country/liberia.

Partners In Health. (2017d). *Blindness*. Retrieved November 17, 2017, from http://www.pih.org/blog/blindness-uveitis-partners-in-health-sierra-leone-eye-care.

Pieterse, P., & Lodge, T. (2015). When free healthcare is not free. Corruption and mistrust in Sierra Leone's primary healthcare system immediately prior to the Ebola outbreak. *International Health, 7*(6), 400–404.

Raykar, N. P., Kralievits, K. E., Greenberg, S. L., Gillies, R. D., Roy, N., & Meara, J. G. (2015). The blood drought in context. *Lancet Global Health, 3*(Suppl 2), S4–S5. Retrieved December 22, 2017, from http://www.thelancet.com/journals/langlo/article/PIIS2214-109X(14)70351-1/abstract.

Ribacke, K. J. B., van Duinen, A. J., Nordenstedt, H., Höijer, J., Molnes, R., Froseth, T. W., et al. (2016). The impact of the West Africa Ebola outbreak on obstetric healthcare in Sierra Leone. *PLoS One, 211*(2), e0150080. Retrieved December 22, 2017, from http://journals.plos.org/plosone/article?id=10.1371/journal.pone.0150080.

Sack, K., Fink, S., Belluck, P., & Nossiter, A. (2014). How Ebola roared back. *New York Times*. Retrieved November 17, 2017, from www.nytimes.com/2014/12/30/health/how-ebola-roared-back.

Save the Children. (2015). *A wake-up call: Lessons from Ebola for the world's health systems*. Retrieved November 17, 2017, from https://www.savethechildren.net/sites/default/files/libraries/WAKE%20UP%20CALL%20REPORT%20PDF.pdf.

Sharkey, A., Yansaneh, A., Bangura, P. S., Kabano, A., Brady, E., Yumkella, F., et al. (2016). Maternal and newborn care practices in Sierra Leone: A mixed methods study of four underserved districts. *Health Policy and Planning, 32*(2), 151–162.

Sheku, M. (2018). Koidu Government Hospital. Personal communication.

Strong, A., & Schwartz, D. A. (2016). Sociocultural aspects of risk to pregnant women during the 2013-2015 multinational Ebola virus outbreak in West Africa. *Health Care for Women International, 37*(8), 922–942. https://doi.org/10.1080/07399332.2016.1167896.

Strong, A., & Schwartz, D. A. (2019). Effects of the West African Ebola epidemic on health care of pregnant women—Stigmatization with and without infection. In D. A. Schwartz, J. N. Anoko, & S. Abramowitz (Eds.), *Pregnant in the time of Ebola: Women and their children in the 2013-2015 West African Ebola epidemic*. New York: Springer. ISBN 978-3-319-97636-5.

Tavernise, S. (2016). Maternal mortality rate in U.S. rises, defying global trend, study finds. *The New York Times*. Retrieved November 17, 2017, from https://www.nytimes.com/2016/09/22/health/maternal-mortality.html.

Treacy, L., & Sagbakken, M. (2015). Exploration of perceptions and decision-making processes related to childbirth in rural Sierra Leone. *BMC Pregnancy and Childbirth, 15*, 87. Retrieved December 21, 2017, from https://bmcpregnancychildbirth.biomedcentral.com/articles/10.1186/s12884-015-0500-9.

U.S. Centers for Disease Control and Prevention. (2016). *CDC's ongoing work to contain Ebola in West Africa*. Retrieved November 17, 2017, from https://www.cdc.gov/vhf/ebola/pdf/cdcs-ongoing-work.pdf.

United Nations. (2016). *Global Ebola Response. UN Mission for Ebola Emergency Response (UNMEER)*. Retrieved November 17, 2017, from http://ebolaresponse.un.org/un-mission-ebola-emergency-response-unmeer.

United Nations Development Programme (UNDP). (2017). *About Sierra Leone*. Retrieved December 22, 2017, from http://www.sl.undp.org/content/sierraleone/en/home/countryinfo.html.

United Nations Population Fund (UNFPA). (2014). *Setting standards for emergency obstetric and newborn care*. Retrieved November 17, 2017, from http://www.unfpa.org/resources/setting-standards-emergency-obstetric-and-newborn-care.

VSO. (2015). *Exploring the impact of the Ebola outbreak on routine maternal health services in Sierra Leone*. Retrieved December 22, 2017, from https://www.vsointernational.org/news/press-releases/higher-maternal-and-newborn-death-rates-sierra-leone-due-ebola-fears.

Walker, P. G., White, M. T., Griffin, J. T., Reynolds, A., Ferguson, N. M., & Ghani, A. C. (2015). Malaria morbidity and mortality in Ebola-affected countries caused by decreased health-care capacity, and the potential effect of mitigation strategies: A modelling analysis. *Lancet Infectious Diseases, 15*(7), 825–832. Retrieved December 22, 2017, from http://www.thelancet.com/journals/laninf/article/PIIS1473-3099(15)70124-6/abstract.

WHO Ebola Response Team. (2015). West African Ebola epidemic after one year—Slowing but not yet under control. *New England Journal of Medicine, 372*, 584–587. Retrieved December 22, 2017, from http://www.nejm.org/doi/full/10.1056/NEJMc1414992#t=article.

World Health Organization. (2014). *Statement on the 1st meeting of the IHR Emergency Committee on the 2014 Ebola outbreak in West Africa*. Retrieved November 17, 2017, from http://www.who.int/mediacentre/news/statements/2014/ebola-20140808/en/.

World Health Organization. (2015a). *Ebola situation report—30th September 2015*. Retrieved November 17, 2017, from http://apps.who.int/ebola/current-situation/ebola-situation-report-30-september-2015.

World Health Organization. (2015b). *Factors that contributed to undetected spread of the Ebola virus and impeded rapid containment*. Retrieved November 17, 2017, from http://www.who.int/csr/disease/ebola/one-year-report/factors/en/.

World Health Organization. (2016a). *Latest Ebola outbreak over in Liberia; West Africa is at zero, but new flare-ups are likely to occur*. Retrieved November 17, 2017, from http://www.who.int/mediacentre/news/releases/2016/ebola-zero-liberia/en/.

World Health Organization. (2016b). *Ebola virus disease. Fact sheet*. Retrieved November 17, 2017, from http://www.who.int/mediacentre/factsheets/fs103/en/.

World Health Organization. (2016c). *Ebola data and statistics: Situation summary by sex and age group, 11 May 2016*. Retrieved November 17, 2017, from http://apps.who.int/gho/data/view.ebola-sitrep.ebola-summary-latest-age-sex.

World Health Organization. (2016d). *Global Health Observatory data repository*. Retrieved November 17, 2017, from http://apps.who.int/gho/data/.

World Health Organization. (2017a). *WHO Statement on Caesarean Section Rates*. Retrieved on 17 November 2017, http://apps.who.int/iris/bitstream/10665/161442/1/WHO_RHR_15.02_eng.pdf.

World Health Organization. (2017b). *Ebola virus disease brings opportunities for improved blood systems in West Africa*. Retrieved November 17, 2017, from http://www.who.int/medicines/ebola-treatment/ebola_improved_bs_wa/en/.

Ebola Virus Disease and Pregnancy: Perinatal Transmission and Epidemiology

4

Lisa M. Bebell

4.1 Introduction

The 2013–2015 West African Ebola virus disease (EVD) outbreak led to 28,616 recorded cases (CDC 2016a), ranking it as the largest recorded viral hemorrhagic fever outbreak. Over 5,000 of these cases occurred among reproductive-aged women (WHO 2016a), and 75% of recorded deaths occurred in women (Luginaah et al. 2016). Despite these unprecedented numbers, official case counts likely underestimated the true number of cases and fatalities and may have misclassified some women with early pregnancies as nonpregnant, biasing mortality estimates (Aylward et al. 2014; CDC 2016a). The prolonged EVD outbreak devastated local health systems in Liberia, Sierra Leone, and Guinea and led to many thousands of deaths from other causes as crippled health systems were unable to provide usual care. Though the outbreak was a profound tragedy, the large number of EVD cases allowed for greater insights into EVD epidemiology, clinical course, and management strategies, particularly in smaller subpopulations such as pregnant and breastfeeding women. This discussion utilizes knowledge gained from recent outbreak to focus on the epidemiology of EVD in pregnant women, risk of perinatal EVD transmission to their fetuses and neonates, and subsequent pregnancy outcomes among EVD survivors.

4.2 Epidemiology of Ebola Infection in Pregnancy

Of the 112 cases of pregnant patients with EVD reported in the literature prior to the 2014 outbreak, aggregate maternal mortality was 86% (Akerlund et al. 2015; Baggi et al. 2014; Baize et al. 2014; Bower et al. 2016a; Caluwaerts et al. 2016; Chertow et al. 2014; Maganga et al. 2014; Mupapa et al. 1999; Oduyebo et al. 2015; Schieffelin et al. 2014; WHO 1978). The vast majority of these cases occurred during the first recognized Ebola outbreak in Zaire (now the Democratic Republic of Congo) in 1976, where retrospectively collected data suggested 82 pregnant women had EVD, of whom 73 (89%) died. Mortality among pregnant women was similar to the overall mortality of 88% in this

L. M. Bebell (✉)
Division of Infectious Diseases, Massachusetts General Hospital and Harvard Medical School, Boston, MA, USA
e-mail: lbebell@mgh.harvard.edu

D. A. Schwartz et al. (eds.), *Pregnant in the Time of Ebola*, Global Maternal and Child Health,
https://doi.org/10.1007/978-3-319-97637-2_4

outbreak (WHO 1978). A second outbreak in the Democratic Republic of Congo in 1995 infected 15 pregnant women, of whom 14 (93%) died (Mupapa et al. 1999). During this outbreak, mortality among nonpregnant women was lower (70%) than for pregnant women (Mupapa et al. 1999). Based on these cohorts and extrapolating from the clinical course of other infectious diseases contracted during pregnancy, it was hypothesized that infected pregnant women may fare worse than nonpregnant similarly aged women as a result of pregnancy-related physiologic changes, altered immune state, and placental Ebola virus infection (Mupapa et al. 1999; World Health Organization 1978). An important caveat when interpreting mortality rates in this early epidemic and many outbreaks prior to 2014 is that most patients were never definitively tested for EVD, potentially biasing epidemiologic associations and mortality estimates. In addition, almost all presumed EVD cases in the 1976 Zaire outbreak had received injections using poor infection control practices at a single hospital or one of its clinics, or had close contact with a case that did. In this outbreak, no persons with EVD and a history of having had a parenteral injection as their mode of exposure survived (Mupapa et al. 1999; WHO 1978), suggesting that mortality rates may also be affected by mode of exposure. In addition, despite hypotheses of increased mortality in pregnant women, there is no evidence that pregnant women are more susceptible to Ebola virus infection (CDC 2016c; Jamieson et al. 2014).

During the 2013–2015 West Africa EVD outbreak, mortality among all patients cared for in West Africa varied from 37 to 74% (Aylward et al. 2014; Qin et al. 2015; Yan et al. 2015), an overall mortality rate lower than prior epidemics. The majority of EVD cases were of reproductive age, similar to prior outbreaks (Aylward et al. 2014). The small number of EVD patients managed in Europe and the United States had a lower mortality of 18.5% (Uyeki et al. 2016b), though EVD mortality in West Africa also declined over the course of the outbreak. Though the 2013–2015 EVD outbreak was historically the largest, only five women with EVD and known pregnancies were enumerated in large studies published in 2014 and 2015 (Baize et al. 2014; Chertow et al. 2014; Schieffelin et al. 2014). Additional cases of EVD in pregnancy were later described in case reports and series (Akerlund et al. 2015; Baggi et al. 2014; Bower et al. 2016a; Caluwaerts et al. 2016; Henwood et al. 2017; Maganga et al. 2014; Oduyebo et al. 2015). Accurately assessing the burden of EVD in pregnancy and its epidemiology has been constrained by the paucity of data on attack rates in pregnancy and maternal and neonatal outcomes. This lack of data was exacerbated by the destabilization of health systems and reporting infrastructure during the 2013–2015 outbreak, weakening reporting mechanisms. In addition, it was not routine practice to determine pregnancy status among EVD suspects and cases (Henwood et al. 2017), raising the specter that many pregnancies—especially early pregnancies—may have gone undetected leading to misclassification bias and errors in determining true mortality rates in pregnancy. Pregnant women may also have been underrepresented in publications from the West African outbreak due to decreased mobility in late pregnancy and inability to reach treatment centers in a gravid state. With these caveats in mind, the limited pregnancy-specific outcomes data published from the West Africa EVD outbreak suggest mortality rates among pregnant women were lower than previously reported, with only 11 of 25 (44%) reported EVD-infected pregnant women dying (Akerlund et al. 2015; Baggi et al. 2014; Baize et al. 2014; Bower et al. 2016a; Caluwaerts et al. 2016; Chertow et al. 2014; Henwood et al. 2017; Maganga et al. 2014; Oduyebo et al. 2015; Schieffelin et al. 2014).

4.3 Managing Ebola Virus Disease in Pregnancy–Supportive Care and Obstetric Considerations

Correct triage of pregnant women suspected of having EVD versus those with non-EVD pregnancy symptoms or another infectious or hemorrhagic disorder is a challenging but critical first step in EVD outbreak settings. Many common pregnancy-associated conditions can mimic EVD, especially vaginal bleeding (Deaver and Cohen 2015). Potential clinical overlap between normal or complicated pregnancy and EVD makes pregnant women a particularly complex population to triage and care for during EVD outbreaks, and it is estimated that during the 2014–2016 EVD outbreak in Sierra Leone, only about 1.5% of pregnant women referred to EVD treatment units had EVD (Deaver and Cohen 2015). Once a pregnant woman is suspected of or diagnosed with EVD, early and aggressive supportive care is the mainstay of their clinical management (Bebell and Riley 2015). Pregnant women suspected of EVD should be admitted to an intensive care unit or other highly monitored isolation bed, when resources allow. In general, despite clear differences in disease pathogenesis and clinical course between patients with EVD and other infections, EVD patients can be treated according to guidelines for severe sepsis (Clark et al. 2012) or even dengue (Deen et al. 2015b) management. Maternal vital signs should guide initial treatment, with massive fluid resuscitation as the main support (Bebell and Riley 2015). A summary of therapies given to 27 EVD patients treated in the United States and Europe noted that most patients received antiemetics and broad-spectrum antibiotics; few received antidiarrheal agents (Uyeki et al. 2016b). Transfusion of blood products including transfusion of plasma from convalescent donors may also be beneficial (Lyon et al. 2014); however, a study in Guinea demonstrated no change in survival among convalescent plasma recipients (van Griensven et al. 2016).

The value of common obstetric interventions such as fetal monitoring, cesarean delivery, induction of labor, or pregnancy interruption must be considered on a case-by-case basis, because pregnant women with EVD are at risk of vascular collapse (Bebell and Riley 2015). Pregnant EVD patients are also at high risk of spontaneous, often preterm, labor, and thus, healthcare providers should be prepared for delivery at any time. Pregnancy outcomes in women with EVD can include spontaneous abortion, stillbirth, or delivery of a live fetus, and all products of delivery including the placenta should be considered potentially infectious with Ebola virus and handled accordingly. Neonatal survival is almost universally poor, and high fetal loss rates underscore the need to focus efforts on treatment of the pregnant mother (CDC 2014). Furthermore, pregnant women with decompensated EVD may not survive surgical delivery. Given the apparent improvement in survival for infected mothers in the most recent outbreak as compared with previous outbreaks, the use of obstetric interventions will depend on the mother's stability, available resuscitation options, and the obstetric intervention in question. Expectant management of labor seems the most appropriate current strategy (CDC 2014), and case reports of successful care of pregnant women suggest this is feasible (Oduyebo et al. 2015).

4.4 Transplacental and Perinatal Ebola Virus Transmission

Filoviruses, those viruses belonging to the family *Filoviridae*, include the Ebola virus and are negative-stranded, lipid-enveloped ribonucleic acid (RNA) viruses (Martines et al. 2014). They are classic zoonotic diseases, with index transmission occurring from an animal to human host (Martines et al. 2014). Bats are thought to be the primary asymptomatic reservoir for Ebola virus (Towner et al. 2009; Vogel 2014), but outside of laboratory and bush settings Ebola virus is transmitted by contact with the body fluids of an infected individual (Martines et al. 2014; Smith et al. 1982). Human blood, amniotic fluid, vaginal secretions, urine, saliva, sweat, feces, vomit, breast milk, and semen of an individual with EVD are potentially infectious (CDC 2015b). Ebola virus may be transmitted from mother to baby in utero, during delivery, or through contact with maternal body fluids after birth including breast milk.

Hematogenous spread of filovirus by the maternal bloodstream through the placenta and into the fetus and amniotic fluid is likely, as high titers of Ebola and other hemorrhagic fever viruses have been detected in placental tissue both during acute infection and after recovery (Baggi et al. 2014;

Caluwaerts et al. 2016; Jamieson et al. 2014; Oduyebo et al. 2015; Walker et al. 1982). One case report of a pregnant Guinean woman whose fetus died describes high titers of Ebola virus RNA detected via amniocentesis even after full maternal recovery (Baggi et al. 2014). This case and others support theories that the placenta is an Ebola viral reservoir, transplacental spread of filovirus occurs, and in utero fetal infection may result from altered immune defenses during pregnancy allowing filoviruses to cross the placenta. Fetal loss rates are nearly 100% for pregnant women EVD, with or without maternal death (Baggi et al. 2014; Jeffs 2006; Mehedi et al. 2011; Schieffelin et al. 2014).

No fetal autopsy or placental studies have been published from the 2014–2016 West Africa EVD outbreak (Schwartz 2017b), likely a result of international guidelines recommending against handling products of conception from EVD-affected pregnancies (Caluwaerts et al. 2016; CDC 2015c; WHO 2014). Placental histology and fetal autopsy could provide important additional insights into the mechanisms of transplacental Ebola virus transmission as it has for such other recent vertically transmitted emerging infections such as Zika virus (Rosenberg et al. 2017; Schwartz 2017a, b). However, even in the absence of tissue-based studies, the predilection of Ebola virus for fetal tissues can be inferred from epidemiologic studies of pregnancy outcomes in EVD-affected women. Multiple pregnant women who survived acute EVD went on to deliver stillborn fetuses testing positive for Ebola virus with high viral loads days to weeks after maternal convalescence (Baggi et al. 2014; Bower et al. 2016a; Caluwaerts et al. 2016; Caluwaerts and Lagrou 2014; Muehlenbachs et al. 2017). Histological specimens obtained from products of conception during two other non-Zaire strain EVD outbreaks in Uganda (in the year 2000 with the Sudan virus) and Democratic Republic of Congo (in the year 2012 with the Bundibugyo virus) provide some insight into the findings that may have been seen in fetuses and placentas from the 2014–2016 Zaire EVD outbreak. One placenta demonstrated mild subchorionitis, malaria pigmentation, and hemozoin crystals seen by electron microscopy, though all tissues were negative for Ebola virus antigen by immunohistochemical stain (Muehlenbachs et al. 2017). The second placenta had scattered atypical maternal macrophages in the intervillous space with degenerate-appearing nuclei, cytoplasmic blebs, and small eosinophilic cytoplasmic granules suggestive of viral inclusions. In this placenta, immunohistochemical analysis revealed Ebola virus antigen in circulating large atypical maternal mononuclear cells and villous syncytiotrophoblasts, staining intensely on the basal aspect (Muehlenbachs et al. 2017). The lining cells of the maternal blood vessels of the basal plate also stained positive for Ebola virus. No fetal or placental tissue necrosis and no virions were seen (Muehlenbachs et al. 2017). The presence of Ebola virus in the fetal trophoblast cells is strongly suggestive of transplacental virus transfer across the epithelial barrier to effect vertical Ebola virus transmission. The transplacental route of transmission is supported by evidence of Ebola virus in amniocentesis, fetal blood, and fetal swab specimens (Baggi et al. 2014; Caluwaerts et al. 2016). Ebola virus enters cells via macrominocytosis and clathrin-mediated endocytosis (Aleksandrowicz et al. 2011), which are molecular transport routes for fetal nutrition, making it plausible that these trafficking routes could be co-opted by Ebola virus and lead to fetal infection. Additionally, Ebola virus cellular infection requires the NPC1 gene (Carette et al. 2011), which is expressed in placental syncytiotrophoblast cells (Uhlen et al. 2015). In future outbreaks, safely sampling and analyzing products of conception including placental and fetal tissues will be critical to better understanding the pathogenesis of vertical Ebola virus transmission. Further analysis of the coexistence and potential interactions of placental malaria and Ebola virus should be a research priority, given the frequent epidemiological co-occurrence of these two pathogens, their placental tissue tropism, and the clinical interactions between malaria and EVD observed during the 2014–2016 West Africa EVD outbreak (Gignoux et al. 2016).

Though clinical and pathological findings support transplacental EVD transmission from mother to baby *in utero*, it is likely that transmission also occurs during delivery and through contact with maternal body fluids after birth including breast milk (Bausch et al. 2007). Ebola virus has been cultured from breast milk 15 days after symptom onset and Ebola virus RNA has been detected in breast milk by reverse transcription polymerase chain reaction (RT-PCR) 26 days after disease onset (Bausch et al. 2007; Nordenstedt et al. 2016). This evidence suggests that Ebola virus can be transmitted through breast milk, though other modes of mother-to-child transmission are also possible, given the frequent intimate contact between mothers and their children. Cumulative and repeated exposure to Ebola virus likely occurs in neonates surviving until birth and breastfeeding, similar to vertical transmission of HIV (Muehlenbachs et al. 2017). However, a study of EVD survivors in Sierra Leone demonstrated no excess risk of EVD transmission to infants during breast feeding (Bower et al. 2016b). Despite this finding, it is recommended that if safe alternatives to breastfeeding exist, mothers with suspected or confirmed EVD avoid close contact with their infants (CDC 2016d). This includes cessation of breastfeeding, resuming only when breast milk is shown to be Ebola virus-free by laboratory testing (CDC 2016d).

4.5 Outcomes of Ebola Virus Disease-Affected Pregnancies

Fetal and neonatal outcomes reported in the literature are consistent across EVD epidemics. Unfortunately, the vast majority of pregnancies affected by EVD terminate in fetal and neonatal loss. To date, outcomes data have been published for 60 confirmed or suspected Zaire Ebola virus cases in pregnancy, which resulted in 47 stillbirths or miscarriages (78%) and 13 live births (22%), of whom all except one died within 19 days of life (Akerlund et al. 2015; Baggi et al. 2014; Baize et al. 2014; Bower et al. 2016a; Caluwaerts et al. 2016; Chertow et al. 2014; Maganga et al. 2014; Mupapa et al. 1999; Nelson et al. 2016; Oduyebo et al. 2015; Schieffelin et al. 2014; WHO 1978). A case report published in 2017 described a single neonate born to a mother infected with EVD during pregnancy. The mother subsequently died, but the neonate received experimental EVD therapy with GS 5734 and the monoclonal antibody ZMapp (Dornemann et al. 2017). The child survived congenital EVD for at least 1 year after birth, meeting normal growth and developmental milestones (Dornemann et al. 2017). Beyond fetal demise and high neonatal mortality, knowledge of the effects of EVD on fetuses and neonates is limited by international guidelines recommending against pathology examination of fetal and placental tissue due to risk of exposure of staff to Ebola virus while performing autopsies and placental pathology evaluations (Caluwaerts et al. 2016; CDC 2015c; WHO 2014). Recently, visual cataracts have been described among children and adults surviving EVD (Grady 2017), and it is possible that children surviving congenital EVD may suffer similar ocular or neurologic complications and should be monitored closely.

4.6 Nosocomial Ebola Virus Disease Transmission

Many common pregnancy-associated conditions as well as endemic febrile illnesses can mimic EVD (Deaver and Cohen 2015). This clinical overlap makes pregnant women suspected of having EVD a particularly complex population to triage and care for, while awaiting the results of EVD testing. Using mathematical modeling studies of the 2014–2016 EVD outbreak in Sierra Leone, researchers estimated that of all suspected EVD cases in pregnancy referred to Ebola treatment units, only about 1.5% would have had EVD (Deaver and Cohen 2015). The authors speculate that the remainder of EVD suspect cases would have had common pregnancy-related infections including chorioamnionitis, septic abortion, and urinary tract infections (Deaver and Cohen 2015). An additional percentage likely presented with endemic infections such as *Salmonella typhi*, other hemorrhagic viral infections, placental abnormalities with bleeding, and even malaria (Deaver and Cohen 2015). The implication of these estimates is that EVD-negative women meeting criteria for suspected EVD could be placed in holding areas with EVD-positive women for hours to days while awaiting confirmatory EVD testing. Such cohorting practices may be necessary to efficiently utilize limited resources, but also has the potential to expose EVD-uninfected women to the risk of nosocomial EVD transmission and divert medical attention away from their true diagnosis, thus delaying treatment. Though EVD suspect case definitions are not specific enough to distinguish between EVD and other pregnancy-related diagnoses and endemic infections, case definitions that are too specific can also carry the risk of incorrectly triaging unidentified EVD patients to non-Ebola treatment facilities and leading to nosocomial transmission to unsuspecting healthcare workers and patients. One small study from the 2014–2016 West Africa EVD outbreak provided evidence of nosocomial EVD transmission on both maternity and pediatric wards in Ebola treatment units in Sierra Leone. The authors report that unrelated pediatric and maternal patients were admitted to separate non-ETU wards with illnesses not meeting EVD case definitions and then ultimately died of EVD after transmitting EVD to nosocomial contacts (Dunn et al. 2016), with an attack rate of 3% in the pediatric case and 13% in the maternal case (Dunn et al. 2016). EVD suspect case definitions should be revisited for future outbreaks to ensure that triage of pregnant women is optimized to minimize nosocomial transmission risks.

4.7 Sexual Transmission

Once recovered from EVD, a patient is likely immune to future infection by the same viral strain and is considered noninfectious to the nonimmune public. However, Ebola virus has been found in semen for 1 to 24 months after convalescence (Fallah 2016; Fischer et al. 2017; Green et al. 2016; Mayor 2015; Sow et al. 2016; Uyeki et al. 2016a). For this reason, abstinence or use of condoms for all sexual activity is recommended until male survivors know their semen is negative for Ebola virus (CDC 2015a; Deen et al. 2015a). Two separate studies following EVD survivors in Sierra Leone and Guinea detected Ebola virus RNA in semen for up to 9 months in 26% of male survivors studied (Deen et al. 2015a; Sow et al. 2016), with decreasing viral loads decrease over time, though the infectivity of semen with detectable Ebola virus is still unknown and being investigated (Sow et al. 2016). One case of sexual EVD transmission through penile-vaginal intercourse has been published, where transmission occurred from an EVD survivor whose semen was PCR-positive for Ebola virus more than 6 months after the onset of his illness (Christie et al. 2015). The WHO reports 15 additional cases of suspected male-to-female sexual transmission of Ebola virus in Liberia, 12 in Sierra Leone, and 3 in Guinea, though mortality rates of sexually acquired EVD are low, suggesting that mode of exposure to Ebola virus plays a role in pathogenesis (WHO 2015b). Better understanding the infectivity of semen from men surviving EVD is relevant to reproductive-aged women who may wish to become pregnant, but wish to reduce their risk of sexually acquired EVD. Though it is unclear how long viable Ebola virus is present in semen, condoms may help prevent Ebola virus transmission. However, condoms have not been studied clinically as precautions against sexual transmission of EVD, and latex condom pore size may be large enough to permit viral particle translocation. Despite case reports of sexually transmitted Ebola virus, the scarcity of new EVD cases since the official end of the outbreak suggests that sexual transmission is relatively rare. Using polymerase chain reaction (PCR) techniques, Ebola viral RNA has also been detected in vaginal secretions as late as 33 days after symptom onset (WHO 2016c). A paucity of data on Ebola virus in vaginal fluid limits interpretations of its infectiousness (WHO 2016c). Currently, no evidence exists to evaluate whether persistent Ebola virus in semen and vaginal fluid affects future birth outcomes among survivors. WHO recommends close monitoring of survivors who later become pregnant, but does not recommend enhanced precautions at delivery (WHO 2015b).

4.8 Impact of Ebola Virus Disease on Routine Pregnancy Care in West Africa

In Liberia, the EVD outbreak was associated with a decline in births that were accompanied by skilled birth attendants (SBAs) from 52% in the year prior to the outbreak to 38% during the outbreak (Bernstein September 20, 2014). Though the decline in skilled birth attendance may be multifactorial, it is thought that the perceived high mortality and risk of caring for pregnant EVD patients led to negative treatment given to pregnant women by healthcare providers and internalized stigma among pregnancy women seeking care. Perceived or actual lack of care for pregnant women during the epidemic may have led to lingering distrust between pregnant women and healthcare workers and a subsequent decline in care-seeking behavior among pregnant women in EVD-affected areas (Luginaah et al. 2016; Strong and Schwartz 2016, 2019). A Liberian study conducted in Bong demonstrated that facility-based births decreased by more than 75% over the course of 2014, as the EVD epidemic took hold (Lori et al. 2015). Similar findings were reported in Sierra Leone, comparing healthcare utilization by pregnant and peripartum women over a 10-month period during the EVD outbreak to a 12-month period before the outbreak (Jones et al. 2016). The authors found an 18% decrease in the number of women attending antenatal clinic, 11% decrease in women giving birth at a health care facility, and a 22% decrease in attendance of postnatal care visits, despite stable numbers of skilled healthcare workers and availability of supplies and cesarean delivery during the outbreak (Jones et al. 2016). Maternal mortality also increased by 34% and stillbirths by 24% during the EVD outbreak (Jones et al. 2016). These increases in mortality and stillbirth could be explained by delayed access to care by women during the outbreak, poorer care delivery, or even undiagnosed EVD in obstetric patients. Researchers surveyed 1,298 reproductive-aged rural-dwelling women in Liberia during the outbreak about their attitudes towards facility-based deliveries. Women reporting a belief that health facilities are or may be sources of Ebola transmission were significantly less likely to deliver at a facility (AOR = 0.59, 95%CI 0.36–0.97, $p = 0.038$), compared to women who did not report such a belief (Ly et al. 2016). These findings reinforce the notion that stigma likely played a role in delaying pregnant women's decisions to access care (Strong and Schwartz 2019).

4.9 Pregnancy Outcomes After Recovery from Ebola Virus Disase

As of May 2016, it was estimated that 17,306 persons would have survived the 2014–2016 West Africa EVD outbreak, including approximately 5,000 women of childbearing age (Kamali et al. 2016; WHO 2016b). The WHO recommends close monitoring of survivors who later become pregnant, but does not recommend enhanced precautions at subsequent delivery. There is a paucity of data on whether EVD infection impacts future pregnancies in terms of birth defects or obstetric outcomes, and additional research is needed to understand the effect of prior EVD on subsequent pregnancies and fertility among survivors.

Two case reports and a larger case series illustrate the range of outcomes that can occur in women who have recovered from EVD. The first case report is of an Ebola virus-positive stillborn fetus born from a woman who tested EVD-negative at the time of birth (Bower et al. 2016a). The woman was approximately 7 months' pregnant when she was exposed to EVD through nonsexual contact with her EVD-positive husband. She subsequently developed a relatively mild illness with nausea, fatigue, loss of appetite and abdominal pain, had vaginal leakage of some non-bloody fluid, and recovered. She later noticed the absence of fetal movements and went on to deliver an EVD-positive macerated fetus at a local healthcare center in Sierra Leone (Bower et al. 2016a). This case suggests that EVD during pregnancy can be variable in presentation and have devastating effects on the fetus even when maternal illness is relatively mild, and further argues for predilection of Ebola virus for placental and fetal tissues.

In the second case report, a West African EVD survivor became pregnant 22 weeks after her last negative PCR test for EBOV (designation for the Zaire ebolavirus—the Makona variant caused the West African epidemic) and delivered a healthy baby in the United States, 14 months after acute EBOV infection (Kamali et al. 2016). The authors describe working closely with public health officials in the state of California for weeks before the patient's due date, agreeing that at the time of delivery no additional precautions were needed to protect health workers above and beyond the procedures employed for normal deliveries in their hospital. The patient tested negative for Ebola virus prior to delivery, and neonatal skin swab, colostrum, blood, placental tissue, and umbilical cord blood samples all tested negative for Ebola virus (Kamali et al. 2016). The only laboratory evidence of prior maternal EVD was a high Ebola virus IgG antibody titer, indicating past infection. However, colostrum from the baby tested negative for EBOV IgG. No histological abnormalities were seen in the placenta or umbilical cord, and both stained negative for Ebola virus using immunohistochemistry (Kamali et al. 2016).

A recent case series published pregnancy outcomes among 70 Liberian EVD survivors (Fallah et al. 2016). Of these, 15 (22%) pregnancies ended in miscarriage, 4 (6%) in stillbirth, and 2 (3%) in elective pregnancy terminations (Fallah et al. 2016). Notably, EVD survivors who became pregnant within 2 months of discharge after EVD treatment had a higher proportion of miscarriages (50%). However, most miscarriages presented in women who became pregnant between 4 and 15 months after discharge, with an overall miscarriage rate of 22%, higher than the expected range of 11–13% for healthy West African women (Fallah et al. 2016). The overall prevalence of adverse pregnancy outcomes was 28%, suggesting a long-lasting negative impact of prior EVD on future pregnancy outcomes (Fallah et al. 2016). While these findings are concerning, further prospective evaluation and larger studies of EVD survivors and age-matched women without EVD will help confirm the prevalence of adverse future pregnancy outcomes and reduce the chance of miscarriage recall bias among EVD survivors. Current WHO guidelines state that women who have recovered from EVD are not infectious, should receive routine prenatal care, and that their deliveries should be attended by health workers employing the same routine protective equipment used for other deliveries (WHO 2015a). In contrast, women who survive EVD and remain pregnant should have their deliveries attended to by healthcare workers employing full PPE (WHO 2015a). After maternal recovery from EVD, there is thought to be no risk of transmission of Ebola virus through breastfeeding for infants born to survivors of EVD and the CDC recommends breastfeeding for EVD survivors who later give birth (CDC 2016b).

4.10 Summary

Pregnant women appear equally biologically susceptible to acquiring Ebola virus compared to nonpregnant individuals. It remains unclear whether risk of death is higher in pregnancy, though mounting evidence from the 2013–2015 epidemic suggests lower mortality than historical epidemics and comparable mortality between pregnant and nonpregnant reproductive-aged women with EVD (Henwood et al. 2017). Supportive care is the mainstay of EVD management during pregnancy and delivery until effective Ebola virus-specific vaccines and therapeutics are available. Decisions about mode of delivery and other obstetric interventions in pregnancy should be individualized, based on a range of factors including the status of the mother, gestational age and status of the fetus, and available resources including trained personnel and protective equipment. Maternal EVD is almost universally fatal to the developing fetus, and limited fetal autopsy data prevent inferences on risk of birth defects. In survivors, Ebola virus is detectable for months to years in semen and weeks in vaginal secretions. Though sexual transmission of Ebola virus has been documented from male survivors to female partners, this is a relatively rare occurrence. Birth outcomes among EVD survivors are now being published and may indicate a higher risk of miscarriage and stillbirth in this population. Additional research documenting pregnancy outcomes among survivors will be important to appropriately counsel women on likely future pregnancy outcomes.

References

Akerlund, E., Prescott, J., & Tampellini, L. (2015). Shedding of Ebola virus in an asymptomatic pregnant woman. *New England Journal of Medicine, 372*(25), 2467–2469. https://doi.org/10.1056/NEJMc1503275. Retrieved December 23, 2017, from http://www.nejm.org/doi/full/10.1056/NEJMc1503275#t=article.

Aleksandrowicz, P., Marzi, A., Biedenkopf, N., Beimforde, N., Becker, S., Hoenen, T., et al. (2011). Ebola virus enters host cells by macropinocytosis and clathrin-mediated endocytosis. *Journal of Infectious Diseases, 204*(Suppl 3), S957–S967. https://doi.org/10.1093/infdis/jir326. Retrieved December 22, 2017, from https://www.ncbi.nlm.nih.gov/pmc/articles/PMC3189988/.

Aylward, B., Barboza, P., Bawo, L., Bertherat, E., Bilivogui, P., Blake, I., et al. (2014). Ebola virus disease in West Africa—The first 9 months of the epidemic and forward projections. *New England Journal of Medicine, 371*(16), 1481–1495. https://doi.org/10.1056/NEJMoa1411100. Retrieved December 23, 2017, from http://www.nejm.org/doi/full/10.1056/NEJMoa1411100#t=article.

Baggi, F., Taybi, A., Kurth, A., Herp, M. V., Caro, A. D., Wölfel, R., et al. (2014). Management of pregnant women infected with Ebola virus in a treatment Centre in Guinea, June 2014. *Eurosurveillance, 19*(49), 2–5. Retrieved December 24, 2017, from http://www.eurosurveillance.org/images/dynamic/EE/V19N49/V19N49.pdf.

Baize, S., Pannetier, D., Oestereich, L., Rieger, T., Koivogui, L., Magassouba, N., et al. (2014). Emergence of Zaire Ebola virus disease in Guinea. *New England Journal of Medicine, 371*(15), 1418–1425. https://doi.org/10.1056/NEJMoa1404505. Retrieved December 24, 2017, from http://www.nejm.org/doi/full/10.1056/NEJMoa1404505#t=article.

Bausch, D. G., Towner, J. S., Dowell, S. F., Kaducu, F., Lukwiya, M., Sanchez, A., et al. (2007). Assessment of the risk of Ebola virus transmission from bodily fluids and fomites. *Journal of Infectious Diseases, 196*(Suppl 2), S142–S147. https://doi.org/10.1086/520545. Retrieved December 23, 2017, from https://academic.oup.com/jid/article/196/Supplement_2/S142/858852.

Bebell, L. M., & Riley, L. E. (2015). Ebola virus disease and Marburg disease in pregnancy: A review and management considerations for filovirus infection. *Obstetrics & Gynecology, 125*(6), 1293–1298. https://doi.org/10.1097/aog.0000000000000853. Retrieved December 23, 2017, from https://www.ncbi.nlm.nih.gov/pmc/articles/PMC4443859/.

Bernstein, L. (2014, September 20). With Ebola crippling the health system, Liberians die of routine medical problems. *The Washington Post*. Retrieved December 24, 2017, from https://www.washingtonpost.com/world/africa/with-ebola-crippling-the-health-system-liberians-die-of-routine-medical-problems/2014/09/20/727dcfbe-400b-11e4-b03f-de718edeb92f_story.html?utm_term=.65dae977831e.

Bower, H., Grass, J. E., Veltus, E., Brault, A., Campbell, S., Basile, A. J., et al. (2016a). Delivery of an Ebola virus-positive stillborn infant in a rural community health center, Sierra Leone, 2015. *American Journal of Tropical Medicine and Hygiene, 94*(2), 417–419. https://doi.org/10.4269/ajtmh.15-0619.

Bower, H., Johnson, S., Bangura, M. S., Kamara, A. J., Kamara, O., Mansaray, S. H., et al. (2016b). Effects of mother's illness and breastfeeding on risk of Ebola virus disease in a cohort of very young children. *PLoS Neglected Tropical Diseases, 10*(4), e0004622. https://doi.org/10.1371/journal.pntd.0004622. Retrieved December 24, 2017, from http://journals.plos.org/plosntds/article?id=10.1371/journal.pntd.0004622.

Caluwaerts, S., & Lagrou, D. (2014). *Guidance paper Ebola treatment Centre (ETC): Pregnant and lactating women*. Retrieved December 24, 2017, from https://www.rcog.org.uk/globalassets/documents/news/etc-preg-guidance-paper.pdf.

Caluwaerts, S., Fautsch, T., Lagrou, D., Moreau, M., Modet Camara, A., Gunther, S., et al. (2016). Dilemmas in managing pregnant women with Ebola: 2 case reports. *Clinical Infectious Diseases, 62*(7), 903–905. https://doi.org/10.1093/cid/civ1024. Retrieved December 24, 2017, from https://academic.oup.com/cid/article/62/7/903/2462754.

Carette, J. E., Raaben, M., Wong, A. C., Herbert, A. S., Obernosterer, G., Mulherkar, N., et al. (2011). Ebola virus entry requires the cholesterol transporter Niemann-Pick C1. *Nature, 477*(7364), 340–343. https://doi.org/10.1038/nature10348. Retrieved December 24, 2017, from https://www.nature.com/articles/nature10348.

CDC. (2014). *Guidance for screening and caring for pregnant women with Ebola virus disease for healthcare providers in U.S. hospitals*. Retrieved November 20, 2014, from http://www.cdc.gov/vhf/ebola/hcp/guidance-maternal-health.html.

CDC. (2015a). *Ebola (Ebola virus disease)—Q&As on transmission*. Retrieved October 17, 2016, from http://www.cdc.gov/vhf/ebola/transmission/qas.html.

CDC. (2015b). *Ebola (Ebola virus disease): Transmission*. Retrieved December 19, 2017, from https://www.cdc.gov/vhf/ebola/transmission/index.html.

CDC. (2015c). *Guidance for safe handling of human remains of Ebola patients in U. S. Hospitals and mortuaries*. Retrieved October 17, 2016, from http://www.cdc.gov/vhf/ebola/healthcare-us/hospitals/handling-human-remains.html.

CDC. (2016a). *2014 Ebola outbreak in West Africa: Case counts*. Retrieved October 9, 2017, from http://www.cdc.gov/vhf/ebola/outbreaks/2014-west-africa/case-counts.html.

CDC. (2016b). *Care of a neonate born to a mother who is confirmed to have Ebola, is a person under investigation, or has been exposed to Ebola*. Retrieved October 17, 2016, from http://www.cdc.gov/vhf/ebola/healthcare-us/hospitals/neonatal-care.html.

CDC. (2016c). *Guidance for screening and caring for pregnant women with Ebola virus disease for healthcare providers in U.S. hospitals*. Retrieved October 17, 2016, from http://www.cdc.gov/vhf/ebola/healthcare-us/hospitals/pregnant-women.html.

CDC. (2016d). *Recommendations for breastfeeding/infant feeding in the context of Ebola*. Retrieved October 17, 2016, from http://www.cdc.gov/vhf/ebola/hcp/recommendations-breastfeeding-infant-feeding-ebola.html.

Chertow, D. S., Kleine, C., Edwards, J. K., Scaini, R., Giuliani, R., & Sprecher, A. (2014). Ebola virus disease in West Africa—Clinical manifestations and management. *New England Journal of Medicine, 371*, 2054–2057. https://doi.org/10.1056/NEJMp1413084. Retrieved December 23, 2017, from http://www.nejm.org/doi/full/10.1056/NEJMp1413084#t=article.

Christie, A., Davies-Wayne, G. J., Cordier-Lasalle, T., Blackley, D. J., Laney, A. S., Williams, D. E., et al. (2015). Possible sexual transmission of Ebola virus—Liberia, 2015. *MMWR Morbidity and Mortality Weekly Report, 64*(17), 479–481. Retrieved December 23, 2017, from https://www.cdc.gov/mmwr/preview/mmwrhtml/mm6417a6.htm.

Clark, D. V., Jahrling, P. B., & Lawler, J. V. (2012). Clinical management of filovirus-infected patients. *Viruses, 4*(9), 1668–1686. https://doi.org/10.3390/v4091668.

Deaver, J. E., & Cohen, W. R. (2015). Ebola virus screening during pregnancy in West Africa: Unintended consequences. *Journal of Perinatal Medicine, 43*(6), 649–655. https://doi.org/10.1515/jpm-2015-0118.

Deen, G. F., Knust, B., Broutet, N., Sesay, F. R., Formenty, P., Ross, C., et al. (2015a). Ebola RNA persistence in semen of Ebola virus disease survivors—Preliminary report. *The New England Journal of Medicine, 377*, 1428. https://doi.org/10.1056/NEJMoa1511410.

Deen, J., Dondorp, A. M., & White, N. J. (2015b). Treatment of Ebola. *The New England Journal of Medicine, 372*(17), 1673–1674. https://doi.org/10.1056/NEJMc1500452#SA1. Retrieved December 24, 2017, from http://www.nejm.org/doi/full/10.1056/NEJMc1500452#t=article.

Dornemann, J., Burzio, C., Ronsse, A., Sprecher, A., De Clerck, H., Van Herp, M., et al. (2017). First newborn baby to receive experimental therapies survives Ebola virus disease. *Journal of Infectious Diseases, 215*(2), 171–174. https://doi.org/10.1093/infdis/jiw493. Retrieved December 24, 2017, from https://academic.oup.com/jid/article/215/2/171/2877903.

Dunn, A. C., Walker, T. A., Redd, J., Sugerman, D., McFadden, J., Singh, T., et al. (2016). Nosocomial transmission of Ebola virus disease on pediatric and maternity wards: Bombali and Tonkolili, Sierra Leone, 2014. *American Journal of Infection Control, 44*(3), 269–272. https://doi.org/10.1016/j.ajic.2015.09.016.

Fallah, M. (2016). *A cohort study of survivors of Ebola virus infection in Liberia (PREVAIL III)*. Paper presented at the Conference on Retroviruses and Opportunistic Infections, Boston.

Fallah, M. P., Skrip, L. A., Dahn, B. T., Nyenswah, T. G., Flumo, H., Glayweon, M., et al. (2016). Pregnancy outcomes in Liberian women who conceived after recovery from Ebola virus disease. *Lancet Global Health, 4*(10), e678–e679. https://doi.org/10.1016/s2214-109x(16)30147-4. Retrieved December 24, 2017, from http://www.thelancet.com/journals/langlo/article/PIIS2214-109X(16)30147-4/fulltext.

Fischer 2nd, W. A., Loftis, A. J., & Wohl, D. A. (2017). Screening of genital fluid for Ebola virus. *Lancet Global Health, 5*(1), e32. https://doi.org/10.1016/s2214-109x(16)30300-x. Retrieved December 24, 2017, from http://www.thelancet.com/journals/langlo/article/PIIS2214-109X(16)30300-X/fulltext.

Gignoux, E., Azman, A. S., de Smet, M., Azuma, P., Massaquoi, M., Job, D., et al. (2016). Effect of artesunate-amodiaquine on mortality related to Ebola virus disease. *New England Journal of Medicine, 374*(1), 23–32. https://doi.org/10.1056/NEJMoa1504605. Retrieved December 23, 2017, from http://www.nejm.org/doi/full/10.1056/NEJMoa1504605#t=article.

Grady, D. (2017). Ebola's legacy: Children with cataracts. *The New York Times*. Retrieved from https://www.nytimes.com/2017/10/19/health/ebola-survivors-cataracts.html?emc=edit_th_20171020&nl=todaysheadlines&nlid=52019166&_r=2

Green, E., Hunt, L., Ross, J. C., Nissen, N. M., Curran, T., Badhan, A., et al. (2016). Viraemia and Ebola virus secretion in survivors of Ebola virus disease in Sierra Leone: A cross-sectional cohort study. *Lancet Infectious Diseases, 16*(9), P1052–P1056. https://doi.org/10.1016/s1473-3099(16)30060-3. Retrieved December 23, 2017, from http://www.thelancet.com/journals/laninf/article/PIIS1473-3099(16)30060-3/abstract.

Henwood, P. C., Bebell, L. M., Roshania, R., Wolfman, V., Mallow, M., Kalyanpur, A., et al. (2017). Ebola virus disease and pregnancy: A retrospective cohort study of patients managed at five Ebola treatment units in West Africa. *Clinical Infectious Diseases, 65*(2), 292–299. https://doi.org/10.1093/cid/cix290. Retrieved December 23, 2017, from https://academic.oup.com/cid/article/65/2/292/3097901.

Jamieson, D. J., Uyeki, T. M., Callaghan, W. M., Meaney-Delman, D., & Rasmussen, S. A. (2014). What obstetrician-gynecologists should know about Ebola: A perspective from the Centers for Disease Control and Prevention. *Obstetrics & Gynecology, 124*(5), 1005–1010. https://doi.org/10.1097/aog.0000000000000533. Retrieved December 24, 2017, from http://ldh.louisiana.gov/assets/oph/Center-PHCH/Center-CH/infectious-epi/EpiManual/ObstetricsEbola.pdf.

Jeffs, B. (2006). A clinical guide to viral haemorrhagic fevers: Ebola, Marburg and Lassa. *Tropical Doctor, 36*(1), 1–4. https://doi.org/10.1258/004947506775598914.

Jones, S. A., Gopalakrishnan, S., Ameh, C. A., White, S., & van den Broek, N. R. (2016). 'Women and babies are dying but not of Ebola': The effect of the Ebola virus epidemic on the availability, uptake and outcomes of maternal and newborn health services in Sierra Leone. *BMJ Global Health, 1*(3), e000065. https://doi.org/10.1136/bmjgh-2016-000065. Retrieved December 24, 2017, from https://www.ncbi.nlm.nih.gov/pmc/articles/PMC5321347/.

Kamali, A., Jamieson, D. J., Kpaduwa, J., Schrier, S., Kim, M., Green, N. M., et al. (2016). Pregnancy, labor, and delivery after Ebola virus disease and implications for infection control in obstetric services, United States. *Emerging Infecioust Dieases, 22*(7), 1156–1161. https://doi.org/10.3201/eid2207.160269. Retrieved December 23, 2017, from https://www.ncbi.nlm.nih.gov/pmc/articles/PMC4918171/.

Lori, J. R., Rominski, S. D., Perosky, J. E., Munro, M. L., Williams, G., Bell, S. A., et al. (2015). A case series study on the effect of Ebola on facility-based deliveries in rural Liberia. *BMC Pregnancy and Childbirth, 15*, 254. https://doi.org/10.1186/s12884-015-0694-x. Retrieved December 24, 2017, from https://bmcpregnancychildbirth.biomedcentral.com/articles/10.1186/s12884-015-0694-x.

Luginaah, I. N., Kangmennaang, J., Fallah, M., Dahn, B., Kateh, F., & Nyenswah, T. (2016). Timing and utilization of antenatal care services in Liberia: Understanding the pre-Ebola epidemic context. *Social Science & Medicine, 160*, 75–86. https://doi.org/10.1016/j.socscimed.2016.05.019.

Ly, J., Sathananthan, V., Griffiths, T., Kanjee, Z., Kenny, A., Gordon, N., et al. (2016). Facility-based delivery during the Ebola virus disease epidemic in rural Liberia: Analysis from a cross-sectional, population-based household survey. *PLoS Medicine, 13*(8), e1002096. https://doi.org/10.1371/journal.pmed.1002096. Retrieved December 24, 2017, from http://journals.plos.org/plosmedicine/article?id=10.1371/journal.pmed.1002096.

Lyon, G. M., Mehta, A. K., Varkey, J. B., Brantly, K., Plyler, L., McElroy, A. K., et al. (2014). Clinical care of two patients with Ebola virus disease in the United States. *New England Journal of Medicine, 371*(25), 2402–2409. https://doi.org/10.1056/NEJMoa1409838. Retrieved December 23, 2017, from http://www.nejm.org/doi/full/10.1056/NEJMoa1409838#t=article.

Maganga, G. D., Kapetshi, J., Berthet, N., Kebela Ilunga, B., Kabange, F., Mbala Kingebeni, P., et al. (2014). Ebola virus disease in the Democratic Republic of Congo. *New England Journal of Medicine, 371*(22), 2083–2091. https://doi.org/10.1056/NEJMoa1411099. Retrieved December 24, 2017, from http://www.nejm.org/doi/full/10.1056/NEJMoa1411099#t=article.

Martines, R. B., Ng, D. L., Greer, P. W., Rollin, P. E., & Zaki, S. R. (2014). Tissue and cellular tropism, pathology and pathogenesis of Ebola and Marburg viruses. *Journal of Pathology, 235*, 153. https://doi.org/10.1002/path.4456.

Mayor, S. (2015). Ebola virus persists and may be transmitted in survivors' semen, studies show. *British Medical Journal, 351*, h5511. https://doi.org/10.1136/bmj.h5511. Retrieved December 23, 2017, from https://search.proquest.com/openview/aff82d3e75433bbeb2b82af331bd079b/1?pq-origsite=gscholar&cbl=2043523.

Mehedi, M., Groseth, A., Feldmann, H., & Ebihara, H. (2011). Clinical aspects of Marburg hemorrhagic fever. *Future Virology, 6*(9), 1091–1106. https://doi.org/10.2217/fvl.11.79.

Muehlenbachs, A., de la Rosa Vazquez, O., Bausch, D. G., Schafer, I. J., Paddock, C. D., Nyakio, J. P., et al. (2017). Ebola virus disease in pregnancy: Clinical, histopathologic, and immunohistochemical findings. *Journal of Infectious Diseases, 215*(1), 64–69. https://doi.org/10.1093/infdis/jiw206.

Mupapa, K., Mukundu, W., Bwaka, M. A., Kipasa, M., De Roo, A., Kuvula, K., et al. (1999). Ebola hemorrhagic fever and pregnancy. *Journal of Infectious Diseases, 179*(Suppl 1), S11–S12. https://doi.org/10.1086/514289.

Nelson, J. M., Griese, S. E., Goodman, A. B., & Peacock, G. (2016). Live neonates born to mothers with Ebola virus disease: A review of the literature. *Journal of Perinatology, 36*(6), 411–414. https://doi.org/10.1038/jp.2015.189.

Nordenstedt, H., Bah, E. I., de la Vega, M. A., Barry, M., N'Faly, M., Crahay, B., et al. (2016). Ebola virus in breast milk in an Ebola virus-positive mother with twin babies, Guinea, 2015. *Emerging Infectious Diseases, 22*(4), 759–760. https://doi.org/10.3201/eid2204.151880.

Oduyebo, T., Pineda, D., Lamin, M., Leung, A., Corbett, C., & Jamieson, D. J. (2015). A pregnant patient with Ebola virus disease. *Obstetrics & Gynecology, 126*(6), 1273–1275. https://doi.org/10.1097/aog.0000000000001092.

Qin, E., Bi, J., Zhao, M., Wang, Y., Guo, T., Yan, T., et al. (2015). Clinical features of patients with Ebola virus disease in Sierra Leone. *Clinical Infectious Diseases, 61*(4), 491–495. https://doi.org/10.1093/cid/civ319.

Rosenberg, A. Z., Yu, W., Hill, D. A., Reyes, C. A., & Schwartz, D. A. (2017). Placental pathology of Zika virus: Viral infection of the placenta induces villous stromal macrophage (hofbauer cell) proliferation and hyperplasia. *Archives of Pathology & Laboratory Medicine, 141*(1), 43–48. https://doi.org/10.5858/arpa.2016-0401-OA. Retrieved December 24, 2017, from http://www.archivesofpathology.org/doi/full/10.5858/arpa.2016-0401-OA.

Schieffelin, J. S., Shaffer, J. G., Goba, A., Gbakie, M., Gire, S. K., Colubri, A., et al. (2014). Clinical illness and outcomes in patients with Ebola in Sierra Leone. *New England Journal of Medicine, 371*(22), 2092–2100. https://doi.org/10.1056/NEJMoa1411680. Retrieved December 24, 2017, from http://www.nejm.org/doi/full/10.1056/NEJMoa1411680#t=article.

Schwartz, D. A. (2017a). Autopsy and postmortem studies are concordant: Pathology of Zika virus infection is neurotropic in fetuses and infants with microcephaly following transplacental transmission. *Archives of Pathology & Laboratory Medicine, 141*(1), 68–72. https://doi.org/10.5858/arpa.2016-0343-OA. Retrieved December 22, 2017, from http://www.archivesofpathology.org/doi/full/10.5858/arpa.2016-0343-OA.

Schwartz, D. A. (2017b). Viral infection, proliferation, and hyperplasia of Hofbauer cells and absence of inflammation characterize the placental pathology of fetuses with congenital Zika virus infection. *Archives of Gynecology and Obstetrics, 295*(6), 1361–1368. https://doi.org/10.1007/s00404-017-4361-5. Retrieved December 22, 2017, from https://www.ncbi.nlm.nih.gov/pmc/articles/PMC5429341/.

Smith, D. H., Johnson, B. K., Isaacson, M., Swanapoel, R., Johnson, K. M., Killey, M., et al. (1982). Marburg-virus disease in Kenya. *Lancet, 1*(8276), 816–820.

Sow, M. S., Etard, J. F., Baize, S., Magassouba, N., Faye, O., Msellati, P., et al. (2016). New evidence of long-lasting persistence of Ebola virus genetic material in semen of survivors. *Journal of Infectious Diseases, 214*(10), 1475–1476. https://doi.org/10.1093/infdis/jiw078. Retrieved December 24, 2017, from https://academic.oup.com/jid/article/214/10/1475/2514585.

Strong, A., & Schwartz, D. A. (2016). Sociocultural aspects of risk to pregnant women during the 2013-2015 multinational Ebola virus outbreak in West Africa. *Health Care for Women International, 37*(8), 922–942. https://doi.org/10.1080/07399332.2016.1167896.

Strong, A., & Schwartz, D. (2019). Effects of the West African Ebola epidemic on health care of pregnant women—Stigmatization with and without infection. In D. A. Schwartz, J. N. Anoko, & S.A. Abramowitz (Eds.), *Pregnant in the time of Ebola: Women and their children in the West African Ebola Epidemic*. New York: Springer.

Towner, J. S., Amman, B. R., Sealy, T. K., Carroll, S. A., Comer, J. A., Kemp, A., et al. (2009). Isolation of genetically diverse Marburg viruses from Egyptian fruit bats. *PLoS Pathogens, 5*(7), e1000536. https://doi.org/10.1371/journal.ppat.1000536. Retrieved December 24, 2017, from http://journals.plos.org/plospathogens/article?id=10.1371/journal.ppat.1000536.

Uhlen, M., Fagerberg, L., Hallstrom, B. M., Lindskog, C., Oksvold, P., Mardinoglu, A., et al. (2015). Proteomics. Tissue-based map of the human proteome. *Science, 347*(6220), 1260419. https://doi.org/10.1126/science.1260419. Retrieved December 22, 2017, from http://science.sciencemag.org/content/347/6220/1260419.long.

Uyeki, T. M., Erickson, B. R., Brown, S., McElroy, A. K., Cannon, D., Gibbons, A., et al. (2016a). Ebola virus persistence in semen of male survivors. *Clinical Infectious Diseases, 62*(12), 1552–1555. https://doi.org/10.1093/cid/ciw202. Retrieved December 21, 2017, from https://academic.oup.com/cid/article/62/12/1552/1745446.

Uyeki, T. M., Mehta, A. K., Davey Jr., R. T., Liddell, A. M., Wolf, T., Vetter, P., et al. (2016b). Clinical management of Ebola virus disease in the United States and Europe. *The New England Journal of Medicine, 374*(7), 636–646. https://doi.org/10.1056/NEJMoa1504874. Retrieved December 21, 2017, from http://www.nejm.org/doi/full/10.1056/NEJMoa1504874#t=article.

van Griensven, J., Edwards, T., de Lamballerie, X., Semple, M. G., Gallian, P., Baize, S., et al. (2016). Evaluation of convalescent plasma for Ebola virus disease in Guinea. *New England Journal of Medicine, 374*(1), 33–42. https://doi.org/10.1056/NEJMoa1511812. Retrieved December 23, 2017, from http://www.nejm.org/doi/full/10.1056/NEJMoa1511812#t=article.

Vogel, G. (2014). Infectious disease. Are bats spreading Ebola across sub-Saharan Africa? *Science, 344*(6180), 140. https://doi.org/10.1126/science.344.6180.140. Retrieved December 24, 2017, from http://www.chinacdc.cn/jkzt/crb/ablcxr/zstd_6214/201408/P020140818534819230508.pdf.

Walker, D. H., McCormick, J. B., Johnson, K. M., Webb, P. A., Komba-Kono, G., Elliott, L. H., et al. (1982). Pathologic and virologic study of fatal Lassa fever in man. *American Journal of Pathology, 107*(3), 349–356. Retrieved December 23, 2017, from http://europepmc.org/backend/ptpmcrender.fcgi?accid=PMC1916239&blobtype=pdf.

WHO. (1978). Ebola haemorrhagic fever in Zaire, 1976. *Bulletin of the World Health Organization, 56*(2), 271–293.

WHO. (2014). *Infection prevention and control guidance for care of patients in health-care settings, with focus on Ebola interim guidance*. Retrieved December 24, 2017, from http://www.who.int/csr/resources/publications/ebola/filovirus_infection_control/en/.

WHO. (2015a). *Ebola virus disease in pregnancy: Screening and management of Ebola cases, contacts and survivors*. Retrieved October 8, 2017, from http://www.who.int/csr/resources/publications/ebola/pregnancy-guidance/en/.

WHO. (2015b). *WHO meeting on survivors of Ebola virus disease: Clinical care of EVD survivors, Freetown, 3–4 August 2015: Meeting report*. Retrieved November 27, 2016, from http://www.who.int/mediacentre/events/2015/meeting-on-ebola-survivors/en/.

WHO. (2016a). *Ebola data and statistics*. Retrieved October 17, 2016, from http://apps.who.int/gho/data/view.ebola-sitrep.ebola-summary-20160511?lang=en.

WHO. (2016b). *Ebola data and statistics: Situation summary 11 May, 2016*. Retrieved October 8, 2017, from http://apps.who.int/gho/data/view.ebola-sitrep.ebola-summary-latest?lang=en.

WHO. (2016c). *Interim advice on the sexual transmission of the Ebola virus disease*. Retrieved October 17, 2016, from http://www.who.int/reproductivehealth/topics/rtis/ebola-virus-semen/en/.

World Health Organization. (1978). Ebola haemorrhagic fever in Sudan, 1976. Report of a WHO/International Study Team. *Bulletin of the World Health Organization, 56*(2), 247–270.

Yan, T., Mu, J., Qin, E., Wang, Y., Liu, L., Wu, D., et al. (2015). Clinical characteristics of 154 patients suspected of having Ebola virus disease in the Ebola holding center of Jui Government Hospital in Sierra Leone during the 2014 Ebola outbreak. *European Journal of Clinical Microbiology & Infectious Diseases, 34*(10), 2089–2095. https://doi.org/10.1007/s10096-015-2457-z.

Comprehensive Clinical Care for Infants and Children with Ebola Virus Disease

5

Indi Trehan, Peter Matthew George, and Charles W. Callahan

5.1 Introduction

In addition to the devastation that the 2013–2015 Ebola epidemic wrought on the health of pregnant women and their fetuses, infants and children throughout Guinea, Liberia, and Sierra Leone also suffered immensely during the outbreak. Those under 5 years of age, as is often the case when it comes to neglected infectious diseases that disproportionately affect the impoverished, were particularly vulnerable (Frieden 2015; WHO Ebola Response Team 2015). Due to the nature of the transmission of the virus, children during this outbreak generally had lower rates of infection than adults (Bower et al. 2016; Glynn 2015; McElroy et al. 2014; WHO Ebola Response Team 2015). Nevertheless, the Ebola outbreak created havoc at all levels of the child health care system (Hermans et al. 2017; UNICEF 2014), disrupted innumerable family and social structures, and posed a direct threat to child health even among those who were not infected (Nebehay 2015; Peacock et al. 2014). Given the recognition that Ebola may even occur without overt clinical manifestations, children may even be infected by breast milk from a mother who was never known to have Ebola (Sissoko et al. 2017).

Neonates born to mothers with Ebola are highly unlikely to survive (Nelson et al. 2016) without experimental therapies (Dornemann et al. 2017), and we will thus focus on children who acquire Ebola through traditional horizontal transmission from a close contact or family member. The specifics of the complicated pathophysiologic cascade and societal barriers to health care that led to high rates of morbidity and mortality in children with Ebola are outside the scope of this review, as here we will focus on the highly challenging care of children at the bedside in the context of an Ebola outbreak. The clinical protocols described are based on published literature from the outbreak,

I. Trehan (✉)
Lao Friends Hospital for Children, Luang Prabang, Lao PDR

Washington University in St. Louis, St. Louis, MO, USA
e-mail: indi@alum.berkeley.edu

P. M. George
Sierra Leone Ministry of Health, Port Loko, Sierra Leone

University of Sierra Leone, Freetown, Sierra Leone

C. W. Callahan
University of Maryland, Baltimore, MD, USA

D. A. Schwartz et al. (eds.), *Pregnant in the Time of Ebola*, Global Maternal and Child Health,
https://doi.org/10.1007/978-3-319-97637-2_5

modifications of standard pediatric critical and emergency care based on first principles, and our own experiences caring for large numbers of suspect and infected children with Ebola (Trehan et al. 2016).

5.2 Protocol Development

Our clinical experience with providing care for Ebola patients came at the Maforki Ebola Holding and Treatment Centre (Fig. 5.1), operated by the Sierra Leone Ministry of Health in partnership with the international nongovernmental organization (NGO) Partners In Health and the Cuban Medical Brigade coordinated by the World Health Organization (Baez et al. 2015). Located in Port Loko,[1] Sierra Leone, this 106-bed facility was repurposed from a former vocational school (rather than being specifically purpose-built) (Fig. 5.2) due to the urgency of providing care for Ebola patients in this geographic area. The center operated from October 2014 to March 2015, consisting of a holding ward for persons with suspected Ebola virus disease (EVD) pending confirmation of their infection. It also had a treatment ward for persons with confirmed EVD (World Health Organization 2015), as well as a separate holding room for suspected pediatric patients (Fitzgerald et al. 2016a) (Fig. 5.3). Of the 910 persons admitted during the unit's operation, 29% were under 18 years old and nearly 10% were under 5 years of age. The reported case fatality rates (CFRs) were reported to vary from 50 to 80%

Fig. 5.1 Front entrance of Maforki Ebola Holding and Treatment Centre, Port Loko, Sierra Leone

[1]Port Loko is the capital city and the second largest town in the Port Loko District of the Northern Province of Sierra Leone. It has a population of approximately 24,000 inhabitants, the majority of whom are from the *Temne* ethnic group, speaking the *Temne* and *Krio* languages. Lying 45 miles east of Freetown, Port Loko is an importance bauxite mining and trade center for the country.

Fig. 5.2 View of interior courtyard of the suspect area of the Maforki Ebola Holding and Treatment Centre

among young children by others during the outbreak (Damkjaer et al. 2017; Frieden 2015; Fitzgerald et al. 2016b; Nebehay 2015; Peacock et al. 2014; Shah et al. 2016; Smit et al. 2017; WHO Ebola Response Team 2015) and were consistent with our experience.

The general principle of the protocols is that aggressive critical care, initiated early in the course of the illness, should decrease the mortality rate from EVD in all patients, including children (Chertow et al. 2014; Eriksson et al. 2015; Farmer 2015; Fowler et al. 2014; Lamontagne et al. 2014; Lyon et al. 2014; Schibler et al. 2015; Wolf et al. 2015). Given the limitations in the ability to fully examine patients, together with the challenging infection prevention and control requirements for this virus in resource-constrained tropical settings, many of the interventions are empirical and not able to necessarily be individualized to a given patient's condition as might be done in a pediatric ETU in a well-resourced setting (DeBiasi et al. 2016).

This aggressive and comprehensive approach was most remarkably demonstrated at the Hastings Ebola Treatment Unit (ETU) near Freetown, Sierra Leone, where among 581 patients they reported a case fatality rate approaching 30% (Ansumana et al. 2015). It is unclear how well this approach worked for children in their center as these data were not reported, but did serve as a foundation for our own protocols (Trehan et al. 2016), those of other treatment centers (Smit et al. 2017), and ultimately the revised World Health Organization (WHO) management guidelines (World Health Organization 2016a). One of the more challenging elements is the recognition that in impoverished tropical areas, most children admitted for care will have some degree of chronic malnutrition (often superimposed with acute malnutrition) and that this requires more careful attention to fluid and nutritional balance (Iannotti et al. 2015; WHO/UNICEF/WFP 2014).

We acknowledge that these are challenging protocols to follow and will in many cases be aspirational rather than completely feasible in all settings. At the very least, we hope that they can serve as

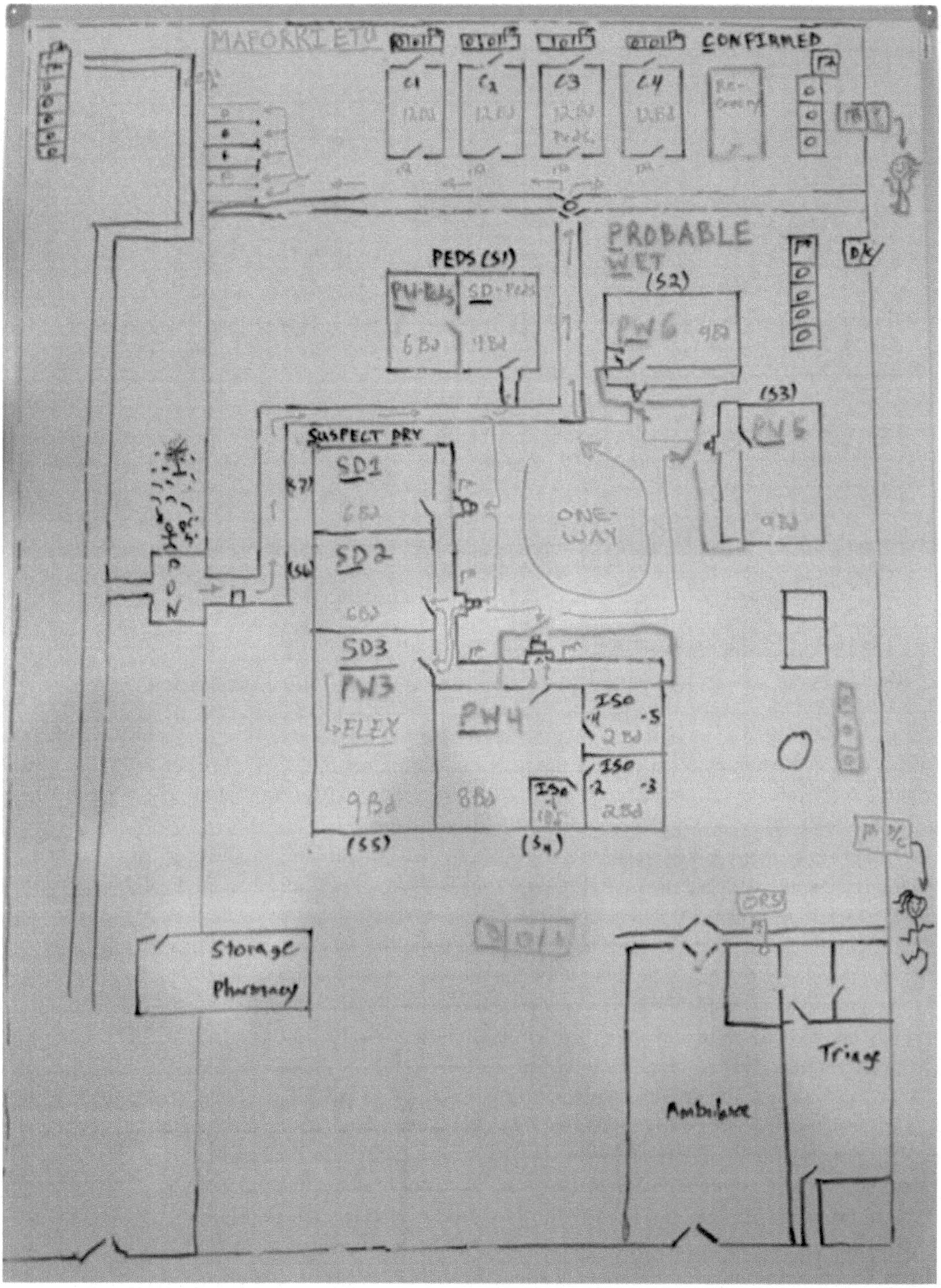

Fig. 5.3 Schematic layout of Maforki Ebola Holding and Treatment Centre, demonstrating patient and staff flow and the separation between "suspect dry" (SD1-SD3), "probable wet" (PW3-PW6), and "confirmed" (C1-C4) patient rooms

a framework for future clinical protocols used in an Ebola Treatment Unit, saving the next generation of frontline Ebola care providers from having to "reinvent the wheel" and thus be able to provide more effective care sooner to reduce morbidity and mortality among children with Ebola. Most, if not all, of the principles and treatments recommended can and should be provided to adults as well and need to be modified mostly with regard to dosing.

While several clinical trials took place during the course of the outbreak, most focused on experimental therapies, novel vaccines, and diagnostic testing algorithms. Rigorous trials to demonstrate efficacy of most of the interventions recommended here could not be conducted, both from a practical as well as an ethical standpoint. Instead, these protocols are based on clinical experience accumulated over decades from other critical and emergency contexts, adapted, and iteratively implemented and tested during the 2013–2015 Ebola outbreak. Unfortunately, they do not (and may never) have the solid empirical footing that is the hallmark of modern evidence-based clinical practice.

5.3 Admission Criteria for Ebola Treatment Units

In the context of any given Ebola outbreak, clear criteria for ETU admission should be developed, based upon the predominant manifestation of Ebola virus disease occurring at that time. These criteria need to be highly sensitive so as to keep infected individuals isolated from the rest of the community, but this will necessarily make any suspect case definition less specific and may lead to a high rate of ETU admissions among those with illnesses that mimic Ebola.

During the 2013–2015 West African Ebola epidemic, the clinical criteria for admission used in most screening algorithms consisted of:

- *Fever plus at least one clinical symptom* for children under 5 years of age, or
- *Fever plus at least two clinical symptoms* for children over 5 years of age, or
- *Close contact with an Ebola suspect or case within the preceding 3 weeks.*

The clinical symptoms included headache, anorexia, fatigue, musculoskeletal pain, diarrhea, nausea or vomiting, abdominal pain, difficulty swallowing, difficulty breathing, hiccups, and unusual or unexplained bleeding (World Health Organization 2016a). Unfortunately, among children living in tropical areas, this is of particular concern, given that fever, diarrhea, vomiting, abdominal pain, and generalized malaise and weakness are common findings in many acute childhood illnesses, such as malaria, measles, and gastroenteritis (Fitzgerald et al. 2017b), not to mention other possibly co-endemic viral hemorrhagic fevers (MacDermott et al. 2016).

These criteria, and similar ones used in practice (Smit et al. 2017; World Health Organization 2015), ultimately were not as sensitive or specific as would have been desired (Lado et al. 2015; Schieffelin et al. 2014). However, they still proved practical and feasible to implement in diverse health centers in the affected areas, and given the limited resources available, were likely the best approach at the time. The clustering of suspected EVD cases in often overcrowded Ebola holding units while awaiting confirmatory diagnostic testing also places both patients and health care providers at risk for acquiring infection from the environment or infected patients within the holding center. Although such nosocomial transmission certainly occurred, the rate of this transmission fortunately may have been lower than had been originally feared (Fitzgerald et al. 2017a). It is anticipated that improved rapid point-of-care testing will continue to be developed and hopefully available to diagnose Ebola quickly and accurately for use in future triage situations (Broadhurst et al. 2016), as well as to help decrease the spread of future outbreaks (Dhillon et al. 2015).

Among children admitted to the ETU, a number of schemas and prediction rules were developed for identifying those most likely to die during admission (Hartley et al. 2017; Shah et al. 2016). Those covariables which were most strongly linked to pediatric mortality were higher viral loads, disorientation, confusion, bleeding, diarrhea, and significant weakness.

5.4 Accompaniment and Challenges in Pediatric Ebola Care

In crisis circumstances, it is unlikely that most health care providers working in an ETU will have a significant background in providing medical care to infants and children. Thus, it is important for ETU leadership to proactively pair those caregivers with more experience together with those new to pediatric care. Weight-based medication dosing and nutritional requirements, the use of smaller-sized intravenous (IV) catheters for smaller veins, and the different cardiorespiratory physiological reserves of children are all examples of lessons that will need to be taught to those new to pediatric care in an ETU—in a short time frame. ETU staff will also benefit from constant accompaniment and attention to their own mental health as they care for potentially large numbers of ill and dying children, while at the same time understandably overcoming constant fear for their own safety (Mobula 2014).

Similar to staff accompaniment, the continuous accompaniment of children is of vital importance. In addition to the medical needs of antimicrobials, fluid resuscitation, cleaning and hygiene, and medications, psychological support for frightened children (often separated from their caregivers as well as having lost family members to disease) is an important component of good care. Ideally, pediatric ETU wards will have plentiful medical supplies of various sizes, as well as toys, diapers, and other items necessary for pediatric care.

If it is possible to keep families together inside the ETU, obviously that is ideal, but compromises should not be made in standard infection prevention and control protocols (World Health Organization 2014a). If a family is together in the suspect ward, and any single family member is confirmed to have EVD, then that person (adult or child) should be separated and moved to the confirmed ward, even if it means separating children from their only familial caregivers. Similarly, if a suspect case is confirmed to be negative, that person should be discharged from the ETU so as to minimize their own risk of nosocomial acquisition of Ebola. Nursing mothers should also be asked to stop breastfeeding in order to reduce the risk of infection through breast milk; alternative nutritional products must be supplied and caregivers educated on how to use them.

Given that children may often be left unattended in an ETU, adaptations need to be made to the standard ward architecture for safety. These precautions include keeping medical supplies locked securely, placing mattresses and beds on the ground while keeping sharps receptacles on higher tables, and restricting movement so unattended children cannot wander off to other areas of the ETU.

One of the greatest challenges in caring for children with suspected or confirmed EVD (as with adults) is the only intermittent presence of care providers at the bedside due to the need for protecting their own safety and limiting the risk for heat stroke while working in personal protective equipment (PPE) in extraordinarily hot and humid conditions. Most current PPE (World Health Organization 2016c) limits care providers to 1–2 h at a time and requires at least an equivalent amount of time out of PPE to rehydrate and recover. This is a barrier that must be overcome in future outbreaks if major improvements in patient care are to be achieved (Sprecher et al. 2015). Increased staffing, cooler PPE, and climate-controlled environments will be extremely helpful in overcoming these barriers and providing better care in the future.

5.5 Protocols for the Management of Ebola Virus Disease in Children

5.5.1 Overview

A summary list of the key elements of the medical management in an ETU of children infected, or suspected to be infected, with Ebola virus disease in a resource-limited setting is presented in Box 5.1. Detailed protocols and their rationale are outlined below.

Box 5.1 Key elements in the medical management of children in an Ebola Treatment Unit in resource-limited settings

- Initial Assessment and Management
  - Vital signs including temperature and blood pressure
  - Anthropometry including weight and mid-upper arm circumference (MUAC)
  - Assessment for degree of dehydration and hemorrhage
  - Mental status assessment
  - Obtain IV access, ideally securing two IV lines
- Fluid and Electrolyte Management
  - Oral rehydration solution (ORS) for all children above 6 months of age
  - Ondansetron as needed for nausea and vomiting
  - Loperamide as needed for non-bloody diarrhea in confirmed cases
  - Intravenous fluids

    Initial bolus of lactated ringers with 5% dextrose (D_5LR)

      Reassess hydration status and repeat half boluses until euvolemic state achieved on clinical exam or by ultrasound examination

    Continued aggressive rehydration throughout hospitalization

    Potassium and magnesium supplementation for those with significant diarrhea
  - Oral zinc daily
- Nutritional Supplementation
  - F-100 formula or ready-to-use therapeutic food (RUTF) or BP-100 biscuits
- Empiric Antimalarials
  - IV artesunate daily initially if vomiting
  - Complete treatment course with PO artemisinin combination therapy (ACT), preferably artesunate-amodiaquine
- Empiric Antimicrobials
  - IV ceftriaxone or PO cefixime or PO ciprofloxacin
  - Add metronidazole if particularly voluminous or bloody diarrhea
- Vitamin K
  - Oral or intramuscular dose on admission
  - Consider additional doses for those with active bleeding

Table 5.1 Age-adjusted thresholds for identifying acute malnutrition in children

Age	MUAC (cm)
6 months to 5 years	<12.5
6–7 years	<14.0
8–9 years	<15.5
10–11 years	<17.0
12–13 years	<18.5
14–15 years	<20.0
16–17 years	<21.0

5.5.2 Initial Assessment and Management

All children should have their temperature, weight, and automatic blood pressure measured upon admission. The mid-upper-arm circumference (MUAC) should be assessed as well; if the MUAC is below the age-adjusted threshold (Tang et al. 2013) (Table 5.1), the child should be provided additional therapeutic feeding. Children should be assessed for their degree of dehydration and whether any bleeding suggestive of Ebola virus disease has occurred.

Unless the child is extremely well-appearing or has an obvious alternative diagnosis such as uncomplicated malaria, IV access should be obtained at the time of admission. Children with any recent vomiting, diarrhea, or evidence of moderate (or worse) dehydration should ideally have two IV lines placed. If consistent IV access is not available in those who are significantly dehydrated, intraosseous access in the proximal tibia should be sought (Paterson and Callahan 2015); subcutaneous access can also be considered (Rouhani et al. 2011; Spandorfer 2011). If an ultrasound machine is available, this can be used as an aid to obtaining IV access, as well as for comparing the sizes of the inferior vena cava and descending aorta to assess hydration status and to assess for pulmonary edema or ascites as late signs of fluid overload after rehydration has begun (Shah et al. 2011).

Vitamin K should be provided at admission for all patients, as hepatic damage (Lyon et al. 2014) and diminished synthetic function may contribute to prolonged prothrombin time and worsened coagulation function (Liddell et al. 2015). Oral administration is preferred over intramuscular administration; the dose is 5 mg for children under 12 years old, or 10 mg for older children.

5.5.3 Fluid Management

Ebola, at least as manifested during the 2013–2015 West African outbreak, is a fundamentally dehydrating disease due to the profound gastrointestinal fluid losses, sometimes together with hemorrhage, in almost all infected patients. Emphasis should thus be on early and aggressive fluid and electrolyte resuscitation, even among those who appear initially well-hydrated. The major exception to this principle would be for children who were confirmed or suspected to have malaria or malnutrition. The risk of fluid overload and cardiovascular collapse in these populations may lead to increased mortality (Maitland et al. 2011), although this must of course be balanced with the risk of continued fluid losses due to Ebola infection. Limitations in staff time at the bedside makes it particularly difficult to provide sufficient fluid resuscitation, both because encouraging oral rehydration in a child takes significant time and because IV lines generally need to be disconnected when staff leave the bedside.

5.5.3.1 Oral Rehydration

Oral rehydration solution (ORS) should be provided in liberal quantities; if flavored solutions are available, these may increase consumption. Limits should not be placed on the maximum amount a child should be allowed to drink, but is instead guided by thirst, unless a nasogastric (NG) tube is used for hydration, in which case overhydration is possible. Children will benefit from frequent positive reinforcement for drinking ORS, and all care providers should be encouraged to provide this encouragement frequently. Children may consume up to 20 mL/kg of ORS per hour when initially dehydrated. Fluids other than ORS or water should not be provided to children, especially those with added sugar as they may serve to increase diarrhea.

Due to the high rates of nausea and vomiting experienced by Ebola patients, we also advocate the use of ondansetron as a means of improving the intake of ORS and other fluids (Tomasik et al. 2016). The best route of administration—IV, oral solution, oral tablet, or oral dissolving tablet—will vary from patient to patient. Children over 12 years of age may be given 4–8 mg every 8–12 h; the dose should be lowered to 2–4 mg for younger children.

Although there is some controversy about its safety in patients who may have severe diarrhea due to bacterial pathogens, we do believe that loperamide is warranted in this context to decrease diarrheal fluid losses at a dose of 2 mg initially, then in subsequent doses of 1 mg, up to a maximum of 8 mg/day (Chertow et al. 2014, 2015; Smit et al. 2017). If used, care should be taken that the child does not have bloody diarrhea, as this may be a sign of bacterial enteritis, and also that their abdomen is examined frequently for any rigidity that may be a marker of ileus or paresis. Loperamide should be discontinued in these cases. It is also important to stay aware of local epidemiology of other diarrheal disease outbreaks, most notably cholera, and to be particularly cautious about the use of loperamide when there is such an outbreak.

5.5.3.2 Parenteral Rehydration

Very few children will be able to drink sufficient quantities of ORS to keep up with the gastrointestinal losses due to Ebola virus disease, and thus parenteral access should be sought in all children. We recommend using Lactated Ringers' solution supplemented with 5% dextrose (D_5LR). Children who are "dry" at the time of presentation (*i.e.*, without vomiting, diarrhea, or bleeding) should receive an initial bolus of 20 mL/kg of D_5LR. "Wet" children should receive 40 mL/kg, unless there is significant concern for malaria or acute malnutrition, in which case the bolus should be limited to 20 mL/kg.

After the initial bolus, hydration status should be reassessed by evaluating capillary refill, skin turgor, the warmth of the extremities, and mental status. Children who are still dehydrated after this initial bolus should receive repeated boluses of half the volume as the initial bolus, reassessing hydration each time. Boluses may be repeated 2–3 times or more as needed, but an increase in respiratory or heart rate during these boluses may be a marker of intravascular fluid overload.

After admission, children will require continued aggressive fluid rehydration to account for insensible losses (due to fever, acidosis, tachypnea, and high environmental temperatures), even in the absence of significant gastrointestinal losses or hemorrhage. Dextrose should continue to be added to these fluids in almost all cases, especially younger children and those with significant vomiting or anorexia. Hydration status should be reassessed frequently to monitor for fluid overload. Care will need to be individualized, but in general "dry" children and those with significant concern for malaria or malnutrition should receive 150 mL/kg/day for the first 10 kg of body weight plus 75 mL/kg/day for the next 10 kg of body weight plus 40 mL/kg/day of remaining weight. "Wet" children with significant concern for malaria or malnutrition should additionally receive at least 100 mL of fluid for each loose stool. "Wet" children without obvious concern for malaria or malnutrition should receive at least 200 mL for each loose stool.

Rapid infusion, either with manual pressure on the IV fluid bags or with the use of pressure bags, is helpful in providing large amount of fluid to Ebola patients during the short spans of time that clinicians are generally able to be at the bedside. However, caution should be advised if these approaches are used with children as it may be quite easy to lose IV access due to the high pressures "blowing" the vein.

5.5.4 Electrolyte Supplementation

Supplementation of fluids with glucose, potassium, and magnesium are important, as these are all likely to be at relatively low levels in infants and children with profuse vomiting and diarrhea. If point-of-care electrolyte monitoring is available, this can greatly help with targeted and rational supplementation (Palich et al. 2016).

5.5.4.1 Potassium

Potassium supplementation of IV fluids should generally be included for children with significant diarrhea, unless there is evidence of decreased urine output and renal insufficiency or failure. Titration of this supplementation is best guided by bedside measurements due to the high risk of prerenal failure and consequent hyperkalemia. Nevertheless, empirically adding 10 mEq/L of potassium supplementation to IV fluids (when replacing stool losses liter-for-liter) is generally safe. IV fluids containing potassium should be infused by gravity rather than by pressure bag or manually squeezing the bag. Given that this slow infusion would limit the amount of fluids a child would receive, LR without potassium should be administered via a second IV line at a more aggressive rate to achieve the fluid goals needed for adequate resuscitation. The goal of overall volume resuscitation likely supersedes the goal of electrolyte supplementation, but both goals must be balanced and individualized for each child and also balanced with the time available for bedside patient care, given the constraints of PPE.

5.5.4.2 Magnesium

If available, magnesium should also be supplemented into IV fluids for children with significant diarrhea unless there is evidence of significant hypotension. If the systolic blood pressure is low for age or if the pulse pressure is wide, the child should first be resuscitated with fluids without magnesium. For ease, minimum systolic blood pressures before giving magnesium can be considered to be 90 mmHg for adolescents, 80 mmHg for school-aged children, 70 mmHg for toddlers, and 60 mmHg for infants. Rapid infusions of magnesium can lead to significant hypotension, thus fluids supplemented with magnesium should also be run over gravity rather than aggressively infused, as with potassium. The typical dose of magnesium sulfate is 25 mg/kg, up to 100 mg/kg/day in total (maximum 2 g).

5.5.4.3 Zinc

Given that zinc decreases the duration and quantity of diarrhea in children due to a variety of infectious causes (Lazzerini and Wanzira 2016), zinc should be provided to all children. For children under 6 months of age, the standard dose is 10 mg by mouth per day; for children over 6 months of age, the dose is 20 mg/day.

5.5.5 Antimicrobial Therapy

5.5.5.1 Antibacterials

Due to the risk of severe bacterial infections due to organisms such as *Samonella* and pneumococcus complicating or mimicking Ebola in children, it is often impossible to clinically distinguish between Ebola and bacterial sepsis. The profound diarrhea that was often the most significant clinical manifestation of Ebola during the West African outbreak also placed children at an extremely high risk for bacterial translocation across the denuded gut mucosa, thus leading to superimposed Gram-negative sepsis (Kreuels et al. 2014). Further, given the severely hampered ability to perform a thorough physical examination and relatively little ability to perform blood cultures or other microbiologic testing in the context of an Ebola outbreak, it is prudent to empirically treat all patients with suspected or confirmed Ebola with antibiotics.

The regimen of choice used in West Africa varied from ETU to ETU, but the general theme was that the choice of agent should be one that had effective coverage against enteric Gram-negative organisms. At Maforki, we generally gave patients intravenous or intramuscular ceftriaxone. Other units preferred ciprofloxacin, or preferred oral therapy with amoxicillin-clavulanate, or a second- or third-generation cephalosporin. Bowel wall edema and gut translocation are more likely to occur late in the course of EVD during the period of most severe gastrointestinal injury, and patients may develop an acute abdomen. Some ETUs thus included metronidazole as part of their empiric management (Ansumana et al. 2015). We generally used metronidazole if the child's abdominal exam was specifically worrisome, or if the patient had particularly voluminous or bloody diarrhea in order to provide coverage for anaerobes, *Giardia*, and *Cryptosporidium*.

The duration of therapy is similarly unclear and variable; a balance must be struck between prudent antimicrobial stewardship and caring for critically ill children with very high mortality in a setting where the usual physical and laboratory diagnostics are unavailable. Our approach was generally to treat children throughout their hospitalization in the ETU, while others provided a fixed duration to their empiric treatment courses.

5.5.5.2 Antimalarials

West Africa is endemic for malaria, and the most likely sites for future outbreaks of Ebola are likely to be malaria hotspots as well. In an absolutely fascinating and unexpected finding, coinfection with malaria correlated with an *increased* likelihood of survival, even after controlling for age, Ebola viral load, treatment for malaria; the effect was even more pronounced among those with the highest levels of parasitemia (Rosenke et al. 2016). While this immunomodulatory effect induced by malaria may help a patient recover from Ebola, this does not obviate the need for treating patients for malaria. What remains clinically challenging in an ETU is that the clinical presentation of falciparum malaria overlaps significantly with the case definition for suspected and confirmed Ebola cases, particularly in children whose predominant clinical syndrome is a febrile gastroenteritis.

We thus strongly recommend empiric antimalarial therapy for all children admitted to the ETU, even those with a negative malaria test at the time of admission since they may well become infected (or clinically manifest an infection that was incubating) while in the ETU. Modern artemisinin-combination therapy (ACT) also provides some short-term prophylaxis against malaria, and thus can also be considered helpful in prophylaxing against malaria during their post-Ebola convalescence when they are already likely to be weak, undernourished, and deconditioned (Nosten and White 2007).

In order to most rapidly clear *Plasmodium* parasitemia, IV artesunate or IM artemether for at least 24 h is preferred at the time of admission, unless the child appears particularly well and clearly able to ingest oral medications without vomiting, in which case oral therapy with ACT can be used. We recommend that all children complete a full 3-day course of ACT, either at the ETU or at home after discharge. There is significant evidence to suggest that the choice of which ACT to use does indeed matter, with artesunate-amodiaquine leading to lower Ebola-related mortality than artemether-lumefantrine (Gignoux et al. 2016).

5.5.6 Nutritional Support

Ever since Nevin Scrimshaw's pivotal studies in Guatemala from the 1960s demonstrating a strong bidirectional link between malnutrition and infection (Scrimshaw et al. 1968; Scrimshaw 2003), a large body of literature has emerged demonstrating that undernourished children fare far worse than their well-nourished counterparts when faced with a severe infection. Similarly, survivors of severe infectious diseases often find themselves newly malnourished and with lingering weakness and long-term disability (Katona and Katona-Apte 2008). Malnutrition is by far the biggest cause of immunodefi-

Table 5.2 Sample minimum nutritional intake goals for children in an ETU

Age	RUTF sachets per day	BP-100 biscuits per day
1–4 years	2–3	3–5
5–9 years	3–4	5–7
10–14 years	4–5	6–8
≥15 years female	4–5	6–8
≥15 years male	5–6	8–10

ciency worldwide and acts through a number of mechanistic pathways; those most relevant to Ebola include compromised immune defenses at epithelial barriers (respiratory mucosa, skin, and most importantly, the gut mucosa), reduced availability of complement components, and decreased leptin levels (needed for T cell activation) (Guerrant et al. 2008).

Recognizing that most children with Ebola virus disease in the midst of an epidemic are at least chronically malnourished to some degree (as manifested by them being underweight and/or stunted), aggressive attention should be paid to their nutritional status while in the ETU. During the West African outbreak in 2013–2015, WHO issued interim guidelines for nutritional support of patients with Ebola (WHO/UNICEF/WFP 2014). However, these were never studied symptomatically or updated based upon the experience implementing them in ETUs. In our own experience, we found that most patients, especially children, were not particularly fond of locally prepared meals served at fixed times of the day, given how much anorexia, nausea, vomiting, sore throat, and abdominal pain they suffered from. These meals would arrive at the bedside and, if not eaten right away, remain at the bedside and lose their palatability in the heat and humidity of our tropical setting, especially as they quickly attracted insects.

Our approach to nutritional care thus evolved to make use of energy-dense food that can provide significant quantities of micronutrients and macronutrients even when only small amounts are eaten at a time. These food also need to retain their taste and palatability while remaining at the bedside for many hours at a time. They must also be easy for weak patients to eat by themselves and should be able to be fed quickly by caregivers and ETU staff to young children. Ready-to-use therapeutic food (RUTF) and BP-100 biscuits are two examples of nutritional food products that achieved these goals and objectives.

All children who were able to tolerate solid or semisolid food should be provided with RUTF and/or BP-100 in ample quantities. Children, caregivers, and staff should be educated that nutritional care will be optimized if children eat all of their intended RUTF and/or BP-100 each day (Table 5.2). The intake of these should always be prioritized over standard hospital food, which generally does not provide as complete of a daily diet of fats, proteins, and micronutrients. Patients should always be provided with more supplements if they consume all of their minimum recommended intake. Since these products do not contain any water, children are likely to want more water or ORS than they otherwise would. Young children may not be able to eat RUTF without some mixing with water in a spoon or cup. BP-100 can be crushed and mixed with water to make a porridge.

Children over 6 months of age who are only able to tolerate liquids should be provided with F-75 or F-100 formula instead of giving ready-to-use formula milk, as these formulas provide much higher nutritional content (World Health Organization 1999). F-100 is likely a better nutritional choice for most children, as it contains more energy and macronutrients, but has a higher likelihood of causing diarrhea due to its higher osmotic load, and thus F-75 may be a better choice for some children. These formula milks do spoil if unrefrigerated and thus may need to be reconstituted multiple times per day. Children under 6 months of age should be provided with ready-to-use formula milk instead, as the safety of F-75 and F-100 has not been fully established in this age group. Nasogastric tubes may be considered for children without vomiting who are too weak to eat or drink on their own; these may be used for water and ORS in addition to milk feeds.

Children with acute malnutrition should be identified at the time of admission based on their MUAC and by checking for the presence of the bilateral pitting edema diagnostic for kwashiorkor; for the purposes of ETU management, severely and moderately malnourished children can be treated together rather than separately (Trehan and Manary 2015). The MUAC thresholds for children above 5 years of age are suggested guidelines, based on extrapolated estimates and not from clear international consensus. Clinical judgment, incorporating signs of visible wasting and history of recent weight loss, should be used in all circumstances. These children, especially those under 5 years of age, should be fed only F-75, F-100, RUTF, or BP-100; breast milk, water, ORS, and ReSoMal should be the only additional liquids they drink.

Malnourished children will require particularly aggressive therapeutic feeding during their hospitalization and special arrangements should be made to provide frequent directly observed feedings around the clock. All malnourished children should be prioritized for antibiotics at the time of admission (Trehan et al. 2013), in case antibiotics are not being used for all patients as described in Sect. 5.5.5.1.

5.5.7 Psychological Support

Particular attention should be paid to providing psychological support and accompaniment for children in an ETU as well. They are likely to be extremely frightened by the ETU environment, which can be exacerbated by health care providers dressed in PPE and often performing painful procedures on them. Their sense of abandonment is also likely to be acutely heightened, given the separation from their families and home environments, potentially compounded further by the death of close family members as well. As in all health care settings, children will benefit from frequent encouragement and positive reinforcement for cooperating with their own medical care (Fig. 5.4). We encourage the recruitment and integration of adult Ebola survivors as bedside caretakers to assist with feeding and oral fluids and also to provide accompaniment and psychological support for children. If feasible, children should ideally have a caregiver present at the bedside all times.

Fig. 5.4 A new bicycle tire requested by an Ebola patient and used as a reward to motivate him to drink as much ORS as possible during his hospitalization

5.5.8 Management of Constitutional Symptoms and Clinical Complications

Patients with EVD can develop a number of constitutional symptoms and clinical complications that should be addressed using standard clinical management, tailored to each patient's symptoms and to the medications available in a given ETU's formulary—a starting point for developing such an ETU formulary is available from the WHO (World Health Organization 2014b). Symptoms and complications that patients are likely to suffer from include fever, myalgias, gastritis, hypoglycemia, abdominal pain, anxiety, confusion, agitation, seizures, rhabdomyolysis, and fluid overload (Cournac et al. 2016; Smit et al. 2017).

Considered thought should be given to palliative care as well, and protocols should be developed to incorporate the rational and ethical use of benzodiazepines and opiates for the relief of air hunger, pain, anxiety, and suffering in terminally ill patients.

5.5.9 Experimental Therapies

A number of experimental therapies, including antiviral medications, monoclonal antibodies, and serum from Ebola survivors were developed and attempted during the course of the 2013–2015 West African Ebola outbreak (Mendoza et al. 2016; Uyeki et al. 2016). It remains unclear which therapies (if any) will be approved and available for use for future Ebola cases, and their use in children is not nearly as well-studied as it has been in adults. In the future, if such resources are available for use, then it seems quite prudent to provide these therapies as well, as their use in this outbreak was strongly linked to improved survival. The triple monoclonal antibody marketed under the name ZMapp seems particularly promising in clinical trials (Prevail II Writing Group 2016).

5.5.10 Discharge Criteria

Virologic testing protocols may vary in future Ebola outbreaks, given further developments in Ebola tests (Broadhurst et al. 2016), but will likely remain the key decision point when it comes to discharging patients from an ETU. During the 2013–2015 West African outbreak, reverse-transcription polymerase chain reaction (RT-PCR) testing was the method of choice for confirmation of Ebola virus infection in nearly all care centers. However, given the possible delay in RT-PCR positivity from the time of clinical symptom onset, at least 72 h should have elapsed from the symptom onset before a negative RT-PCR test should be considered truly negative (Chertow et al. 2014) in a suspect case.

Due to the possibility of false-negative RT-PCR testing, confirmed cases should not be discharged until at least two RT-PCR tests are obtained, taken from 48 to 72 h apart. Discharge can then be considered if the child is clinically stable. If further medical care is needed, a balance must be struck between continuing to care for the child in the ETU (with all of the limitations of testing and treatment available there) and transferring the child to a non-Ebola hospital (with the concern of potentially further infectious body fluid transmission such as from urine).

Once discharged, all Ebola patients, including children, will benefit from continued medical care due to the high rate of lingering medical problems (World Health Organization 2016b). Post-discharge nutritional supplementation is of particularly high importance to include in their medical care, as survivors are generally profoundly weak, emaciated, and deconditioned from their illness. Discharged patients

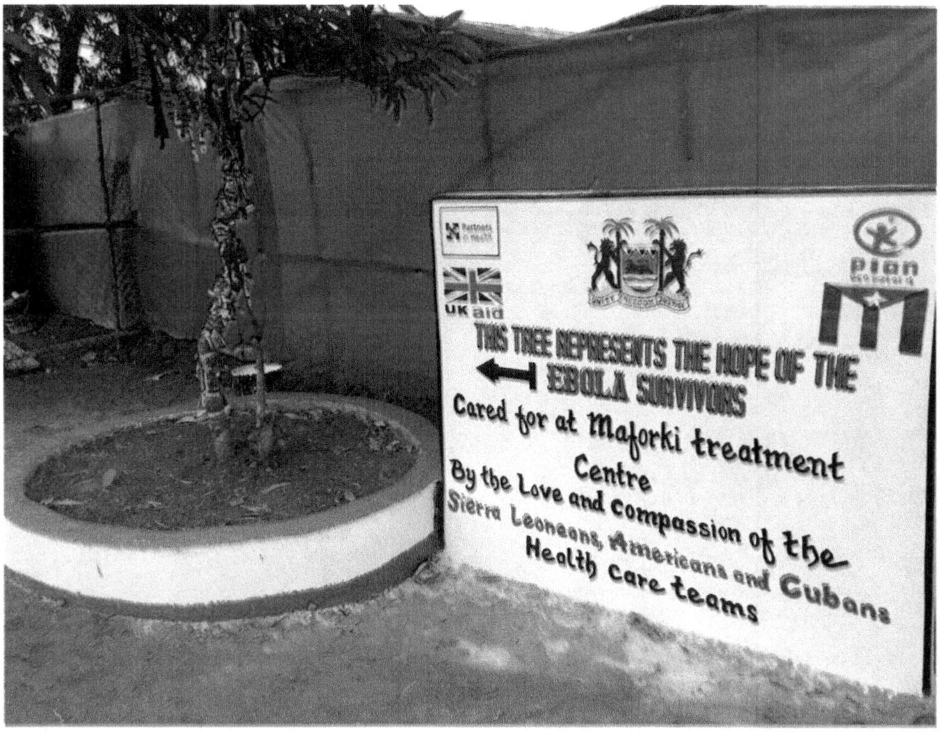

Fig. 5.5 Survivors' tree at Maforki Ebola Holding and Treatment Centre. At the time of discharge, each survivor was publicly celebrated and allowed to tie a piece of cloth to the tree as a symbol of their resilience and reintegration back into their community

should also receive assistance reintegrating into their communities and helping them overcome any stigma associated with their disease; ceremonies celebrating their perseverance and survival may be helpful in this regard (Fig. 5.5). The disruption of routine primary care during an Ebola epidemic (Delamou et al. 2017; Walker et al. 2015) is likely to lead to significant delays in routine, critical vaccination (Takahashi et al. 2015), and thus, children being discharged from an ETU should ideally be caught up on any vaccines they may have missed (Fig. 5.6).

5.6 Conclusions

Caring for children in an ETU is extraordinarily challenging, rewarding work. Rigorous implementation of standardized treatment protocols emphasizing aggressive fluid resuscitation, electrolyte repletion, antimicrobial prophylaxis, nutritional supplementation, and symptomatic management should drastically reduce morbidity and mortality from Ebola in both adults and children. The global medical establishment was not prepared for the scale or severity of the 2013–2015 West African Ebola epidemic and the state of knowledge on how best to diagnose and treat Ebola was relatively limited; thus, the recommendations made here are likely to need updating should another outbreak ever occur. However, these serve as an initial point from which care providers and ETU directors can plan to start their care for children with Ebola.

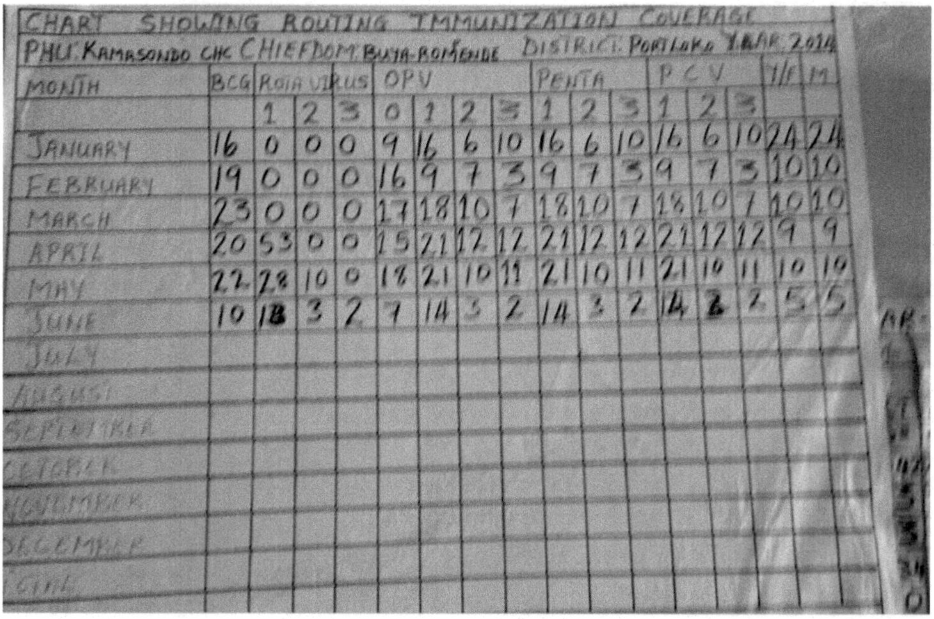

Fig. 5.6 Photo taken in January 2015 of the immunization administration log book from a clinic in Port Loko District, Sierra Leone, demonstrating the collapse of the routine health care system in mid-2014 as the Ebola epidemic spread through the area

References

Ansumana, R., Jacobsen, K. H., Sahr, F., Idris, M., Bangura, H., Boie-Jalloh, M., et al. (2015). Ebola in Freetown area, Sierra Leone--A case study of 581 patients. *New England Journal of Medicine, 372*(6), 587–588. https://doi.org/10.1056/NEJMc1413685.

Baez, F., Perez, J., & Reed, G. (2015). Meet Cuban Ebola fighters: Interview with Felix Baez and Jorge Perez. *MEDICC Review, 17*(1), 6–10.

Bower, H., Johnson, S., Bangura, M. S., Kamara, A. J., Kamara, O., Mansaray, S. H., et al. (2016). Exposure-specific and age-specific attack rates for Ebola virus disease in Ebola-affected households, Sierra Leone. *Emerging Infectious Diseases, 22*(8), 1403–1411. https://doi.org/10.3201/eid2208.160163.

Broadhurst, M. J., Brooks, T. J., & Pollock, N. R. (2016). Diagnosis of Ebola virus disease: Past, present, and future. *Clinical Microbiology Reviews, 29*(4), 773–793. https://doi.org/10.1128/CMR.00003-16.

Chertow, D. S., Kleine, C., Edwards, J. K., Scaini, R., Giuliani, R., & Sprecher, A. (2014). Ebola virus disease in West Africa--Clinical manifestations and management. *New England Journal of Medicine, 371*(22), 2054–2057. https://doi.org/10.1056/NEJMp1413084.

Chertow, D. S., Uyeki, T. M., & DuPont, H. L. (2015). Loperamide therapy for voluminous diarrhea in Ebola virus disease. *Journal of Infectious Diseases, 211*(7), 1036–1037. https://doi.org/10.1093/infdis/jiv001.

Cournac, J. M., Karkowski, L., Bordes, J., Aletti, M., Duron, S., Janvier, F., et al. (2016). Rhabdomyolysis in Ebola virus disease. Results of an observational study in a treatment center in Guinea. *Clinical Infectious Diseases, 62*(1), 19–23. https://doi.org/10.1093/cid/civ779.

Damkjaer, M., Rudolf, F., Mishra, S., Young, A., & Storgaard, M. (2017). Clinical features and outcome of Ebola Virus disease in pediatric patients: A retrospective case series. *Journal of Pediatrics, 182*(3), 378–381.e371. https://doi.org/10.1016/j.jpeds.2016.11.034.

DeBiasi, R. L., Song, X., Cato, K., Floyd, T., Talley, L., Gorman, K., et al. (2016). Preparedness, evaluation, and care of pediatric patients under investigation for Ebola virus disease: Experience from a pediatric designated care facility. *Journal of the Pediatric Infectious Diseases Society, 5*(1), 68–75. https://doi.org/10.1093/jpids/piv069.

Delamou, A., Ayadi, A. M., Sidibe, S., Delvaux, T., Camara, B. S., Sandouno, S. D., et al. (2017). Effect of Ebola virus disease on maternal and child health services in Guinea: A retrospective observational cohort study. *Lancet Global Health, 5*(4), e448–e457. https://doi.org/10.1016/S2214-109X(17)30078-5.

Dhillon, R. S., Srikrishna, D., Garry, R. F., & Chowell, G. (2015). Ebola control: Rapid diagnostic testing. *Lancet Infectious Diseases, 15*(2), 147–148. https://doi.org/10.1016/S1473-3099(14)71035-7.

Dornemann, J., Burzio, C., Ronsse, A., Sprecher, A., De Clerck, H., Van Herp, M., et al. (2017). First newborn baby to receive experimental therapies survives Ebola virus disease. *Journal of Infectious Diseases*. https://doi.org/10.1093/infdis/jiw493.

Eriksson, C. O., Uyeki, T. M., Christian, M. D., King, M. A., Braner, D. A., Kanter, R. K., et al. (2015). Care of the child with Ebola virus disease. *Pediatric Critical Care Medicine, 16*(2), 97–103. https://doi.org/10.1097/PCC.0000000000000358.

Farmer, P. (2015). *The secret to curing West Africa from Ebola is no secret at all*. Retrieved November 5, 2017, from http://wapo.st/14IXlAZ.

Fitzgerald, F., Awonuga, W., Shah, T., & Youkee, D. (2016a). Ebola response in Sierra Leone: The impact on children. *Journal of Infection, 72*(Suppl), S6–S12. https://doi.org/10.1016/j.jinf.2016.04.016.

Fitzgerald, F., Naveed, A., Wing, K., Gbessay, M., Ross, J. C., Checchi, F., et al. (2016b). Ebola virus disease in children, Sierra Leone, 2014-2015. *Emerging Infectious Diseases, 22*(10), 1769–1777. https://doi.org/10.3201/eid2210.160579.

Fitzgerald, F., Wing, K., Naveed, A., Gbessay, M., Ross, J. C., Checchi, F., et al. (2017a). Risk in the "Red Zone": Outcomes for children admitted to Ebola holding units in Sierra Leone without Ebola virus disease. *Clinical Infectious Diseases*. https://doi.org/10.1093/cid/cix182.

Fitzgerald, F., Wing, K., Naveed, A., Gbessay, M., Ross, J. C. G., Checchi, F., et al. (2017b). Refining the paediatric Ebola case definition: A study of children in Sierra Leone with suspected Ebola virus disease. *Lancet, 389*, S19. https://doi.org/10.1016/s0140-6736(17)30415-4.

Fowler, R. A., Fletcher, T., Fischer II, W. A., Lamontagne, F., Jacob, S., Brett-Major, D., et al. (2014). Caring for critically ill patients with ebola virus disease. Perspectives from West Africa. *American Journal of Respiratory and Critical Care Medicine, 190*(7), 733–737. https://doi.org/10.1164/rccm.201408-1514CP.

Frieden, T. (2015). *Eight ways Ebola threatens the children of West Africa*. Retrieved October 23, 2017, from http://fxn.ws/1G6fSnZ.

Gignoux, E., Azman, A. S., de Smet, M., Azuma, P., Massaquoi, M., Job, D., et al. (2016). Effect of artesunate-amodiaquine on mortality related to Ebola virus disease. *New England Journal of Medicine, 374*(1), 23–32. https://doi.org/10.1056/NEJMoa1504605.

Glynn, J. R. (2015). Age-specific incidence of Ebola virus disease. *Lancet, 386*(9992), 432. https://doi.org/10.1016/S0140-6736(15)61446-5.

Guerrant, R. L., Oria, R. B., Moore, S. R., Oria, M. O., & Lima, A. A. (2008). Malnutrition as an enteric infectious disease with long-term effects on child development. *Nutrition Reviews, 66*(9), 487–505. https://doi.org/10.1111/j.1753-4887.2008.00082.x.

Hartley, M. A., Young, A., Tran, A. M., Okoni-Williams, H. H., Suma, M., Mancuso, B., et al. (2017). Predicting Ebola severity: A clinical prioritization score for Ebola virus disease. *PLoS Neglected Tropical Diseases, 11*(2), e0005265. https://doi.org/10.1371/journal.pntd.0005265.

Hermans, V., Zachariah, R., Woldeyohannes, D., Saffa, G., Kamara, D., Ortuno-Gutierrez, N., et al. (2017). Offering general pediatric care during the hard times of the 2014 Ebola outbreak: Looking back at how many came and how well they fared at a Medecins Sans Frontieres referral hospital in rural Sierra Leone. *BMC Pediatrics, 17*(1), 34. https://doi.org/10.1186/s12887-017-0786-z.

Iannotti, L. L., Trehan, I., Clitheroe, K. L., & Manary, M. J. (2015). Diagnosis and treatment of severely malnourished children with diarrhoea. *Journal of Paediatrics and Child Health, 51*(4), 387–395. https://doi.org/10.1111/jpc.12711.

Katona, P., & Katona-Apte, J. (2008). The interaction between nutrition and infection. *Clinical Infectious Diseases, 46*(10), 1582–1588. https://doi.org/10.1086/587658.

Kreuels, B., Wichmann, D., Emmerich, P., Schmidt-Chanasit, J., de Heer, G., Kluge, S., et al. (2014). A case of severe Ebola virus infection complicated by Gram-negative septicemia. *New England Journal of Medicine, 371*(25), 2394–2401. https://doi.org/10.1056/NEJMoa1411677.

Lado, M., Walker, N. F., Baker, P., Haroon, S., Brown, C. S., Youkee, D., et al. (2015). Clinical features of patients isolated for suspected Ebola virus disease at Connaught Hospital, Freetown, Sierra Leone: A retrospective cohort study. *Lancet Infectious Diseases, 15*(9), 1024–1033. https://doi.org/10.1016/S1473-3099(15)00137-1.

Lamontagne, F., Clement, C., Fletcher, T., Jacob, S. T., Fischer II, W. A., & Fowler, R. A. (2014). Doing today's work superbly well--Treating Ebola with current tools. *New England Journal of Medicine, 371*(17), 1565–1566. https://doi.org/10.1056/NEJMp1411310.

Lazzerini, M., & Wanzira, H. (2016). Oral zinc for treating diarrhoea in children. *Cochrane Database of Systematic Reviews*, (12), CD005436. https://doi.org/10.1002/14651858.CD005436.pub5.

Liddell, A. M., Davey Jr., R. T., Mehta, A. K., Varkey, J. B., Kraft, C. S., Tseggay, G. K., et al. (2015). Characteristics and clinical management of a cluster of 3 patients eith Ebola virus disease, including the first domestically acquired cases in the United States. *Annals of Internal Medicine, 163*(2), 81–90. https://doi.org/10.7326/M15-0530.

Lyon, G. M., Mehta, A. K., Varkey, J. B., Brantly, K., Plyler, L., McElroy, A. K., et al. (2014). Clinical care of two patients with Ebola virus disease in the United States. *New England Journal of Medicine, 371*(25), 2402–2409. https://doi.org/10.1056/NEJMoa1409838.

MacDermott, N. E., De, S., & Herberg, J. A. (2016). Viral haemorrhagic fever in children. *Archives of Disease in Childhood, 101*(5), 461–468. https://doi.org/10.1136/archdischild-2014-307861.

Maitland, K., Kiguli, S., Opoka, R. O., Engoru, C., Olupot-Olupot, P., Akech, S. O., et al. (2011). Mortality after fluid bolus in African children with severe infection. *New England Journal of Medicine, 364*(26), 2483–2495. https://doi.org/10.1056/NEJMoa1101549.

McElroy, A. K., Erickson, B. R., Flietstra, T. D., Rollin, P. E., Nichol, S. T., Towner, J. S., et al. (2014). Biomarker correlates of survival in pediatric patients with Ebola virus disease. *Emerging Infectious Diseases, 20*(10), 1683–1690. https://doi.org/10.3201/eid2010.140430.

Mendoza, E. J., Qiu, X., & Kobinger, G. P. (2016). Progression of Ebola therapeutics during the 2014-2015 Outbreak. *Trends in Molecular Medicine, 22*(2), 164–173. https://doi.org/10.1016/j.molmed.2015.12.005.

Mobula, L. M. (2014). Courage is not the absence of fear: Responding to the Ebola outbreak in Liberia. *Global Health: Science and Practice, 2*(4), 487–489. https://doi.org/10.9745/GHSP-D-14-00157.

Nebehay, S. (2015). *High rates of child deaths from Ebola, special care needed: WHO*. Retrieved March 25, 2017, from http://reut.rs/1D5rmbR.

Nelson, J. M., Griese, S. E., Goodman, A. B., & Peacock, G. (2016). Live neonates born to mothers with Ebola virus disease: A review of the literature. *Journal of Perinatology, 36*(6), 411–414. https://doi.org/10.1038/jp.2015.189.

Nosten, F., & White, N. J. (2007). Artemisinin-based combination treatment of falciparum malaria. *American Journal of Tropical Medicine and Hygiene, 77*(6 Suppl), 181–192.

Palich, R., Gala, J. L., Petitjean, F., Shepherd, S., Peyrouset, O., Abdoul, B. M., et al. (2016). A 6-Year-old child with severe Ebola virus disease: Laboratory-guided clinical care in an Ebola treatment center in Guinea. *PLoS Neglected Tropical Diseases, 10*(3), e0004393. https://doi.org/10.1371/journal.pntd.0004393.

Paterson, M. L., & Callahan, C. W. (2015). The use of intraosseous fluid resuscitation in a pediatric patient with Ebola virus disease. *Journal of Emergency Medicine, 49*(6), 962–964. https://doi.org/10.1016/j.jemermed.2015.06.010.

Peacock, G., Uyeki, T. M., & Rasmussen, S. A. (2014). Ebola virus disease and children: What pediatric health care professionals need to know. *JAMA Pediatrics, 168*(12), 1087–1088. https://doi.org/10.1001/jamapediatrics.2014.2835.

Prevail II Writing Group. (2016). A randomized, controlled trial of ZMapp for Ebola virus infection. *New England Journal of Medicine, 375*(15), 1448–1456. https://doi.org/10.1056/NEJMoa1604330.

Rosenke, K., Adjemian, J., Munster, V. J., Marzi, A., Falzarano, D., Onyango, C. O., et al. (2016). Plasmodium parasitemia associated with increased survival in Ebola virus-infected patients. *Clinical Infectious Diseases, 63*(8), 1026–1033. https://doi.org/10.1093/cid/ciw452.

Rouhani, S., Meloney, L., Ahn, R., Nelson, B. D., & Burke, T. F. (2011). Alternative rehydration methods: A systematic review and lessons for resource-limited care. *Pediatrics, 127*(3), e748–e757. https://doi.org/10.1542/peds.2010-0952.

Schibler, M., Vetter, P., Cherpillod, P., Petty, T. J., Cordey, S., Vieille, G., et al. (2015). Clinical features and viral kinetics in a rapidly cured patient with Ebola virus disease: A case report. *Lancet Infectious Diseases, 15*(9), 1034–1040. https://doi.org/10.1016/S1473-3099(15)00229-7.

Schieffelin, J. S., Shaffer, J. G., Goba, A., Gbakie, M., Gire, S. K., Colubri, A., et al. (2014). Clinical illness and outcomes in patients with Ebola in Sierra Leone. *New England Journal of Medicine, 371*(22), 2092–2100. https://doi.org/10.1056/NEJMoa1411680.

Scrimshaw, N. S. (2003). Historical concepts of interactions, synergism and antagonism between nutrition and infection. *Journal of Nutrition, 133*(1), 316S–321S.

Scrimshaw, N. S., Taylor, C. E., & Gordon, J. E. (1968). Interactions of nutrition and infection. *Monograph Series of the World Health Organization, 57*, 3–329.

Shah, S., Price, D., Bukham, G., Shah, S., & Wroe, E. (2011). *The partners in health manual of ultrasound for resource-limited settings*. Boston: Partners In Health.

Shah, T., Greig, J., van der Plas, L. M., Achar, J., Caleo, G., Squire, J. S., et al. (2016). Inpatient signs and symptoms and factors associated with death in children aged 5 years and younger admitted to two Ebola management centres in Sierra Leone, 2014: A retrospective cohort study. *Lancet Global Health, 4*(7), e495–e501. https://doi.org/10.1016/S2214-109X(16)30097-3.

Sissoko, D., Keita, M., Diallo, B., Aliabadi, N., Fitter, D. L., Dahl, B. A., et al. (2017). Ebola virus persistence in breast milk after no reported illness: A Likely source of virus transmission from mother to child. *Clinical Infectious Diseases*. https://doi.org/10.1093/cid/ciw793.

Smit, M. A., Michelow, I. C., Glavis-Bloom, J., Wolfman, V., & Levine, A. C. (2017). Characteristics and outcomes of pediatric patients with Ebola virus disease admitted to treatment units in Liberia and Sierra Leone: A retrospective cohort study. *Clinical Infectious Diseases, 64*(3), 243–249. https://doi.org/10.1093/cid/ciw725.

Spandorfer, P. R. (2011). Subcutaneous rehydration: Updating a traditional technique. *Pediatric Emergency Care, 27*(3), 230–236. https://doi.org/10.1097/PEC.0b013e31820e1405.

Sprecher, A. G., Caluwaerts, A., Draper, M., Feldmann, H., Frey, C. P., Funk, R. H., et al. (2015). Personal protective equipment for filovirus epidemics: A call for better evidence. *Journal of Infectious Diseases, 212(Suppl 2)*, S98–S100. https://doi.org/10.1093/infdis/jiv153.

Takahashi, S., Metcalf, C. J., Ferrari, M. J., Moss, W. J., Truelove, S. A., Tatem, A. J., et al. (2015). Reduced vaccination and the risk of measles and other childhood infections post-Ebola. *Science, 347*(6227), 1240–1242. https://doi.org/10.1126/science.aaa3438.

Tang, A. M., Dong, K., Deitchler, M., Chung, M., Maalouf-Manasseh, Z., Tumilowicz, A., et al. (2013). *Use of cutoffs for mid-upper arm circumference (muac) as an indicator or predictor of nutritional and health-related outcomes in adolescents and adults: A systematic review*. Washington, DC: FHI 360/FANTA.

Tomasik, E., Ziolkowska, E., Kolodziej, M., & Szajewska, H. (2016). Systematic review with meta-analysis: Ondansetron for vomiting in children with acute gastroenteritis. *Alimentary Pharmacology and Therapeutics, 44*(5), 438–446. https://doi.org/10.1111/apt.13728.

Trehan, I., & Manary, M. J. (2015). Management of severe acute malnutrition in low-income and middle-income countries. *Archives of Disease in Childhood, 100*(3), 283–287. https://doi.org/10.1136/archdischild-2014-306026.

Trehan, I., Goldbach, H. S., LaGrone, L. N., Meuli, G. J., Wang, R. J., Maleta, K. M., et al. (2013). Antibiotics as part of the management of severe acute malnutrition. *New England Journal of Medicine, 368*(5), 425–435. https://doi.org/10.1056/NEJMoa1202851.

Trehan, I., Kelly, T., Marsh, R. H., George, P. M., & Callahan, C. W. (2016). Moving towards a more aggressive and comprehensive model of care for children with Ebola. *Journal of Pediatrics, 170*(28–33), e21–e27. https://doi.org/10.1016/j.jpeds.2015.11.054.

UNICEF. (2014). *Guidelines for a revised implementation of integrated community case management of childhood illnesses (ICCM) during the Ebola outbreak*. New York: UNICEF.

Uyeki, T. M., Mehta, A. K., Davey Jr., R. T., Liddell, A. M., Wolf, T., Vetter, P., et al. (2016). Clinical management of Ebola virus disease in the United States and Europe. *New England Journal of Medicine, 374*(7), 636–646. https://doi.org/10.1056/NEJMoa1504874.

Walker, P. G., White, M. T., Griffin, J. T., Reynolds, A., Ferguson, N. M., & Ghani, A. C. (2015). Malaria morbidity and mortality in Ebola-affected countries caused by decreased health-care capacity, and the potential effect of mitigation strategies: A modelling analysis. *Lancet Infectious Diseases, 15*(7), 825–832. https://doi.org/10.1016/S1473-3099(15)70124-6.

WHO Ebola Response Team. (2015). Ebola virus disease among children in West Africa. *New England Journal of Medicine, 372*(13), 1274–1277. https://doi.org/10.1056/NEJMc1415318.

WHO/UNICEF/WFP. (2014). *Interim guideline: Nutritional care of children and adults with Ebola virus disease in treatment centres*. Geneva: World Health Organization.

Wolf, T., Kann, G., Becker, S., Stephan, C., Brodt, H. R., de Leuw, P., et al. (2015). Severe Ebola virus disease with vascular leakage and multiorgan failure: Treatment of a patient in intensive care. *Lancet, 385*(9976), 1428–1435. https://doi.org/10.1016/S0140-6736(14)62384-9.

World Health Organization. (1999). *Management of severe malnutrition: A manual for physicians and other senior health workers*. Geneva: World Health Organization.

World Health Organization. (2014a). *Interim infection prevention and control guidance for care of patients with suspected or confirmed filovirus haemorrhagic fever in health-care settings, with focus on Ebola*. Geneva: World Health Organization.

World Health Organization. (2014b). *Interim list of WHO essential medicines necessary to treat Ebola cases based on existing guidelines*. Geneva: World Health Organization.

World Health Organization. (2015). *Manual for the care and management of patients in Ebola Care Units/Community Care Centres: Interim emergency guidance*. Geneva: World Health Organization.

World Health Organization. (2016a). *Clinical management of patients with viral haemorrhagic fever: A pocket guide for front line health workers*. Geneva: World Health Organization.

World Health Organization. (2016b). *Interim guidance: Clinical care for survivors of Ebola virus disease*. Geneva: World Health Organization.

World Health Organization. (2016c). *Personal protective equipment for use in a filovirus disease outbreak: Rapid advice guideline*. Geneva: World Health Organization.

Ebola and Pregnant Women: Providing Maternity Care at MSF Treatment Centers

6

Severine Caluwaerts and Patricia Kahn

6.1 Introduction

As one of the main responders to the 2014–2016 West Africa Ebola epidemic (Fig. 6.1), Médecins Sans Frontières/Doctors Without Borders (MSF) faced many challenges, among them the complicated question of how to care for pregnant women who were infected with the Ebola virus. When the West African epidemic began, the scarce data available from past outbreaks suggested an extremely pessimistic prognosis: earlier reports documented 89–100% mortality levels among pregnant women (Report of an International Comission 1978; Bebell and Riley 2015) and 100% mortality among neonates born to infected mothers (Mupapa et al. 1999; Bebell and Riley 2015).

During the 2-year West African epidemic, MSF sought solutions for the many clinical, operational, and ethical challenges these patients presented. From early on, we began developing protocols and best practice guidelines for managing pregnant women, and these tools evolved as we gained experience and knowledge. By the end of the outbreak, Ebola virus disease (EVD) was no longer seen as an almost inevitable death sentence for patients, or for pregnant women specifically—although the nearly 40% overall mortality rate in this epidemic (WHO 2016b) was still appalling high. The exception was neonates, where no such improvement was seen: there was only one documented case of an infant who recovered from Ebola contracted *in utero*. This tiny survivor was among the last Ebola patients in an MSF treatment center—a small ray of hope at the very end of the epidemic that took over 11,000 lives.

This chapter describes the evolution of MSF's response to managing these highly vulnerable patients, told by recounting some key time periods, seminal events on the ground, and unforgettable patients that helped shape our thinking. It also discusses how we addressed some of the difficult, often very complicated dilemmas that arose along the way.

All names have been changed to protect patient privacy. The exception is the infant Nubia, whose remarkable survival was widely reported in the media in late 2015, with her father's consent.

S. Caluwaerts
Médecins Sans Frontières, Operations Center, Brussels, Belgium

P. Kahn (✉)
Médecins Sans Frontières, New York, NY, USA
e-mail: patricia.kahn@newyork.msf.org

D. A. Schwartz et al. (eds.), *Pregnant in the Time of Ebola*, Global Maternal and Child Health,
https://doi.org/10.1007/978-3-319-97637-2_6

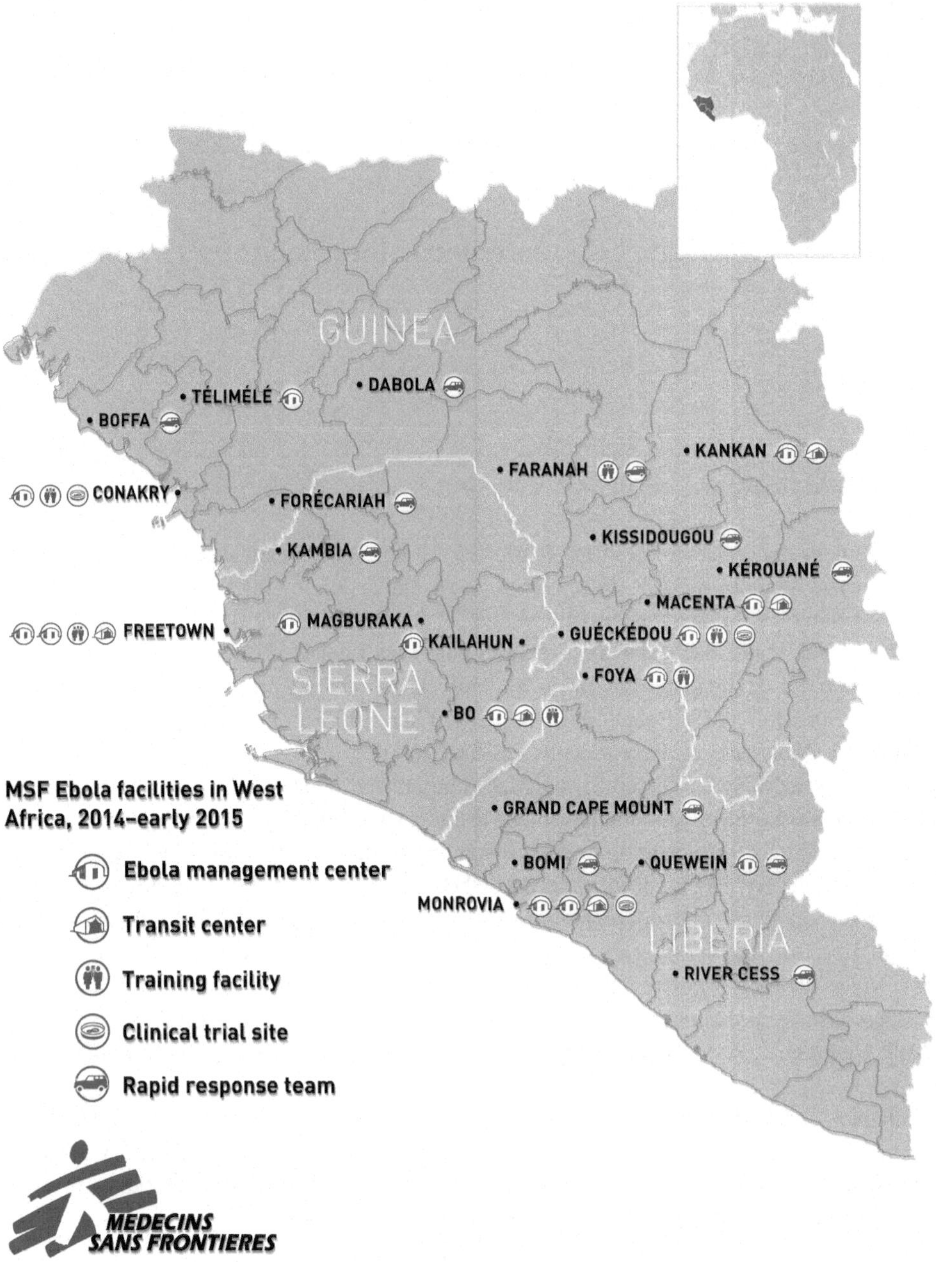

Fig. 6.1 Map of MSF Ebola activities. Published with permission of © Doctors Without Borders/Médecins Sans Frontières 2018. All Rights Reserved

6.2 The Outbreak Begins, Maternity Care Breaks Down: Sierra Leone in Mid-2014

MSF had already been working in West Africa for more than 25 years when the Ebola epidemic unexpectedly spread through three countries in the region. In early 2014, MSF was managing a human immunodeficiency virus (HIV) project and conducting malaria surveillance in Guinea as well as running a large maternity hospital, the Gondama Referral Centre (GRC), in Sierra Leone.

The GRC program was an important provider of emergency obstetrical care (EmOC) within Sierra Leone, which had the highest maternal mortality of any country in the world even prior to the Ebola virus outbreak (WHO 2014b); in 2014, MSF staff at GRC provided more than 10% of the country's total Caesarean sections. Although the hospital's official catchment area was only Bo Province (one of Sierra Leone's 14 provinces), in practice it served patients with serious obstetrical complications coming from over one third of the country, due to the scarcity of quality obstetrical care (Brolin Ribacke et al. 2016). GRC's provision of pediatric and obstetrical care (since 2003 and 2007, respectively) was a lifeline for many sick mothers and children.

By the time Ebola crossed the border from Guinea to Sierra Leone in May 2014, all alarm signs were going off in the MSF program. Hemorrhagic fevers are not unknown in this region: Lassa virus infection, an acute viral hemorrhagic fever related to Ebola but less virulent and with higher survival rates if treated early enough, has long been endemic. But Ebola and its extremely high mortality rate was something completely new and far more dangerous. Published data for the Ebola Zaire strain (the strain behind the West Africa outbreak) showed overall mortality rates ranging from 47 to 90% (WHO 2014a), as compared to 1% for Lassa fever in the community and 15% in hospitalized Lassa patients (WHO 2016a).

Anyone who practices obstetrics knows that deliveries come with lots of blood and body fluids getting spread around during most procedures, presenting an important potential source of contamination in cases of Ebola infection. Even in the high-risk zone of Ebola Treatment Centers (ETCs), where all personnel were required to wear personal protective equipment (PPE), pregnant women were a source of anxiety both for staff and other patients. Despite these worries, the MSF obstetrics team at GRC tried to prepare for the arrival of infected patients. But the possibility of carrying out deliveries and Caesarean sections in Ebola-positive women was quite frightening for many health care workers (HCWs). And the danger was not limited to them, but also extended to other patients and caretakers, and potentially to the families of HCWs. There was also the diagnostic challenge of distinguishing Ebola from common pregnancy complications, which all share some symptoms and ultimately rely on results from Ebola laboratory tests.

MSF considered, and ultimately rejected, two strategies for keeping the hospital running in the safest possible way.

The first strategy was to perform Ebola testing on all pregnant women presenting to the hospital and to admit only Ebola-negative patients; women who tested positive would be sent to the ETC, where they could get care in a safe setting and from specialized staff. But this approach would create new problems. First, it was likely to add a significant workload, since GRC was a referral center for complicated pregnancies and was therefore likely to receive many patients with symptoms that could indicate either Ebola or obstetrical complications. Testing would require sending blood samples to a laboratory in Kailahun, in the north, which usually took 24 h to return results. As a result, definitive diagnosis sometimes took longer, since false-negative results are not uncommon in the first 72 h after exposure to the virus. The MSF team felt it was inhumane to keep pregnant women (most of whom were not infected with the Ebola virus) waiting for a laboratory result while possibly hemorrhaging in a tent or losing their baby. Moreover, this strategy risked exposing the women to Ebola from other patients awaiting test results, and it could frighten other pregnant women away from coming to the hospital for delivery.

Alternatively, we could admit all pregnant women who presented to GRC and treat everyone as potentially Ebola-positive. While in theory this was the better option, it also raised serious issues. We would need to manage the maternity as an ETC, and within this structure, would face difficult decisions about performing surgery (especially Caesarean sections), when indicated, on patients who might be Ebola-positive. Surgery under these conditions would require a significant increase in hospital staff (twice as many obstetrician-gynecologists and triple the number of nurses and cleaners, by our assessment), since an individual staff member is allowed to work in full PPE only for a maximum of 3 h daily. We simply did not have those human resources available for GRC at a time when we were investing heavily in finding doctors and nurses to work inside ETCs across all three affected countries. And even using PPE, there would still be a (much smaller) risk of infecting staff. Another factor was that the entire operating theater could become contaminated with Ebola-positive body fluids, a considerable risk even with rigorous infection control measures in place. What would happen then? In an ETC, basically every object that comes in contact with Ebola patients must be burned, but there were no clear procedures for potentially contaminated operating tables or anesthesia material. And there was an ominous new example of the potentially fatal risk posed by surgery: the 2014 Ebola outbreak in the Democratic Republic of Congo (DRC), which infected 66 people in total (49 deaths) between August and November, had started with a pregnant woman as an index case. After she died, the fetus was removed by surgery, so as not to bury it inside her (according to her community's custom). All four staff members who took care of her or had contact—including doctor, nurse, and cleaner—died and spread the deadly virus to their families (Maganga et al. 2014).

The difficult decision of how to manage and deliver pregnant Ebola patients ultimately forced MSF staff to assess the risks they were personally willing to take. Every national and expatriate colleague had to ask him- or herself: Am I willing to risk my life operating on Ebola-positive patients who have a high risk of mortality even before surgery—at that point it was thought to be over 90%— and perhaps even higher with the additional stress of surgery? Is it worth it? If a physician dies from Ebola virus infection in a country with so few doctors, how does this affect other patients, both Ebola-positive and -negative? Is this something I can discuss with my already-worried family and loved ones? The strain of working in this environment took a considerable toll on every staff member. We were all in emergency mode (Fig. 6.2).

Fig. 6.2 An MSF obstetrician/gynecologist at the Gondama Referral Centre for children and women with complicated pregnancies just before the Center's closure. Published with permission of © Doctors Without Borders/Médecins Sans Frontières 2018. All Rights Reserved

The heartbreaking decision to close the MSF maternity hospital at the GRC (Fig. 6.2) came a few weeks after the first Ebola-positive pregnant patients were received in July 2014. The triage system had worked: the first patient was taken to the ETC without infecting other GRC staff or patients. More patients with Ebola infection followed, and since the epidemic was just beginning, we knew that the case load would only increase. One patient who was later confirmed with Ebola infection arrived in the Gondama Clinic next to GRC and was not correctly triaged; she sat with 39 °C fever among other patients in the antenatal care room before alarm signs went off. Fortunately, no one became infected. After further assessment of our options, the team on the ground decided that continuing to provide obstetric care posed too many risks for patients and staff, and they took the very painful decision to close.

The pediatric ward remained open for another 3 months, but was then also shut down. In caring for sick children, the risk of contact with body fluids is less, and triage seemed more straightforward in children than in pregnant women with complications. Since the health system was collapsing, we had expected an increase in pediatric admissions during that period, but instead we saw a decrease (Delamou et al. 2017). Parents were afraid to bring their children to hospitals because they feared Ebola in the community—so sometimes they preferred for their children to die at home. MSF was putting all possible resources into fighting the burgeoning Ebola epidemic and just did not have the additional human resources and means to keep the hospital safely open.

Facilities that continued to offer health care paid a heavy price. At least 10 of only 130 doctors in Sierra Leone died of Ebola (Evans et al. 2015), along with many more nurses and midwives. MSF lost community health officers we had been trained in doing caesarean sections 1 or 2 years before and who had been working in rural hospitals. These deaths are a huge loss for the country and will be felt for many years to come.

6.3 Caring for Pregnant Patients in ETCs: Early Days, Epidemic Peak, and New Guidelines for Care

Meanwhile, in Gueckedou, Guinea, the first confirmed pregnant Ebola patient was admitted to the MSF ETC in May 2014 and died 3 days later. Two others admitted in June 2014 survived, but lost their pregnancies.

Medically speaking, we learned something vitally important from these two survivors: their amniotic fluid remained Ebola-positive even several days after the mother's Ebola test became negative (Baggi et al. 2014), which told us that pregnancy continues to pose a potential risk of Ebola exposure even in survivors. This finding raised some difficult dilemmas around what to do if a pregnant woman recovers from Ebola with an intact pregnancy and living fetus. One option was to discharge her and advise her to return to an ETC for delivery. But if she miscarried or delivered at home, she might infect her family or the traditional birth attendant (TBA) who assisted her. The alternative was to keep her at the ETC or another safe facility until she delivered—which meant a potentially long separation from her family and daily exposure to the death of other patients. Survivors with other children were also needed at home. For those with intact first-trimester pregnancies, termination of pregnancy was another option. These issues, and especially their ethical dimensions, continued to be discussed throughout the epidemic, both in decisions that had to be made with individual patients in our ETCs and in beginning to develop protocols for managing pregnant women with Ebola.

Protocol development was an important early objective for MSF. Drawing on early lessons learned from caring for pregnant Ebola patients, in August 2014 MSF published online the first field-adapted guidelines for pregnant women; these guidelines were updated five times during the epidemic, reflecting

Table 6.1 Principles for managing pregnant women with Ebola virus disease

At admission to an ETC, assess all women of childbearing age for pregnancy and breastfeeding
Ensure that all pregnant women with Ebola and all Ebola survivors deliver in designated high-risk areas within an ETC
Avoid all invasive procedures, including Caesarean sections and other surgery, episiotomy, removal of retained placenta, and use of vacuum
Favor oral treatment over injectable or intravenous ones (e.g., misoprostol over oxytocin)
Apply strict PPE and waste management protocols, including for materials used during delivery and which cannot be disinfected and sterilized
Treat new born babies as Ebola-positive and test them at birth. Apply the same PPE and safety protocols
Before discharge, counsel surviving women and provide them with family planning and nutritional support

our ongoing learning (MSF 2014). The key principles underpinning these guidelines are summarized below in Table 6.1.

The guidelines included procedures to mitigate certain clinical risks associated with delivery; for instance, since some patients had an increased tendency to bleed, the protocol emphasized measures for preventing and treating postpartum hemorrhage. Protocols also stressed the importance of protecting health care workers and, therefore, saw no place in an ETC for surgery or even suturing, the latter due to the high risk linked to needle stick injuries a HCW could sustain while suturing wounds in agitated patients. Also for staff protection, the guidelines recommended that three people assist at each delivery (even though this was ultimately not always feasible due to the immense workloads and understaffing) and that one staff person serves as a timekeeper to limit all staff to 1½ h in the ETC, for their own safety. Another key point was to recommend routine pregnancy testing for women of childbearing age admitted to an ETC, although we were ourselves not able to implement this consistently until later in the outbreak.

At the time when these first guidelines were published, the epidemic was at its most disastrous point in Liberia—during August to November 2014. One of our MSF colleagues who worked in ELWA3 in Monrovia—the biggest ETC ever constructed (shown below in Fig. 6.3), with a capacity of 250 patients at its peak—said it was how she could imagine Dante's circles of hell. The number of patients was overwhelming, and the staff was continuously confronted with death and horrendous suffering. At one stage, MSF had to refuse admission to sick patients at the door due to the dangerous overcrowding.

During this peak period in Monrovia, 10 women with known pregnancies were admitted to ELWA3; nine of them died, five with their baby still within the uterus (*in utero)*. The single survivor, Hawa, had an intact pregnancy, raising the question posed above about where she should go after she was discharged from the hospital. Given the extreme shortage of beds, there was no space to keep a cured patient in the ETC between follow-up appointments. Hawa was therefore sent home with a follow-up appointment, plus instructions to return if labor began. In case she was unable to return and went through labor at home, she was given a safe delivery kit with sterile material and chlorine. All HCWs involved knew this was far from ideal, but the circumstances left little choice. A few days later, Hawa delivered a macerated 5-month stillborn fetus at home, returning to ELWA3 for an appointment a few days later. She told the MSF staff that no one had touched the fetus or placenta, and that she chlorinated the floor after delivery. Although anxiety levels about infection of family members were high, fortunately no Ebola infections related to this delivery were identified during follow-up in the subsequent weeks.

Fig. 6.3 Aerial view of ELWA 3, the Ebola Case Management Center in Monrovia, Liberia. Published with permission of © David Darg/Médecins Sans Frontières 2018. All Rights Reserved

6.4 Managing the Surviving Pregnancies: Difficult Choices

Patient stories from ETCs that were less overwhelmed than ELWA3 further illustrate the dilemmas that arose in providing care for survivors with live pregnancies. In particular, while the guidelines called for offering these women the option of terminating their pregnancy (Fig. 6.4), some declined, in the hopes of having a healthy baby. Finding ways to care for these survivors while minimizing the infection risk to others was challenging and required flexibility to adapt to the local and family circumstances.

In one instance in Guéckédou, Guinea, an Ebola-positive woman, Fatmata, was admitted at 5-months gestation to an MSF ETC. She had high fever and general body weakness, but recovered after a few days. The fetus also survived. This was Fatmata's eleventh pregnancy, and she and her husband had 10 other children at home; the baby was very much wanted by the family.

MSF offered to assist the mother in terminating the pregnancy. As mentioned above, in such cases there is a high chance of miscarriage at home, posing an infection risk for family or any traditional birth attendant (TBA) assisting at the birth. But Fatmata and the family declined. So what should be done?

To spare Fatmata the distressing experience of staying inside the ETC for the remainder of her pregnancy, the ETC's logistics team came up with a novel solution: they built a small wooden hut next to the ETC where she could stay until labor began or her membranes ruptured. In Brussels, where MSF has one of its five operational centers, this unlikely surviving 5-month old fetus became known as "the next president of Guinea." Everyone at MSF working on Ebola seemed to know about this baby and to feel inspired by the hope of new life and a better future. In an epidemic where no one yet knew how and when the Ebola virus epidemic in West Africa would ever end, Fatmata's fetus became very special.

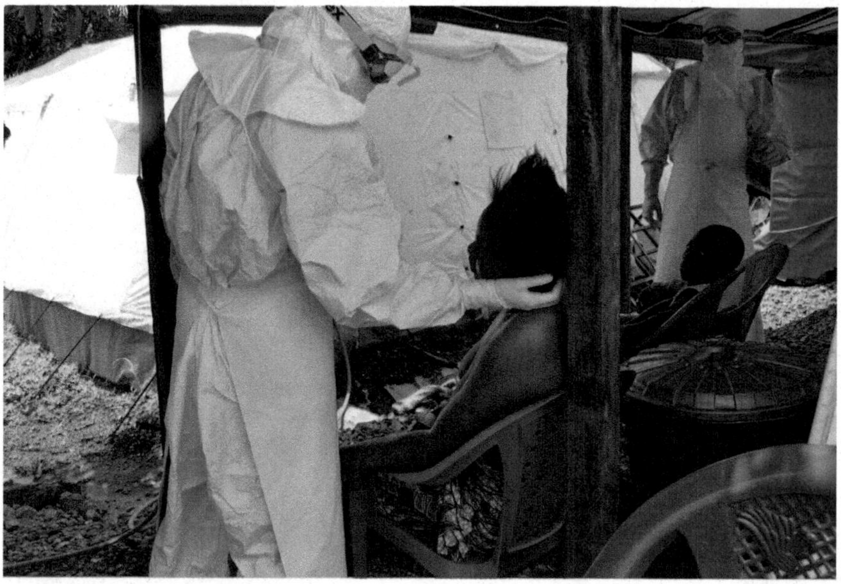

Fig. 6.4 Caretaker and patient inside an MSF Ebola case management center. Published with permission of © Julien Rey/Médecins Sans Frontières 2018. All Rights Reserved

Sadly, 32 days after her cure, Fatmata experienced vaginal bleeding in her wooden hut and was transferred back to the ETC. A few hours later, she delivered a macerated 5-month old stillborn fetus. Swabs were tested and found to be highly Ebola-positive. The mother recovered and left the ETC 24 h later, while the MSF team ensured a safe burial for the highly infectious fetus (Fig. 6.4).

The same situation arose in Freetown in March 2015. A young woman, Aminata, recovered from EVD at an MSF facility, with a live 4-month old fetus. Like Fatmata, Aminata felt strongly about continuing her pregnancy; it was her third, but she had no living children. The medical team discussed different options, preferring that she stay at the ETC. But she chose to go home, and the team continued to support her by phoning twice a day and by checking her weekly at the ETC. But 3 weeks after discharge, the baby died *in utero*. Aminata agreed to have labor induced in the ETC, where we determined from swabs after delivery that the fetus was highly Ebola-positive.

These two tragic, but exceptional, cases taught us that amniotic fluid remains potentially infectious not just for a few days after the mother's cure, as we had learned from our earliest patients, but even more than 1 month later—a finding that further shaped our protocols going forward. At the same time, MSF managed to find creative, safe, and culturally adapted solutions for the mother and family without jeopardizing the safety of other family members and/or HCWs.

But over time, even as more pregnant women survived—occasionally with their pregnancies intact—live births remained rare, and survival beyond the neonatal period was unheard of. It was known from previous Ebola virus outbreaks that babies had been born alive despite having Ebola virus. During the first-ever Ebola outbreak in the Democratic Republic of Congo (then called Zaire) in 1976, there were 11 live births, but none of the babies survived for more than 3 weeks (Mupapa et al. 1999). During the entire West Africa outbreak, only two babies were born alive at MSF ETCs. The first one was born in September 2014 in Gueckedou, Guinea, and had a good Apgar score at birth, but tested positive for Ebola. After 24 h, he developed a high fever and died 48 h later. His mother survived. Transplacental congenital Ebola is such an extreme stress on a neonate's body that the prospect of ever having a survivor past this early stage seemed impossible throughout most of the epidemic. We learned otherwise only at the very end of the epidemic, when Baby Nubia became the first documented survivor (see Sect. 6.7 below).

6.5 Asymptomatic Ebola Presentation at Hospitals, and Community Access to Maternal Care

In December 2014, Mary, a patient in her fifth pregnancy and with four previous normal vaginal births, presented at the MSF ETC in Monrovia in latent labor. She was at full term with a live fetus and no signs of illness. The government hospital had refused to admit her because she did not have a certificate stating that she was Ebola-negative, a requirement to be admitted for delivery (Akerlund et al. 2015). The hospital staff were understandably extremely afraid of Ebola, but on a practical level the testing requirement was an enormous barrier to normal obstetric care during the epidemic.

MSF tested Mary for the virus so that she could get a certification of Ebola-negative status and then return to the hospital for delivery. But to everyone's surprise, the test came back as highly positive. Initially, MSF thought the result was an error, and that the sample had been mixed-up with one from another patient. So Mary was retested, and once again was highly positive. Still, apart from having slight uterine contractions, she looked healthy and had no clinical signs and no fever.

The MSF team admitted her to the ETC and gave her the available supportive care: antibiotics, antimalarial drugs, an intravenous drip, and painkillers. On day 3, she developed a fever and reported no fetal movements; the fetus had died *in utero*. She deteriorated gradually over the next few days and died 7 days after admission, with the undelivered fetus still inside her.

This case left the medical team with a lot of questions and frustrations. It was a painful experience to see a perfectly healthy patient deteriorate so quickly and die in agony with hemorrhagic signs, and for the clinicians to have so little to offer.

But it also triggered internal reflection within our team and across MSF. We couldn't blame the national hospitals for not admitting untested patients: an asymptomatic patient like Mary, who unexpectedly proved to be positive, was a stark reminder that there was no 100% safe way to manage a delivery or perform a Caesarean section in an Ebola epidemic without testing all patients before admission. MSF also saw Mary's case as an after-the-fact validation of our difficult decision to close the GRC maternity hospital in Sierra Leone early in the outbreak. "We were right to close, given the circumstances," we said, while ignoring the terrible feelings in our stomachs as we thought of all the patients who died—not of Ebola but of perfectly treatable obstetrical complications (Strong and Schwartz 2016; UNPFA 2015).

Towards the end of the epidemic, our ETC had another case that illustrated the difficulties of providing routine obstetric care during an Ebola outbreak without testing all patients before admission. The case was discovered due to a requirement in Sierra Leone to test all corpses for Ebola with an oral swab, as community surveillance. So when a burial team went to bury a macerated stillborn preterm fetus in July 2015, they did a routine swab, which unexpectedly came back highly positive (Bower et al. 2016).

The mother of this stillborn baby was alive and well. She had never been admitted to an ETC, but had been followed up during her pregnancy by a contact-tracing team because the father of the stillborn infant was a known Ebola survivor who had been discharged from an ETC several months earlier. She had no fever and did not meet the case definition for suspicion of Ebola. Following the stillbirth, she was tested and found to have antibodies (IgG and IgM) against Ebola, meaning that she had certainly had a recent Ebola infection. In a follow-up interview, she reported that shortly after her husband became ill she had experienced symptoms not considered as EVD warning signs, including intense fatigue, loss of appetite, abdominal pain, jaundice, eye pain, sensitivity to light, and confusion. The authors hypothesize that she was most likely infected around the time of her husband's exposure or illness and that pregnancy might alter the ways that EVD presents and progresses—which would present significant challenges for case detection and for protection of other patients and health care workers during outbreaks.

So this was apparently a case of a completely asymptomatic woman delivering an Ebola-positive stillbirth in a health care center. Of course, MSF was very concerned about the staff who assisted at the delivery wearing only gloves and a mask, rather than the full PPE as in an ETC. Fortunately, neither the nurse who did the delivery nor the traditional birth attendant who was also present developed any signs of Ebola over a 21-day follow-up, the incubation time of the virus.

6.6 Ebola Clinical Trials and Pregnancy: More Dilemmas

In December 2014, after negotiations that had begun the previous July, MSF took a step into new territory for the organization with a decision to collaborate on conducting clinical trials of experimental treatments for Ebola. The collaborations involved three separate trials to be carried out in MSF-managed ETCs. Two sites were in Guinea: one would study favipiravir (an antiviral drug developed against influenza virus) and another would assess the use of plasma from Ebola survivors (convalescent plasma); a third would evaluate the broad antiviral drug brincidofovir, in Liberia. All three were on the WHO list of priority therapeutics to be tested in Ebola-positive patients during the outbreak. Separately, MSF also participated in WHO's vaccination trial, which took place towards the end of the epidemic.

As soon as MSF began to seriously consider collaborating in these studies, questions arose over the inclusion of pregnant women. Pharmaceutical companies always have strong reservations about including pregnant women in trials testing new drugs, mostly due to fear of birth defects—including potentially disastrous ones (such as the severe malformations caused by Softenon-thalidomide)—and of legal liability. But at the time the Ebola drug trials were being planned, the mortality rate of pregnant women appeared to be even higher than that for Ebola patients overall, and there was as yet no documented case of a neonate surviving transplacental (vertical) Ebola virus infection. Thus, the issue of teratogenicity seemed a theoretical problem. MSF also considered it particularly unfair that pregnant women with Ebola virus infection, already such a highly vulnerable group, would be denied access to potentially life-saving drugs, and we argued that the patients should have a voice in these discussions and decisions concerning their care.

6.6.1 Brincidofovir Study

For this study, no negotiations were possible: the drug's manufacturer was very clear about not having pregnant women in the trial. But this became a moot point when the trial was stopped prematurely after only a few patients had been enrolled, due to the lack of new Ebola patients—it was already the end of the epidemic in Liberia (Dunning et al. 2016).

6.6.2 Favipiravir Study

The company producing this drug, which was already approved in Japan for use against resistant influenza in adults, was more open for discussion. Through negotiations they agreed to allow MSF to give favipiravir for monitored emergency use in pregnant women, outside of the clinical study, since they could not be included in the trial. This small trial (JIKI Trial 2016), which ended after 126 patients had received the drug, demonstrated that favipiravir is safe and well-tolerated. But it did not resolve the question of efficacy; findings showed only that further clinical studies in patients with low or medium Ebola viral load—but not in the highest viral load group—might be warranted (Sissoko et al. 2016).

6.6.3 Convalescent Plasma Study

Inclusion of pregnant patients in this study posed no problem, since treatment used a blood-derived natural product rather than a drug. In total, eight pregnant patients were included in the trial, which took place in March-July 2015 (van Griensven et al. 2016), and an additional three received convalescent plasma in monitored emergency use, while awaiting the final efficacy results of the trial. Eight of the 11 women (but no fetuses) survived, but these numbers were too small to give insight into whether the treatment had contributed to their recovery.

6.6.4 ZMapp Study

Another trial at the time (The PREVAIL II Writing Group 2016), one that MSF declined to be involved with, tested a product called ZMapp, which is based on monoclonal antibodies. Although, in principle, pregnant women were included, this study raised another difficult ethical issue: it was a randomized controlled trial (RCT) where ZMapp was compared to standard supportive treatment, so each patient had only a 50% chance of receiving ZMapp (the other 50% were given standard supportive care alone). While this trial design might give scientifically clearer results, MSF felt very strongly that it was not ethical to take this approach, i.e., to withhold the experimental treatment from some trial participants, for a disease with such high mortality, especially given the circumstances of the West Africa epidemic. For this reason, we participated only in trials (the three described above) where *all* eligible patients could receive the potentially beneficial drug. The communities of the affected countries supported MSF's decision and the Guinean, Sierra Leonean, and Liberian ethical committees also preferred nonrandomized designs. Inclusion of pregnant women was allowed under the Zmapp protocol, but in practice none of the included patients were pregnant.

6.6.5 rVSV-ZEBOV Vaccine Study

This experimental vaccine became available for efficacy testing towards the end of the epidemic. The trial was called "Ebola ça suffit" and was conducted in Guinea starting in March 2015 (Henao-Restrepo et al. 2017). But it excluded pregnant women and children below age 18, because a live attenuated vaccine (which has very limited, weak viral replication) carries a theoretical risk of transmission to the fetus. However, since pregnancy testing during the epidemic was not routine, some women who did not realize they were pregnant were vaccinated. There have been no reports of malformations in the offspring of these unintentionally vaccinated pregnant women (Henao Restrepo 2016).

6.7 The First Neonatal Survivor: Nubia in Nongo, Guinea[1]

In late October 2015, as the outbreak was ending, a 25-year old woman named Zeinabu with confirmed Ebola disease was admitted to the Nongo ETC in Guinea. According to the history she provided, she was 7 months into her third pregnancy and was being clinically followed as a household contact of an Ebola patient who had died of the disease. At this stage of the epidemic, the rVSV-ZEBOV vaccine had shown some preliminary success in clinical trials and was being offered, within

[1] The father specifically agreed to the use of his daughter's name in MSF publications.

the trial parameters, to known contacts. But because Zeinabu was pregnant, she had been ineligible for vaccination.

A few days after exposure to Ebola, Zeinabu developed symptoms and tested positive for Ebola in Forecariah, the province in Guinea where the ZMapp clinical trial was ongoing. The Forecariah ETC did not want to keep her because they thought that caring for a pregnant woman with Ebola was too complex, so they referred her to MSF in Nongo.

As soon as Zeinabu was admitted to the MSF facility, we tried to obtain ZMapp for her outside the randomized clinical trial. But this was refused. The decision was then made to treat her with favipiravir, the experimental antiviral drug that at that point appeared to be showing potential limited success in small clinical studies (Sissoko et al. 2015). Three days later, Zeinabu went into spontaneous labor and delivered an unexpectedly healthy baby girl of 2800 g, but who tested positive for Ebola. The family decided to call the baby girl Nubia in order to thank the MSF team and the Brazilian nurse named Nubia, who took care of their infant daughter. Zeinabu deteriorated rapidly after delivery and died 7 h later of postpartum hemorrhage caused by an Ebola-related complication, despite having received the appropriate medical treatment for a hemorrhage (oxytocin, ergometrin, and misoprostol in high doses).

The baby's survival and initial good health came as a complete surprise to the MSF team. And even as they barely dared to hope, they sought and were able to obtain ZMapp outside the clinical trial for the newborn. Nubia received her first dose the day after birth, followed by three more doses 3 days apart. After an initial improvement, her viral load increased again, indicating that ZMapp was having little effect, so on day 11 she was also given a transfusion of an Ebola survivor's buffy coat (containing white blood cells and platelets) to supplement her immature immune system's response. On day 15, she developed convulsions, raising fear that Ebola virus might have migrated in the cerebrospinal fluid, causing neurologic damage. However, a single sample from a lumbar puncture showed no Ebola virus. She was then (day of life 19) given another experimental medicine, the small molecule drug GS-5734 (Warren et al. 2015)—only the second person in the world to receive this drug for treating Ebola (the first is described in Jacobs et al. 2016).

Nubia recovered and survived (Dörnemann et al. 2017). She left the ETC 33 days after she was born and was followed up by MSF until January 2017 (Fig. 6.5). She is a beautiful child who is developing normally and is the first documented survivor of transplacental (vertically transmitted) Ebola virus infection.

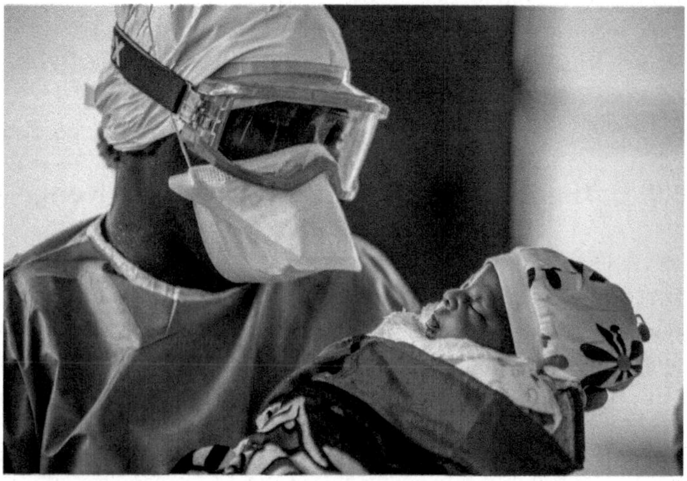

Fig. 6.5 The infant Nubia, shown here at 3 weeks old, is carried outside the MSF ETC to meet her uncle, who said: "We didn't think she could survive. She is a warrior." Nubia is the only known case of a newborn surviving Ebola. Published with permission of © Tommy Trenchard/Médecins Sans Frontières 2018. All Rights Reserved

6.8 What Did We Learn?

For MSF, our response to the West Africa outbreak was a case of learning while doing, under extraordinarily difficult circumstances. Although at the start of the epidemic, we were among the most experienced organizations globally in responding to Ebola, the West Africa outbreak was overwhelming, particularly in the early stages. Nearly all previous outbreaks had occurred in isolated, rural settings and at a fraction of the scale—the largest previous outbreak (occurring in Uganda from 2000 to 2001) had a total of 425 cases, compared with over 28,600 people infected in three contiguous countries of West Africa (CDC 2016), many in crowded urban settings. In the beginning, many ETCs did not have midwives, or physicians or nurses experienced in obstetrics. Despite huge deployments of material and people, at some critical points workers faced situations when there were simply no beds left to accommodate new patients, and clinics had to refuse dying people at the door, placing teams under extreme stress. Especially at such times, pregnant Ebola-positive women suffered even more. This suffering was compounded by the lack of knowledge and, early in the epidemic, the absence of protocols for managing these more complicated patients.

MSF, therefore, quickly began developing specific protocols for pregnant women, and subsequently adapted them continuously over the course of the epidemic as we gained experience and knowledge. This process involved the ongoing dilemma of finding a balance between the best possible care for patients versus keeping caregivers safe (and minimizing their fear for their own safety).

Data on pregnancy status of patients were unfortunately not collected consistently during the outbreak, so it is not possible to make a rigorous estimate of mortality in this group. Still, MSF has documentation on 36 known pregnant patients who survived Ebola virus infection out of a total of 77 documented pregnant patients in this epidemic, giving a survival rate of 47% (article under review)—considerably better than the figure of 10% or less (i.e., 90% mortality) in previously published data. At MSF, we made the mental transition from an almost palliative clinical care situation for pregnant women early in the outbreak to one where patients have a better chance of survival.

However, many pregnancies were not detected during this epidemic because routine pregnancy testing only became part of clinical care after December 2014. Particularly in those patients who died shortly after arrival at an ETC, clinicians would probably not have thought about or tested for possible pregnancy. We have no information about what happened to these many pregnancies. Did they all abort? Did any health care workers or family get infected at delivery? Or did any fetuses survive Ebola, to be born healthy? We saw that this is indeed possible with Nubia, the baby born at the end of the epidemic in Guinea and who became the first documented survivor of transplacental Ebola. However, we don't know *why* she survived. It is possible that she was infected only very late, either during labor or delivery (her initial viral load was quite low, which is a good prognostic sign). She also received extremely dedicated, around-the-clock care from medical staff (a level of care possible because she was the only Ebola-positive patient in the ETC at the time), as well as treatment with three experimental products. Was it any (or all) of these drugs, the extremely dedicated staff, a combination of these factors, or just luck?

Many other questions about best care practices also remain unanswered. If a pregnant woman has Ebola and her fetus is still alive, is it better to induce labor near term or to wait until the woman is cured and tests Ebola-negative? In our protocol, we opted to wait for cure, both to enhance the mother's chance of surviving the delivery (less risk of postpartum hemorrhage) and to protect HCWs against the risk of infection in case of accidental blood contact. But it is possible that shorter exposure to maternal virus might reduce the neonatal mortality risk for a near-term fetus. These questions reflect our lack of understanding about when and how the fetus of an Ebola-positive mother becomes infected: does this occur mainly in the antenatal phase or also during delivery?

The infectiousness of amniotic fluid is another unknown. We know that the level of virus in amniotic fluid remains very high even after the mother is cured. But without virus culture studies, there are no answers as to whether it is as highly infectious as blood or less infectious, like urine.

We remain frustrated that we did not manage to provide better family planning to women survivors, even at the end of the epidemic. Condoms were given to all survivors in abundance, but we never managed to offer women long-term methods, such as implants. For future epidemics, provision of family planning to women who survived Ebola must be a priority. It should also be a priority for all women in affected communities during outbreaks. There was virtually no obstetrical care available during the epidemic, especially in Sierra Leone and Liberia; therefore, by preventing some pregnancies, access to family planning could likely have saved more women from dying of untreated pregnancy complications.

Despite the various clinical trials conducted late in the epidemic, and although MSF managed to include a few pregnant women, the outbreak ended without clear evidence of effective treatment. There were encouraging hints: ZMapp had some promising results, and GS-5734 has shown positive findings in studies with nonhuman primates, while two individuals who received the drug both survived. During the next Ebola outbreak, follow-up research on these drugs will be needed, along with new studies on emerging experimental treatments. As for vaccines, pregnant women cannot (yet) benefit from the development of what appears to be an effective vaccine against the Ebola virus Zaire strain: as of this writing (February 2017) they are still excluded from receiving this potentially life-saving measure.

So for now, the only gain is that we will at least be better prepared to provide supportive care to pregnant women, thanks to new knowledge and guidelines that have advanced best practice.

Dedication This article is dedicated to the 14 MSF staff who lost their lives in this Ebola epidemic.

References

Akerlund, E., Prescott, J., & Tampellini, L. (2015). Shedding of Ebola virus in an asymptomatic pregnant woman. *New England Journal of Medicine, 372*(25), 2467–2469.

Baggi, F. M., Taybi, A., Kurth, A., Van Herp, M., Di Caro, A., Wolfel, R., et al. (2014). Management of pregnant women infected with Ebola virus in treatment center in Guinea. *Eurosurveillance, 19*(49). Retrieved from http://www.eurosurveillance.org/ViewArticle.aspx?ArticleId=20983.

Bebell, L. M., & Riley, L. E. (2015). Ebola virus disease and Marburg disease in pregnancy: A review and management considerations for filovirus infection. *Obstetrics and Gynecology, 125*(6), 1293–1298.

Bower, H., Grass, J. E., Veltus, E., Brault, A., Campbell, S., Basile, A. J., et al. (2016). Delivery of an Ebola Virus-positive stillborn infant in a rural community health center, Sierra Leone, 2015. *The American Journal of Tropical Medicine and Hygiene, 94*(2), 417–419.

Brolin Ribacke, K. J., van Duinen, A. J., Nordenstedt, H., Höijer, J., Molnes, R., Froseth, T. W., et al. (2016). The impact of the West Africa Ebola outbreak on obstetric health care in Sierra Leone. *PLoS One, 11*(2), e0150080. https://doi.org/10.1371/journal.pone.0150080.

CDC. (2016). *Outbreaks chronology: Ebola virus disease. Known cases and outbreaks of Ebola virus disease, in reverse chronological order*. Retrieved from https://www.cdc.gov/vhf/ebola/outbreaks/history/chronology.html.

Delamou, A., Hammonds, R. M., Caluwaerts, S., Utz, B., & Delvaux, T. (2017). Ebola in Africa: Beyond epidemics, reproductive health in crisis. *The Lancet, 384*(9960), 2105. https://doi.org/10.1016/S0140-6736(14)62364-3.

Dörnemann, J., Burzio, C., Ronsse, A., Sprecher, A., De Clerck, H., Van Herp, M., et al. (2017). First newborn baby to receive experimental therapies survives Ebola virus disease. *The Journal of Infectious Diseases, 215*(2), 171–174. https://doi.org/10.1093/infdis/jiw493.

Dunning, J., Kennedy, S. B., Antierens, A., Whitehead, J., Ciglenecki, I., Carson, G., et al. (2016). Experimental treatment of Ebola virus disease with brincidofovir. *PLoS One, 11*(9), e0162199. https://doi.org/10.1371/journal.pone.0162199.

Evans, D. K., Goldstein, M., & Popova, A. (2015). Health-care worker mortality and the legacy of the Ebola epidemic. *The Lancet Global Health, 3*(8), e439–e440. https://doi.org/10.1016/S2214-109X(15)00065-0.

Henao-Restrepo, A. M., Camacho, A., Longini, I. M., Watson, C. H., Edmunds, W. J., Egger, M., et al. (2017). Efficacy and effectiveness of an rVSV-vectored vaccine in preventing Ebola virus disease: Final results from the Guinea ring vaccination, open-label, cluster-randomised trial (Ebola ça Suffit!). *The Lancet, 389*(10068), 505–518. https://doi.org/10.1016/S0140-6736(16)32621-6.

Henao Restrepo, A. (2016). Personal communication with PI.

Jacobs, M., Rodger, A., Bell, D. J., Bhagani, S., Cropley, I., Filipe, A., et al. (2016). Late Ebola virus relapse causing meningoencephalitis: A case report. *Lancet, 388*(10043), 498–503. https://doi.org/10.1016/S0140-6736(16)30386-5.

JIKI Trial. (2016). *Efficacy of favipiravir against Ebola (JIKI trial).* Retrieved from https://clinicaltrials.gov/ct2/show/NCT02329054.

Maganga, G. D., Kapetshi, J., Berthet, N., Ilunga, B. K., Kabange, F., Kingebeni, P. M., et al. (2014). Ebola virus disease in the Democratic Republic of Congo. *The New England Journal of Medicine, 371*, 2083–2089. https://doi.org/10.1056/NEJMoa1411099.

MSF. (2014). *Guidance paper for Ebola Treatment Center (ETC): Pregnant and lactating women.* Retrieved from https://www.rcog.org.uk/globalassets/documents/news/etc-preg-guidance-paper.pdf.

Mupapa, K., Mukundu, W., Bwaka, M. A., Kipasa, M., De Roo, A., Kuvula, K., et al. (1999). Ebola hemorrhagic fever and pregnancy. *The Journal of Infectious Diseases, 179*(Suppl 1), S11–S12. https://doi.org/10.1086/514289.

Report of an International Commission. (1978). Ebola haemorrhagic fever in Zaire, 1976. *Bulletin of the World Health Organization, 56*(2), 271–293 Retrieved from http://www.ncbi.nlm.nih.gov/pmc/articles/PMC2395567.

Sissoko, D., Folkesson, E., Abdoul, M., Beavogui, A. H., Gunther, S., Shepherd, S., et al. (2015). Favipiravir in patients with Ebola virus disease: Early results of the JIKI trial in Guinea. In *Conference on Retroviruses and Opportunistic Infections*. Retrieved from http://www.croiconference.org/sessions/favipiravir-patients-ebola-virus-disease-early-results-jiki-trial-guinea.

Sissoko, D., Laouenan, C., Folkesson, E., M'Lebing, A.-B., Beavogui, A.-H., Baize, S., et al. (2016). Experimental treatment with favipiravir for Ebola virus disease (the JIKI Trial): A historically controlled, single-arm proof-of-concept trial in Guinea. *PLOS Medicine, 13*(3), e1001967. https://doi.org/10.1371/journal.pmed.1001967.

Strong, A., & Schwartz, D. A. (2016). Sociocultural aspects of risk to pregnant women during the 2013-2015 multinational Ebola virus outbreak in West Africa. *Health Care for Women International, 37*(8), 922–942.

The PREVAIL II Writing Group. (2016). A randomized, controlled trial of ZMapp for Ebola virus infection. *New England Journal of Medicine, 375*(15), 1448–1456. https://doi.org/10.1056/NEJMoa1604330.

UNPFA. (2015). *Rapid assessment of Ebola impact on reproductive health services and service seeking behaviour in Sierra Leone.* Retrieved from http://reliefweb.int/sites/reliefweb.int/files/resources/UNFPAstudy _synthesis_March 25_final.pdf.

van Griensven, J., Edwards, T., de Lamballerie, X., Semple, M. G., Gallian, P., Baize, S., et al. (2016). Evaluation of convalescent plasma for Ebola virus disease in Guinea. *New England Journal of Medicine, 374*(1), 33–42. https://doi.org/10.1056/NEJMoa1511812.

Warren, T., Jordan, R., Lo, M., Soloveva, V., Ray, A., Bannister, R., et al. (2015). Nucleotide prodrug GS-5734 is a broad-spectrum filovirus inhibitor that provides complete therapeutic protection against the development of Ebola virus disease (EVD) in infected non-human primates. *Open Forum Infectious Diseases, 2*(Suppl_1), LB-2. https://doi.org/10.1093/ofid/ofv130.02.

WHO. (2014a). *Ebola virus disease: Fact sheet.* Retrieved from http://www.who.int/mediacentre/factsheets/fs103/en/.

WHO. (2014b). *Trends in maternal mortality: 1990 to 2013.* Retrieved from http://www.who.int/reproductivehealth/publications/monitoring/maternal-mortality-2013/en/.

WHO. (2016a). *Lassa fever: Fact sheet.* Retrieved from http://www.who.int/mediacentre/factsheets/fs179/en/.

WHO. (2016b). *Situation report. Ebola virus disease.* Retrieved February 17, 2017, from http://apps.who.int/iris/bitstream/10665/208883/1/ebolasitrep_10Jun2016_eng.pdf?ua=1.

7 Understanding the Personal Relationships and Reproductive Health Changes of Female Survivors of Ebola Infection in Liberia

Christine L. Godwin, Alexandria Buller,
Margaret Bentley, and Kavita Singh

7.1 Introduction

The Ministry of Health and Social Work in Liberia estimates that 997 of the 1,558 documented Ebola survivors living in the country (64%) are female (Gbarmo-Ndorbor 2017). Studies of previous Ebola outbreaks have demonstrated that women tend to disproportionately experience many of the social and economic consequences of Ebola survivorship, such as stigma and social rejection (De Roo et al. 1998; Hewlett and Amola 2003). In addition, press reports and anecdotal accounts from the West Africa Ebola epidemic suggest that female survivors have experienced changes in menstruation, sexual behavior, and subsequent pregnancies since recovering from Ebola (World Health Organization 2015a). Despite these reports, much of the research conducted in West Africa over the last 3 years has focused on the reproductive health of males due to the documented persistence of the Ebola virus in their semen (Mate et al. 2015; Deen et al. 2017). As described elsewhere in this book, the importance of understanding the potential for sexual transmission of Ebola from male survivors to their female partners cannot be overstated. However, it is also important to document and understand the biological and social mechanisms through which Ebola can uniquely affect the sexual and reproductive health of female survivors.

The Ebola virus has been demonstrated to persist in the semen of male survivors for from 9 months up to 2 years or more after an episode of acute Ebola infection (Ghose 2017; Fischer et al. 2017; Mate et al. 2015; Deen et al. 2017; Sissoko et al. 2017). Although no evidence currently exists regarding sexual transmission of Ebola from female survivors to their male partners, the rapidly changing guidelines

C. L. Godwin (✉) · K. Singh
Department of Maternal and Child Health, Gillings School of Global Public Health, University of North Carolina at Chapel Hill, Chapel Hill, NC, USA

A. Buller
Global Studies Department, Abilene Christian University, Abilene, TX, USA

M. Bentley
Department of Nutrition, Gillings School of Global Public Health, University of North Carolina at Chapel Hill, Chapel Hill, NC, USA

D. A. Schwartz et al. (eds.), *Pregnant in the Time of Ebola*, Global Maternal and Child Health,
https://doi.org/10.1007/978-3-319-97637-2_7

regarding condom use for survivors (World Health Organization 2015b), combined with preexisting stigma and fear, create an atmosphere where rejection of female survivors by their sexual partners is plausible. The World Health Organization's (WHO's) *Interim advice on the sexual transmission of the Ebola virus disease* (World Health Organization 2015b) fails to distinguish between male and female survivors when stating that, "*Ebola survivors and their sexual partners should either (a) abstain from all types of sex or (b) observe safer sex through correct and consistent condom use…*" In addition, the same document describes specific cases of sexual transmission of Ebola virus disease (EVD) from male survivors to their sexual partners and adds, "*Less probable, but theoretically possible, is female to male transmission*," (World Health Organization 2015b). The uncertainty regarding the persistence of the Ebola virus in survivors and potential for sexual transmission can contribute to the stigmatization of survivors (Center for Disease Control and Prevention 2015).

In April 2016, WHO addressed female reproductive health, stating that "*menorrhagia/metrorrhagia, and amenorrhea are all frequently reported, although the causal link of these conditions with EVD remains to be determined*," (World Health Organization 2016a). In addition, the guidelines describe that, "*numerous anecdotal reports exist of stillbirths in women who have conceived after recovering from EVD but it is still uncertain whether the rate of stillbirth in EVD survivors is higher than that of the general population. Until more evidence is available, pregnancy in EVD survivors should be considered at-risk for fetal complications and providers should consider performing intermittent assessments of fetal wellbeing via exam or ultrasound when available*," (World Health Organization 2015c, 2016a). This statement was supported by the study of Fallah et al. that included a small sample of 70 female EVD survivors in Montserrado and Margibi Counties in Liberia (Fallah et al. 2016). The study found that incidence of miscarriage and stillbirth (22.1%) was higher for female survivors who became pregnant after Ebola Treatment Unit (ETU) release, compared to figures for healthy women in West Africa (11–13%) (Fallah et al. 2016).

In our formative study described in this chapter, we utilize qualitative methods to complete some of the gaps in survivor research by describing female survivors' experiences with sexual relationships and behavior, menstruation, and pregnancy following the Ebola virus epidemic and infection (Fig. 7.1).

7.2 Methods

7.2.1 Setting

This study was conducted in three counties within Liberia: Montserrado, Lofa, and Bong. Montserrado County, home to the country's capital of Monrovia, was included in order to acquire perspectives from survivors living in an urban area. Lofa County, which borders Guinea, represents the origin of the Ebola outbreak in Liberia as the first two cases were diagnosed in Foya District. Lastly, Bong County, which borders both Lofa County and shares a small border with Guinea, was also hard hit by Ebola. Of the estimated 10,600 Ebola cases in Liberia, 6,000 cases were in Montserrado County, compared to 700 identified cases in Lofa County and 670 cases in Bong County (World Health Organization 2016b). The inclusion of Lofa and Bong Counties was designed to gain an understanding of survivor experiences in more rural communities (Fig. 7.2).

Upon being released from the ETU, survivors were often given a small care package, including soap, sandals, a bucket, condoms, and a mobile phone (Qureshi et al. 2015). These supplies were intended to help survivors as they transitioned back into their homes. Unfortunately, many survivors returned to their homes to find that all of their belongings had been burnt by community members, in an effort to prevent the spread of Ebola. As has been noted in other studies of Ebola survivors (Arwady et al. 2014; Lee-Kwan et al. 2014; Rabelo et al. 2016), there is a wide range in the way in which survivors were received

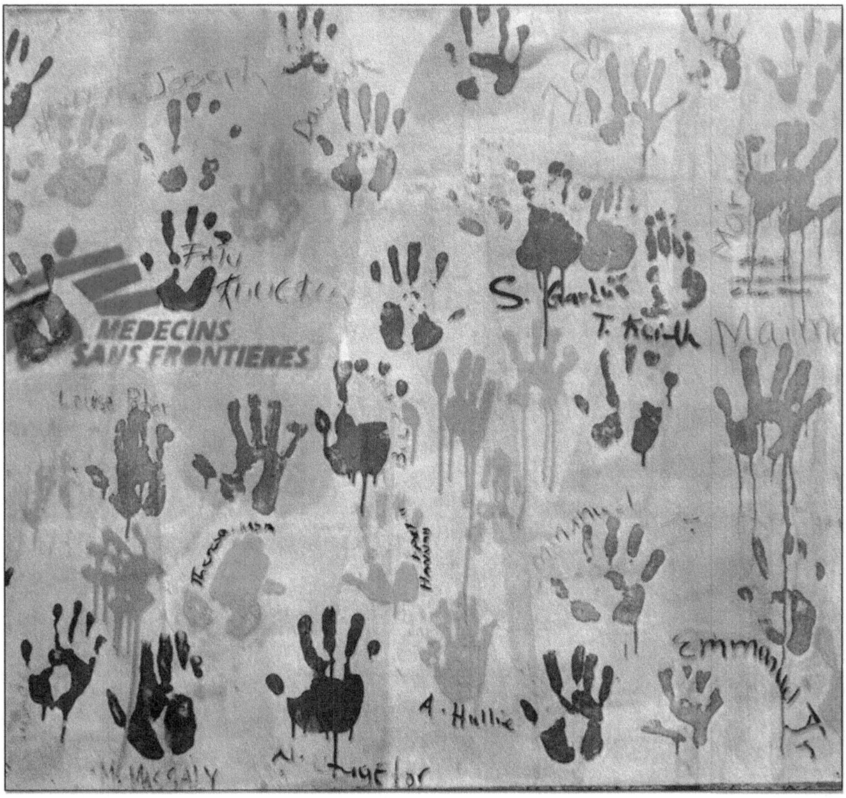

Fig. 7.1 Photo of sign outside of Ebola Treatment Unit administered by Medecins Sans Frontieres (Doctors Without Borders) in Monrovia, Liberia. Upon being released from the ETU, each survivor would place his or her handprint on the sign. Photo taken by Alexandria Buller

by their communities upon their release from the ETU. Some communities celebrated the return of recovered individuals from the ETU. However, many communities begrudgingly accepted the return of survivors to their community, while other communities refused to allow Ebola survivors to return. These actions reflected the tight grip of fear that Ebola had on these communities, as well as the poor understanding of how survivors could not transmit the virus through casual contact. Many survivors also lost their source of income, due to stigma from employers and peers, or from physical Ebola-caused disabilities that prevented them from working. To combat this stigmatization and rejection of survivors, several organizations launched campaigns to educate communities on how they should accept and seek to help Ebola survivors. Various signs and murals were erected throughout the country, encouraging people to accept, and not fear, Ebola survivors (Fig. 7.3).

As an increasing number of Ebola survivors were released from ETUs, organizations began to recognize the need to support survivors and would provide cash transfers to individuals who could offer proof of their survivorship, usually through a facility-issued Ebola survivor certificate. Survivors also began to organize themselves into groups, forming both the Liberia Survivors Association and the Survivors Network. Elected leaders of these groups could then liaise with the Government of Liberia and other organizations to communicate the needs of survivors and advocate for policies and programs that would support them. In addition, regular meetings held by these groups allow survivors a chance to gather together and socialize with people who could understand the unique struggles that came with Ebola survivorship. Within these associations, survivors could find comradery and acceptance from other survivors.

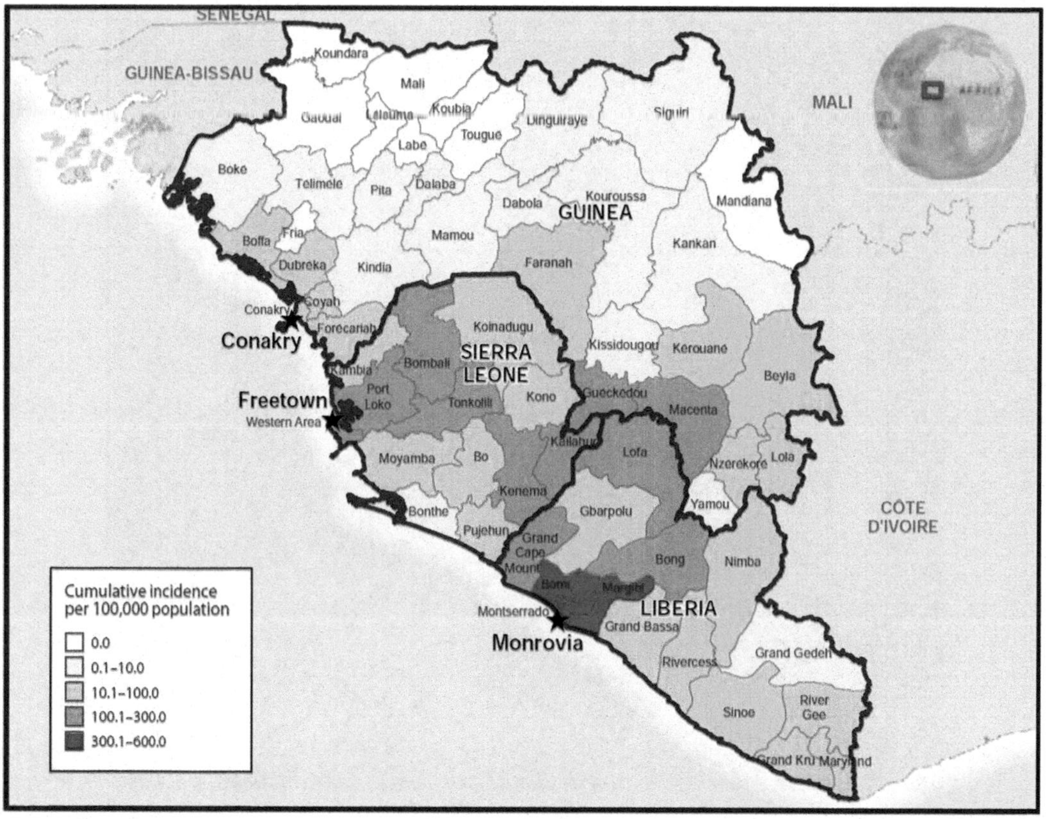

Fig. 7.2 Cumulative incidence of Ebola virus diseases—Guinea, Liberia, and Sierra Leone, November 30, 2014 (Dahl 2016)

After the Ebola outbreak subsided, several health facilities that once hosted ETUs began offering clinical support specific to the needs of Ebola survivors. These "survivor clinics" would offer screening and follow-up on the myriad of health-related complications of Ebola survivorship that became known as Post-Ebola Syndrome (Wilson et al. 2018).

It is in this context, an environment of both support and stigmatization of Ebola survivors, that this study was launched. This study sought to document and describe female survivors' experiences with sexual relationships and behavior, menstruation, and pregnancy following Ebola, in an effort to better inform supportive services and programs for this unique population (Fig. 7.4).

7.2.2 Data Collection

The study consisted of in-depth interviews with 69 female survivors of Ebola virus infection who were 18 years and older at the time of the study. A registry of female survivors in each county was provided by the Liberian Ministry of Health and Social Work. This registry was used as a sampling frame from which women were randomly selected to participate in the study. The targeted sample sizes of 30 in Montserrado County and 20 each in Lofa and Bong Counties were designed to ensure saturation of thematic content given the expected wide range of experiences among survivors.

Study recruitment, screening, and interviews were conducted by research assistants who are themselves female Ebola survivors. These assistants were trained in research ethics and the methods to be

Fig. 7.3 Photo of anti-stigma campaign advertisement in Monrovia, Liberia. Photo by Alexandria Buller

Fig. 7.4 Female Ebola survivor having blood drawn at an Ebola survivor clinic in Monrovia, Liberia. Photo by David Wohl

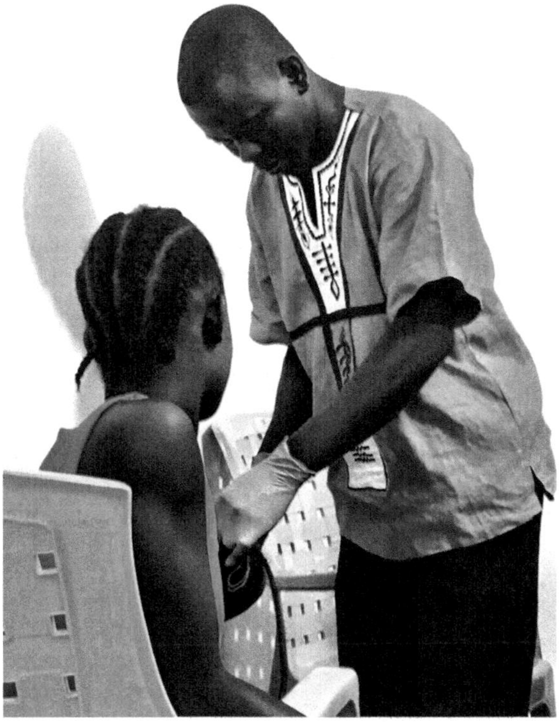

utilized for this particular study. Two of the three assistants had previously worked in healthcare settings as nurses. All three assistants were actively involved in the Liberian Survivors Association and were already known by many of the other survivors in their county.

In recruiting study participants, the research assistants would either call or travel to the home of the potential participant, ask to speak privately with the woman, and then explain the purpose of the study and gauge interest and availability for participation. Women who expressed interest in participating in the study were then screened to ensure that they were 18 years or older and to confirm that they were, in fact, an Ebola survivor. Additional screening was used to ensure safety: women who were currently experiencing emotional distress or who perceived they might be in danger if someone learned of their status as an Ebola survivor were not eligible to participate in the study. Verbal consent was obtained from the women and a time was set for the first of two interviews.

Each research participant was interviewed twice over a period of 3 weeks. Initially, interviews took place in one of three designated health facilities that had survivor clinics (one in each of the three counties). However, due to unanticipated concerns regarding participants being reluctant to come to the health facility, the study protocol was amended to allow interviews to take place in a location of the participant's choosing, as long as the venue was private. The first interview focused on topics related to general experiences as an Ebola survivor, including current living situation, impact of Ebola on family structure and dynamics, and stigma. A major objective of the first interview was to establish rapport between the interviewer and participant, as well as to gain a general understanding of the woman's experiences since being released from the ETU. The second interview centered on topics related to sexual partnerships, sexual behavior, perceptions of sexual transmission, menstruation, and pregnancy. Basic demographic information, including age and parity, were also collected from most participants. Each interview lasted approximately 30 min. Participants were compensated at the conclusion of each of the two interviews with a small cash amount, designed to cover the cost of their time and any travel. The study protocol was reviewed for ethical considerations and approved by institutional review boards at both the University of North Carolina at Chapel Hill and the University of Liberia Pacific Institute for Research and Evaluation.

7.2.3 Data Analysis

All interviews were audio-recorded and transcribed, verbatim, by native Liberian English speakers. Transcripts were uploaded into NVivo 11 software for qualitative analysis. After initial reading of all transcripts for general familiarity, the transcripts were coded based on a series of a priori codes. These codes included the woman's relationship status at the time of falling ill with Ebola, incidence of new relationships since recovering from Ebola, reported menstrual irregularities, and incidence and outcome of pregnancies conceived since Ebola. Relationship status at the time of Ebola included being in a relationship (self-described boyfriend, fiancée, or husband). In Liberia, the terms boyfriend, fiancée, and husband are sometimes used interchangeably and so any of these descriptions were used to fill the criteria of being in a relationship.

The outcome of this original relationship was defined as "still together," abandonment by partner, or death of partner due to Ebola. For the purpose of classification, abandonment was defined as the partner no longer having sex with and/or financially supporting the female survivor. For example, if a woman's husband no longer lived at home but would occasionally visit and send money to the woman, then the two were considered to be in a relationship. Death of partner due to Ebola was as reported by the female survivor. No attempt to verify the cause of death was made by the researchers.

For women whose relationship ended due to death or abandonment, incidence of any new intimate relationship was coded. The outcome of this relationship was also coded as either "still together" or "abandonment due to Ebola," as perceived and reported by the woman.

Increases or decreases in libido, as defined as changes in the desire or "feeling for" sex and as reported by the woman, were coded. Incidence of any menstrual irregularities (i.e., a time of amenorrhea, change in consistency or color of menses, etc.) since release from the ETU was also coded. This was self-reported and based on the woman's perceptions of any changes in her menstruation, compared to before she became sick with Ebola. Any self-reported pregnancies conceived post-Ebola, along with their outcomes (pending, live birth, miscarriage, or stillbirth) were also coded.

After all transcripts were coded, the data were reduced into matrices for further summary and analysis (Tolley et al. 2016). These matrices were developed in Excel and also allowed for analysis of the data by county or age of the woman. Key quotes summarizing or characterizing each theme were also extracted from transcripts.

7.3 Intimate Partner Relationships

Approximately 90% of the women interviewed were in an intimate relationship at the time that they became sick with Ebola. It was expected that Ebola infection would have a disruptive effect on many relationships, due to partners either dying of Ebola or abandoning the relationship due to fear of Ebola. Of the 62 women who were in relationships at the time they fell ill with Ebola, 27 (44%) of the women reported that their relationship ended due to Ebola, through either death or abandonment. This loss of a male partner could have a wide range of destabilizing effects for the surviving woman, including loss of income, economic hardship, psychological stress, and loss of social standing (Sossou 2002) (Fig. 7.5).

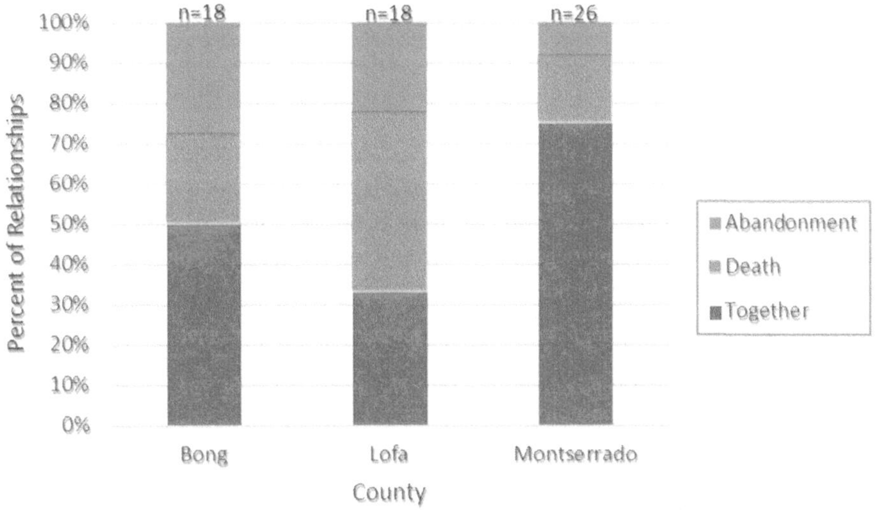

Fig. 7.5 Relationship outcomes of female Ebola survivors who were in a relationship at the time that they became infected with Ebola

7.3.1 Death of Partner

As seen in Fig. 7.5, the proportion of relationships that resulted in the death of a male partner due to Ebola varied greatly across the three counties. In Lofa County, nearly 50% of women in relationships experienced the death of a male partner from Ebola, compared to only 17% in Montserrado County. Many women reported that they and their partner went to the ETU either at the same time or within days of each other. Having lived together with their partners at the time one of them became ill, the women often said that they had gotten the sickness from their husbands or vice versa. Often times, several family members would be transported to the ETU together, after a single-family member fell ill and/or died. The experience of Ebola-stricken individuals witnessing their family members die of Ebola while in the ETU has been documented in other studies as a source of great stress and trauma for survivors (Rabelo et al. 2016). Several participants from this study also reported how seeing their family members die had affected them.

> *"I was thinking, now I coming die because the man died, four children died. So myself I was thinking I coming die."*—Lofa County Ebola survivor, age 35
> *"The way I get the sickness, I not know. That God know everything. Beside my husband died, that the time I was get the sickness... I was thinking, I say, I coming die or all my people them die. My husband die, my children them, three children them die. Now I coming die."*—Bong County Ebola survivor, age 34
> *"My husband na [did not] make it up from the Ebola so that's how kind we all contracted with the virus."*—Lofa County Ebola survivor, age 35

The level of acceptance and support offered by relatives of the deceased partner varied greatly and was, at times, influenced by the religious and cultural norms of the community. Lofa County also has a greater Muslim population than the other two counties, a factor that might have been protective for Ebola widows, as Muslim tradition in some communities dictates that a surviving brother of the deceased marry his late brother's widow. This practice was observed among some of the study participants in Lofa County and is believed to be a source of economic and social stability for both the widow and her children. The woman is protected against the potential of isolation and stigma that might otherwise prevent her from finding a new partner and, in some cases, financial provider (Fig. 7.6).

> *"When my husband leave [died] in the Ebola so after that time, that's how the family say you have to be with this man because that your husband little brother. So that's how I decide to be with the man."*—Lofa County Ebola survivor, age 33

In contrast, many of the study participants who were widowed by Ebola lamented their lack of support. In many instances, survivors also had several family members die of Ebola, effectively leaving them alone to try to find money for housing, food, and school fees for their children. Moreover, the participants experienced varying degrees of post-EVD sequelae, with some women reporting that they could do little to no work due to health complications.

> *"Yes no body to help me, their pa, die from Ebola, I na get money... I get nobody to helping me oh."*—Bong County Ebola survivor, age unknown
> *"I can't get money the children them they not go to school. They not go to school... My...my husband not living. Life hard for me."*—Bong County Ebola survivor, age 34

7.3.2 Abandonment by Partner

Many women reported being abandoned by their male partners, either upon falling ill with Ebola, or upon being released from the ETU. In some cases, the woman simply never heard from or saw her

Fig. 7.6 Photo of Ebola cemetery in Foya, Lofa County, Liberia. Photo by Samaritan's Purse International Relief

partner again after being taken to the ETU. In other cases, the couple would have some communication after the woman was released from the ETU, but the man would have moved to a different community. Participants from Bong County were most likely to report being abandoned by a male partner due to Ebola, with 28% of women in relationships reporting this outcome. Women in Montserrado County were least likely to report being abandoned by their male partners, with only 8% reporting abandonment. The women conveyed a variety of emotion when describing their abandonment—many explained their stories in a matter-of-fact manner, while others communicated hurt, anger, and betrayal.

> *"When we go in the camp [ETU], the boy called me he say 'what place you are?' I say 'I to the camp [ETU]'... Since that time the boy na used to come we move from there the boy na used to speak to me... since the boy do that to me, I start getting vexed. I say oh so you was scare of me because I have, I was having this sickness?"*—Lofa County Ebola survivor, age 18
>
> *"I had a relationship before Ebola and during the Ebola crisis, we breakup... We broke away before even going home before I could leave from the ETU."*—Lofa County Ebola survivor, age 28

According to the study participants, the man's abandonment was often due to fear of the survivor being contagious and passing Ebola on to the partner. Abandonment of the survivor was sometimes encouraged by family and friends. Some women also described peers of the partner as suggesting that the man could "do better" than being with an Ebola survivor—suggesting that survivors are of lower social status than women who were not infected with the virus.

> *"I was in a relationship but due to the Ebola virus my partner left he went, when he went after I got well he came back and family, friends told him that if you keep around that girl you will get the same sickness. So since then he left me I have not getting any other relationship."*—Bong County Ebola survivor, age unknown
>
> *"He move in different community self... His people [family] carry him far off. Since I came back [from ETU] self I na lay eyes on him yea."*—Bong County Ebola survivor, age 21

7.3.3 Maintained Partnership

An estimated 53% of study participants who were in a relationship at the time of their Ebola infection reported that the relationship continued after Ebola. Relationships in Montserrado County were more likely to be intact after Ebola than relationships in the other two study counties, as 75% of women were still with their male partners after Ebola. Many of these cases were instances where both the male and female in a partnership became infected with Ebola and survived. This scenario was quite common as individuals living in the same home were likely to have unintentionally passed the virus, leading to multiple household members becoming infected with Ebola.

In other cases, the man remained uninfected by Ebola even after his wife or girlfriend fell ill. Some women described how their partners took care of them while they were sick, prior to going to the ETU. The women discussed how the men would clean up after them, cook, and feed them—despite the personal risk of becoming infected with Ebola.

> *"My husband taking care of me… when wash me, he bath me, he will wash his self with chloride… but God was with him, he never got down with that sickness."*—Bong County Ebola survivor, age 45

In some instances, the man stayed with his partner after her ETU release, despite facing stigma and social rejection for remaining with an Ebola survivor. Participants described how their partners' coworkers, friends, and other peers would urge the men to leave their survivor girlfriends or wives. Men who stayed with their survivors would endure jokes, insults, and discrimination from other men.

> *"His Pa still vex, his Ma she understand she say me I woman because that the woman my son want I will not say no… the people say Ebola people can't give people sickness when they come from there again but his Pa, he can just call me Ebola woman. That your Ebola woman I na want see her here."*—Bong County Ebola survivor, age 21

While some couples' relationships remained intact through the Ebola epidemic, the nature of the relationship was changed. Some women described that their male partners were no longer willing to engage in sexual intercourse with them, due to fear of contracting the Ebola virus. Of these, some men no longer lived at home and would only return to bring financial support to the women. Other participants described that, although they were still in a relationship with their men, the pair rarely had sex as the woman's desire and interest in sex had decreased dramatically since Ebola. The women discussed how these changes in dynamics and sexual activity had placed stress on the relationship, causing disagreement within the relationship. Many women described how they assumed their men were engaging in sex with other women, to compensate for the lack of sex in the existing relationship.

> *"My husband and I are not in good time because he say I am a survivor… he myself no sexual desire between us. Now, you know whole day I grieving in my heart."*—Montserrado County Ebola survivor, age 41

7.4 Subsequent New Partnerships

Of the 27 women who had a relationship end due to abandonment or death related to Ebola, 17 (63%) had experienced a subsequent relationship with a man by the time of the study, nearly 3 years after the West Africa Ebola Epidemic. This outcome is notable as being in a relationship with a man could offer some degree of financial and social stability to the female survivor (Fig. 7.7).

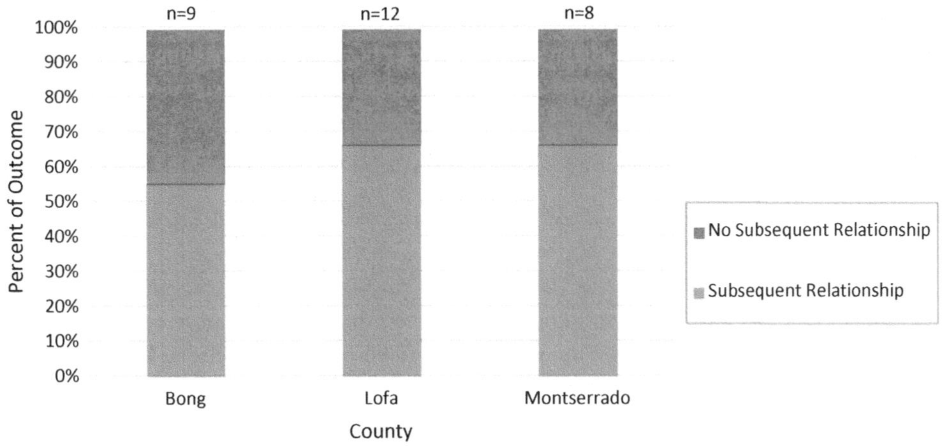

Fig. 7.7 Percent of female Ebola survivors who had a subsequent intimate relationship after Ebola

7.4.1 No New Partnerships

The proportion of women who had a relationship end due to Ebola (from death or abandonment) but have had a subsequent relationship varied greatly across the study sites. Female survivors in Bong County were slightly less likely than survivors from the other two counties to enter into new relationships following Ebola. This could reflect a difference in how female Ebola survivors are perceived and accepted in the two counties, especially given that women in Bong County were also more likely to be abandoned by their partners. Female survivors across all three counties consistently described their perception that they are unable to enter into relationships with men because they are Ebola survivors. As discussed above, some men do not want to be in relationships with female survivors, as they are afraid of contracting Ebola and/or view survivors as lower in social status.

> *"No man, because the people are scared."*- Bong County Ebola survivor, age unknown
>
> *"Because since I came from the ETU. You know I can't just get man like that... Because the stigma behind me. They say I survivor."*—Lofa County Ebola survivor, age 18
>
> *"One other man came around... he said "oh but I like that girl", other people them say "oh she survivor oh." I heard it from them clear... The man since that day he left. When I even calling he can't respond my called."*—Bong County Ebola survivor, age 38

7.4.2 Formation of New Partnerships

Several survivors described how they had started relationships with new men after recovering from Ebola. In many instances, the man came from the same community or neighborhood as the woman, and so knew she was an Ebola survivor before they began the relationship.

> *"You know during the Ebola time my husband died. So after, after the Ebola I came, I fall in relationship and up to now I still in that relationship."*—Lofa County Ebola survivor, age 33
>
> *"After my people died he the one who came and we were together. He was giving me some encouraging words."*—Montserrado County Ebola survivor, age 18

A handful of participants described that they had had relationships with men who were themselves Ebola survivors. At least one woman described that she had met her boyfriend at a meeting for Ebola survivors. The same woman also explained the benefits of being in a relationship with a survivor, as she didn't have to fear stigma when she is with him.

> *"The man came to me, he approach me. He say you mon na [shouldn't] be scared, me too I Ebola survivor."*—Lofa County Ebola survivor, age 26
> *"Because when I love to him no stigma no nothing, and he good, when he and myself together no problem."*—Lofa County Ebola survivor, age 26

A small number of study participants also detailed how their new partners are accepting and supportive of their status as Ebola survivors. One woman described how her boyfriend had helped provide counsel as she recovered from the trauma and stigma. Another participant explained that her boyfriend demonstrated pride in her for being an Ebola survivor:

> *"So that how that day I told him that you know that I Ebola survivor he say wow... the man was so proud he called me I when to his bank and he was so proud of me he even told his workmate that the girl I loving to, that Ebola survivor."*—Montserrado County Ebola survivor, age unknown

When study participants were asked as to whether or not they would tell a new boyfriend that they are an Ebola survivor, opinions were divided. Those women who reported that they would want to disclose their status as an Ebola survivor to their boyfriends gave several reasons for this answer, including wanting to be honest with the man and fearing the man would find out from others in the community and be angry. Participants who supported the idea of telling new boyfriends that they are Ebola survivors also tended to be women who themselves took pride in having survived Ebola. In contrast, the many participants who suggested they might keep their Ebola survivor status a secret from new boyfriends were also more likely to express feelings of shame and experiences with stigma. These women explained that they would be afraid that the boyfriend would quickly abandon them if they found out that they were a survivor early into the relationship.

> *"But still, I always make them to know say I am a survivor because, I will [not] like to hide my status from you then tomorrow while we [are] together... you say, 'so you a survivor you didn't tell me say you a survivor? You hide it from me?'... and then it will bring another problem. So, I can make them to know that I am a survivor."*—Montserrado County Ebola survivor, age unknown
> *"They can be shame; they na [don't] want tell their boyfriend them because they scared. Maybe the boy will leave them—and that how most of the boy them can do. As soon they know that you Ebola person, they na [don't] want come around you again."*—Bong County Ebola survivor, age 21
> *"[When asked if she would tell a boyfriend she is an Ebola survivor] No... the person will cut off from me."*—Montserrado County Ebola survivor, age 30

7.4.3 Abandonment by New Partners

The study found that nearly one third of relationships that have begun since the woman's recovery from Ebola have ended with the woman being abandoned by her partner. In some cases, the man entered into the relationship without knowing that the woman was an Ebola survivor. Then, upon finding out—often times through friends or community members telling the man—he would terminate the relationship. The exact reasons why men sometimes choose to end these relationships is unknown; however, it is likely a combination of fearing transmission of Ebola, fearing being stigmatized themselves due to their partner being an Ebola survivor, and/or being upset that their partner did not disclose her status as an Ebola survivor.

"He and myself met we decided to have fine (date) time, but unfortunately he decided to check in my bag he wanted something from there. The Ebola survivor ID card dropped so he saw it he just put on his trousers and left.... since that day I called his line, now, now I call his line you see he will not answer. It will ring, ring.. no answer. It so sorrowful on my part."—Montserrado County Ebola survivor, age 41

"One man a come and say he want me. We be together four month. The people talk it to him a say I...I Ebola survivor, then [he] run away."—Lofa County Ebola survivor, age 35

"Yeah when I come, the person abandon me. He say he not want me because I survivor... I told the people I told friends but the people too say they not want to bother survivor."—Lofa County Ebola survivor, age 18

7.5 Sexual Behavior

7.5.1 Short-Term Abstinence and/or Condom Use

Study participants consistently described receiving counseling prior to their leaving the ETU. They were instructed that they should remain abstinent or use condoms for 90 days after their release from the ETU. Many participants described that they were advised to abstain from having sex for 3 months, but then decided to add an additional period of some months (often 2–6 months). Women described their rationale for extending the abstinence period as their way of being extra cautious in avoiding Ebola transmission. In addition, some women described that they wanted to give their bodies some time to recover from Ebola, as many were still physically weak after they returned home from the ETU. This idea of increasing the abstinence period was not consistently described as always being the man's idea or the woman's idea, it seemed to be something both parties were amenable to. This finding is consistent with data reported from a small, mixed methods study by Karafillakis et al. that reported both male and female survivors from Sierra Leone deciding to voluntarily extend their abstinence period beyond the recommended 3 months (Karafillakis et al. 2016).

"Because I was sick I was not doing man business... it can make one year I can't do man business [sex]."—Lofa County Ebola survivor, age 35

"When I came from the center {ETU] I can say it stay long before I start doing man business [sex] because the man too was afraid but it stay long...before we started something [having sex] and he used to used condom on me. Yes. Because he himself, he was still afraid. He used condom for over two months. After the ninety days now oo. Mm hmm. He used condom for almost two months. Yeah. And I used to be talking to him I say 'oh, but this thing here it not in me again but don't be afraid you not use condom self you will not get Ebola' but still his heart was not satisfied. His heart was not satisfied."—Montserrado County Ebola survivor, age 30

7.5.2 Decreased Libido

Approximately one half of interviewed female survivors reported a decrease in their libido after recovering from Ebola. Most research on the sexual outcomes of survivors after Ebola have focused on male survivors, as erectile dysfunction is a well-documented symptom of Post-Ebola Syndrome (Bausch 2015; World Health Organization 2016c). In addition, other studies have found that male survivors experience decreased libido (Karafillakis et al. 2016), yet no studies have examined the impact of Ebola on the libido of female survivors. When female participants were asked about "man-woman business," many women simply stated that the "feeling is not there." These comments seemed independent of the woman's current relationship status; however, it was often not clear if the women were referring to a physical desire to have sex, versus an emotional desire for sexual intimacy. Given that the mechanisms through which Ebola can affect sexual desire are unknown, it is uncertain how much of these changes in libido can be attributed to physical versus psychological causes.

> *"I got no feeling for man business [sex]."*—Lofa County Ebola survivor, age 35
> *"The reason ain't have time for man business again, my husband died from me, go now, my daughter left [died] in the Ebola. So I can be thinking too much. So ain't...my heart not for man business again."*—Montserrado County Ebola survivor, age 40

Additionally, some women described experiencing back or stomach pain during sexual intercourse. Women reported that this pain was new since recovering from Ebola and they attributed it to Ebola.

> *"When I come from the ETU my back was hurting. It can be hurting me small, small. But when we having sex now even that it can be hurting me to come down to my foot."*—Lofa County Ebola survivor, age 35
> *"What happened some time when me and my husband ready to meet some time my stomach can hurting me, sometime infection [laugh] some time my stomach can hurting me."*—Lofa County Ebola survivor, age 24

Several study participants reported that their lack of interest in sex—either due to low libido or pain during intercourse—had caused problems in their relationship. The women described getting into arguments with their partners over sex, and the decrease in sexual activity causing an ongoing tension in the relationship. Many women commented that their partners accused them of having affairs on the side, as this seemed a plausible reason for the women's lack of interest in sex. A few women explained that their partners were now engaging in affairs with women outside of the relationship, since they felt the survivors were unwilling or unable to meet their sexual needs.

> *"...but now he can complain on me. My husband get girlfriend now... because the sex business is not really in me."*—Montserrado County Ebola survivor, age unknown

7.6 Menstruation

As mentioned in the chapter introduction, the WHO's *Interim Guidance for Clinical Care for Survivors of Ebola Virus Disease*, released in April 2016, states that, "*menorrhagia/metrorrhagia, and amenorrhea are all frequently reported, although the causal link of these conditions with EVD remains to be determined,*" (World Health Organization 2016a). Apart from this mention, there is no literature on the topic of menstruation issues in female Ebola survivors. Despite this, approximately 46% of study participants reported some form of menstrual irregularity following their recovery from Ebola. A few types of menstrual changes were reported by female survivors. Many women stated that they experienced a period of amenorrhea following their release from the ETU. Of these women, most reported their menstruation resuming after anywhere from 3 to 9 months after Ebola. A handful of women had, at the time of the interview 3 years after Ebola, still not resumed menstruation.

> *"The time I came back first from Ebola... it was almost nine month I was not use to see my time, sometime when it come like that it will just be small!"*—Bong County Ebola survivor, age 21

In addition, a large number of women reported that they experience their menses only once every few months. They describe the timing and frequency of their menstruation as unpredictable.

> *"Sometime two months will pass, I can't bleed. Sometime I will be sitting down—not expecting—and I just start bleeding."*—Bong County Ebola survivor, age 38

Other women reported a change in the menses itself, following Ebola. Some women reported having heavier or longer periods. Interestingly, many women described their menses as being black in color for at least the first day of their menstruation, as well as having stomach pain. A few of these women described being given antibiotics or unspecified treatment to resolve this issue. Given the women's descriptions of their symptoms, it seems possible that the women were suffering from endometritis—an infection of the endometrial lining of the uterus. However, it is not clear why female Ebola survivors would be more susceptible to an endometrial infection than other women.

> *"The first day, it can be black, the second day na [then] it can be it can change. Then the third day, it can just be coming with water na. Like the way it used to come before, [it] is not coming like that. I na [have] take[n] medicine, they say infection, infection. I don't know what kind of infection that [is]."*—Montserrado County Ebola survivor, age 24
>
> *"My stomach was hurting, my body was hurting, my menstruation was, it was getting, it was black. I go to the hospital they gave me medicine today, today, I thank God for that one."*—Lofa County Ebola survivor, age 40

7.7 Pregnancy

Since the end of the West African Ebola epidemic, many reports have been circulated in the media regarding female survivors' experiences with miscarriages and stillbirths for those pregnancies conceived post-Ebola. A 2016 study by Fallah et al. examined the pregnancy outcomes of female survivors in Margibi and Montserrado Counties in Liberia. The study reported that, of 70 EVD survivors who had become pregnant after Ebola infection, 15 of the women miscarried and 4 had stillbirths (Fallah et al. 2016). The fact that 27.9% of pregnancies resulted in an adverse outcome is significant, relative to general figures regarding miscarriages and stillbirths for women in West Africa (Grout et al. 2015) (Fig. 7.8).

When asked about any pregnancies conceived after recovering from Ebola, a total of 17 survivors reported having been pregnant since Ebola. Of these pregnancies, 10 resulted in live births, 3 resulted in stillbirth, 0 resulted in miscarriage, 3 were ongoing, and 1 ended with an elective abortion.

Fig. 7.8 Photo of female Ebola survivor and her healthy baby boy conceived after Ebola. Photo by Samaritan's Purse International Relief

Women seemed to hold a variety of perceptions regarding pregnancy outcomes for Ebola survivors. Several women described knowing female survivors who had become pregnant since Ebola, and their opinions of Ebola's impact on pregnancies seemed to be influenced heavily by these personal accounts. Women who were acquainted with survivors who had successful live births felt that Ebola was not a problem for pregnancies. Other women who knew of survivors who had experienced miscarriages or stillbirths perceived Ebola to have a large impact on pregnancies. Some of the women who had experienced pregnancies post-Ebola reported experiencing intense swelling in their feet.

> *"Our friend get pregnant... the baby not live... their foot swell up, all their body swell up, they not able to do small lil' work."*—Montserrado County Ebola survivor, age 40
> *"The child is one year, one month [old]... I used [to] be worrying because I say maybe when I deliver the child will be having Ebola oh, because the thing was in my blood."*—Bong County Ebola survivor, age 23
> *"The problem that I was facing because during that time I was pregnant all my foot was puffed up, I was just swelling up. Time for me to deliver, I faced whole lots of problems because that time when the baby came I was bleeding... I delivered but the child not live"*—Bong County Ebola survivor, age 38

7.8 Conclusions

In terms of relationships and reproductive health issues, the experiences of female Ebola survivors in the years since the West African Ebola outbreak can best be described as highly varied. According to the results of this study, nearly half of women who were in a relationship at the time of Ebola had that relationship end due to either death or abandonment of the male partner. Yet, these figures vary greatly by county with only 33% of relationships in Lofa County being maintained after Ebola, compared to 75% in Montserrado County. It is hypothesized that the increased access to both treatment and education for Ebola in Montserrado contributed to its lower numbers of Ebola deaths and male partner abandonment. In contrast, half of relationships in Lofa County ended due to death, a reflection of the higher mortality rates experienced by that county. Abandonment of women by male partners was highest in Bong County, as was the percent of women who were unable to form a new, subsequent partnership after the loss of their first relationship. These findings might reflect a greater persistence of stigma in Bong County.

An estimated 46% of study participants reported some menstrual changes, following their recovery from Ebola. These changes ranged from months of amenorrhea to reported black menses and stomach pain, indicating infection. To date, no other studies have examined the potential impact of Ebola on the reproductive system of female survivors. More research is needed to understand the underlying cause(s) behind these menstrual changes and any impact Ebola might have on the functioning of the immune system in the female reproductive tract.

The findings of this study can be used to inform policymakers and program implementers of the unique challenges faced by female survivors of Ebola in Liberia. In this way, interventions for this population can be based on scientific findings.

References

Arwady, M. A., Garcia, E. L., Wollor, B., Mabande, L. G., Reaves, E. J., Montgomery, J. M., et al. (2014). Reintegration of Ebola survivors into their communities—Firestone District, Liberia, 2014. *MMWR Morbidity Mortality Weekly Report, 63*(50), 1207–1209. Retrieved November 4, 2017, from https://www.cdc.gov/mmwr/preview/mmwrhtml/mm6350a7.htm.

Bausch, D. G. (2015). Sequelae after Ebola virus disease: Even when it's over it's not over. *Lancet Infectious Diseases, 15*(8), 865–866. Retrieved March 31, from, http://www.thelancet.com/journals/laninf/article/PIIS1473-3099(15)70165-9/fulltext.

Center for Disease Control and Prevention. (2015). *Ebola survivors questions and answers*. Retrieved November 5, 2017, from https://www.cdc.gov/vhf/ebola/outbreaks/2014-west-africa/survivors.html.

Dahl, B. A. (2016). CDC's response to the 2014–2016 Ebola epidemic—Guinea, Liberia, and Sierra Leone. *MMWR Supplements, 65*(3), 12–20. Retrieved October 30, from, https://www.cdc.gov/mmwr/volumes/65/su/su6503a3.htm.

De Roo, A., Ado, B., Rose, B., Guimard, Y., Fonck, K., & Colebunders, R. (1998). Survey among survivors of the 1995 Ebola epidemic in Kikwit, Democratic Republic of Congo: Their feelings and experiences. *Tropical Medicine & International Health, 3*(11), 883–885.

Deen, G. F., Broutet, N., Xu, W., Knust, B., Sesay, F. R., McDonald, S. L. R., et al. (2017). Ebola RNA persistence in semen of Ebola virus disease survivors - Final report. *New England Journal of Medicine, 377*(15), 1428–1437. Retrieved March 31, 2018, from http://www.nejm.org/doi/full/10.1056/NEJMoa1511410.

Fallah, M. P., Skrip, L. A., Dahn, B. T., Nyenswah, T. G., Flumo, H., Glayweon, M., et al. (2016). Pregnancy outcomes in Liberian women who conceived after recovery from Ebola virus disease. *Lancet Global Health, 4*, e678–e679. Retrieved March 31, 2018, from, http://www.thelancet.com/journals/langlo/article/PIIS2214-109X(16)30147-4/fulltext.

Fischer, W.A., Brown, J., Wohl, D.A., Loftis, A.J., Tozay, S., Reeves, E., et al., (2017). Ebola virus ribonucleic acid detection in semen more than two years after resolution of acute Ebola virus infection. *Open Forum Infectious Diseases, 4*(3), ofx155. Retrieved November 19, 2018, from https://academic.oup.com/ofid/article/4/3/ofx155/4004818.

Gbarmo-Ndorbor, A. (2017). *Number of Liberian Ebola survivors* [online]. E-email to Christine Godwin (chrissylgodwin@gmail.com) 2017 March 13 [cited 2017 November 5].

Ghose, T. (2017). Ebola may linger in men's semen for more than 2 years. *Live Science*. Retrieved November 18, 2018, from https://www.livescience.com/59996-ebola-virus-stays-in-semen-for-years.html.

Grout, L., Martinez-Pino, I., Ciglenecki, I., Keita, S., Diallo, A. A., Traore, B., et al. (2015). Pregnancy outcomes after a mass vaccination campaign with an oral cholera vaccine in Guinea: A retrospective cohort study. *PLoS Neglected Tropical Diseases, 9*, e0004274. Retrieved November 5, 2017, from http://journals.plos.org/plosntds/article?id=10.1371/journal.pntd.0004274.

Hewlett, B. S., & Amola, R. P. (2003). Cultural contexts of Ebola in northern Uganda. *Emerging Infectious Diseases, 9*(10), 1242–1248. Retrieved March 31, 2018, from https://wwwnc.cdc.gov/eid/article/9/10/02-0493_article.

Karafillakis, E., Jalloh, M. F., Nuriddin, A., Larson, H. J., Whitworth, J., Lees, S., et al. (2016). Once there is life, there is hope Ebola survivors' experiences, behaviours and attitudes in Sierra Leone, 2015. *British Medical Journal Global Health, 1*(3), e000108. Retrieved March 31, 2018, from https://www.ncbi.nlm.nih.gov/pmc/articles/PMC5321361/.

Lee-Kwan, S. H., DeLuca, N., Adams, M., Dalling, M., Drevlow, E., Gassama, G., et al. (2014). Support services for survivors of Ebola virus disease-Sierra Leone, 2014. *MMWR. Morbidity and Mortality Weekly Report, 63*, 1205–1206. Retrieved March 31, 2018, from https://www.cdc.gov/mmwr/preview/mmwrhtml/mm6350a6.htm.

Mate, S. E., Kugelman, J. R., Nyenswah, T. G., Ladner, J. T., Wiley, M. R., Cordier-Lassalle, T., et al. (2015). Molecular evidence of sexual transmission of Ebola virus. *New England Journal of Medicine, 373*(25), 2448–2454. Retrieved March 31, 2018, from http://www.nejm.org/doi/full/10.1056/NEJMoa1509773.

Qureshi, A. I., Chughtai, M., Loua, T. O., Pe Kolie, J., Camara, H. F., Ishfaq, M. F., et al. (2015). Study of Ebola virus disease survivors in Guinea. *Clinical Infectious Diseases, 61*(7), 1035–1042.

Rabelo, I., Lee, V., Fallah, M. P., Massaquoi, M., Evlampidou, I., Crestani, R., et al. (2016). Psychological distress among Ebola survivors discharged from an Ebola treatment unit in Monrovia, Liberia–A qualitative study. *Frontiers in Public Health, 4*, 142. Retrieved November 1, 2017, from, https://www.ncbi.nlm.nih.gov/pmc/articles/PMC4931229/.

Sissoko, D., Duraffour, S., Kerber, R., Kolie, J.S., Beavogui, A.H., Camara, A-M., et al. (2017) Persistence and clearance of Ebola virus RNA from seminal fluid of Ebola virus disease survivors: a longitudinal analysis and modelling study. *The Lancet Global Health, 5*(1), e80-e88. Retrieved November 19, 2018, from https://www.thelancet.com/journals/langlo/article/PIIS2214-109X(16)30243-1/fulltext?code=lancet-site.

Sossou, M. A. (2002). Widowhood practices in West Africa: The silent victims. *International Journal of Social Welfare, 11*, 201–209.

Tolley, E. E., Ulin, P. R., Mack, N., Robinson, E. T., & Succop, S. M. (2016). *Qualitative methods in public health: A field guide for applied research* (2nd ed.). Hoboken: Wiley.

Wilson, H.W., Amo-Addae, M., Kenu, E., Ilesanmi, O.S., Ameme, D.K., Sackey, S.O. (2018). Post-Ebola syndrome among Ebola virus disease survivors in Montserrado County, Liberia 2016. *BioMed Research International*, Article ID 1909410, https://doi.org/10.1155/2018/1909410. Retrieved November 19, 2018, from https://www.hindawi.com/journals/bmri/2018/1909410/.

World Health Organization. (2015a). *WHO meeting on survivors of Ebola virus disease: Clinical care of EVD survivors. Meeting Report, Freetown, Sierra Leone*. Geneva: World Health Organization. Accessed 8 December 2017.

World Health Organization. (2015b). *Interim advice on the sexual transmission of the Ebola virus disease*. Retrieved October 30, 2017, from http://www.who.int/reproductivehealth/topics/rtis/ebola-virus-semen/en/.

World Health Organization. (2015c). *Ebola virus disease in pregnancy: Screening and management of Ebola cases, contacts and survivors: Interim guidance*. Retrieved October 20, 2017, from http://www.who.int/csr/resources/publications/ebola/pregnancy-guidance/en/.

World Health Organization. (2016a). *Interim guidance: Clinical care for survivors of Ebola virus disease*. Geneva: WHO. Retrieved October 20, 2017, from apps.who.int/iris/bitstream/10665/204235/1/WHO_EVD_OHE_PED_16.1_eng.pdf.

World Health Organization. (2016b). *Ebola situation reports—Archive*. Retrieved October 25, 2017, from http://www.who.int/csr/disease/ebola/situation-reports/archive/en/.

World Health Organization. (2016c). *Clinical care for survivors of Ebola virus disease. Interim guidance*. Retrieved October 30, 2017, from http://www.who.int/csr/resources/publications/ebola/guidance-survivors/en/.

8 Gender-Based Violence Among Adolescent Girls and Young Women: A Neglected Consequence of the West African Ebola Outbreak

Monica Adhiambo Onyango, Kirsten Resnick, Alexandra Davis, and Rupal Ramesh Shah

8.1 Introduction

During the 2013-2015 Ebola virus disease (EVD) outbreak in West Africa, response efforts from the government, healthcare groups, and nongovernmental organizations (NGOs) focused on containing the disease and bringing the number of cases to zero. Despite each country eventually being declared Ebola-free, the outbreak severely impacted security, economic and social activity, previous gains in development, provision of healthcare and education, and the essential services of government. With organizations intently focused on outbreak containment, protocols were never established to protect adolescent girls and young women during the outbreak, even though studies have indicated that women and girls are more likely to become victims of gender-based violence (GBV) in disaster and humanitarian crises such as the one created by the Ebola epidemic (McKay 2015). GBV is a term

Gender-based violence is an umbrella term for any harmful act that is perpetrated against a person's will and that is based on socially ascribed (i.e., gender) differences between males and females. It includes acts that inflict physical, sexual, or mental harm. These acts can occur in public or in private (Inter-Agency Standing Committee [IASC] 2015:5).

M. A. Onyango (✉)
Department of Global Health, Boston University School of Public Health, Boston, MA, USA
e-mail: monyango@bu.edu

K. Resnick
Boston University, Boston, MA, USA

A. Davis
Department of Global Health, Boston University School of Public Health, Boston, MA, USA
e-mail: ardavis1@bu.edu

R. R. Shah
Partners in Health, Boston, MA, USA

Zanmi Lasante, Mirebalais, Haiti

D. A. Schwartz et al. (eds.), *Pregnant in the Time of Ebola*, Global Maternal and Child Health,
https://doi.org/10.1007/978-3-319-97637-2_8

commonly used to highlight the systemic inequality between males and females—which exists in every society in the world (Interagency Standing Committee [IASC] 2015). Importantly, there is a growing recognition that populations affected by humanitarian crises can and do experience various forms of GBV. The West Africa Ebola outbreak was no exception.

It has been widely acknowledged that during the West African Ebola outbreak, GBV against adolescent girls and women was common and had severe consequences including an increase in sexual violence and rape (McKay 2015). NGOs and other humanitarian programs waited until the spread of the virus had slowed to focus on adolescent girls and young women. By then, the impact of violence had already resulted in negative outcomes and consequences for many (McKay 2015; Yasmin 2016).

This chapter examines how GBV increases during times of crisis, and how the needs of adolescent girls and young women were overlooked and neglected during the Ebola outbreak in Sierra Leone and Liberia. An analysis of the rise in GBV during the Ebola outbreak is an opportunity to review the areas in which the humanitarian community can improve and apply lessons learned to future epidemics to protect adolescent girls and women. Since this vulnerable population already faced disparities in opportunities and health outcomes prior to the outbreak, this analysis will review GBV through each country's history including the civil wars experienced in both countries, postwar periods, and the Ebola outbreak. Although GBV that occurred during the wars and post-conflict periods in both Sierra Leone and Liberia have been extensively documented, it must be noted that few changes to humanitarian response have been made for young women and girls (Liebling-Kalifani et al. 2011).

8.2 The Vulnerability of Adolescents and Young Women

The Ebola outbreak of 2013–2015 spread quickly across three West African countries: Guinea, Liberia, and Sierra Leone. While this was not the first outbreak of Ebola virus in history, it was by far the largest and longest, the first to occur in densely populated areas, the first to extend across national boundaries and the first to be considered an epidemic. By the end of 2016, there were 28,616 total cases (suspected, probable, and confirmed) and 11,310 deaths over the course of the 3 years (Centers for Disease Control and Protection [CDC] 2016; Coltart et al. 2017).

The EVD outbreak in West Africa occurred within a context of social and political inequities, structural violence, and an already weak health system infrastructure. The intersections of long-term economic, sociocultural, political exclusions, and injustices that plagued communities in these countries for a long time were obvious through the duration of the epidemic (Coltart et al. 2017; McKay 2015; Wilkinson and Leach 2015). As quarantines were put in place to contain the spread of disease, neighborhood by neighborhood, women and adolescent girls were vulnerable to violence, especially sexual violence (McKay 2015). In addition, school closings provided teenage girls free time that made them more vulnerable to coercion, exploitation, and sexual abuse (Labous 2014). Economic vulnerabilities can drive young girls into sexual relationships which are sometimes exploitative. For example, young girls who were not able to financially support themselves formed relationships with older men to obtain money for food and clothing. In this case, girls voluntarily seek out older men for economic reasons (Yasmin 2016).

GBV impacts women's economic productivity, disrupts educational attainment, and further reinforces gender inequalities, which eventually reinforces the cycle of female poverty. In humanitarian crises, for example, GBV rates are often underreported, largely due to economic dependence, fear, and stigma. For example, women risk divorce or loss of future marriage prospects if identified as victims of GBV, and as a result, they don't report (Gurman et al. 2014).

In Liberia and Sierra Leone, women are traditionally the caregivers, and as such, during the Ebola outbreak they carried an additional burden of caring for the sick and dying. Young women and adolescent girls were often at home and unable to attend community meetings where education and instructions were given about how to protect themselves from contracting the disease. Women and girls who survived having EVD, as well as those who cared for the ill and had exposure to the virus but remained uninfected, faced stigma and were turned away from their families and community. Many were left without homes or means to provide support for themselves and their children (Korkoyah and Wreh 2015). Women and adolescent girls were also more at risk because they were charged with preparing the bodies for burial. Although warnings were made during the crises to not touch dead bodies for fear of infection (Abramowitz et al. 2017), women continued to do so, while also facing greater inequities in access to healthcare (Korkoyah and Wreh 2015). Adolescent girls and women occupied a liminal space during the outbreak, and as such, they were often forgotten during relief and recovery efforts (Korkoyah and Wreh 2015; Harman 2015).

8.3 Gender-Based Violence in Liberia and Sierra Leone: What Do We Know?

Long before the Ebola epidemic spread in West Africa, the women of Liberia and Sierra Leone were well-versed in the harsh realities that accompany war and disease. Both countries experienced long civil wars—Liberia from 1989 to 1996 and again from 1999 to 2003 and Sierra Leone from 1991 to 2002. These wars gravely impacted development and left little social and medical infrastructure in their wake (Friedman-Rudovsky 2013; Strong and Schwartz 2016). Rape, forced marriage, and kidnapping were used as weapons of fear and intimidation in both countries and affected both boys and girls (Friedman-Rudovsky 2013). For the generations of children that grew up during civil war, GBV was reinforced and normalized as a part of everyday life and became the foundation for violence that continued post-civil war and became magnified during the Ebola outbreak (Abramowitz and Moran 2012).

Adolescent girls and women were already vulnerable within the culture and exploited during the civil war. It was estimated that about 275,000 women and girls experienced some form of sexual violence during the war. In the years following the war, The Sierra Leonean Truth and Reconciliation Commission documented a list of 1012 victims of sexual violence and forced conscription, with particular focus on adolescent girls and women (Truth & Reconciliation Commission, Sierra Leone 2004a, b).

From a 2006 study, years after the conflict ended, researchers found that in two Liberian counties, including the capital Monrovia, more than one-half of the women interviewed had survived at least one violent sexual attack during an 18-month period (Jones 2008). In addition to GBV, teenage pregnancy has long been a problem in the region. In 2013, Sierra Leone ranked among the ten highest countries in the world for teenage pregnancy, with 28% of girls aged 15–19 years pregnant or already having given birth at least once (Denney et al. 2015). These data reveal that postwar violence against women in each country was still frequent and continuing; not nearly enough programs or policies had been implemented to help women and adolescent girls.

It is not the first time a nation has witnessed this type of cyclical violence against women. Olujic studied the genocide in Croatia and Bosnia-Herzegovina, believing that gendered violence is not only used during war—that its roots are well-established during peacetime (Olujic 1998). Olujic argues that sexual violence in peacetime is often left out of the conversation regarding sexual coercion. The author asserts that the literature treats rape during peace as a crime against an individual woman and only rape in wartime as a tactic of terror against women in general (Olujic 1998). The war in the former Yugoslavia shed light on using rape as a war tactic and revealed how the underlying sociocultural dynamics, such as honor, shame, and sexuality attached to women's bodies, make war rape such an effective weapon (Olujic 1998). Scott et al. (2013) found a similar trend in South Sudan. Their assessment of GBV revealed that violence during war time is a reflection of violence during peace time, and that GBV is more prevalent in those communities where some forms of violence are readily acceptable within households (Scott et al. 2013). In post-conflict Uganda, one-third of the women experiencing GBV reported being raped in their past, and several had children as a result (IASC 2015). When Ebola spread through West Africa, conditions that left women and girls vulnerable to sexual coercion and sexual violence were exacerbated. Consequently, as the impacts of Ebola intensified, the cycle of violence as a tactic in disaster and conflict increased, resulting in high rates of pregnancy and poor health outcomes (Friedman-Rudovsky 2013; Yasmin 2016).

8.4 The "Other" Ebola Victims

8.4.1 Why Focus on Adolescents and Young Girls?

Public health infrastructure, which was still in development in Liberia and Sierra Leone prior to the epidemic, completely halted during the Ebola crisis as it had during each country's civil wars (O'Brien and Tolosa 2016). Epidemic relief efforts were heavily overburdened and they did not account for particularly vulnerable groups—mainly adolescent girls and young women. In Sierra Leone and Liberia, adolescents and young people 15–24 years of age make approximately 20% of the women in both countries (UNFPA 2017).

Although the World Health Organization (WHO) did not provide sex-disaggregated data for the EVD mortality, sufficient evidence has been found to support that more women and girls were affected than men and boys in regard to EVD- and non-EVD-related mortality and morbidity (O'Brien and Tolosa 2016). The Ebola epidemic revealed the deeply rooted gender inequities in each country. Across Guinea, Liberia, and Sierra Leone, it is estimated that collectively 55–60% of those who died were women (Lai 2015). Furthermore, in Liberia, 75% of deaths were estimated to be women, further revealing the disproportionate impact Ebola has on female populations (O'Brien and Tolosa 2016).

While women were more likely to contract Ebola, our review focuses on a particularly vulnerable subgroup, teenage girls between 15 and 19 years of age. The impact of emergencies on teenage girls is well-known. Girls are always the most vulnerable population to sexual exploitation and abuse (Harman 2015). However, despite this knowledge, the international community failed to recognize the specific needs of girls during the outbreak and in recovery planning. For example, in response to the Ebola outbreak, quarantines, curfews, and school closures were enacted—all public health measures to prevent the spread of the disease (Coltart et al. 2017; DuBois et al. 2015). However, these same measures actually escalated the risk of violence and rape against girls (Yasmin 2016).

8.4.2 Uncounted, Unrecognized, and Unattended

While the number of Ebola-infected patients and deaths were recorded as accurately as possible, victims of violence during the outbreak went uncounted, unrecognized, and unattended. Various factors, such as separation (temporary or permanent) from immediate family members and/or caregivers, rising poverty rates, lack of communication, and limited or no access to education all compounded the situation of violence against women (Harman 2015).

Although for obvious reasons precise data were difficult to obtain, the United Nations Children's Fund (UNICEF) reported that 401 of the 450 rape cases recorded in Liberia since the beginning of the EVD outbreak were perpetrated against children between the ages of 0 and 17 years (Korkoyah and Wreh 2015). In an assessment of the differing impact of Ebola virus disease on men and women in Liberia, of the 1562 respondents surveyed, 22.9% reported that cases of GBV were still occurring during the Ebola crisis in numerous forms including domestic violence, sexual abuse, and rape (Korkoyah and Wreh 2015). The assessment also found that 52.6% of respondents recognized that women and girls had been bearing a greater burden in the household when the Ebola outbreak began, particularly due to increased work at home and men not contributing to the family income. This may have also contributed to exploitation and/or sexual violence by men taking advantage of the situation (Korkoyah and Wreh 2015).

Due to the outbreak, many girls became the heads of households. The need for food, water, and basic necessities led some girls into transactional sex[1]—this was especially true for girls who lost parents to Ebola disease. Furthermore, girls whose parents and relatives died from Ebola were forced into transactional sex work to buy food or pay for housing (Werber 2015; Yasmin 2016).

8.4.3 Increased Pregnancy During the Ebola Outbreak

As the Ebola epidemic spread throughout Liberia and Sierra Leone, public places such as schools voluntarily shut down or were ordered closed to help curb transmission. Sexual exploitation, including rape, was more pronounced during the epidemic. Many girls stated they were fearful of rape, and many told stories of peers who were raped, even in quarantined houses (Werber 2015).

Save the Children's survey of 1100 boys and girls in Sierra Leone indicated that most girls believed that teenage pregnancy was rising, and 10% stated that more girls were being forced into prostitution/transactional sex due to the loss of family members and financial insecurity (Minor 2017; Werber 2015). Although empirical data is limited, in Sierra Leone, the United Nations Development Program (UNDP) estimates that teenage pregnancy increased by 65% due to the socioeconomic conditions imposed by Ebola (Werber 2015). The increase in transactional sex due to economic concerns is supported by both children and adults, as illustrated by the following quote:

> *We are encountering lots of teenage pregnancy. Girls get pregnant because they are not going to school and some because they want money ... Prostitution is rampant, girls don't eat unless they go and sleep with older men for money ... Now, we girls do have sex with our father's age group, because we need money and men don't give money for nothing. (Selection of quotes from a girls group, Mile 47, Sierra Leone, 16 December)* (Plan International 2015).

First-hand accounts and surveys from Sierra Leone clearly support the impact the EVD outbreak had on teenage girls and pregnancy. While less is known about the situation in Liberia, UN Women Liberia, Oxfam Liberia, and the Minister of Gender, Children and Social Protection all recognize the sharp increase in teenage pregnancy rates and have also identified that the numbers of girls expected to return to school will be low (Korkoyah and Wreh 2015). West Africa will deal with the impact of increased teenage pregnancy for years to come and could see the impact in education levels and livelihoods in general.

8.4.4 Girls Twice Victims

Additional laws put teenagers who became pregnant as a result of the epidemic at even greater disadvantage from pursuing an education. In Liberia, pregnant teenagers are prohibited from attending daytime classes, and in Sierra Leone, pregnant girls are forbidden from attending school altogether

[1]Transactional sex is a form of sex-for-exchange. Among girls and women in the Sub-Saharan region that are typically young and poor, it involves acceptance of sexual proposals from men as a form of exchange. In some cases it may also be referred to as *survival sex*, and is considered by some to be a coping strategy for economically disadvantaged women, often occurring in areas of disaster or conflict.

(Walsh and Mulhern 2019; Yasmin 2016). This governmental policy only made the cycle of victimization worse for these young mothers. If girls were suspected to be pregnant, teachers often did public tests, feeling the already victimized girl's breasts and stomach. Some feel that pregnant girls' presence in school is only encouraging others to become pregnant. Due to the degrading and humiliating treatment of pregnant, or suspected pregnant, girls across Sierra Leone, many girls said they would simply never return to school (Werber 2015).

These laws have made girls to become victims twice. First, these girls became pregnant due to the failure of social institutions (Plan International 2015). Furthermore, pregnancies are often more difficult as young women are more likely to experience common pregnancy complications including obstructed and prolonged labor, fistula, and bleeding which are all more likely to lead to higher death rates among teenage girls (Yasmin 2016). Second, these country-wide school closures, although a public health necessity, may have created a generation of girls who will miss out on education and never be able to break out of the cycle of poverty (Plan International 2015). By stigmatizing and preventing pregnant girls from receiving an education, it is likely they will never return to school, putting them at greater risk of exploitation in the future (Bruce 2016; Minor 2017). Whether by force, or by choice, girls who were leaving the scars of the war years behind and forming a new educated generation are finding their progress retarded by the devastating virus (Labous 2014).

8.5 The Feedback Loop

8.5.1 How Ebola Influenced Gender-Based Violence

Evidenced from humanitarian crises around the world, teenage girls face particularly heightened risks of violence and rape due to their age and gender (IASC 2015; Stark and Ager 2011). While the international community acknowledges these risks in relation to war or natural disaster, organizations failed to view the EVD outbreak under a similar lens. The West Africa EVD outbreak confirmed that epidemics are indeed disasters and they leave women vulnerable to GBV (DuBois et al. 2015; IASC 2015). All of the measures put into place to protect society from the spread of disease—quarantines, curfews, and school closures—put women at risk of violence. Gender was overlooked during the response and left young girls highly vulnerable, similar to the civil wars in each country (Yasmin 2016).

8.5.2 Feedback Loop: Structural Violence

Due to years of civil war in both countries, many Liberians and Sierra Leoneans had already lived through the complete breakdown of their government, health services, and social structure. Many had witnessed first-hand the rampant increase in GBV, rape, and unintended pregnancies brought on by disaster. The WHO estimates that 90% of Liberian women had suffered physical or sexual violence during the war, and three of every four had been raped (Jones 2008). While the full impact of the Ebola epidemic is not yet clear on teenager's livelihoods, the possible ramifications from GBV, rape, and failed social structure could potentially be similar to the outcomes women suffered during the war (Jones 2008).

By comparing the state of women's health and safety during the war, peacetime, and the Ebola outbreak, we are attempting to illustrate the impact that years of violence had on the citizens of these countries. Having lived through years of war, violence became embedded into their social structure and continued to harm individuals; in this case, mainly adolescent and teenage girls and women. Once the Ebola outbreak began, society again felt the social fabric around them dissipate, resulting in a second wave of heightened violence, rape, and GBV. The impact of losing social structure had already been experienced, and therefore, when all structure was lost again, violence reemerged.

This analysis reinforces that disasters of any kind can result in violence against women due to existing political, economic, and social inequalities based only on gender. Specific to Liberia and Sierra Leone, we postulate further and maintain that a feedback loop was in place for the populations of these countries. Having both lived through tragic civil wars marked by violence and rape, the opening for violence remained prevalent within these countries. Even during times of peace, violence was not only commonplace but even considered normal, and perpetrators were rarely penalized (Denov 2006; Liebling-Kalifani et al. 2011). When the EVD outbreak occurred, society already had the underpinnings of GBV against women. Thus, the outbreak recreated an opening for the increase in gender-based violence (Liebling-Kalifani et al. 2011). Throughout the civil war, during peace, and the EVD outbreak, girls were either ignored or forgotten by some relief and recovery organizations which has allowed for the cycle of violence to continue, and the lasting consequences are still not fully known.

Structural violence is one way of describing social arrangements that put individuals and populations in harm's way. The arrangements are structural because they are embedded in the political and economic organization of our social world; they are violent because they cause injury to people (typically, not those responsible for perpetuating such inequalities) (Farmer et al. 2006).

8.6 Impact and Plan for Recovery

After 17 months, Liberia was declared Ebola-free on January 14, 2016, and Sierra Leone on March 17, 2016 (Sirleaf 2016; WHO 2016). The President of Sierra Leone, Dr. Ernest Bai Koroma, acknowledged that any gains since the civil war that previously ravaged the country had been largely undone by the Ebola outbreak (Government of Sierra Leone 2015). The President subsequently outlined a 24-month recovery strategy with the first 6–9 months focusing on (1) restoring basic access to healthcare; (2) getting children back to school; (3) social protection; and (4) restoring growth through the private sector and agriculture (Government of Sierra Leone 2015).

Similarly, in Liberia, The Economic Stabilization and Recovery Plan (ESRP) aimed to revitalize the country to precrisis levels, specifically rebuilding and strengthening the capacity to deliver social services including education, social welfare, and healthcare with better coverage to rural areas (International Labour Organization 2015). Liberia's plan did make note to strengthen resilience and reduce vulnerability for the poor and other at-risk groups. However, international organizations and the governments of the two countries failed both during the epidemic and in recovery to focus on highly at-risk groups, specifically, teenage girls.

8.7 Suggestions for a Way Forward

As witnessed during each country's civil war and the Ebola outbreak, teenage girls and their needs are overlooked in times of chaos, resulting from either conflict or epidemic. Girls experience loss and trauma while also dealing with the impact of schools being closed and their family structure drastically changing. Teenage pregnancy resulted from this chaos. The international community and each country's government must begin to evaluate the impact of these events to see these 'other' victims of Ebola. Teenage girls must be included in the epidemic response program and their voices must be heard to restore structure so that they are able to transition into adulthood in a more healthy and empowered manner (UNFPA 2016). Recent research indicates that, during the Ebola outbreak, local communities were able to process health messages and adapt quickly to the new information (Abramowitz et al. 2017). It may be possible to use this same type of messaging systems to influence behavior that may be based on gender.

8.7.1 Ensure Equitable Health Systems

Equitable health systems are defined by the absence of systematic or potentially remediable differences in health status, access to healthcare and health-enhancing environments, and treatment in one or more aspects of health across populations or population groups defined socially, economically, demographically, or geographically within and across countries (WHO 2011).

While each country was on a path to develop and improve their health care systems, the epidemic nullified any progress made, and each country's systems were already considered weak. Prior to the outbreak, each country already had some of the highest maternal and infant mortality rates globally as well as high percentages of teenage pregnancies. Hence, when the outbreak occurred, the gaps in health care revealed themselves further. Systems must be put into place to strengthen services in equitable ways to not only deal with future outbreaks, but also sexual and reproductive health for teens and women (International Labour Organization 2015). A priority must be to maintain SRH services during outbreaks, but to restore and strengthen immediately following (International Labour Organization 2015). Providing more equitable services would ensure a quicker recovery as well as improve the health outcomes of vulnerable citizens (Diez Roux 2011).

8.7.2 Create Safe Spaces

It is increasingly recognized as best practice to create and allocate specific safe spaces during humanitarian crises (UNFPA 2016). In the case of the EVD outbreak, women, men, girls, and boys were all quarantined together resulting in GBV, rape, and although less discussed, incest (Denney et al. 2015). Organizations focused on epidemic recovery could prioritize creation of spaces for vulnerable populations. These places, whether formal or informal, allow girls to feel physically and emotionally safe (UNFPA 2016). While everyone's lives drastically changed during the outbreak, many girls found themselves heads of households responsible for the well-being of their family members. These spaces could have helped girls psychologically as well as prevented them from turning to transactional sex as economic resources and support would have been available. Additionally, since schools were closed, safe spaces could have provided girls with safe socialization as well as education on Ebola, best practices to avoid infection and to have sexual and reproductive health services that are necessary but ignored during epidemics.

Safe spaces provide adolescent girls with livelihood skills, psychosocial counselling for gender-based violence, and access to sexual and reproductive health information and referral services (UNFPA 2016).

Safe spaces have been used in previous disasters including Malawi in 2015 after severe flooding occurred. Named 'youth clubs,' adolescents had access to recreation opportunities as well as age-appropriate SRH information and services (UNFPA 2016). From January to June of that year, the 32 youth clubs provided 18,000 internally displaced adolescents and reduced the incidence of sexually transmitted infections (UNFPA 2016). Although Malawi experienced a different type of disaster, adolescent girls were similarly put into risks due to idle time (UNFPA 2016). Similar to Sierra Leone and Liberia, due to school closures and disruption to SRH services, girls in Malawi also experienced fear of rape and were vulnerable to unwanted pregnancies and GBV (UNFPA 2016). As seen from Malawi, safe spaces work and are effective in times of disaster and chaos to provide adolescents with the structure and services they require. Within the context of an outbreak, precaution considerations should be observed in safe spaces so as not to further spread the disease.

8.7.3 Remove Restrictions on Schooling for Pregnant Girls

The schooling of an estimated five million children and youth were impacted by the outbreak across Liberia, Sierra Leone, and Guinea (UNDP 2015). Getting kids back to school should be a priority for each government to return to normalcy and continue to advance the younger generations. Instead, governments added restrictions, all together preventing pregnant girls from attending school in Sierra Leone and preventing attendance at day classes in Liberia. Education should be mandatory for all in these countries which already suffer from low attendance rates. Furthermore, schools should be a priority of the government, making sure they are safe, sanitary, and equipped with psychosocial care post-humanitarian crisis. To ensure health for teenage girls, sexual and reproductive health (SRH) services should also be provided along with referrals to local health centers (UNDP 2015). Preventing girls from attending school will only add to the cycle of low education and high poverty in each country. Providing child care for girls who had babies as a result of Ebola so that they can return to school should be considered.

8.7.4 Empowering Teenage Girls

Sierra Leone and Liberia have been detrimentally impacted by years of conflict, and now from the Ebola epidemic. During the outbreak, the youth were especially hard hit taking on new familial roles, being orphaned, removed from school, and in the case of many girls, becoming pregnant. In order to end the feedback loop of violence, it is essential to involve young girls in recovery assessments and future planning. By empowering girls to channel this experience to something that can effectively transform the future of recovery for generations to come, they can contribute to society to be regarded as part of the solution, not part of the problem (UNFPA 2016). An approach which can be considered for this age group is using arts for public health messages. Sonke and Pesata (2015) describe how West African musicians wrote Ebola songs which became popular in Liberian radio stations and dance clubs. The songs demonstrated the power of popular music to convey Ebola messages. Dozens of songs and music videos were created through collaborations between artists and health professionals to deliver more targeted messages. If packaged correctly, such an initiative can target and empower adolescents and young people.

8.8 Conclusions

Increases in GBV, rape, and teenage pregnancy have been seen time and again, yet recovery efforts during the West African Ebola outbreak ignored the populations most at risk. International organizations and governments alike must begin to view all disasters, whether conflict or disease outbreak, as threats to teenage girls and, furthermore, must build systems that facilitate women's specific needs. Focusing on the younger generation can prevent the feedback loop from occurring again, and instead, these countries can strive to empower their girls and have them prosper in environments without targeted violence, fear, and stigma (O'Brien and Tolosa 2016).

References

Abramowitz, S., & Moran, M. (2012). International human rights, gender-based violence and local discourses of abuse in postconflict Liberia: A problem of "culture"? *African Studies Review, 55*(2), 119–146. https://doi.org/10.1353/arw.2012.0037.

Abramowitz, S., McKune, S., Fallah, M., Monger, J., Tehoungue, K., & Omidian, P. (2017). The opposite of denial: Social learning at the onset of the Ebola emergency in Liberia. *Journal of Health Communication, 22*(Suppl 1), 59–65. https://doi.org/10.1080/10810730.2016.1209599.

Bruce, J. (2016). The difficulties of 'living while girl'. *Journal of Virus Eradication, 2*(3), 177–182. Retrieved December 5, 2017, from https://www.ncbi.nlm.nih.gov/pubmed/27482459.

Centers for Disease Control and Prevention. (2016). *Ebola (Ebola virus Disease)*. Retrieved December 10, 2017, from https://www.cdc.gov/vhf/ebola/outbreaks/2014-west-africa/case-counts.html.

Coltart, C. E. M., Lindsey, B., Ghinai, I., Johnson, A. M., & Heymann, D. L. (2017). The Ebola outbreak, 2013–2016: Old lessons for new epidemics. *Philosophical Transactions of The Royal Society B, 372*, 20160297. https://doi.org/10.1098/rstb.2016.0297.

Denney, L., Gordon, R., & Ibrahim, A. (2015). *Teenage pregnancy after Ebola in Sierra Leone: Mapping responses, gaps and ongoing challenges*. Retrieved December 5, 2017, from https://www.odi.org/publications/10396-teenage-pregnancy-after-ebola-sierra-leone-mapping-responses-gaps-and-ongoing-challenges.

Denov, M. S. (2006). Wartime sexual violence: Assessing a human security response to war-affected girls in Sierra Leone. *Security Dialogue, 37*(3), 319–342. Retrieved December 12, 2017, from http://citeseerx.ist.psu.edu/viewdoc/download?doi=10.1.1.473.8651&rep=rep1&type=pdf.

Diez Roux, A. V. (2011). Complex systems thinking and current impasses in health disparities research. *American Journal of Public Health, 101*(9), 1627–2634. https://doi.org/10.2105/AJPH.2011.300149.

DuBois, M., Wake, C., Sturridge, S., & Bennett, C. (2015). *The Ebola response in West Africa exposing the politics and culture of international aid*. Humanitarian policy group working paper. Retrieved December 15, 2017, from https://www.odi.org/sites/odi.org.uk/files/odi-assets/publications-opinion-files/9903.pdf.

Farmer, P. E., Nizeye, B., Stulac, S., & Keshavjee, S. (2006). Structural violence and clinical medicine. *PLoS Medicine, 3*(10), 1686–1691. https://doi.org/10.1371/journal.pmed.0030449.

Friedman-Rudovsky, J. (2013). The women who bear the scars of Sierra Leone's civil war. *The Telegraph*. Retrieved December 5, 2017, from http://www.telegraph.co.uk/news/worldnews/africaandindianocean/sierraleone/10450619/The-women-who-bear-the-scars-of-Sierra-Leones-civil-war.html.

Government of Sierra Leone. (2015). *National Ebola recovery strategy for Sierra Leone 2015-2017, 1–58*. Retrieved December 5, 2017, from http://ebolaresponse.un.org/sites/default/files/sierra_leone_recovery_strategy_en.pdf.

Gurman, T., Trappler, R., Acosta, A., McCray, P., Cooper, C., & Goodsmith, L. (2014). By seeing with our own eyes, it can remind in our mind: Qualitative evaluation findings suggest the ability of participatory video to reduce gender-based violence in conflict-affected settings. *Health Education Research, 29*(4), 690–701. https://doi.org/10.1093/her/cyu018.

Harman, S. (2015). Ebola, gender and conspicuously invisible women in global health governance. *Third World Quarterly, 37*(3), 524–541. https://doi.org/10.1080/01436597.2015.1108827.

Inter-Agency Standing Committee [IASC]. (2015). *Guidelines for integrating gender-based violence interventions in humanitarian action camp coordination and camp management food security and agriculture. Reducing risk, promoting resilience and aiding recovery*. Retrieved December 5, 2017, from http://gbvguidelines.org/wp/wp-content/uploads/2015/09/2015-IASC-Gender-based-Violence-Guidelines_lo-res.pdf.

International Labour Organization. (2015). *Recovery of the world of work in Guinea, Liberia and Sierra Leone.* Retrieved December 5, 2017, from http://www.ilo.org/wcmsp5/groups/public/%2D%2D-africa/%2D%2D-ro-addis_ababa/documents/publication/wcms_381359.pdf.

Jones, A. (2008). A war on women. *Los Angeles Times*. Retrieved December 5, 2017, from http://www.latimes.com/la-op-jones17feb17-story.html.

Korkoyah, D. T., & Wreh, F. F. (2015). *Ebola impact revealed.* Retrieved December 5, 2017, from https://www.oxfam.org/sites/www.oxfam.org/files/file_attachments/rr-ebola-impact-women-men-liberia-010715-en.pdf.

Labous, J. (2014). Ebola shutdown brings new fears of rape and teenage pregnancy. *Thomas Reuters Foundation News*. Retrieved December 5, 2017, from http://news.trust.org//item/20141117091228-3n0cq/.

Lai, D. (2015). Women must be at the center of the Ebola response. *Care Insights*. Retrieved December 5, 2017, from http://insights.careinternational.org.uk/development-blog/women-must-be-at-the-centre-of-the-ebola-response.

Liebling-Kalifani, H., Ojiambo-Ochieng, R., Were-Oguttu, J., & Kinyanda, E. (2011). Women war survivors of the 1989-2003 conflict in Liberia: The impact of sexual and gender-based violence. *Journal of International Women's Studies, 12*(1), 1–21.

McKay, B. (2015). West Africa struggles to rebuild its ravaged health-care system. *Wall Street Journal*, 1–8. Retrieved December 9, 2017, from https://www.wsj.com/articles/africa-struggles-to-rebuild-its-ravaged-health-care-system-1433457230.

Minor, O. M. (2017). Ebola and accusation: Gender dimensions of stigma in Sierra Leone's Ebola response. *Anthropology in Action, 24*(2), 25–35. https://doi.org/10.3167/aia.2017.240204.

O'Brien, M., & Tolosa, X. (2016). The effect of the 2014 West Africa Ebola virus disease epidemic on multi-level violence against women. *International Journal of Human Rights in Healthcare, 9*(3), 151–160. https://doi.org/10.1108/IJHRH-09-2015-0027.

Olujic, M. B. (1998). Embodiment of terror: Gendered violence in peacetime and wartime in Croatia and Bosnia-Herzegovina. *Medical Anthropology Quarterly, 12*(1), 31–50. https://doi.org/10.1525/maq.1998.12.1.31.

Plan International. (2015). *Teenage pregnancy rates rise in Ebola-stricken West Africa*. Retrieved December 5, 2017, from https://plan-international.org/news/2014-11-17-teenage-pregnancy-rates-rise-ebola-stricken-west-africa#.

Scott, J., Averbach, S., Modest, A. M., Hacker, M. R., Cornish, S., Spencer, D., et al. (2013). An assessment of gender inequitable norms and gender-based violence in South Sudan: A community-based participatory research approach. *Conflict and Health, 7*, 4. https://doi.org/10.1186/1752-1505-7-4. Retrieved December 9, 2017, from https://conflictandhealth.biomedcentral.com/articles/10.1186/1752-1505-7-4.

Sirleaf, E. J. (2016). Two years after Ebola: Liberia is a changed nation. *Time Magazine*. Retrieved December 5, 2017, from http://time.com/4440771/liberian-president-ebola-recovery/.

Sonke, J., & Pesata, V. (2015). The arts and health messaging: Exploring the evidence and lessons from the 2014 Ebola outbreak. *British Medical Journal*. Retrieved December 9, 2017, from https://www.researchgate.net/publication/277312649_The_arts_and_health_messaging_Exploring_the_evidence_and_lessons_from_the_2014_Ebola_outbreak.

Stark, L., & Ager, A. (2011). A systematic review of prevalence studies of gender-based violence in complex emergencies. *Trauma, Violence & Abuse, 12*(3), 127–134. https://doi.org/10.1177/1524838011404252. Retrieved December 5, 2017, from https://www.researchgate.net/publication/277312649.

Strong, A., & Schwartz, D. A. (2016). Sociocultural aspects of risk to pregnant women during the 2013-2015 multinational Ebola virus outbreak. *Health Care for Women International, 37*(8), 922–942.

Truth & Reconciliation Commission, Sierra Leone. (2004a). *Witness to truth: Report of the Sierra Leone Truth and Reconciliation Commission*, (Vol. 3B, pp. 1–520).

Truth & Reconciliation Commission, Sierra Leone. (2004b). *Witness to truth: Report of the Sierra Leone Truth and Reconciliation Commission*, (Vol. 2, pp. 1–503).

United Nations Development Programme (UNDP). (2015). *Recovering from the Ebola crisis*. Retrieved December 5, 2017, from http://www.undp.org/content/undp/en/home/librarypage/crisis-prevention-and-recovery/recovering-from-the-ebola-crisis%2D%2D-full-report.html.

United Nations Population Fund (UNFPA). (2016). *Adolescent girls in disaster and conflict: Interventions for improving access to sexual and reproductive health services* (pp. 1–92). Retrieved December 5, 2017, from http://www.unfpa.org/sites/default/files/pub-pdf/UNFPA-Adolescent_Girls_in_Disaster_Conflict-Web.pdf.

United Nations Population Fund (UNFPA). (2017). *Statistics*. Retrieved December 5, 2017, from http://www.unfpa.org/data.

Walsh, S., & Mulhern, F. (2019). A step in the rights direction: Advocacy, negotiation and money as tools for realising the right to education for pregnant girls in Sierra Leone during the Ebola epidemic. In D. A. Schwartz, J. A. Anoko, & S.A. Abramowitz (Eds.), *Pregnant in the time of Ebola: Women and their children in the 2013-2015 West African epidemic*. New York: Springer.

Werber, C. (2015). How Ebola led to more teenage pregnancy in West Africa. *Quartz Africa*. Retrieved December 5, 2017, from https://qz.com/543354/how-ebola-led-to-more-teenage-pregnancy-in-west-africa/.

Wilkinson, A., & Leach, M. (2015). Briefing: Ebola-myths, realities, and structural violence. *African Affairs, 114*(454), 136–148. https://doi.org/10.1093/afraf/adu080.

World Health Organization (WHO). (2011). *Health systems strengthening: Glossary*. Retrieved December 5, 2017, from http://www.who.int/healthsystems/Glossary_January2011.pdf.

World Health Organization (WHO). (2016). *Ebola situation report*. Retrieved December 5, 2017, from http://apps.who.int/ebola/current-situation/ebola-situation-report-16-march-2016.

Yasmin, S. (2016). The Ebola rape epidemic no one's talking about. *Foreign Policy*. Retrieved December 5, 2017, from http://foreignpolicy.com/2016/02/02/the-ebola-rape-epidemic-west-africa-teenage-pregnancy/.

9 Translating Models of Support for Women with Chronic Viral Infection to Address the Reproductive Health Needs of West African Ebola Survivors

Caroline Crystal, Laura A. Skrip, Tolbert Nyenswah, Hilary Flumo, Alison P. Galvani, David P. Durham, and Mosoka P. Fallah

9.1 Introduction

The 2013–2015 epidemic of Ebola virus disease (EVD) devastated the West Africa region with an estimated 28,610 cases and 11,308 deaths documented in Liberia, Guinea, and Sierra Leone. The region was last declared free of known Ebola transmission in March 2016. However, the social and

C. Crystal · D. P. Durham
Center for Infectious Disease Modeling and Analysis (CIDMA), Yale School of Public Health, New Haven, CT, USA

L. A. Skrip
Center for Infectious Disease Modeling and Analysis (CIDMA), Yale School of Public Health, New Haven, CT, USA

National Public Health Institute of Liberia, Monrovia, Liberia

T. Nyenswah
National Public Health Institute of Liberia, Monrovia, Liberia

H. Flumo
National Institute of Allergy and Infectious Diseases, PREVAIL-III Study, Monrovia, Liberia

A. P. Galvani
Center for Infectious Disease Modeling and Analysis (CIDMA), Yale School of Public Health, New Haven, CT, USA

A.M. Dogliotti College of Medicine, University of Liberia, Monrovia, Liberia

M. P. Fallah (✉)
Center for Infectious Disease Modeling and Analysis (CIDMA), Yale School of Public Health, New Haven, CT, USA

National Public Health Institute of Liberia, Monrovia, Liberia

National Institute of Allergy and Infectious Diseases, PREVAIL-III Study, Monrovia, Liberia

A.M. Dogliotti College of Medicine, University of Liberia, Monrovia, Liberia

D. A. Schwartz et al. (eds.), *Pregnant in the Time of Ebola*, Global Maternal and Child Health, https://doi.org/10.1007/978-3-319-97637-2_9

economic consequences of the outbreak continue to affect tens of thousands of Ebola survivors, orphans, caregivers, and families of the deceased. Female Ebola survivors have emerged as a particularly vulnerable group due to perceptions of infectiousness (Rabelo et al. 2016). In addition to psychosocial needs associated with the trauma during their time in Ebola treatment centers and social exclusion upon their return home, female survivors experience physical ailments characteristic of post-Ebola syndrome (Mattia et al. 2016). It has been hypothesized that sequelae reported by survivors is related, in part, to viral persistence. Ebola virus RNA has been detected in ocular aqueous humor (Varkey et al. 2015), as well as in several fluids produced by reproductive organs—including breast milk and amniotic fluid (Bausch et al. 2007; Sissoko et al. 2017; Black et al. 2015). In one cluster of EVD, molecular analysis revealed that a Liberian woman who suvived an episode of Ebola infection in 2014 had viral peristence or recurent infection, and transmitted the infection to three family members one year later (Dokubo et al. 2018). Accordingly, the physical consequences of surviving acute EVD among women of child-bearing age have implications for conception, pregnancy, and newborn health.

The reproductive health needs of female Ebola survivors are varied and significant. A study on pregnancy outcomes among Ebola survivors has suggested that women recovering from the disease may be at heightened risk of fetal loss, particularly in the absence of specialized care (Fallah et al. 2016). In addition, the possibility of viral persistence in reproductive organs has implications not only for health care workers and others engaged with survivors during prenatal care and delivery, but also for sexual partners of the survivors. Lastly, post-Ebola sequelae include various physical ailments, and the stigma surrounding survivorship has been associated with psychological health issues. Such health concerns may impair the ability and/or interest of women to conceive and care for children.

The uncertainty surrounding pregnancy, sexual and general health, and transmission potential associated with viral persistence may lead to both formal policies and personal decisions about having children. For instance, Ebola survivors may need to follow policies such as those in place for HIV (human immunodeficiency virus) infection; integrated programs combining prevention and treatment, as one example, are currently used in Liberia to reduce sexual and vertical transmission of HIV/acquired immunodeficiency syndrome (AIDS). Psychosocial support and peer groups could simultaneously address the nonmedical needs of survivors opting to delay pregnancy or experiencing pregnancy complications.

Due to the unprecedented nature of the Ebola epidemic in West Africa and the subsequent number of survivors, programs to meet the reproductive health needs of females recovering from the disease will be critical. Unfortunately, there is a lack of historical models. We use this chapter to review successfully implemented models for other communicable diseases. Specifically, we identify peer-led support group initiatives and efforts targeting health care workers to address the needs of young adult women with HIV and hepatitis C virus (HCV) infections.

9.2 Review of Programs Intended to Prevent Transmission

Transmission of Ebola virus during acute disease was associated with care-giving, both in community and health care settings (Dunn et al. 2016; Richards et al. 2015), as well as intimate contacts, such as sharing of bedding and eating utensils and funerary preparation of corpses (Bausch et al. 2007; Richards et al. 2015). While less is known about the transmission of persistent virus during convalescence, sexual transmission from a male Ebola survivor was hypothesized to have precipitated the last disease cluster in Liberia's primary epidemic (Mate et al. 2015). Fears surrounding transmission potential of bodily fluids from Ebola survivors have led to stigmatization in sexual relationships and refusal of care, thus jeopardizing the prospects for conception as well as health of pregnant survivors. Interventions and guidelines around conception and care for female patients infected with HIV or hepatitis may provide a precedent for ensuring safe conception and quality antenatal care for pregnant women with chronic viral infections.

9.2.1 Current Knowledge of Transmission Potential Among Female Ebola Survivors

It is suspected that Ebola may be transmitted from male-to-female through sexual activity; however, unlike HIV/AIDS and hepatitis, there is less evidence for female-to-male sexual transmission (WHO 2016). The live Ebola virus can be isolated from semen for up to 82 days after symptom onset, and Ebola RNA persists for much longer – greater than 2 years after the onset of disease in one study (Fischer et al. 2017). There are several reported and suspected cases of male-to-female transmission of the Ebola virus following exposure to infected semen, but the exact mode of infection (ex. sexual transmission, contact) is not yet clear. Genetic analysis of Ebola virus from the semen of EVD survivors has demonstrated that active viral replication/transcription continues to occur during the period of clinical convalescence, decreasing over time, which is consistent with viral persistence (Whitmer et al. 2018). On the other hand, Ebola RNA—but not the live virus—has been detected in vaginal fluid for up to 33 days after symptom onset. The duration of RNA persistence is uncertain, as is whether the virus can be sexually transmitted from female to male.

Although greater surveillance data and research are needed, the WHO has offered interim advice regarding the risks of sexual transmission of Ebola. These recommendations include (1) safe sexual practices or abstinence between Ebola survivors and sexual partners, until either infected semen has tested negative by reverse transcription polymerase chain reaction (RT-PCR) twice, or 12 months have elapsed since infection if semen has not been tested; (2) semen testing for male Ebola survivors at 3-month intervals after the onset of disease and, if positive, then every month until negative twice by RT-PCR, with at least an interval of 1 week between tests; (3) thorough hand and personal hygiene after physical contact with semen until semen tested negative twice or 12 months after symptom onset; and (4) consistent condom use to prevent HIV, other sexually transmitted infections (STIs), and unwanted pregnancy even in the absence of Ebola transmission risk.

Although Ebola infection during pregnancy is associated with a high rate of obstetric complications, poor maternal outcomes and a neonatal mortality approaching 100%, the obstetrical risks are far lower for Ebola survivors than for those acutely infected. There is no evidence that Ebola survivors who subsequently become pregnant pose a risk for Ebola virus transmission or that there is persistent Ebola virus infection in their fetus, amniotic fluid, or placenta (WHO 2015). These women and their health care practitioners should, therefore, observe standard obstetric IPC (infection prevention and control) precautions during childbirth and/or management of obstetrical complications. Pregnant women who survive Ebola with an ongoing pregnancy, however, may transmit the virus during delivery and/or management of obstetric complications and should, therefore, use comprehensive Ebola IPC precautions during these times to prevent exposure to infectious intrauterine contents. The neonates of these women should also be managed using Ebola IPC precautions for 21 days following birth.

9.2.2 Model Transmission-Preventing Interventions for Other Chronic Viral Infections

Transmission-reducing interventions have been well-studied for HIV and viral hepatitis. Among sexually active women of child-bearing age, both the decision to plan for pregnancy and the decision to actively prevent it are supported by established guidelines. At the preconception phase, guidance for HIV-positive women opting to delay pregnancy involves a combination of female-controlled contraceptive products, including female condoms, vaginal gels, oral birth control, and transmission prevention methods, including the nonpharmaceutical methods of male condoms and male circumcision (Laga and Piot 2012). Among sero-discordant heterosexual couples, pharmaceutical approaches are being increasingly adopted. Preexposure prophylaxis (PrEP) has been shown to be 62% efficacious at reducing HIV transmission to the seronegative partner (McMahon et al. 2014). Furthermore, adherence to an antiretroviral therapy (ART) regimen by the seropositive partner suppresses HIV RNA in the genital tract, and consequently, has implications for reduced sexual transmission risk (Cohen 2007).

For HIV-positive women planning a pregnancy, preconception planning guidelines have been suggested to enhance the odds of conception and to prevent sexual transmission to HIV-negative male partners (Money et al. 2014). For enhancing the odds of conception, folic acid and other supplements are recommended, while for preventing sexual transmission, PrEP and treatment as prevention afford an HIV-positive woman the option to conceive with reduced chance of transmitting virus through sexual contact unprotected by physical barrier (Money et al. 2014).

The efficiency of sexual contact as a transmission route varies across types of viral hepatitis (Centers for Disease Control and Prevention 2015). In the case of hepatitis C virus (HCV), sexual transmission is much less frequent than it is for HIV or hepatitis B virus (HBV); HCV RNA in body fluids other than blood is usually absent or of very low titer (Myung et al. 2003). For sexually active women infected with hepatitis B, recommendations include use of antivirals and vaccination of partners (Chen et al. 2014).

Culturally competent public awareness campaigns to stress the importance of prevention, screening, and management have been effective at reducing the population-level burden of hepatitis B (Chen et al. 2014). Such systems-level approaches have incorporated targeted messaging both for pregnant women and women interested in becoming pregnant (Evans et al. 2015). One such educational tool, "Hepatitis B and You," is designed to communicate essential information about hepatitis B in a way that is accessible to HBV-infected pregnant women (Wilson 2003). Available both as an online slide set on the CDC website or as a booklet, "Hepatitis B and You" is a valuable educational resource for both health care professionals and the general public in preventing the perinatal transmission of HBV. "Hepatitis B and You" uses educational strategies to most effectively reach its patient audience. Information is presented at a sixth-grade reading level, using images and graphics to depict complex concepts and conveying information in short, bulleted lists. After completing the online slide set, viewers have the option to fill out a web-based survey assessing knowledge, beliefs, and what information they would like to know more about. The content of "Hepatitis B and You" is continuously modified on the basis of those areas respondents would like more information on. After exposure to the tool, participants in the survey demonstrated increased knowledge about the prevention of perinatal transmission of HBV infection, and almost all respondents said that the information was helpful and easy to follow. The provision of readily accessible information on any disease is critical in preventing transmission, and that this tool is designed for pregnant HBV-positive women to become proactive in the care necessary to prevent HBV transmission to their infants makes this study particularly relevant for combating perinatal transmission among female Ebola survivors.

While the efficiency and frequency of Ebola virus transmission via sexual contact requires further study, guidelines for HIV and HBV could direct investigations into transmission-reducing efforts among sexual-active survivors interested in becoming pregnant or deciding to delay pregnancy. For example, continued study of the duration of Ebola persistence (Chughtai et al. 2016; Rodriguez et al. 1999) and the impact of antivirals on reducing Ebola viral RNA (Huggins et al. 1999; Haque et al. 2015) in the female genital tract would elucidate the potential effectiveness of translating a treatment as prevention approach.

9.3 Review of Programs Intended to Educate the Broader Community

Due to fear among the community of ongoing risk for transmission, Ebola survivors have experienced significant stigma affecting opportunities for employment, education, social engagement, and health care. Programs to educate communities about Ebola transmission, as well as the health and emotional needs of survivors and their families, could correct misconceptions and promote discussion about fears surrounding Ebola survivors. Community-based stigma-reduction interventions that have been developed for HIV could be adopted for use in Ebola-affected areas.

9.3.1 HIV Intervention Education in South Africa

HIV educational programs have focused on improving the lives and relationships and fostering a deeper understanding of HIV stigma, not only in people living with HIV but also among those persons living close to them and in their broader communities (French et al. 2015). Implemented for 1 month in both urban and rural settings in North West Province in South Africa, one HIV education intervention was targeted at both people living with HIV (PLWH) who are negatively affected by stigma and persons living close to them (PLC), who also experience stigma due to their association with PLWH. The PLC group included partners, children, close family members, close friends, spiritual leaders, and community members.

The program design included an initial 2-day workshop with only PLWH, followed by six 3-day workshops with PLWH and PLC together. Focused on understanding stigma, self-empowerment, and coping with HIV stigma and disclosure, the workshops consisted of presentations and small-group discussions and activities. Afterwards, the participants planned and implemented a stigma-reduction project in their communities for 1 month. The project had wide reach and resulted in greater awareness of and knowledge about HIV stigma. Project participants reported improved relationships, greater sense of empowerment, and increased knowledge and understanding of HIV and stigma. Moreover, PLWH reported feeling less stigmatized and a greater willingness to disclose their status, while PLC reported greater awareness of their stigmatizing behaviors. A stigma-reduction and wellness-enhancement intervention similar to the one described above may improve the lives of female Ebola survivors. The possibility of viral persistence in reproductive organs heightens the importance of Ebola survivors' willingness to disclose their status, particularly with sexual partners, and demands an aware and accepting audience for such information.

9.3.2 HIV Health Education in Nicaragua

Another HIV-focused health education campaign targeting psychologists, doctors, and teachers was implemented in Managua, Nicaragua (Pauw et al. 1996). The intervention incorporated a brief presentation on HIV transmission and ways to avoid infection, distribution of leaflets and condoms, educational activities in schools and public meeting places, and communication of HIV and AIDS-related information through vehicle loudspeakers. Following the education program, both intervention and control groups reported increased knowledge regarding transmission and prevention of HIV infection, increased condom usage, and decreased levels of anxiety about HIV and AIDS. The effect of the health education campaign was only statistically significantly different between control and intervention samples for a minority of items, however, likely due to the control group's exposure to materials (e.g., billboards, radio broadcasts, articles) meant for the experimental group and the higher proportion residents with formal education within the control group. It is noted that while knowledge about transmission routes and ways to prevent infection is necessary for behavior change, such knowledge alone does not guarantee behavior change. Any health promotion intervention,—including one among female Ebola survivors—therefore, must attempt to ensure that any changes in knowledge and attitudes are reflected in practice.

9.3.3 HIV Education in Saudi Arabia

An alternative approach to education about HIV/AIDS has involved targeting high-risk groups versus the general community. One such health education program was administered to secondary school students in Buraidah City, Saudi Arabia (Saleh et al. 1999). The program sought to clarify misperceptions about transmission of HIV. The intervention involved health talks as well as distribution of posters, booklets, and pamphlets. Exposure to the health education program was associated with increases in participants' knowledge about the disease in general and transmission specifically, while misperception scores decreased. Given the lack of knowledge surrounding viral persistence and transmission potential among EVD survivors, the implementation of a similar school-based education program may not only reduce the risk of Ebola transmission, but also give young female Ebola survivors and potential sexual partners the knowledge to make more informed reproductive health decisions.

9.3.4 Education for Viral Hepatitis in the United States: "Know More Hepatitis"

Educational programs have likewise been employed for creating improved understanding of viral hepatitis transmission and disease progression. In 2012, to accompany its recommendations calling for one-time testing for HCV infection of people born between 1945 and 1965 (termed baby boomers), the United States Centers for Disease Control and Prevention launched "Know More Hepatitis", a multimedia national education campaign (Jorgensen et al. 2016). The program was aimed at (1) the baby boomer generation, who account for 75% of all HCV infections and are at highest risk for HCV-related liver disease; and (2) primary care medical providers, who play a critical role in testing baby boomer patients and linking infected patients to medical care and treatment. As approximately one-half of HCV-infected baby boomers are unaware of their infection, education to promote testing was particularly important.

From 2011 to 2014, the CDC held focus groups to assess baby boomers' and primary care providers' knowledge, attitudes and practices regarding hepatitis C and to test proposed messages for the intervention. The central message of the campaign recommended testing for high-risk population of baby boomers, highlighted the consequences of untreated HCV infection, relayed general facts about the disease and its consequences, and attempted to dispel commonly held myths about HCV infection. The program relied largely on donated time and space from broadcast and print outlets, including magazine, newspaper, digital, YouTube and Google display advertisements; printed posters, infographics, and billboards; airport dioramas; video and radio PSAs; a website; and messages through social media sites such as Facebook and Twitter. For primary care providers, information was disseminated through email updates, clinical news announcements, social media, medical and professional news outlets, and the CDC website. The campaign resulted in more than 1.2 billion documented audience impressions.

Certain aspects of the "Know More Hepatitis" campaign can be applied to an education campaign regarding Ebola survival; for example, in addition to targeting baby boomers, the "Know More Hepatitis" information campaign also targeted and provided practical tools for primary care providers, who play a crucial role in the implementation of any recommendation. Additionally, in its research-based formulation of campaign messages, the program aimed to overcome audience stigma which exists for hepatitis. Lastly, because the campaign succeeded in reaching a wide audience, its strategies for information dissemination may be applied to an intervention for Ebola.

9.3.5 Continuing Medical Education in Australia

While community-focused programs have provided education on HCV to the population, educational campaigns targeting health care workers have facilitated improved general practitioners' hepatitis B knowledge and practices. Continuing medical education (CME), programs have been used as a way of educating general practitioners (GPs) about hepatitis B (Robotin et al. 2013). For example, all local GPs in Sydney, Australia, were invited to participate in a CME program which consisted of four sessions offered over a 12-month period and focused on hepatitis B epidemiology, diagnosis, and management. The CME activities included three components: (1) a predisposing activity, which included a prereading and brief questionnaire; (2) a face-to-face seminar, which was a mixture of interactive and instructive activities; and (3) a post-seminar activity, which encouraged reflection on the skills and knowledge necessary to manage chronic hepatitis B (CHB) and associated liver diseases. The sessions were delivered by academic hepatologists and gastroenterologists and facilitated by an academic GP and the staff of B-Positive, a program which aims to support local GPs in southwest Sydney to identify and manage their patients with CHB.

The use of multimedia CME interventions, multiple instructional techniques, and multiple exposures to CME content has been recommended (Moores et al. 2009). CME programs are important for maintaining current knowledge of any disease, including Ebola. To ensure proper care of female Ebola survivors, therefore, a CME program—perhaps using the recommended flexible, individualized approach—should be considered.

While education programs provide information to both women with chronic viral infection and the health care providers who are working with them, programs that offer emotional and social support, particularly among women having difficulty conceiving or choosing to not have children, are likewise important for well-being. Several interventions have been shown to effectively offer support to HIV-infected women of reproductive age.

9.4 Review of Programs Intended to Offer Support

9.4.1 Project Masihambisane in KwaZulu-Natal, South Africa

Standard programs focused on prevention of mother-to-child transmission frequently do not address the daily challenges faced by pregnant women living with HIV regarding their physical and mental health. Project Masihambisane ("let's walk together") conducted in eight rural primary health care clinics in KwaZulu-Natal, South Africa, utilized peer mentors in helping pregnant women living with HIV to cope with physical and mental health challenges (Rotheram-Borus et al. 2011; Richter et al. 2014). The program was evaluated using a cluster randomized controlled trial in which pregnant women living with HIV were randomly assigned by clinic to either a Masihambisane intervention program or a standard care program on mother to child transmission. The intervention clinics received standard care in addition to optional participation in four antenatal and four postnatal small-group sessions led by a peer mentor: a trained pregnant woman living with HIV who had undergone training on mother-to-child transmission and was perceived as a positive role model. Supplemented by the distribution of informational media and piloted through focus groups and interviews, the Masihambisane sessions were 60–90 min long. They were attended by four to ten pregnant women living with HIV who discuss, roleplay, and practice strategies to address issues including mental and physical health for the mother and the child and plans for the future. A trained independent assessment team held in-person interviews with women living with HIV at baseline and at times coinciding with their clinic visits—during pregnancy and within 6 weeks, 6 months, and 12 months post-birth—to evaluate the mother's and baby's health and well-being. Data were collected and uploaded via mobile phone. At 6 weeks post-birth, the intervention, despite low uptake and mixed results, was found to have significant improvements in maternal and child outcomes. Given the impact of this peer-led program on improving health among women living with HIV and their babies, an adapted version may be beneficial for pregnant Ebola survivors. Additionally, the use of inexpensive and widely available mobile devices for data collection could be adopted for any future intervention with pregnant Ebola survivors.

9.4.2 Education Intervention Among the AmaXhosa People in South Africa

Another peer-led exercise and education intervention focused on pain among AmaXhosa women living with HIV/AIDS in South Africa (Parker et al. 2016). The education intervention consisted of a 6-week peer-led physical exercise and education program. Peer leaders were bilingual in English and isiXhosa and recruited directly from the target community. The pain of all participants was assessed using the International Classification of Functioning, Disability and Health over a 4-month period. Both groups experienced reductions in pain, yet there were no significant differences between the intervention and control groups in terms of pain severity, pain interference, and self-efficacy. The evaluation of the intervention, therefore, suggests that the provision of a workbook alone or engagement in a more intensive education intervention are both more effective ways to reduce pain compared to standard care. The lack of difference between groups may be due to the formation of a therapeutic relationship during the monthly data collection between control subjects and the interviewer, who, unlike most health care professionals, spoke isiXhosa. The efficacy of the program coupled with its low-cost components makes it applicable to pain-reduction endeavors in resource-poor settings, including those typical of female Ebola survivors.

Because high levels of education may have aided the control group in alleviating pain, in communities with high levels of education, the workbook alone may be sufficient to reduce pain in Ebola survivors, whereas in communities with lower levels of education, the peer-led intervention may be more effective. In addition, the "care factor" and the possible curative effects of the formation of a therapeutic relationship may also be considered in developing a program to help relieve pain and suffering among female survivors of child-bearing age.

9.4.3 Maternal-Child HIV Program in New Orleans, United States

Encouraging utilization of health care services, particularly after sensitization of providers to the needs of survivors and transmission-reduction practices, could have important benefits for improved maternal and newborn outcomes. As an example of a program aimed at enhanced health care utilization, a US-based HIV clinic undertook a maternal-child program to improve the attendance rates of women at follow-up clinic visits (Kissinger et al. 1995). The study was set in an outpatient clinic that offers multidisciplinary HIV care in New Orleans, Louisiana. Women attending the clinic were asked to identify barriers to their receiving optimal health care. On the basis of those recommendations, the hospital made modifications that included its provision of free childcare and transportation, separate waiting and examination rooms for mothers and their children, more female health care providers specially trained in female and pediatric needs, merged scheduled visits for both mother and child, daily availability of health care professionals for urgent visits, and on-site gynecologic services. An evaluation of the program suggested that responding to sex-specific needs can improve women's attendance for clinic visits. In diseases that have the possibility of mother-to-child transmission, such as HIV and Ebola, attendance at follow-up clinic visits is crucial for both the mother and the child's health. This study is particularly relevant for Ebola programs because they are most likely to be carried out in low-resource settings, and the clinic in this study reported that the modifications were relatively simple and low-cost to implement, with many reforms simply consisting of reorganizing existing staff, changing the scheduling scheme, and moving examination rooms.

9.4.4 The Toronto Community Hepatitis C Program

A program that attempted to provide low-barrier HCV treatment and support to people who use illicit drugs and/or have mental health issues could also be studied to inform programs for Ebola survivors, as both groups are similarly stigmatized and increasingly hidden (Mason et al. 2015). In the Toronto Community Hepatitis C Program, established in 2006, the HCV study participants were marginalized former or current drug users who had difficulties accessing treatment and care. Centering around a group support model, the participants would also attend weekly psycho-educational support group meetings led by peer support workers, as well as receiving medical care for HCV. The peer support workers were current or former clients of the program who provided informational counseling, group facilitation, and outreach and patient accompaniments. In the meetings, participants would receive accessible information about HCV and access to HCV care through on-site practitioners and counsellors. After 3 years of the program, significantly more participants had been assessed by an HCV specialist, increasing numbers of study participants initiated HCV treatment at rates comparable to those seen for specialist-based HCV care in tertiary settings, and housing status and income showed significant improvement. The housing and income improvements were probably due to clinicians and support staff's working with clients to help find and maintain housing and to obtain income support, particularly through receiving provincial disability benefits. Thus, the program accomplished successful provision of HCV care and support to highly marginalized individuals who would be less likely to receive treatment elsewhere. It is, therefore, highly relevant as a potential application to female Ebola survivors, who are also marginalized and have limited access to health care, particularly surrounding their reproductive needs. This type of program may be especially useful in identifying and addressing key social determinants of health and health care utilization.

9.5 Recent Efforts to Help Ebola Survivors in Liberia

The programs reviewed in this chapter offer approaches for educating female patients, the general community, and health care providers about transmission and prevention of chronic viral disease. Throughout West Africa, pregnant Ebola survivors and those intending to become pregnant may particularly benefit from sensitization efforts to ensure optimal social support and health care during their potentially high-risk terms. In Liberia, efforts have been recently undertaken to address some of the needs of pregnant survivors.

Despite the ongoing threat of localized resurgences, the Ebola epidemic in Liberia has been successfully controlled (Nyenswah et al. 2016). The total number of persons infected with Ebola have been estimated at 10,678 in Liberia alone, with approximately 5800 survivors (Centers for Disease Control and Prevention 2016). In addition to the physical sequelae termed "post-Ebola syndrome," survivors have expressed concerns about stigma in their day-to-day interactions with community

Table 9.1 Experiences of pregnant Ebola survivors in Liberia and impact of birth cohort study

Prior to birth cohort study	"I was in labor and went to a clinic in Caldwell community to give birth, as soon as they knew I was a Survivor, they asked me to go out of the clinic within 30 min after giving birth." "I was 2 months pregnant and slipped when I went to fetched water. When I started to bleed, I call on the mobile team from Refuge Place to take me to the hospital. They were quick in getting to me because they were all Ebola survivors. We went to three renowned hospitals in Monrovia but I was rejected at all of them and I lost my pregnancy."
After initiation of birth cohort study	"I was 2 month pregnant when I went to enroll into the PREVAIL study. They did my pregnant test and it proves positive. I was asked to choose one of the four hospital in Monrovia to attend the antenatal care. Since that time, I was followed by a Tracker and when I got into labor around 1:00 am, the Tracker brought an ambulance and took me to ELWA hospital where I gave birth safely without paying and money." "When I was 5 months pregnant, I was informed about the birth cohort in the PREVAIL study during a meeting organized by the leadership of the Survivor Network. I was asked to go to one of the four hospital in Monrovia as I wish. From that day, the PREVAIL Tracker help me to always go to get good care until I gave birth. I did not pay any money during delivery at Redemption Hospital. Since I delivered, when the baby get sick, the tracker is also helping me to make sure my baby get treatment at JFK or Duport Road Hospital…" "I was followed by the tracker from the day I agreed to be part of the birth cohort at PREVAIL.But because I did not have trust in the health facilities, I stayed home until the baby was almost coming outside during labor. When I called the trackers around 3:00 am on February 22, 2016, they still came to help me but still I delivered in the ambulance because I informed them very late…"

dwellers. Otherwise common activities, such as sharing spoons, buckets, and towels, generate discomfort among family and friends who fear ongoing potential for transmission. Likewise, survivors have reported discrimination during their access of health care services (Table 9.1). From the perspectives of health care workers, there are concerns about their limited training on infection protection and control, limited information about transmission from survivors, and ongoing discontent about low salaries and lack of insurance for workers during outbreak. With the fears among health care workers not properly addressed by authorities during the outbreak, discrimination may continue against EVD survivors. Furthermore, in the event of another outbreak, many health care workers may not be willing to participate or work in ETUs if concerns remain unaddressed.

To better understand the health and social challenges experienced by survivors and to answer questions surrounding transmission potential, Liberia and its partners have undertaken several initiatives: specifically, the Ebola Survivor Network and a series of research studies under the clinical partnership termed PREVAIL, or the Partnership for Research on Ebola Virus in Liberia. The birth cohort study under PREVAIL has sought to ensure that appropriate care be provided for all pregnant EVD survivors immediately upon their identification. The Survivors' Network, representing over 1300 EVD survivors, provides opportunities for emotional and medical support through partnerships with several NGOs and international agencies.

9.5.1 PREVAIL Birth Cohort

The PREVAIL Natural History Study, of which the birth cohort is a sub-study, monitors the clinical needs of participating survivors through administration of a medical history, a brief physical examination, and blood sample collection at each of their regular study visits. Individuals with vision problems or other medical complications are referred for further examination. At the baseline examination,

EVD survivors are asked about household contacts since the time of the diagnosis with EVD and sexual partners since their discharge from an ETU. Close contacts of survivors (i.e., household members, family members and friends who provided care, and sexual partners) may be invited to attend a clinic examination at which demographic information are collected, a blood sample is obtained, a brief questionnaire on contact with the EVD survivor is administered, and an assessment of potential comorbidities, including a visual screen, are assessed. Close contacts are used as controls for comparing the prevalence and incidence of medical complications with EVD survivors. Initially, priority was given to enrolling survivors over close contacts and other individuals with negative serology for Ebola serve as controls. Survivors have been recruited using information from the Liberian Ministry of Health as well as referrals from local physicians, from clinics currently evaluating EVD survivors, and from support groups. Demographic information and dates of hospitalization for Ebola obtained from the registry at the Ministry of Health are used to assess how representative the group of EVD survivors are compared to all EVD survivors in the registry.

The Birth Cohort Study specifically falls under the PREVAIL III Natural History Study and aims to assess potential for mother-to-child transmission of EVD by collecting cord blood and placenta samples for virologic studies, by assessing developmental milestones, including growth parameters of body weight, body length, and head circumference, and by measuring EVD antibody levels for children born to EVD survivors. This initiative was necessitated by repeated reports of negative outcomes (stillbirths, abortion, and miscarriages) among survivors in Liberia (Fallah et al. 2016). Another explanation is the fact that survivors were not receiving adequate antenatal care at health centers, possibly because of stigmatization due to their Ebola-survivor status. On December 24, 2015, the birth cohort officially started with the first delivery on December 28, 2015 and still continues.

Since October 2015 up to present, there has being a well-coordinated collaboration between PREVAIL and four referral hospitals in Montserrado and Margibi counties to provide appropriate antenatal care to ensure safe delivery for those enrolled in the birth cohort. This effort has contributed to the delivery of more than 54 children, including 10 delivered by cesarean section. In addition, discussions are underway to ensure free health care services for all EVD survivors including females of child-bearing age at major hospitals. This initiative is expected to be supported by the United States Agency for International Development (USAID) and implemented by John Snow International in Liberia.

9.5.2 Observations During Support Groups at Refuge Place International

In addition to efforts by international partners, local groups have supported female EVD survivors. Refuge Place International, a local nongovernmental organization (NGO), focused on maternal and child health in the Chicken Soup Factory[1] community of Montserrado County, has advocated for quality health care for Ebola survivors, and has supported such initiatives as the birth cohort. Of the 40 EVD survivors who delivered after the commencement of birth cohort, at least 20 of them were taken to the hospital by an ambulance owned by Refuge Place International. The majority (75%) were escorted to the hospital between 12:00 AM and 3:00 AM, a time when it would have been very difficult to access public transportation. Additionally, a mobile clinic for all Ebola survivors in Montserrado including pregnant women was sponsored by Refuge Place from August 2015 to February 2016. Monthly stipends for

[1] The "Chicken Soup Factory" community in the township of Gardnersville, a suburb of Monrovia, refers to a neighborhood of approximately 50,000 persons that is adjacent to a plant that had been designed to produce chicken soup, but was never built.

Physician Assistants and Nurses on the team were also paid by Refuge Place. Efforts across local NGOs, such as Refuge Place, and international partners, such as the U.S. National Institutes of Health (NIH), improve the safety of pregnant survivors.

9.6 Recommendations and Conclusions

While ongoing studies evaluate the transmission potential of EVD survivors and the implications of Ebola viral persistence on pregnancy outcomes, it is important to guide women, their health care providers, and their support systems based on available information about both Ebola and other viral infections. Providing supportive environments, both during the antenatal visits and in day-to-day social interactions, will promote health of convalescent mothers and their children. Here, we summarize recommendations from our review of existing models and guidelines.

9.6.1 Transmission

Encouraging the implementation of culturally competent public awareness campaigns, the use of contraceptives, and pharmaceutical approaches such as use of antiretroviral therapy (ART) as pre-exposure prophylaxis (PrEP) are recommended as effective ways to help prevent sexual transmission of HIV. Guidelines for women who are pregnant or intend to become pregnant should be established to inform healthy practices and effective health care.

To prevent mother-to-child transmission, supportive guidance regarding reproductive choices, antenatal testing with relevant counselling, prenatal antiretroviral therapy, use of appropriate delivery and infant feeding methods, and infant prophylaxis and long-term follow-up are advised. To prevent mother-to-child transmission, health care workers should be educated concerning risk for and prevention of blood-borne infections and implement standard precautions and protocols against exposure.

9.6.2 Education

Education interventions are recommended to both reduce stigma among the general public and to inform those infected as well as health care practitioners how to best manage the disease; optimally, all three populations should be addressed in some manner. Ideal interventions might involve multimedia platforms, distribution of accessible informative material, repeated exposure to information, the provision of practical tools for health care workers, and the participants' active involvement in some activity.

9.6.3 Support

Providing convenient opportunities for participation in support programs by having support sessions coincide with infant's appointments and adding maternal health programs in clinics providing other services, such as multidisciplinary care and treatment for behaviors linked with high infection rates (such as illicit drug use for hepatitis) is advisable to provide adequate support for pregnant or postpartum-infected women. Additionally, supplying optimal environments, particularly those that provide for gender-related health care needs, may increase attendance at appointments and program sessions.

The formation of therapeutic relationships with either health care practitioners or peer mentors—trained individuals infected with the disease and considered to be positive role models—is recommended to reduce pain and to provide support and information. Peer mentor programs might involve small-group sessions with discussions and activities.

References

Bausch, D., Towner, J., Dowell, S., Kaducu, F., Lukwiya, M., Sanchez, A., et al. (2007). Assessment of the risk of Ebola virus transmission from bodily fluids and fomites. *Journal of Infectious Diseases, 196*(Suppl 2), S142–S147.

Black, B. O., Caluwaerts, S., & Achar, J. (2015). Ebola viral disease and pregnancy. *Obstetric Medicine, 8*(3), 108–113.

Centers for Disease Control and Prevention. (2015). *Sexual transmission and viral hepatitis*. Retrieved February 24, 2017, from https://www.cdc.gov/hepatitis/populations/stds.htm.

Centers for Disease Control and Prevention. (2016). *2014 Ebola outbreak in West Africa—Case counts*. Retrieved April 13, 2016, from https://www.cdc.gov/vhf/ebola/outbreaks/2014-west-africa/case-counts.html.

Chen, G., Block, J., Evans, A., Huang, P., & Cohen, C. (2014). Gateway to care campaign: A public health initiative to reduce the burden of hepatitis B in Haimen City, China. *BMC Public Health, 14*, 754.

Chughtai, A. A., Barnes, M., & Macintyre, C. R. (2016). Persistence of Ebola virus in various body fluids during convalescence: Evidence and implications for disease transmission and control. *Epidemiology and Infection, 144*(8), 1652–1660.

Cohen, M. S. (2007). Preventing sexual transmission of HIV. *Clinical Infectious Diseases, 45*(Suppl 4), S287–S292.

Dokubo, E.K., Wendland, A., Mate, S.E., Ladner, J.T., Hamblion, E.L., Raftery,, P., et al. (2018). Persistence of Ebola virus after the end of widespread transmission in Liberia: an outbreak report. *The Lancet Infectious Diseases, 18*(9), 1015–1024. Retrieved November 15, 2018, from https://www.thelancet.com/journals/lancet/article/PIIS1473-3099(18)30417-1/fulltext.

Dunn, A., Walker, T., Redd, J., Sugerman, D., McFadden, J., Singh, T., et al. (2016). Nosocomial transmission of Ebola virus disease on pediatric and maternity wards: Bombali and Tonkolili, Sierra Leone, 2014. *American Journal of Infection Control, 44*(3), 269–272.

Evans, A., Cohen, C., Huang, P., Qian, L., London, W. T., Block, J., et al. (2015). Prevention of perinatal hepatitis B transmission in Haimen City, China: Results of a community public health initiative. *Vaccine, 33*(26), 3010–3015.

Fallah, M., Skrip, L., Dahn, B., Nyenswah, T., Flumo, H., Glayweon, M., et al. (2016). Pregnancy outcomes in Liberian women who conceived after recovery from Ebola virus disease. *Lancet Global Health, 4*(10), e678–e679. https://doi.org/10.1016/S2214-109X(16)30147-4.

Fischer, W. A., Brown, J., Wohl, D. A., Loftis, A. J., Tozay, S., Reeves, E., et al. (2017). Ebola virus ribonucleic acid detection in semen more than two years after resolution of acute Ebola virus infection. *Open Forum Infectious Diseases, 4*(3). Retrieved September 26, 2018, from https://academic.oup.com/ofid/article/4/3/ofx155/4004818

French, H., Greeff, M., Watson, M., & Doak, C. (2015). A comprehensive HIV stigma-reduction and wellness-enhancement community intervention: A case study. *Journal of the Association of Nurses in AIDS Care: JANAC, 26*(1), 81–96.

Haque, A., Hober, D., & Blondiaux, J. (2015). Addressing therapeutic options for Ebola virus infection in current and future outbreaks. *Antimicrobial Agents and Chemotherapy, 59*(10), 5892–5902.

Huggins, J., Zhang, Z. X., & Bray, M. (1999). Antiviral drug therapy of filovirus infections: S-adenosylhomocysteine hydrolase inhibitors inhibit Ebola virus in vitro and in a lethal mouse model. *The Journal of Infectious Diseases, 179*(Suppl 1), S240–S247.

Jorgensen, C., Carnes, C. A., & Downs, A. (2016). "Know More Hepatitis:" CDC's national education campaign to increase hepatitis C testing among people born between 1945 and 1965. *Public Health Reports, 131*(Suppl 2), 29–34.

Kissinger, P., Clark, R., Rice, J., Kutzen, H., Morse, A., & Brandon, W. (1995). Evaluation of a program to remove barriers to public health care for women with HIV infection. *Southern Medical Journal, 88*(11), 1121–1125.

Laga, M., & Piot, P. (2012). Prevention of sexual transmission of HIV: Real results, science progressing, societies remaining behind. *AIDS, 26*(10), 1223–1229.

Mason, K., Dodd, Z., Sockalingam, S., Altenberg, J., Meaney, C., Millson, P., et al. (2015). Beyond viral response: A prospective evaluation of a community-based, multi-disciplinary, peer-driven model of HCV treatment and support. *The International Journal on Drug Policy, 26*(10), 1007–1013.

Mate, S., Kugelman, J., Nyenswah, T., Ladner, J., Wiley, M., Cordier-Lassalle, T., et al. (2015). Molecular evidence of sexual transmission of Ebola virus. *New England Journal of Medicine, 373*(25), 2448–2454.

Mattia, J., Vandy, M., Chang, J., Platt, D., Dierberg, K., Bausch, D., et al. (2016). Early clinical sequelae of Ebola virus disease in Sierra Leone: A cross-sectional study. *Lancet Infectious Diseases, 16*(3), 331–338.

McMahon, J., Myers, J., Kurth, A., Cohen, S., Mannheimer, S., Simmons, J., et al. (2014). Oral pre-exposure prophylaxis (PrEP) for prevention of HIV in serodiscordant heterosexual couples in the United States: Opportunities and challenges. *AIDS Patient Care and STDs, 28*(9), 462–474.
Money, D., Tulloch, K., Boucoiran, I., Caddy, S., Infectious Diseases Committee, Yudin, M., et al. (2014). Guidelines for the care of pregnant women living with HIV and interventions to reduce perinatal transmission: Executive summary. *Journal of Obstetrics and Gynaecology Canada, 36*(8), 721–751.
Moores, L. K., Dellert, E., Baumann, M. H., Rosen, M. J., & American College of Chest Physicians Health and Science Policy Committee. (2009). Executive summary: Effectiveness of continuing medical education: American College of Chest Physicians Evidence-Based Educational Guidelines. *Chest, 135*(3 Suppl), 1S–4S.
Myung, P., Mallette, C., Taylor, L., Allen, S., & Feller, E. (2003). Sexual transmission of hepatitis C: Practical recommendations. *Medicine and Health, Rhode Island, 86*(6), 168–171.
Nyenswah, T., Kateh, F., Bawo, L., Massaquoi, M., Gbanyan, M., Fallah, M., et al. (2016). Ebola and its control in Liberia, 2014–2015. *Emerging Infectious Diseases, 22*(2), 169.
Parker, R., Jelsma, J., & Stein, D. J. (2016). Managing pain in women living with HIV/AIDS: A randomized controlled trial testing the effect of a six-week peer-led exercise and education intervention. *Journal of Nervous and Mental Disease, 204*(9), 665–672.
Pauw, J., Ferrie, J., Rivera Villegas, R., Medrano Martínez, J., Gorter, A., & Egger, M. (1996). A controlled HIV/AIDS-related health education programme in Managua, Nicaragua. *AIDS, 10*(5), 537–544.
Rabelo, I., Lee, V., Fallah, M., Massaquoi, M., Evlampidou, I., Crestani, R., et al. (2016). Psychological distress among Ebola survivors discharged from an Ebola treatment unit in Monrovia, Liberia—A qualitative study. *Frontiers in Public Health, 4*, 142.
Richards, P., Amara, J., Ferme, M. C., Kamara, P., Mokuwa, E., Sheriff, A. I., et al. (2015). Social pathways for Ebola virus disease in rural Sierra Leone, and some implications for containment. *PLoS Neglected Tropical Diseases, 9*(4), e0003567.
Richter, L., Rotheram-Borus, M. J., Van Heerden, A., Stein, A., Tomlinson, M., Harwood, J. M., et al. (2014). Pregnant women living with HIV (WLH) supported at clinics by peer WLH: A cluster randomized controlled trial. *AIDS and Behavior, 18*(4), 706–715.
Robotin, M., Patton, Y., & George, J. (2013). Getting it right: The impact of a continuing medical education program on hepatitis B knowledge of Australian primary care providers. *International Journal of General Medicine, 6*, 115–122.
Rodriguez, L. L., De Roo, A., Guimard, Y., Trappier, S. G., Sanchez, A., Bressler, D., et al. (1999). Persistence and genetic stability of Ebola virus during the outbreak in Kikwit, Democratic Republic of the Congo, 1995. *Journal of Infectious Diseases, 179*(Suppl 1), S170–S176.
Rotheram-Borus, M.-J., Richter, L., Van Rooyen, H., van Heerden, A., Tomlinson, M., Stein, A., et al. (2011). Project Masihambisane: A cluster randomised controlled trial with peer mentors to improve outcomes for pregnant mothers living with HIV. *Trials, 12*, 2. https://doi.org/10.1186/1745-6215-12-2.
Saleh, M. A., al-Ghamdi, Y. S., al-Yahia, O. A., Shaqran, T. M., & Mosa, A. R. (1999). Impact of health education program on knowledge about AIDS and HIV transmission in students of secondary schools in Buraidah city, Saudi Arabia: An exploratory study. *Eastern Mediterranean Health Journal, 5*(5), 1068–1075.
Sissoko, D., Keïta, M., Diallo, B., Aliabadi, N., Fitter, D. L., Dahl, B. A., et al. (2017). Ebola virus persistence in breast milk after no reported illness: A likely source of virus transmission from mother to child. *Clinical Infectious Diseases, 64*(4), 513–516.
Whitmer, S.L.M., Ladner, J.T., Wiley, M.R, Patel, K., Dudas, G., Rambaut, A., et al. (2018). Active Ebola virus replication and heterogeneous evolutionary rates in EVD survivors. *Cell Reports, 22*(5), 1159–1168. Retrieved November 24, 2018, from https://www.cell.com/cellreports/pdf/S2211-1247(18)30008-1.pdf.
Varkey, J. B., Shantha, J. G., Crozier, I., Kraft, C. S., Lyon, G. M., Mehta, A. K., et al. (2015). Persistence of Ebola virus in ocular fluid during convalescence. *New England Journal of Medicine, 372*(25), 2423–2427.
Wilson, H. R. (2003). Report from the CDC: Hepatitis B and you: A patient education resource for pregnant women and new mothers. *Journal of Women's Health, 12*(5), 437–441.
World Health Organization. (2015). Ebola virus disease in pregnancy: Screening and management of Ebola cases, contacts and survivors, interim guidance.
World Health Organization. (2016). *Interim advice on the sexual transmission of the Ebola virus disease*. Retrieved September 6, 2018, from http://www.who.int/reproductivehealth/topics/rtis/ebola-virus-semen/en/.

10 Maternal and Infant Survival Following Ebola Infection: Their Exclusion from Treatment and Vaccine Trials and "*Primum non nocere*"

David A. Schwartz

10.1 Maternal and Infant Outcomes During Past Ebola Virus Epidemics

Ebola virus disease is one of the most lethal acute infections that occur anywhere in the world. Since the first reported outbreak of Ebola virus disease (EVD in the Democratic Republic of Congo (DRC) (then Zaire) in 1976 (Breman et al. 2016), there have been approximately 27 recorded incidents, clusters and outbreaks of the infection that have occurred. With an average mortality rate in the general population of 65%, there are very few naturally occurring infectious agents that produce death as rapidly and efficiently as does the Ebola virus (Lefebvre et al. 2014).

In the first recognized outbreak of Ebola virus in the area of Yambuku, Zaire in 1976, treatment was restricted to supportive measures, and the overall case fatality rate (CFR) was 88% (WHO 2018). In the clusters and epidemics of infection that followed, the overall CFR has varied from below 50% to as high as 90%, with most outbreaks having greater than one-half of all infected persons dying of Ebola regardless of gender, age, or pregnancy status.

Pregnant women, their unborn fetuses, and neonates are often the most vulnerable segment of a society during infectious disease outbreaks. Although there are only limited data on clinical outcomes of infected pregnant women and their infants from past Ebola epidemics, in cases where these figures were available, the outlook for their survival has not been optimistic. Examining the clinical outcomes of pregnant women during the 1976 Zaire epidemic, there were 73 deaths among the 82 pregnant women infected with Ebola virus, a CFR of 89% (International Commission 1978). During the Kikwit, Zaire, epidemic of 1995, 14 of 15 pregnant women with EVD died—a CFR of 93% (Gomes et al. 2017). In past epidemics, being pregnant with Ebola virus infection also caused poor fetal and neonatal outcomes. During the 1976 Zaire epidemic, spontaneous abortion occurred in 18 of 73 pregnant women who died from EVD and in 1 of 9 infected women who survived. There were 11 live births to mothers infected with Ebola virus—none of the infants survived beyond 19 days of life (Gomes et al. 2017). During the Kikwit epidemic of 1995, in 10 of the 15 women who were pregnant and infected with Ebola virus, the pregnancy resulted in spontaneous abortion (67%), with an additional woman delivered at 32 weeks gestation with a stillborn infant, and four women who died together with their conceptuses during the third trimester. All of the women had uterine or vagi-

D. A. Schwartz (✉)
Department of Pathology, Medical College of Georgia, Augusta University, Augusta, GA, USA

D. A. Schwartz et al. (eds.), *Pregnant in the Time of Ebola*, Global Maternal and Child Health,
https://doi.org/10.1007/978-3-319-97637-2_10

nal bleeding and were at heightened risk for having a spontaneous abortion and pregnancy-related hemorrhage (Gomes et al. 2017; Jamieson et al. 2014; Mupapa et al. 1999). In Gulu, Uganda, a pregnant woman infected with Ebola virus survived, but had a spontaneous abortion at 28 weeks gestation (Muehlenbachs et al. 2016). Following the 2000–2001 EVD outbreak in Uganda, four Ebola virus-infected mother-baby pairs were identified—all of the mothers and infants had died (Francesconi et al. 2003; Gomes et al. 2017; Nelson et al. 2016). A pregnant woman died of EVD at 7 months gestation during the Isiro, DRC outbreak in 2012—the neonate was prematurely delivered on day 6 of the mother's disease and died on the 8th day of life; the mother died the day after delivery (Kratz et al. 2015).

As can be seen, prior to the West African Ebola Epidemic, the prognosis for pregnant women was grim—from 89 to 93% of pregnant women with Ebola virus infection died (Caluwaerts et al. 2016). The prognosis was even worse for their fetuses and infants—no neonates survived for greater than 19 days, representing a 100% infant mortality rate (Caluwaerts et al. 2016).

Garba et al. (2017) evaluated the published literature from 12 studies occurring both prior to the West African epidemic and during it. They identified 108 pregnant women with EVD in five countries—DRC, Liberia, Sierra Leone, Guinea, and Uganda—including six case reports, two cross-sectional studies, three retrospective studies, and one technical report. Among all countries and outbreaks, they found that 91 of the 108 pregnant women died, a case fatality rate of 84.3%. The survival rate among the 15 women who had spontaneous abortions, stillbirths, or induced deliveries was 100%.

10.2 Maternal Survival During the West African Ebola Epidemic

The Ebola epidemic that occurred from 2013 to 2015 in the West African countries of Guinea, Liberia, and Sierra Leone was the largest outbreak of a hemorrhagic fever ever to have occurred. There were 28,652 suspected, probable, and confirmed cases of infection, and 11,325 deaths (CDC 2016, 2017), both from rural and, for the first time, urban areas. Throughout this epidemic, estimates of mortality have shown significant variation depending on the populations studied and the methodology utilized, and it appears likely that the true numbers of infected persons and deaths were higher than those reported (Garske et al. 2017). Following analyses performed after the close of the epidemic, case survival data showed that the overall mortality rate among all ages and genders was 62.9%—it underwent a decline from 69.8% to approximately 39% during the period from July to September 2015. In all three West African countries, the greatest likelihood of mortality were among the elderly (>75 years, 83.8%) and the young (<5 years, 75.6%) (Garske et al. 2017). Both males and females were equally susceptible to Ebola infection, and females had a slight but significant decrease in overall mortality (63%) when compared with men (67.1%) (WHO Ebola Response Team et al. 2016).

Due to the multinational and rapid spread of the epidemic, the emergency response by numerous independent organizations from diverse parts of the world, and the state of the national health care systems in the involved countries, there was unfortunately no organized methodology in place for the surveillance, identification, and registration of women who were pregnant. To add to the confusion, testing of women suspected of or having EVD for their pregnancy status was not routinely performed (Henwood et al. 2017). As a result, estimations for the determination of maternal and infant mortality rests upon independent reports from the multiple independent parties involved.

At the start of the West Africa Ebola epidemic, it appeared that the pregnancy-specific CFR was extremely high—as much as 90% or greater (Caluwaerts and Kahn 2019). It was also believed by some that based upon previous epidemics, pregnant women would be more susceptible to the Ebola virus and have a higher mortality rate (Doucleff 2014; Lang 2014). In an interview conducted early in the outbreak, it was reported that one nongovernmental organization had opined that the survival rate for expectant mothers was virtually zero (Health24 2015). In a report published in 2015, the likelihood of maternal and infant survival of EVD was summarized as follows, "*Present data suggests that maternal mortality remains high (approximately 95%) and peri-natal mortality virtually 100% for infected pregnant women*" (Black 2015).

However, as the epidemic continued, and despite continued maternal deaths occurring around the period of delivery and into the postpartum period, the number of surviving mothers was found to be higher than expected based upon previous Ebola outbreaks. Two pregnant women were reported to have survived an episode of EVD with their 2nd trimester fetuses still alive; following negative maternal PCR tests for Ebola virus the mothers were discharged, and subsequently their fetuses were stillborn (Caluwaerts et al. 2016). By the end of the epidemic, of the 77 pregnant women with EVD treated by Médecins Sans Frontières, there were 36 who survived—a case fatality rate of 47% (Caluwaerts and Kahn 2019). In a review of published reports, Bebell (in print) found that 11 of 25 pregnant women infected with EVD died—a CFR of 44%. According to another review of reports of maternal deaths associated with the Ebola epidemic, Gomes et al. (2017) found the CFR for pregnant women with EVD to be 55% (44 deaths among 80 pregnancies), excluding approximately 20 women who had been administered the rVSV-ZEBOV vaccine (Henao-Restrepo et al. 2017). All pregnant women who managed to survive Ebola infection had poor obstetrical outcomes—spontaneous miscarriage, elective abortion, stillbirth, or a neonatal death. Two women died from EVD with their fetuses still in utero (Gomes et al. 2017).

10.3 Fetal and Infant Survival During the West African Ebola Epidemic

From the very beginning of the West Africa Ebola epidemic, it was not generally expected that any infected fetuses or newborns would survive. In their review, Garba et al. examined the published results of infant outcomes from the 108 pregnancies both before and during the West African Ebola epidemic—there was only one fetal survival (Garba et al., 2017). This sole surviving neonate, Baby Nubia, has been extensively profiled (Caluwaerts 2017; Fox 2015; Médecins Sans Frontières 2015; Schnirring 2015). Her 27-year-old mother was found to be Ebola virus-positive in Forécariah province, Guinea, near the close of the epidemic. The mother was a follow-up household contact of a person who had died of EVD. When she visited the local Ebola clinic very ill, bleeding from her gums and with a high viral load, they declined to treat her because of threat of contagion from the body fluids from pregnant women. At that time, a potentially highly effective vaccine was being tested and she could have received the vaccine along with the other persons who had been in contact with the dead women. However, she was denied access to vaccination because of her pregnant status. The pregnant woman was transferred by ambulance to an Ebola treatment center (ETC) managed by Médecins Sans Frontières (MSF) in Nongo, Conakry. She was eligible to receive a new antiretroviral drug manufactured by Mapp Pharmaceuticals, Zmapp, which at the time was undergoing clinical trials in Guinea as well as other countries - and pregnant women were eligible for inclusion in this trial (Caluwaerts 2017). However, it was only possible to obtain Zmapp through a randomized clinical trial in which she had a 50–50 probability of not receiving the agent for ethical reasons. MSF attempted to obtain Zmapp for her outside of the clinical trial, but was refused (Caluwaerts 2017). At the ETC, the mother was administered favipiravir, an experimental antiviral drug which could be administered to pregnant women in agreement with the manufacturer, Toyama Chemical Company of Japan, under a MEURI (Monitored Emergency Use of Unregistered and Experimental Interventions) protocol. Four days after admission, she went into labor and delivered a liveborn baby girl, Baby Nubia, but the mother subsequently died of postpartum hemorrhage and disseminated intravascular coagulation caused by her EVD (Caluwaerts 2017; Caluwaerts and Baylis 2016). The infant, who tested positive for the Ebola virus, was administered Zmapp a few hours after birth—the drug was obtained outside of the clinical trial by Médecins Sans Frontières, along with GS5734 (Gilead Sciences; an experimental broad-spectrum antiviral), and the white blood cells (buffy coat) taken from an Ebola survivor. Thus, the mother was denied the potentially beneficial ZMapp, yet her newborn was given it after birth. Baby Nubia eventually recovered, the only known infected infant to have survived during the West African epidemic (Caluwaerts and Baylis 2016; Caluwaerts 2017; Dörnemann et al. 2017).

10.4 Pregnant Women in Experimental Clinical Trials

The Latin aphorism *primum non nocere*—first do no harm—has been taught as a maxim to physicians in training for many years. Although thought by some as part of the Hippocratic Oath, an oath taken by medical students at the time of their becoming physicians, the exact phrase does not actually appear in that ancient document. However, a similar concept is present in the Oath of Hippocrates which states "to abstain from doing harm" (ἐπὶ δηλήσει δὲ καὶ ἀδικίῃ εἴρξειν). In whichever form is preferred, the meaning is the same—physicians must be cognizant to carefully consider the potential harm of a treatment versus its potential therapeutic efficacy. In other words, given an illness, it may be better to withhold a specific treatment, or even do nothing at all, rather than to intervene and cause harm to one's patient.

In modern times, pregnancy has always been considered as a potentially hazardous period for the administration of medications. In can be very challenging to be certain of the safety of the mother and developing embryo and fetus of drugs, and thus caution has traditionally guided their usage during pregnancy. The possibility of side effects occurring in the mother may have a deleterious effect on progress of the pregnancy, labor, and delivery; transplacental passage of the drug may result in teratogenic, developmental, and toxic effects to the fetus; intrauterine fetal demise may occur; and there is the danger of passage of the drug to the newborn infant following delivery through breast milk (Smithells 1978). The potential danger of drug administration during pregnancy was perhaps best illustrated during the 1950s and early1960s when congenital limb malformations, termed phocomelia, developed in many thousands of infants from mothers who were given the drug thalidomide as a treatment for morning sickness during pregnancy (Fintel et al. 2009).

No maternal vaccine has been proven to result in birth defects, and inactivated vaccines are usually considered safe to be administered during pregnancy. Even the inadvertent use of live virus vaccines has only rarely been associated with harm to the fetus (Ryan et al. 2008). In 2003 it was reported that 103 women inadvertently received the smallpox vaccine - a live vaccinia virus vaccine - either while pregnant or just before becoming pregnant (CDC 2003). Although two women had early miscarriages, there was no evidence that they were associated with the administration of the vaccine, and there were no other pregnancy complications or incidents of fetal vaccinia infection.

The emergence of the acquired immunodeficiency syndrome (AIDS) pandemic in the 1980s and the recognition of vertical transmission of the human immunodeficiency virus (HIV) provided investigators impetus to enroll pregnant women in the early phases of antiretroviral drug trials, even prior to the completion of experimental animal studies. This was because the life-threatening nature of AIDS was believed to justify an unknown risk to the fetus in order to potentially extend the life of the mother (Merkatz et al. 1993).

The conservative attitude limiting the enrollment of pregnant women in drug and vaccine trials has persisted to the present times. Even the participation of nonpregnant women in experimental trials remained restricted until 1993, when the policy was changed so that both sexes were recruited into drug studies in order to determine gender-based differences (U.S. Food and Drug Administration 2018). This policy shift did not include pregnant women—less than 20 years ago, the United State

Food and Drug Administration (FDA) still had a policy of excluding from experimental drug trials those women "of childbearing potential" (Macklin 2010). More recently, women are being included in drug trials for non-obstetric conditions with the proviso that they are not pregnant and do not intend to become pregnant including the use of birth control (Shields and Lyerly 2013). Perhaps, the clearest and most permissive clearance for the participation of pregnant women in experimental studies originates with the UNAIDS/WHO ethical guidance for HIV prevention trials (UNAIDS, WHO 2012). Guidance Point 9 of this document states

> *Researchers and trial sponsors should include women in clinical trials in order to verify safety and efficacy from their standpoint, including immunogenicity in the case of vaccine trials, since women throughout the life span, including those who are sexually active and may become pregnant, be pregnant or be breastfeeding, should be recipients of future safe and effective biomedical HIV prevention interventions. During such research, women's autonomy should be respected and they should receive adequate information to make informed choices about risks to themselves, as well as to their foetus or breastfed infant, where applicable.*

However, pregnant women are usually still excluded from experimental trial of drugs and vaccines that do not target obstetric conditions.

10.5 Exclusion of Pregnant Women and Infants from Experimental Trials During Ebola: *Primum non nocere*

Prior to the West African Ebola epidemic, there had been no specific or successful forms of therapy available for Ebola virus disease, and medical treatment consisted of supportive measures. At the beginning of the epidemic, there were no specific drugs or vaccines approved for treatment of Ebola infection—no specific therapies had been developed and approved for use from the prior Ebola epidemics. Treatment consisted solely of administration of fluid and electrolyte levels, maintenance of adequate blood pressure, analgesia, usage of blood products, and management of coagulopathy, secondary infections, and other conditions. Although there were vaccines in the early stages of development and testing, only a few had entered the already Phase I stage of testing for safety and efficacy (Alirol et al. 2017; Jones et al. 2005; Qiu et al. 2009)

However, as the West African outbreak progressed, for the first time in almost 40 years since the virus was first discovered in 1976, experimental therapies were available to some persons with EVD either through organized trials or compassionate use. The trials were done through a committee—by the WHO-ERC (World Health Organization Research Ethics Review Committee)—that evaluated and approved proposed studies. They reviewed 24 new and 22 amended protocols for experimental studies that included both interventional (drug, vaccine) and observational studies (Alirol et al. 2017). One of the most problematic decisions faced by the committee members in the design, approval, and implementation of these protocols was the inclusion of pregnant women and infants (Alirol et al. 2017). It was recognized that exclusion of pregnant women and their infants from drug and vaccine trials undermined ethical principles of justice—fairness, equity, and maximization of benefit. In addition, exclusion of these groups from clinical trials would deny them the potential life-saving benefits from an infection with a high mortality rate, at the time believed to be 100% in fetuses and neonates and approximately 90% in pregnant women (WHO Ebola Response Team et al. 2015). These survival data were of immediate relevance in considering the risk/benefit relationship in enrolling pregnant women and infants in clinical trials, as addressed by the WHO Ethics Working Group meeting 20–21 October 2014 (WHO 2014):

> *"It is ethically important to ensure that vulnerable populations such as pregnant women and those with diminished autonomy such as children or those with mental incapacities are not arbitrarily excluded from trials. Instead their inclusion into clinical trials should be guided by a risk benefit analysis and the ability to secure adequate consent."*

Clinical outcomes data from previous epidemics showing the high CFRs for pregnant women and the almost certain deaths of their fetuses and newborns led the WHO committee to support the inclusion of pregnant women together with their fetuses in the planned clinical trials of both drugs and vaccines. This decision was based not only upon the incredibly high CFRs that were to be expected in mothers, fetuses, and neonates, but also because pregnant mothers had a greater interest in and right to decide the fate of themselves and their unborn children than did administrators, sponsors, investigators, or committee members, and pregnant women should be granted the same rights for decision-making as nonpregnant women (Gomes et al. 2017).

As the Ebola epidemic came to a close, the WHO-ERC had reviewed 14 interventional trial protocols and two MEURI proposals (Alirol et al. 2017). All of the protocols for vaccine clinical trials excluded pregnant women. The proposed clinical trials for two promising antiviral drugs—brincidofovir and favipiravir—also excluded women who were pregnant for a variety of reasons. Despite efforts by the WHO-ERC as well as the MSF Ethics Review Board and Inserm Institutional Review Board to the applicants to reconsider the exclusion of pregnant women, the need for rapid implementation of the trials in the field took priority over the delays that would have been encountered in pursuing revision of the protocols to include pregnant women (Alirol et al. 2017; Gomes et al. 2017). And even when the clinical trial of the experimental live attenuated vaccine rVSVΔG/ZEBOV-GP (Merck) demonstrated protective effects in nonpregnant adults, and the WHO-ERC and Data Safety Monitoring Board requested that pregnant women receive the vaccine, 42 pregnant women were denied participation in the trial (Gomes et al. 2017; Henao-Restrepo et al. 2017). Interestingly, because pregnancy tests were not routinely performed, and pregnant women were identified on the basis of self-reporting, greater than 20 pregnant women were administered the vaccine (Gomes et al. 2017; Henao-Restrepo et al. 2017).

Enrollment into clinical trials and access to potentially life-saving drugs and vaccines for pregnant women were difficult in spite of the obvious risk/benefit considerations; in some cases, the pharmaceutical corporations that produced the products would simply not permit their administration to women who were pregnant, or the insurers would not provide insurance to pregnant women (Caluwaerts 2017). By the end of the West African Ebola epidemic, pregnant women, their fetuses, and newborns had been systematically excluded from all drug and vaccine clinical trials. Eventually, pregnant women received access to favipiravir, but only following extensive negotiations between the manufacturer and Médecins Sans Frontières. The sole surviving newborn with EVD, Baby Nubia, received her treatment with ZMapp outside of the clinical trial (Caluwaerts 2017).

10.6 Future Outbreaks

No one can argue that pregnant women, fetuses, and their future children should be protected from the potential effects, either known or unknown, of drugs and vaccines that are in the experimental stages of evaluation. However, compassion and ethics dictate that when a pregnant woman has a life-threatening condition, and especially when it is more likely that she and her unborn infant will die than survive, there must be a paradigm shift in the response of the medical establishment to their treatment. There are multiple reasons to include pregnant women in future drug and vaccine studies for life-threatening infections besides their own survival. Arguably, one of the most important reasons is for the collection of data on clinical efficacy and survival of drugs and vaccines under controlled scientific circumstances.

No mother-baby pair has ever survived an Ebola virus infection, and yet pregnant women remain restricted from participating in EVD drug and vaccine trials. Ethical as well as humanitarian considerations mandate a reevaluation of the exclusion of pregnant women and their fetuses from clinical trials of drugs and vaccines for Ebola virus infection in future outbreaks.

References

Alirol, E., Kuesel, A.C., Guraiib, M.M., dela Fuente-Núñez, V., Saxena, A., Gomes, M.F. (2017). Ethics review of studies during public health emergencies—The experience of the WHO ethics review committee during the Ebola virus disease epidemic. *BMC Medical Ethics, 18*, 43. Retrieved May 16, 2018, from https://www.ncbi.nlm.nih.gov/pmc/articles/PMC5485606/#.

Bebell, L. B. (in print). Ebola virus disease and pregnancy: Perinatal transmission and epidemiology. In D. A. Schwartz, J. Anoko, & S. Abramowitz (Eds.), *Pregnant in the time of Ebola. Women and their children in the 2013-2015 epidemic*. New York: Springer.

Black, B. (2015). *Principles of management for pregnant women with Ebola: A western context*. Retrieved May 12, 2018, from www.rcog.org.uk/globalassets/documents/news/ebola-and-pregnancy-western.pdf.

Breman, J. G., Heymann, D. L., Lloyd, G., McCormick, J. B., Miatudila, M., Murphy, F. A., et al. (2016). Discovery and description of Ebola Zaire virus in 1976 and relevance to the West African epidemic during 2013-2016. *Journal of Infectious Diseases, 214*(Suppl 3), S93–S101. https://www.ncbi.nlm.nih.gov/pubmed/27357339.

Caluwaerts, S. (2017). Nubia's mother: Being pregnant in the time of experimental vaccines and therapeutics for Ebola.*Reproductive Health, 14*(Suppl 3), 17. https://doi.org/10.1186/s12978-017-0429-8. Retrieved May 15, 2018, https://www.ncbi.nlm.nih.gov/pmc/articles/PMC5751508/.

Caluwaerts, S., & Baylis, F. (2016). Nubia—An Ebola survivor orphaned by vaccine policy. *Impact Ethics*. Retrieved May 20, 2018, from https://impactethics.ca/2016/11/25/nubia-an-ebola-survivor-orphaned-by-vaccine-policy/.

Caluwaerts, S. & Kahn, P. (2019). Ebola and pregnant women: Providing maternity care at Médecins Sans Frontières treatment centers. In D. A. Schwartz, J. N. Anoko, & S.A. Abramowitz (Eds.), *Pregnant in the time of Ebola: Women and their children in the 2013-2015 West African epidemic*. New York: Springer.

Caluwaerts, S., Fautsch, T., Lagrou, D., Moreau, M., Camara, A.M., Günther, S., et al. (2016). Dilemmas in managing pregnant women with Ebola: 2 case reports. *Clinical Infectious Diseases*, 62(7), 903–905. https://doi.org/10.1093/cid/civ1024.

CDC. (2016). *2014–2016 Ebola outbreak in West Africa*. Retrieved May 12, 2018, from https://www.cdc.gov/vhf/ebola/outbreaks/2014-west-africa/index.html.

CDC. (2017). *2014–2016 Ebola outbreak in West Africa*. Retrieved May 13, 2018, from https://www.cdc.gov/vhf/ebola/history/2014-2016-outbreak/index.html.

CDC. (2003). Women with smallpox vaccine exposure during pregnancy reported to National Smallpox Vaccine in Pregnancy Registry—United States, 2003. *Morbidity and Mortality Weekly Report*, 52(17), 386–388.

Dörnemann, J., Burzio, C., Ronsse, A., Sprecher, A., De Clerck, H., Van Herp, M., et al. (2017). First newborn baby to receive experimental therapies survives Ebola virus disease. *Journal of Infectious Diseases, 215*(2), 171–174.

Doucleff, M. (2014). Dangerous deliveries: Ebola leaves moms and babies without care. *NPR*. Retrieved May 9, 2018, from https://www.npr.org/sections/goatsandsoda/2014/11/18/364179795/dangerous-deliveries-ebola-devastates-womens-health-in-liberia.

Fintel, B., Samaras, A. T., & Carias, E. (2009). The thalidomide tragedy: Lessons for drug safety and regulation. *Helix Magazine*. Retrieved May 12, 2018, from https://helix.northwestern.edu/article/thalidomide-tragedy-lessons-drug-safety-and-regulation.

Fox, M. (2015). 'A Good Warrior': Newborn beats Ebola in Guinea. *NBC News*. Retrieved May 14, 2018, from https://www.nbcnews.com/storyline/ebola-virus-outbreak/good-warrior-newborn-beats-ebola-guinea-n467331.

Francesconi, P., Yoti, Z., Declich, S., Onek, P. A., Fabiani, M., Olango, J., et al. (2003). Ebola hemorrhagic fever transmission and risk factors of contacts, Uganda. *Emerging Infectious Diseases, 9*(11), 1430–1437.

Garske, T., Cori, A., Ariyarajah, A., Blake, I. M., Dorigatti, I., Eckmanns, T., et al. (2017). Heterogeneities in the case fatality ratio in the west African Ebola outbreak 2013–2016. *Philosophical Transactions of the Royal Society, B: Biological Sciences, 372*(1721), 20160308. Retrieved May 10, 2018, from https://www.ncbi.nlm.nih.gov/pmc/articles/PMC5394646/.

Gomes, M. F., de la Fuente-Núñez, V., Saxena, A., Kuesel, A. C. (2017). Protected to death: Systematic exclusion of pregnant women from Ebola virus disease trials. *Reproductive Health, 14* (Suppl 3), 172. Retrieved May 9, 2018, from https://www.ncbi.nlm.nih.gov/pmc/articles/PMC5751665/#CR19.

Garba, I., Dattijo, L. M., Habib, A. G. (2017). Ebola virus disease and pregnancy outcome: A review of the literature. *Tropical Journal of Obstetrics and Gynaecology, 34*(1), 6–10.

Health24. (2015). *Ebola health workers face life or death decision on pregnant women*. Retrieved May 9, 2018, from https://www.health24.com/Medical/Infectious-diseases/Ebola/Ebola-health-workers-face-life-or-death-decision-on-pregnant-women-20150115.

Henao-Restrepo, A. M., Camacho, A., Longini, I. M., Watson, C. H., Edmunds, W. J., Egger, M., et al. (2017). Efficacy and effectiveness of an rVSV-vectored vaccine in preventing Ebola virus disease: Final results from the Guinea ring vaccination, open-label, cluster-randomised trial (Ebola Ça Suffit!). *The Lancet, 389*(10068), 505–518. https://doi.org/10.1016/S0140-6736(16)32621-6. Retrieved May 10, 2018, from https://www.ncbi.nlm.nih.gov/pmc/articles/PMC5364328/.

Henwood, P. C., Bebell, L. M., Roshania, R., Wolfman, V., Mallow, M., Kalyanpur, A., et al. (2017). Ebola virus disease and pregnancy: A retrospective cohort study of patients managed at five Ebola treatment units in West Africa. *Clinical Infectious Diseases, 65*(2), 292–299. https://doi.org/10.1093/cid/cix290. Retrieved May 15, 2018, from https://academic.oup.com/cid/article/65/2/292/3097901.

International Commission. (1978). Ebola haemorrhagic fever in Zaire, 1976. *Bulletin of the World Health Organization, 56*(2), 271–293.

Jamieson D. J., Uyeki T. M., Callaghan W. M., Meaney-Delman D., & Rasmussen S. A. (2014). What obstetrician–gynecologists should know about Ebola. *Obstetrics & Gynecology,* 124, 1005–1010. https://doi.org/10.1097/AOG.0000000000000533. Retrieved May 10, 2018, from https://journals.lww.com/greenjournal/Fulltext/2014/11000/What_Obstetrician_Gynecologists_Should_Know_About.21.aspx.

Jones, S. M., Feldmann, H., Ströher, U., Geisbert, J. B., Fernando, L., Grolla, A., et al. (2005). Live attenuated recombinant vaccine protects nonhuman primates against Ebola and Marburg viruses. *Nature Medicine, 2005*(11), 786–790. https://doi.org/10.1038/nm1258.

Kratz, T., Roddy, P., Oloma, A. T., Jeffs, B., Ciruelo, D. P., De La Rosa, O., et al. (2015). Ebola virus disease outbreak in Isiro, Democratic Republic of the Congo, 2012: Signs and symptoms, management and outcomes. *PLoS One, 10*(6):e0129333. Retrieved May 10, 2018, from http://journals.plos.org/plosone/article?id=10.1371/journal.pone.0129333.

Lang, J. (2014). Ebola in the maternity ward. *The New Yorker*. Retrieved May 10, 2018, from https://www.newyorker.com/tech/elements/ebola-maternity-ward.

Lefebvre, A., Fiet, C., Belpois-Duchamp, C., Tiv, M., Astruc, K., & Aho Glélé, L. S. (2014). Case fatality rates of Ebola virus diseases: A meta-analysis of World Health Organization data. *Médecine et Maladies Infectieuses, 44*, 412–416.

Macklin, R. (2010). Enrolling pregnant women in biomedical research. *The Lancet, 375*(9715), 632–633. Retrieved May 16, 2018, from https://www.thelancet.com/journals/lancet/article/PIIS0140-6736%2810%2960257-7/fulltext.

Médecins Sans Frontières. (2015). *Nubia. First newborn to survive Ebola*. Retrieved May 15, 2018, from https://msf.exposure.co/nubia.

Merkatz, R. B., Temple, R., Subel, S., Feiden, K., & Kessler, D. A. (1993). Women in clinical trials of new drugs. A change in Food and Drug Administration policy. The working group on women in clinical trials. *New England Journal of Medicine, 329*(4), 292–296.

Muehlenbachs, A., de la Rosa, V. O., Bausch, D. G., Schafer, I. J., Paddock, C. D., Nyakio, J. P., et al. (2016). Ebola virus disease in pregnancy: Clinical, histopathologic, and immunohistochemical findings. *Journal of Infectious Diseases, 215*(1), 64–69. https://doi.org/10.1093/infdis/jiw206. Retrieved May 16, 2018, from https://academic.oup.com/jid/article/215/1/64/2742471.

Mupapa, K., Mukundu, W., Bwaka, M. A., Kipasa, M., De Roo, A., Kuvula, K., et al. (1999). Ebola hemorrhagic fever and pregnancy. *Journal of Infectious Diseases, 179*(Suppl 1), S11–S12.

Nelson, J. M., Griese, S. E., Goodman, A. B., & Peacock, G. (2016). Live neonates born to mothers with Ebola virus disease: A review of the literature. *Journal of Perinatology, 36*(6), 411–414.

Qiu, X., Fernando, L., Alimonti, J. B., Melito, P. L., Feldmann, F., Dick, D., et al. (2009). Mucosal immunization of cynomolgus macaques with the VSVΔG/ZEBOVGP vaccine stimulates strong ebola GP-specific immune responses.*PLoS One, 4*(5), e5547. https://doi.org/10.1371/journal.pone.0005547. Retrieved May 8, 2018, from http://journals.plos.org/plosone/article?id=10.1371/journal.pone.0005547.

Ryan, M. A. K., Smith T. C., Sevick, C. J., Honner, W. K., Loach, R. A., Moore, C. A., et al. (2008). Birth defects among infants born to women who received anthrax vaccine in pregnancy. *American Journal of Epidemiology, 168*(4), 434–442. Retrieved May 10, 2018, from https://academic.oup.com/aje/article/168/4/434/106600.

Schnirring, L. (2015). Youngest Ebola survivor leaves Guinea hospital. *CIDRAP*. Retrieved May 15, 2018, from http://www.cidrap.umn.edu/news-perspective/2015/11/youngest-ebola-survivor-leaves-guinea-hospital.

Shields, K. E., & Lyerly, A. D. (2013). Exclusion of pregnant women from industry-sponsored clinical trials. *Obstetrics & Gynecology, 122*(5), 1077–1081.

Smithells, R. W. (1978). Drugs, infections and congenital abnormalities. *Archives of Disease in Childhood, 53*, 93–99.

U.S. Food & Drug Administration. (2018). *Evaluation of gender differences in clinical investigations—Information sheet*. Retrieved May 9, 2018, from https://www.fda.gov/RegulatoryInformation/Guidances/ucm126552.htm.

UNAIDS, WHO (2012). *Ethical considerations in biomedical HIV prevention trials*. Retrieved May 13, 2018, from http://www.unaids.org/sites/default/files/media_asset/jc1399_ethical_considerations_en_0.pdf.

WHO. (2014). Ethical issues related to study design for trials on therapeutics for Ebola virus disease. *WHO Ethics Working Group Meeting, 20–21 October*. Retrieved May 15, 2018, from http://apps.who.int/iris/bitstream/handle/10665/137509/WHO_HIS_KER_GHE_14.2_eng.pdf;jsessionid=175A553938104AFE15DAB0A3C64A8839?sequence=1.

WHO. (2018). *Ebola virus disease*. Retrieved May 12, 2018, from http://www.who.int/en/news-room/fact-sheets/detail/ebola-virus-disease.

WHO Ebola Response Team, Agua-Agum, J., Ariyarajah, A., Blake, I. M., Cori, A., Donnelly, C. A., et al. (2015). Ebola virus disease among children in West Africa. *New England Journal of Medicine*, 372(13), 1274–1277. Retrieved May 17, 2018, from https://www.nejm.org/doi/full/10.1056/NEJMc1415318.

WHO Ebola Response Team, Agua-Agum, J., Ariyarajah A, Blake, I. M., Cori, A., Donnelly, C. A., et al. (2016). Ebola virus disease among male and female persons in West Africa. *New England Journal of Medicine*, 374(1), 96–98. Retrieved May 18, 2018, from https://www.nejm.org/doi/full/10.1056/NEJMc1510305.

Part II

Liberia

11 Caring for Women in Labor at the Height of Liberia's Ebola Crisis: The ELWA Hospital Experience

Rick Sacra and John Fankhauser

11.1 An Introduction from Rick Sacra, MD

This chapter is written by two medical doctors—John Fankhauser and myself, Rick Sacra. Both John and I provided care to pregnant women at the Eternal Love Winning Africa (ELWA) Hospital in Paynesville, just outside Monrovia, Liberia, during the Ebola crisis.

I grew up in Massachusetts, attended Brown University where I studied Biochemistry, and then the University of Massachusetts Medical School (UMMS) for my Doctor of Medicine degree (MD). I am married, and I have three sons. After an initial term of missionary service in Liberia as a medical student in 1987 (during a leave between my third and fourth years at UMMS), my family and I returned to Liberia in 1995 and spent most of 15 years there before returning to the United States in 2010. My wife, Debbie, and three sons have been very supportive of my work providing medical care in Liberia, both during long periods when we resided together in Liberia as a family, and during periods like 2014 when they were living in the United States.

In 2013, I started making trips to Liberia to continue working at ELWA Hospital. In addition to these "resume items," I should also tell you a little about myself personally. I am a big-hearted, big picture, holistic doctor. I believe in the importance of the physical examination and of meaningful touch as part of the patient encounter. I am not meticulous. I am not a perfectionist; I am a problem solver.

When I started writing this chapter, it had been a little over 2 years since the peak of the Ebola epidemic in West Africa. I marked the 2-year anniversary of my own diagnosis with Ebola virus disease (EVD), evacuation from Liberia, and treatment at the Nebraska Medical Center. As these milestones went by on the calendar, I experienced the full spectrum of emotion: tremendous gratitude

R. Sacra (✉)
University of Massachusetts Medical School, Worcester, MA, USA

ELWA Hospital, Paynesville, Monrovia, Liberia
e-mail: rick.sacra@sim.org

J. Fankhauser
ELWA Hospital, Paynesville, Monrovia, Liberia

David Geffen School of Medicine at UCLA, Los Angeles, CA, USA

D. A. Schwartz et al. (eds.), *Pregnant in the Time of Ebola*, Global Maternal and Child Health,
https://doi.org/10.1007/978-3-319-97637-2_11

that I have been given this chance to continue my life's work in Liberia, and at the same time, deep sorrow for the loss of so many—especially my colleagues in health care—to Ebola. The Ebola virus made its impact in West Africa, not only by infecting over 28,000 individuals (this is widely acknowledged to be a conservative estimate), but also by causing the collapse of the health care system. In this chapter, John and I will look at some basic data from one health facility, the Eternal Love Winning Africa (ELWA) Hospital, and consider the personal reflections, stories, and insights of two expatriate physicians who cared for pregnant women during the Ebola crisis.

I would like to thank the many individuals who played critical roles in responding to Ebola, from those who cared for Ebola victims in Ebola Treatment Units (ETUs), to those who kept hospitals and health facilities open under desperate conditions, those who volunteered to offer logistical support, and those who funded such efforts. I must specifically mention and thank Dr. Jerry Brown, Dr. Kent Brantly, Dr. Afidu Lemfuka, Dr. Patrick Igwilo, Dr. Biligan Korha, Dr. Eric Nyanzeh, Dr. Soka Moses, and Dr. Deborah Eisenhut. I personally knew several physicians who sacrificed their lives serving in Liberia during the Ebola outbreak: Dr. John Dada, Dr. Samuel Brisbane, and Dr. Abraham Borbor; may their souls rest in peace. I hope that my voice can in some way represent these and other people who have dedicated their lives to providing health care in crises.

I especially would like to appreciate Dr. John Fankhauser, my coauthor, who continued to put his own life at risk even after observing colleagues, including me, contract EVD. I count it a privilege to work with him at ELWA Hospital and consider him a colleague of the highest order.

11.2 An Introduction from John Fankhauser, MD

I grew up in a small town in the Pacific Northwest, the only doctor in a family of high-school science teachers. I was raised on the old-fashioned values of hard work and responsibility. When I was not studying, I was playing sports, backpacking, or pursuing the rank of Eagle Scout in the Boy Scouts. I went to college at the University of Washington in Seattle where I ultimately received a Doctor of Medicine degree. My wife and I then transplanted to Southern California for my Family Medicine training. We lived there for most of 20 years, raising a son and two daughters. After completing residency, I stayed on as teaching faculty in the program and eventually became the Medical Director of Ventura County Medical Center. Despite being part of a tremendous organization with amazing colleagues and friends, my wife and I felt a yearning to be involved in mission work in Africa. After months of soul searching, our family decided to make a transition. Soon after, we met Dr. Rick Sacra. Rick's passion for his work and for the people of Liberia was contagious. He is a gifted storyteller, and his love for the people of Liberia was evident with each story. Before long, we had moved out of our home of nearly 20 years, had sold or given away nearly all of our belongings, and we were on our way to Monrovia, Liberia, to work alongside Rick and others at SIM's ELWA Hospital. Little did we realize what lay ahead of us.

11.3 Our Story Begins

In 2014, ELWA Hospital was a small, 50-bed facility, employing seven doctors (three expatriates and four local physicians) and roughly 130 Liberian staff. Founded in 1965, ELWA Hospital is owned and operated by SIM (not an acronym), a nondenominational global Christian mission (www.sim.org). Funding sources for ELWA Hospital's operations include donations from abroad, subsidies from the Liberian government, and the intake from hospital bills that patients pay themselves or are billed to insurance, if available. The 130-acre ELWA Compound hosts several

Christian ministries in addition to the hospital, including Radio ELWA, the ELWA Academy (Kindergarten through High School), and the Liberia offices of Samaritan's Purse (SP), a Christian relief and development organization based in North Carolina. In addition, there are roughly 50 homes, occupied by SIM and Samaritan's Purse missionaries and ELWA staff members serving in key roles.

During the Ebola outbreak, the ELWA Compound became the site of three ETUs. In April of 2014, early in the Ebola crisis, SIM converted the ELWA Hospital chapel into a 6-bed isolation unit for management of Ebola cases (known as ELWA-1). In July of 2014, this unit was closed and its activities transferred to an adjacent building under the direction of the Samaritan's Purse Disaster Assistance Response Team (DART). This unit, ELWA-2, initially had a capacity of 20 patients and gradually was expanded to hold up to 80 patients. On July 29th, 2014, ELWA-2 was turned over to the Ministry of Health (MOH) of Liberia, to manage in partnership with ELWA. Doctors Without Borders (most commonly known by their French name, Médecins Sans Frontières [MSF]) constructed a large ETU at a separate site on the ELWA Compound. Construction at ELWA-3 began on July 19, 2014. Over the following days, members of the adjacent community, fearing that the presence of the ETU would increase their risk of exposure to the Ebola virus, gathered to protest the construction of this large ETU. After protesters threatened to burn or damage the facility, MOH and local government officials met with community leaders, eventually convincing them that the ETU would be a net benefit to them. After a delay of several weeks, construction resumed in early August, and ELWA-3 was opened on August 17th, 2014, with a capacity of 30 beds. Eventually it would grow to 240 beds, the largest ETU that MSF has ever established (Medecins Sans Frontieres 2014).

It goes without saying that much can be learned from a scientific inquiry into the Ebola epidemic which began in 2013, and with the final cases reported in 2016. However, the best understanding may come through a combination of scientific analysis and personal reflection. The crisis was unique, and hopefully, will never be repeated. Furthermore, the process of scientific inquiry can take an intensely human experience and compress it into a data point, a subject for academic analysis, and may obscure ethical and social aspects (Solomon 1985). Instead, we would like to use our memories and records of what took place from July to September of 2014 to communicate the lived experience of health care providers in the midst of crisis, caring for pregnant women and their infants.

Through the fateful timing of their pregnancies, these women faced the prospect of having a complicated, life-threatening birth in a city whose health care system had collapsed—a health system already weakened by fourteen years of civil conflict. Our approach is phenomenological and, therefore, subjective. Phenomenological research and writing have impacted the health care field, with noted examples in nursing research, cross-cultural health care, and health psychology (Smith 1996; Fadiman 1998; Lopez and Willis 2004). The stories and emotions presented will offer insight into the experiences of health care providers faced with risk to themselves as well as their patients. The Ebola epidemic brought out both the best and the worst in health care providers: in some instances, a fierce commitment and unconditional regard for the well-being of patients, and in others, a closed door to those most in need. In the following sections of the chapter, we will present a timeline of events and decision-making points, our basic obstetrical data from July to September 2014, patient care vignettes, and a few memorable staff interactions. The discussion section will shed light on the doctor-patient relationship and how the stresses of the Ebola crisis revealed factors that enhance that relationship amidst difficulties that threaten it.

11.4 Timeline of Events

11.4.1 From John

On March 22nd 2014, the day after I accepted the position of Acting ELWA Hospital Administrator for 6 months, I received a concerning email on the ProMED (Program for Monitoring Emerging Diseases) listserv. Scientists at the Pasteur Institute were reporting six positive tests for Zaire Ebola virus from 12 samples taken from Guinea, where several people had died from suspected viral hemorrhagic fever. That weekend, our ELWA staff physicians began discussing the possibility that the outbreak could affect ELWA Hospital.

On Monday morning, March 24th, Dr. Debbie Eisenhut, a general surgeon from Oregon, offered to be the clinical lead for ELWA's response to this threat. Multiple discussions as a team, with consultation from a group of experienced Ebola responders from MSF, led us to choose our small Chapel, a separate building from the rest of the hospital, as our isolation unit. Three days later, Dr. Eisenhut began training the hospital staff based on a guideline provided by the World Health Organization (WHO) and the United States Centers for Disease Control and Prevention (CDC) (Centers for Disease Control and Prevention and World Health Organization 1998). Over the course of 2 weeks, all of our 130 hospital staff members were trained to protect themselves while providing care (Fig. 11.1). Dr. Eisenhut's quick action and thoughtful preparation likely saved the lives of many of our staff. Throughout the following month, David Writebol (an SIM missionary) led a team of workers to convert the chapel to a six-bed isolation unit. Samaritan's Purse agreed to send several shipments of essential protective equipment and supplies that would be needed in the event a patient with EVD presented to ELWA Hospital.

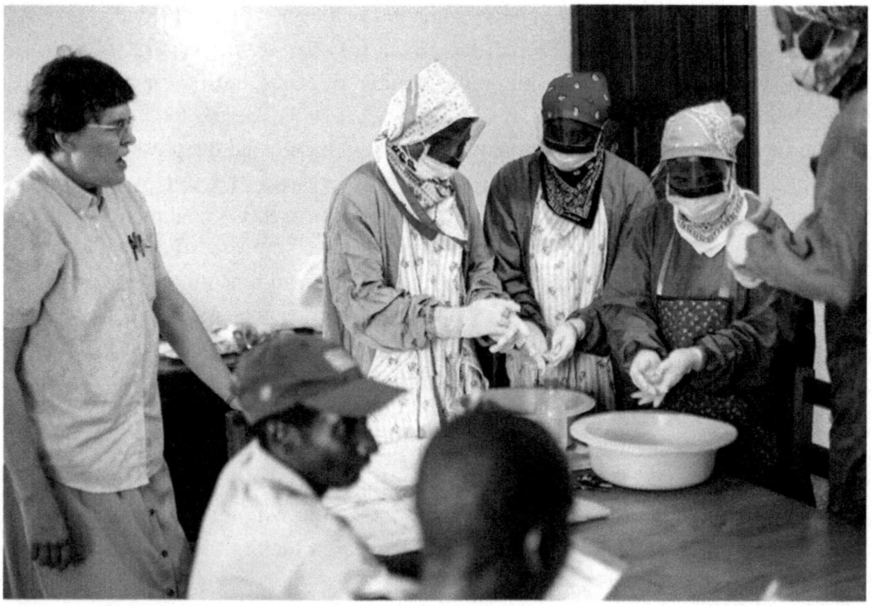

Fig. 11.1 Dr. Debbie Eisenhut (left), an ELWA staff surgeon, trains ELWA staff in infection prevention techniques. Photo by Bethany Fankhauser

11.4.2 From Rick

After my previous trip to Liberia in February/March of 2014, I returned for the month of May. The hospital's 6-bed ETU had yet to host an Ebola patient. There were no confirmed Ebola cases in Monrovia in May. One patient had passed through the city on the way from Lofa County to Margibi County, about an hour outside of Monrovia. We heard she had passed away due to

EVD early in May, but did not hear of any additional cases after that. I returned home to the US and planned my next trip for mid-August.

11.4.3 From John

The first Ebola patients were sent to ELWA by Liberia's Ministry of Health (MOH) on June 10th, 2014. The testing of patient blood samples for Ebola at this time was very limited. Ebola virus testing was performed at the national reference laboratory, an hour's drive from our facility; results could take up to 4 days. There was a gradual increase in cases during June and July (See Fig. 11.2 for the number of Ebola cases in Monrovia each week, June to October 2014). The Liberia Country Director for Samaritan's Purse called in their DART from their international headquarters. By mid-July, more ETU beds were needed, and SP constructed the ELWA-2 unit in collaboration with SIM/ELWA. The personnel for the ELWA-2 ETU were drawn from ELWA Hospital, Samaritan's Purse, and Ministry of Health staff. Personal protective equipment (PPE) and other supplies for the unit had been brought in by SP.

After three staff from the ETU and two from the hospital became ill with EVD, on July 31st ELWA hospital was closed for decontamination and retraining of staff on infection prevention procedures. In addition, hospital leaders perceived that the layout of the facility was problematic for safe and effective

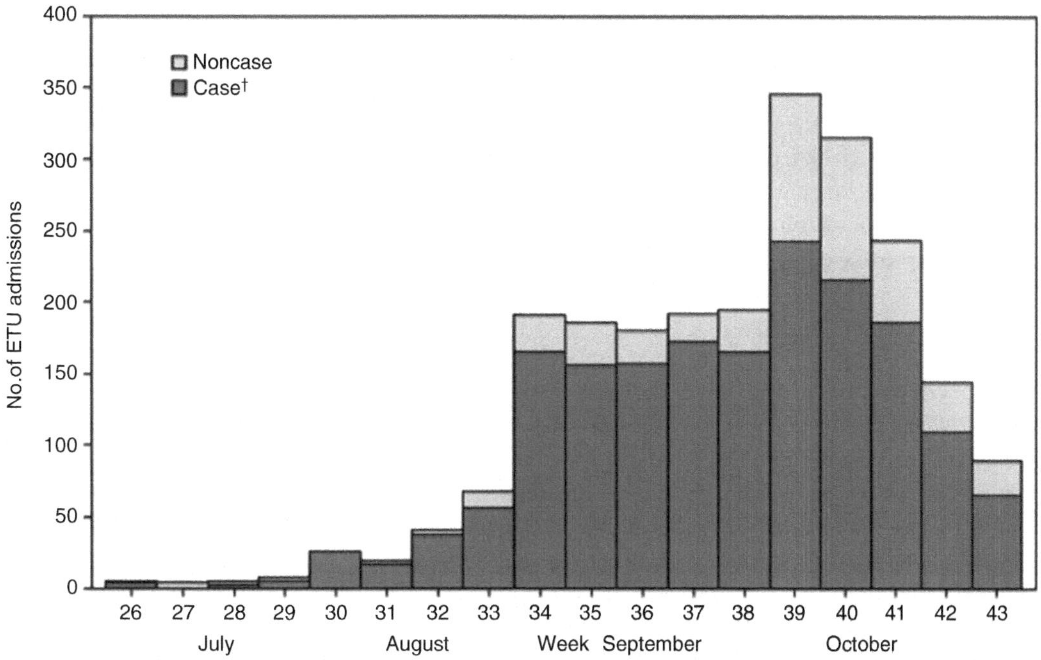

Fig. 11.2 Weekly EVD Cases in Montserrado County, Liberia, June 23 to October 26, 2014 (Nyenswah et al. 2014)

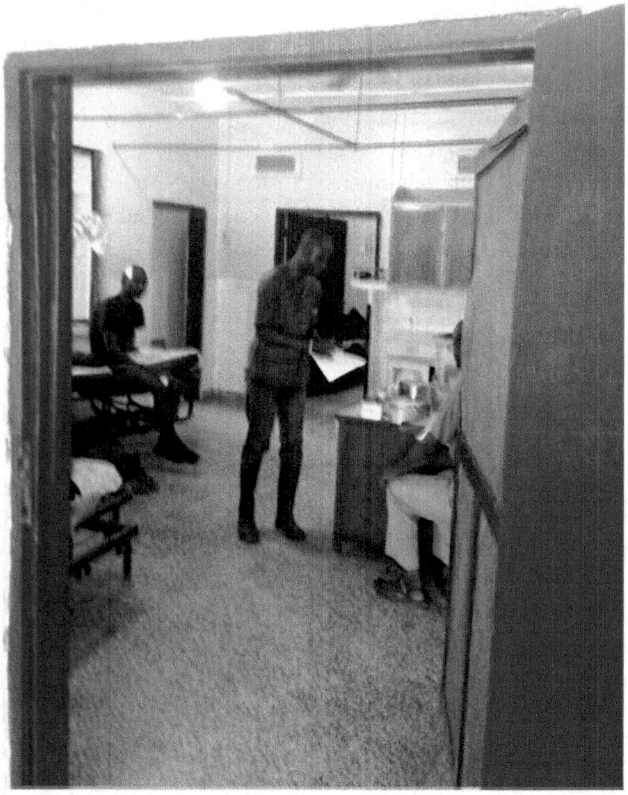

Fig. 11.3 ELWA Hospital's Emergency Department entrance, 2014. The view is from outside the Emergency Room, and patients and visitors could reach this door directly from the parking lot. Photograph courtesy John Fankhauser

patient triage. Each clinical area—Emergency Room (ER), X-ray, dental clinic, business office, visitors' entrance, laboratory, and doctors' office—had its own entrance, which could be accessed directly from the driveways. For example, the ER door opened directly from a drive-up portico into the clinical area (Fig. 11.3). When families arrived with an extremely sick patient, they would sometimes rush through the door with the person in their arms. This design made it impossible to carry out triage of patients, and in particular, to decide whether they should be managed in the hospital or referred to the ETU, which had its own entrance about 75 m away. It was decided that the emergency room would remain closed even after the hospital reopened, until robust procedures could be adopted to make it a safe place to provide care.

11.4.4 From Rick

When Dr. Kent Brantly, a physician serving with Samaritan's Purse, and Nancy Writebol, an SIM Missionary working at the ETU, were diagnosed with Ebola, the need for qualified staff was more urgent than ever. I moved my trip up by a couple weeks. I left Massachusetts on Sunday evening, August 3rd, and arrived in Monrovia on the 4th. I arrived on the ELWA compound just 2 h before Nancy Writebol was taken to the airport for her medical evacuation.

Our hospital was closed. Tuesday was a day of meetings with doctors, midwives, and support staff, preparing to reopen on Wednesday, August 6th. We would initially open only for obstetrics cases: outpatient OB clinic, inpatient deliveries, and emergency surgeries. This would allow us to continue to make changes to the facility and train the staff to provide additional services, while dealing with one acute crisis—women in labor who had nowhere to go for help with a difficult delivery. As we reopened, Dr. Fankhauser turned over the administrators' office to me. My duties included rationing our limited supplies of PPE to the staff, acquiring or ordering additional supplies, and filling a slot in the doctors' call schedule. Dr. Fankhauser was asked by SIM to accompany Dr. Eisenhut and Nancy's husband, David Writebol, on their evacuation flight on August 10th.

The WHO conducted a site visit. They recommended we control access to the facility by creating a single entrance for triage of all patients, visitors, and staff on entry. They provided training in donning and doffing PPE and the use of the WHO triage protocol (Fig. 11.4) to all our staff. The training was focused on patient triage and proper PPE use within the ETU or when caring for patients suspected of having EVD. The initial assumption communicated by health officials was that hospital staff safety could be maintained by keeping any Ebola cases out of the hospital through proper use of the screening protocol. Operating on this assumption, there was little accounting for the possibility that patients harboring Ebola might enter the hospital in spite of proper screening.

While relatively clear guidelines were available (from MSF, WHO and CDC) about how to use PPE in the ETU setting, in August 2014, there was no clear guidance for optimal Ebola infection prevention procedures or use of PPE in the hospital. Due to the high temperatures and high humidity in Liberia (average temperature 26–30 °C/80–86 °F; average humidity 87%), staff could only wear full PPE for about an hour before their condition would be compromised by the heat, perspiration, and dehydration. In light of this, a second level of protective clothing was in use, known as "basic PPE", including various types of gowns and aprons. The leadership and staff were improvising practices and protocols as we went along, basically using whatever we had available in ways that made sense to us. In August, this meant that for procedures involving blood or body fluid exposure, we would generally wear a blue "basic PPE" impermeable plastic gown with full sleeves, gloves, a mask with a face shield attached, and knee-high rubber boots. Decontamination of spills or surfaces was done with a bleach solution. There was a "sprayer" (see Fig. 11.5) who worked in the nearby ETU and could be called upon to come and decontaminate staff members or a zone in the hospital if a death or high risk exposure had taken place, but there was no dedicated sprayer stationed in the hospital at this time.

11.4.5 From John

When Rick was diagnosed with EVD on September 1st, I consulted my family and made immediate plans to return to Liberia. When I arrived, my first task was to make sure Rick's medical evacuation went smoothly. En route to the airport in the back of the ambulance, Rick was still quite lucid, and the two of us reviewed the potential sources of Rick's infection. At that time, there were three entrances to the hospital and triage of patients took place in the waiting area. Rick had drafted a rough plan for establishing a single entrance to the hospital and that had been left in the office that we shared. After his departure, I made some minor modifications, allowing for an ambulance entry and incorporating our isolation unit in the plan. I then worked with our construction team to install the fence, establishing the single entryway with a triage station and a chlorine hand washing station at the front (Fig. 11.6).

Many organizations offered advice on appropriate triage at the hospital (Fig. 11.7). At that time, the common belief was that if a patient reported no ill contacts and did not have a fever, they were not infected with EVD. We were aware of many patients who had presented to ELWA Hospital denying any ill contacts who later admitted to having had contact with an EVD patient. For fear of being turned away, they provided a story that underrepresented their risk. In addition, we had seen several patients who presented without fever, but were eventually diagnosed with EVD. For that reason, all patients who had three symptoms or more were asked to wait until they could be interviewed and examined by a doctor before entering the hospital grounds. They were either moved to the Emergency Department or asked to go to the ETU at the discretion of the physician.

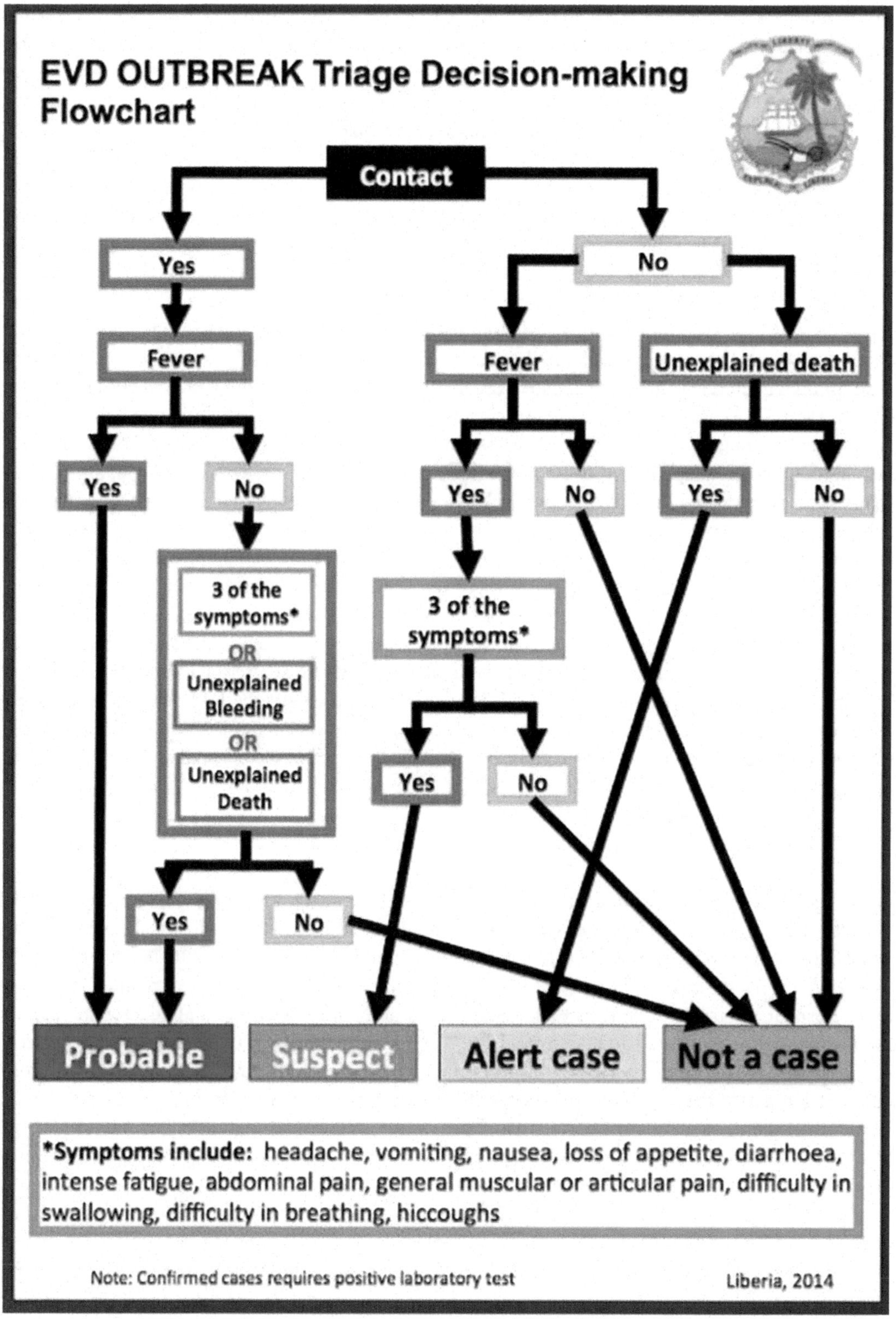

Fig. 11.4 Triage protocol for suspected persons with Ebola virus disease provided by Liberia's Ministry of Health

Fig. 11.5 A sprayer decontaminates an area with 0.5% bleach solution, ELWA Hospital. Photo courtesy of Rick Sacra

Fig. 11.6 ELWA Hospital's front entrance, after enclosing/fencing the facility to provide a single entry for triage. Photo courtesy of John Fankhauser

Fig. 11.7 An ELWA Hospital Nurse triages a patient who has presented for emergency care in September or October 2014. Photograph courtesy of John Fankhauser

11.5 The ELWA Hospital Experience, July to September 2014

In order to provide a numerical summary of the ELWA Hospital's experience, the authors personally reviewed all the charts of patients discharged from the hospital during the 3-month period from July 1, 2014 through September 30, 2014. We present data on all deliveries, whether normal deliveries or cesarean sections, during the period.

During July, August, and September, ELWA Hospital was closed for a period of 6 days (July 31st to August 6th). After August 6th, only obstetric services were open (Fig. 11.8). During this 3-month period, several major hospitals in Monrovia were closed; others were only partly functional. Many of the women we cared for had been to several other facilities before reaching ELWA and had been turned away for various reasons. Reasons given included closure of the facility, lack of staff, or the presence of fever, vomiting, or diarrhea. Some of these patients came with written referral notes; others did not. Several of them had also been cared for by local midwives, at small clinics or private homes, in an attempt to assist them to deliver. During this period, most Ebola virus treatment in Montserrado County took place at the ELWA-1 (June 10 to July 20), ELWA-2 (July 20 onward), or ELWA-3 (August 17 onward) ETUs on ELWA campus. Other ETUs in Monrovia included JFK (John F. Kennedy) and the Island Clinic ETU, which opened on September 20th. Many patients told us that they were hesitant to come to ELWA for care due to the proximity of the ETUs.

During the 3-month period under examination (July to September 2014), ELWA Hospital delivered 201 pregnant women (Table 11.1). Of those, 78 were delivered by cesarean section, along with one cesarean hysterectomy for a ruptured uterus. Normal vaginal deliveries occurred in 119 women, and 3 had assisted deliveries (vacuum or forceps). In 27 women, there were absent fetal heart tones on arrival to the hospital, and they had stillbirths. We delivered an additional four patients with "fresh"

Fig. 11.8 Dr. Sacra makes rounds on the ELWA Hospital Obstetrics Ward, January 2015. Photograph courtesy of Steve King, Worcester Magazine

Table 11.1 ELWA Hospital Statistical Summary July to September 2014

	July	August	September	Total	Registered	Unregistered
Total admissions	152	86	78	**316**		
Total deliveries	71	76	54	**201**	81	120
Cesarean section	17	32	29	**78**	19	59
Cesarean hysterectomy	0	1	0	**1**	0	1
Normal vaginal delivery	53	41	25	**119**	62	57
Instrumented delivery[a]	1	2	0	**3**	0	3
Maternal death	0	3	2	**5**	1	4
Mothers with EVD	0	1[b]	2	**3**	0	3
Known stillbirth[c]	7	13	7	**27**	3	24
Fresh stillbirth	2	2	0	**4**	3	1
Neonatal death	1	2	4	**7**	1	6
Transfer to pediatrics	2	1	1	**4**	2	2

[a]Vacuum extraction or forceps delivery. Forceps were used on one known stillbirth
[b]Inferred due to a case that occurred in expatriate medical staff; laboratory test not performed
[c]Fetal heart tones were absent on arrival at the hospital

stillbirths. There were seven neonatal deaths (one of whom had EVD). An additional four sick neonates were transferred to Pediatrics for ongoing hospital care. Thus, out of 201 deliveries, 159 (79%) had normal fetal/neonatal outcomes (Fig. 11.9). There were five maternal deaths, three of whom most likely had EVD (two proven, one suspected).

ELWA Hospital offers women an opportunity to "register" during their pregnancy, paying a fixed, discounted fee which bundles together the delivery charge with a package of outpatient services, including routine checkups, prenatal vitamins, and other basic medicines. This system incentivizes women to begin prenatal care early and come regularly, hopefully leading to improved rates of important interventions, such as vaccination, intermittent presumptive treatment for malaria, routine testing

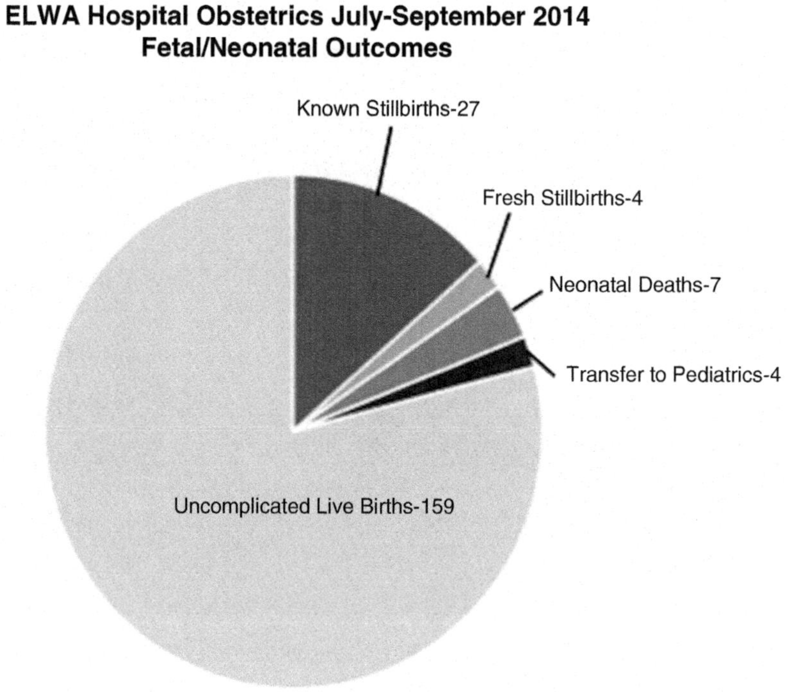

Fig. 11.9 ELWA Hospital fetal/neonatal outcomes, July to September 2014

for human immunodeficiency virus (HIV), hepatitis B, and syphilis, as well as monitoring for complications. Most of the emergency cases in the review period were among "unregistered" patients, while a higher proportion of the normal uncomplicated deliveries were among registered patients. The three columns on the right side of Table 11.1 show ELWA Hospital data, stratified by whether the patient was registered. Sixty percent (120/201) of our total deliveries during the period were unregistered patients, while 40% (81/201) were registered. The cesarean section rate among unregistered patients was 49% (59/120), compared to 23% (19/81) among registered patients. Four out of 5 maternal deaths, all 3 Ebola cases, and 24 of 27 patients presenting with an intrauterine fetal death were unregistered patients. Six out of 7 neonatal deaths, 2 out of 4 of those children too sick to go home with their moms, and 1 out of 4 fresh stillbirths occurred among the unregistered patients.

This stratification brings up several observations about caring for pregnant women during the Ebola crisis. The first was that having an established relationship with a health care facility which remained open benefitted patients and their babies: many of our registered women did come for antenatal care and delivery at ELWA even in the midst of the Ebola epidemic. Secondly, the complication rate was much higher among our unregistered patients. This was most likely because (1) there was significant social pressure for women who could deliver normally outside the hospital to do so; and (2) there were substantial delays in presenting for definitive care, related to reluctance to present to a hospital, closure of many health facilities, and multiple ineffective referrals. Among our unregistered patients, 43% (52/120) had evidence in the chart of having been referred from another facility before reaching ELWA (and this is likely an underestimate), whereas only 1% (1/81) of our registered patients had been referred from elsewhere for definitive care. There was evidence in the medical charts of having been in labor more than 1 day prior to admission in 34% (41/120) of our unregistered parturients (See Fig. 11.10). This group made up only 20% (41/201) of our total deliveries, but accounted for 52% (16/31) of stillbirths, 29% (2/7) of neonatal deaths, and 40% (2/5) of maternal deaths. Delay in reaching definitive care was an important correlate of poor outcomes.

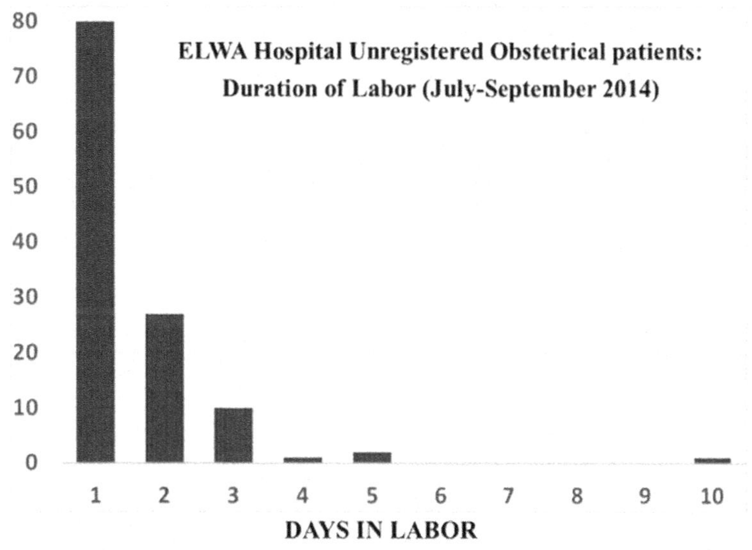

Fig. 11.10 How many days were unregistered obstetrical patients in labor before their arrival at ELWA Hospital (July to September 2014)? There were 41 patients who had evidence in their medical charts that they were already in labor more than 24 h before arrival, with 14 having been in labor more than 48 h

11.6 Patient Care Vignettes by Rick

The statistical data describe one aspect of the ELWA Hospital experience at the height of the Ebola crisis and lead to some relevant conclusions. The following stories illustrate the reality that those statistics represent and the challenges of caring for critical patients in the midst of a health crisis. These vignettes were gathered from our journals and emails and checked against patient charts to add further clinical information. Since John was practicing obstetrics at ELWA in July and September, and Rick was practicing in August, the stories presented belong to those time frames.

11.6.1 Molly

On the first full day we were open, we took care of a young woman whom I'll call Molly. Molly was an unregistered patient with bleeding and a very tender, full-term pregnant uterus. She had visited three other health facilities before reaching ELWA. Fetal heart tones were absent, and she was pale. Her cervix was just open a finger-tip. Blood tests showed she was anemic. We diagnosed her with a placental abruption,[1] and because her cervix was so unfavorable, decided to proceed with a cesarean section even though the baby had already died. The case went according to plan, and she recovered normally, requiring just one unit of blood to restore her blood volume. What struck me about Molly was the knowledge that, if this abruption had happened just 2 or 3 days earlier, there might have been nowhere for her to get a cesarean section and a blood transfusion in the entire city of Monrovia. She would have died.

[1] Placental abruption is a potentially life-threatening event to both the mother and her fetus. It occurs when the placenta either partially or completely prematurely separates from its attachment to the uterus and is typically accompanied by hemorrhage.

Fig. 11.11 Patients waiting to be seen at the ELWA Hospital Clinic and Lab in January 2015. This is the porch where Dr. Sacra first encountered "Naomi" as she was waiting with a family member. Photo courtesy of Steve King, Worcester Magazine

11.6.2 Naomi

A week or so later, in the middle of August, we were working in earnest on setting up the hospital to triage patients, visitors, and staff for symptoms of, or exposure to, Ebola. Appropriate changes to the physical structure of the building were being planned, which would affect the patient and visitor flow and the triage process. We were temporarily running the Obstetric Clinic out of a big classroom in the Administration section of the main hospital, and people were waiting on a porch in front of the lab to be called in to the clinic. The walkway we all used to get into the hospital passed right along under this porch (Fig. 11.11).

One morning, I arrived at the hospital, not wearing any special protective gear yet, because I had not started seeing patients. As I passed by this porch area, a thin, ill-appearing pregnant woman suddenly lunged at me, grabbed my arm and said "*I need help*." It was so shocking to have someone just reach out and grab me. Remember this is Liberia in August of 2014—no one is touching each other at all. No hand shaking, no hugging. I scolded her and told her to sit back down. I told the person who was with her, who brought her to the hospital, that this was not acceptable, that we would get to her soon. I went to a sink and washed my arms.

Later, the midwives called me to see this same patient whom I will call Naomi (not her real name). They had put her in a bed, in a private room, because she was bleeding and a bit confused. Naomi had no fever, and according to the family, no known exposure to anyone ill at home or in the neighborhood, and no vomiting or diarrhea. But she was weak and unsteady on her feet. At one point, she had what appeared to be a brief seizure. She looked to be about 8 months pregnant, but there was no audible fetal heartbeat. The family said she had been having some bleeding and abdominal pain, and a local midwife had tried to deliver her for several hours. On exam, her cervix was very thin but not very dilated—I think her midwife had mistakenly felt she was fully dilated. When we performed her routine labs, she turned out to be HIV-positive, and thus, I felt that the HIV infection explained her being confused and weak. We would try to get her delivered, then see if we could start her on antiretroviral drugs after she stabilized.

Since she was bleeding steadily and appeared to have had a seizure, I felt the best chance we had to help Naomi survive would be to do a cesarean section to end the pregnancy, in case the seizure was a sign of some sort of atypical eclampsia.[2] We had no ability to do a full set of labs or chemistries at that time to help us diagnostically, so we were making most of our decisions based on the clinical situation. I spoke with the staff and we decided to proceed with a cesarean section. We had difficulty obtaining access for the intravenous (IV) line and placing the Foley catheter, because the patient was not cooperative with these uncomfortable procedures. I had to place the Foley catheter myself. The lighting was not ideal, and the patient was moving around a lot. I was wearing the blue "basic PPE" impermeable gown, a face mask, double gloves, and rubber boots.

The first part of the cesarean section went routinely, although the fetus was nonviable. After the delivery of the dead baby, the uterus was boggy and would not contract well. It bled from the sites where the suturing needle passed through the muscle. Oxytocin and ergometrine were administered during the case, along with a rectal dose of misoprostol postoperatively to contract the uterine muscle and stop the bleeding. She seemed to stabilize, but an hour later she began bleeding heavily. In spite of the usual interventions, Naomi went into shock and expired shortly thereafter.

At this time, it took 2–4 days to get Ebola test results, and there was no simple routine to get a patient tested—one had to go over to the ETU and try to locate the lab technician to come over and draw the sample. Therefore, if it didn't make a difference in terms of management of the case (and in this case, the patient had already died), we were not routinely obtaining samples for Ebola testing. Also, contact tracing had not been instituted yet. Within a few weeks, systems would be set up for routine testing of samples from patients who had died, in order to monitor for unknown Ebola cases and chains of transmission.

Looking back on this story, it seems rather horrific. Clearly, this patient should have been tested for Ebola. But the lack of a fever and the positive HIV test seemed to explain her condition and lulled us into thinking we could move on. Though it's impossible to know for sure, in retrospect it seems likely that it was through my contact with Naomi that I contracted EVD.

11.6.3 Fatu

On one Saturday morning, I had come in to distribute PPE to all the staff, and then was hoping to take the rest of the day off, after working 12 days in a row. I went through the PPE, giving out one breathable gown and one heavier impermeable plastic gown for each staff member who was going to be on shift that day, and also gave out another shift's worth of PPE to the supervisors (for the night shift). I was just about ready to go home, when Dr. Eric Nyanzeh came into the administrator's office and told me that an unregistered patient had arrived with a history of two prior cesarean sections who had been in labor pain for a couple of days. The fetal heart rate was normal and the patient was stable. He intended to deliver her surgically, but the anesthetist was not available.

I decided to stay to provide the anesthesia for this case. We gave her a routine spinal anesthetic, and that went smoothly. While the surgeon was prepping and draping the patient, I began a conversation with Fatu (not her real name). I learned she had two living children. I asked her where she had been over the last couple of days while she was in early labor. Fatu told of going to several of the hospitals

[2]Eclampsia is a potentially fatal pregnancy-associated hypertensive disease which affects the central nervous system, including the occurrence of altered mental status, seizures and cerebral hemorrhage. It can also cause liver disease, decreased platelets, and maternal hemorrhage in other organs. It typically develops following the onset of preeclampsia, a milder form of hypertensive disease of pregnancy with no central nervous system manifestations.

in town, but each one turned her away. Then she found a clinic that was open, and the midwife in charge agreed to keep her there with the aim of delivering her vaginally, despite her history of two prior cesarean sections. The midwife surely knew that attempting vaginal birth after two cesarean sections (termed a VBAC) was not recommended, but it was a truly desperate time, and some well-meaning medical providers were doing whatever they thought might work to solve the problems of patients that came to them. Finally, after all these other avenues had failed, she had come to ELWA. She seemed relieved that the baby was still alive. I said something along the lines of "*Wow, you must live way on the other side of Monrovia, to have gone to so many places before reaching here*." Fatu smiled shyly and acted embarrassed. After some coaxing, she finally divulged where she lived—about a mile down the road from us! When I asked her why she hadn't come to ELWA sooner, she indicated that her family members were well aware that two of Monrovia's Ebola management centers were located on our campus, and they advised her to stay away from ELWA, to avoid exposure to Ebola. (What she actually said, in Liberian English, was "*Mah people say da ting too mush on ELWA, so I muh nah go dere*" or, spelled with standard spellings, "*My people say the thing (is) too much on ELWA (compound), so I must not go there*." "The thing" was a euphemism for Ebola—many Liberians didn't even want to speak the term out loud. Only after all these other options had failed had her family decided to bring her to ELWA for evaluation. The surgery went smoothly and this woman carried her healthy baby home 3 days later!

11.6.4 Tasha

I must include another story here, although it is not about maternity care. Tasha's story (not her real name) illustrates the impact of collateral effects from Ebola on families and children. After about 2 weeks in Liberia, I got a phone call from a close friend who told me of a 13-year-old girl in his community who was quite sick. Tasha's father had come to see if my friend could arrange a consultation for his daughter at the hospital. At this time, all hospitals in Monrovia, including ELWA, were closed for medical or surgical patients, due to the difficulty in screening for Ebola virus infection. Tasha had severe abdominal pain and vomiting; she had not passed a bowel movement in about 3 days, and her abdomen was starting to distend. Initially, I explained there was nothing we could do—the hospital was closed to general patients and only open for obstetrics cases. I told him I would try to check with our Medical Director, Dr. Jerry Brown, who is a general surgeon. Over the next day or 2, my friend called me back several times to update me on the girl's condition—she was getting worse—and to request I keep checking on the situation with Dr. Brown.

Finally, on a Saturday, I had a chance to talk with Dr. Brown about the case. At this point, Dr. Brown was extremely busy, as he was both in charge of the medical work at ELWA, playing a significant role providing coverage for cesarean sections needed at the hospital, and also the medical director at the ELWA-2 ETU. I spoke with Dr. Brown, and we reviewed the patient's symptoms. The lack of any diarrhea after more than 5 days of illness, along with the absence of anyone else sick in the house or neighborhood, made Ebola quite unlikely. He said he would be willing to take a look at this patient on Sunday after finishing his work at the ETU. The girl's father brought her to the hospital, and we ran some basic lab tests and performed an abdominal ultrasound, which showed peritonitis (severe inflammation and fluid in the abdomen). The Ebola caseload was very heavy at this time, and Dr. Brown did not complete his work at the ETU until after dark. It was around 8 PM when I got the opportunity to present the case. He agreed that she needed surgery, and that EVD was very unlikely. So Dr. Brown took her to the operating room for an exploratory laparotomy. I performed the role of "circulator" on the case, assisting the technician and anesthetist with needed supplies or instruments. The girl had several perforations in her small bowel caused by typhoid fever. Dr. Brown closed the perforations and washed out the infected fluid in her abdomen, and then closed the abdominal wall and skin. She was taken to the pediatric ward to recover. Over the next few days she seemed to improve, gradually gaining strength. I only learned later, after my own recovery from EVD, that she required three additional surgeries due to recurrent perforations and complications over the next couple months before finally recovering. I did get to see the patient's grandmother during a visit to Liberia about 18 months after Tasha's illness, and she told me that Tasha was doing well, back in school, and behaving like a normal teenager.

Even while I was in the US, these patients were the ones that were most concerning to me over the course of the Ebola crisis: cases that had nothing to do with Ebola that were going untreated because of the collapse of the health system. We were able to help this girl only because of the persistence of her father and my friend, neither of whom would rest until something was done. What of those who did not have such a personal connection with someone in a position to intervene?

11.6.5 Miatta

I have one final story. A woman I'll call Miatta was brought by her family to the hospital during the second week of August. She'd been in labor for 10 days. She could not walk at all; she could barely even move her legs. In an attempt to deliver her 2 days earlier, a midwife had cut an episiotomy, which remained open and unrepaired. By the time of her arrival at ELWA, Miatta's dead baby had become necrotic and smelly. The pressure of the baby's head in Miatta's pelvis for so many days had put pressure on her sciatic nerves, making her legs weak and numb. Miatta was feverish and in shock. The fetal head was visible, right at the vulva, but her vulva had been traumatized and manipulated so much that it was swollen and fragile. I was reluctant to try more manipulation, so we set her up for a cesarean section, even though the risk of sepsis was high. When we got into the operating room, we noticed that the fetal head seemed to be coming down a little lower. I decided that we could try to deliver her vaginally to see if she could be spared a cesarean section. We were able to deliver the head, but then the shoulders were badly impacted. I had to use every maneuver I knew, and was finally able to rotate the baby and get the posterior shoulder out, before delivering the body fully. After completing the delivery of the placenta, we put Miatta on IV antibiotics. We had to leave a Foley catheter in, because she couldn't pass urine on her own. Over the next 2 weeks, Miatta had intermittent fevers and only gradually began to use her legs. She was still in the hospital when my time in the hospital ended, so I only found out later, when reviewing our charts, that Miatta had passed away in mid-September due to an overwhelming infection. Despite our best efforts, we just could not reverse the damage of 10 days of labor.

11.7 Patient Care Vignette by John

11.7.1 Patience

One day in September 2014, as I stepped out of the Labor and Delivery Department, I glanced toward the Emergency Department where two women were helping a young pregnant woman who appeared acutely ill. We'll call her Patience. The older women were supporting most of her weight and her steps were slow and shuffled as they practically carried her down the corridor. My instincts were to reach out and help, but by now, I knew better than to take that risk, so I moved toward them but kept some distance. Within a few feet of the Maternity Department, Patience collapsed and began seizing on the ground. I obtained a temperature with a touch-free thermometer. It was 100.8 °F. At that time, ELWA Hospital was one of the only facilities caring for febrile pregnant patients. The family pleaded with us not to turn her away. I reassured them that we would care for her but let them know that it would take a few minutes for us to get into our protective gear. I instructed them to help Patience into a bed in the

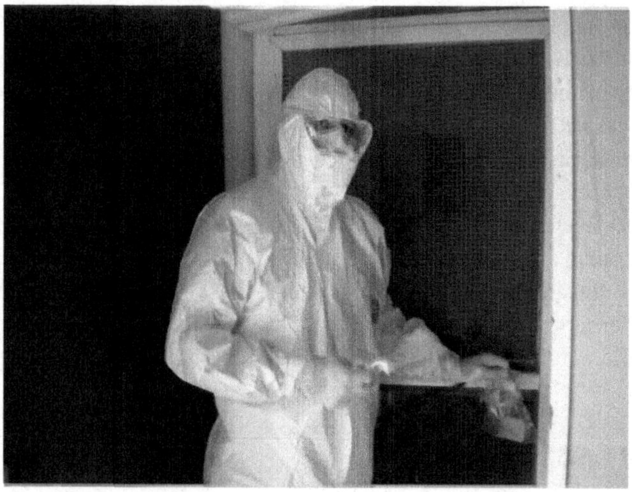

Fig. 11.12 Dr. John Fankhauser enters a patient's room to do a delivery in full PPE

isolation room and we began to prepare. In full PPE, we diagnosed eclampsia and administered magnesium, antibiotics, and malaria medications. We were unable to send her blood to our lab for routine tests until we had a negative Ebola test result. We sent the blood to the CDC lab on the other side of the ELWA compound and waited. For 6 h, we cared for the patient in our PPE. Early in the evening we received the results. Her Ebola test was negative. At that point, we were able to perform an ultrasound, confirming that she was at term or near-term and laboratory testing was ordered. Labor was induced, and several hours later, a healthy male infant was delivered (See Fig. 11.12).

11.8 Staff Interactions

The experience of working in a health facility in the midst of crisis is not solely focused on interactions with patients. The staff are also experiencing high levels of stress and risk. The discussions and conflicts that one has with colleagues, and observes staff having with one another, are just as much a part of the lived experience which we desire to share as the patient care scenarios we have presented.

11.8.1 Rick: Saving the Moms

Some of the midwives were feeling very discouraged by the condition of the patients and the large number of stillbirths. Many women had come with histories of being in labor for 2, 3, or 4 days—some even longer—and often their babies had already died within the uterus (see Table 11.1). I could see the look of disappointment, the heavy sigh as they checked another pregnant woman and found no fetal heartbeat. Normally, a stillbirth is an unexpected loss. It always has an impact on the midwife; but having so many of them seemed too much to bear. I told them at a certain point, "*We are just saving moms now, let's focus on saving the moms.*"

11.8.2 Rick: Limited Resources

The hospital had very scant supplies of personal protective gear during August 2014. We had ordered more supplies for the hospital, which were on the way, but in the meantime the shortage of PPE meant that we had to ration it out carefully. We were reusing items that would be considered "disposable" in a well-resourced setting, washing and disinfecting plastic aprons, or wearing a gown while seeing multiple different patients on the ward if it had not gotten soiled. During August, various local groups arrived at the hospital with donations for the ETU or for the hospital: politicians, churches, business people, or community groups. The donations ranged from food, blankets, bottled water, juices, or soft drinks, to medical supplies and gloves. I was frequently called upon as the acting administrator to join with other hospital supervisors to receive a donation and make a brief remark for the press (who were notified by the donors to come and "cover" the event). We had two main types of "basic PPE" that we were using. One was a blue impermeable plastic gown which was very useful for the operating room or deliveries. It was more comfortable than wearing the full suit of PPE, but provided nearly the same degree of coverage. The second was a breathable yellow gown that was used when one didn't expect to get contaminated, like working in the clinic or making rounds on the ward; it was more comfortable for routine work. On this particular day, I had just opened our very last package of ten of the yellow breathable gowns. I really didn't know what we would do the following day for basic PPE. Later in the morning, I was called to come out to receive a donation. A team had come from the Päynesville City Hall, representing the Mayor's office, with several boxes that had shipped via FedEx, sent by Liberians living in the Philadelphia area to assist medical facilities. I was given two large cartons, still sealed up as they had been sent from the U.S. Imagine my surprise as I opened one of the boxes, and inside were 10 packages (a total of 100) of the yellow breathable gowns! I was really amazed at the timing of this donation and found myself thanking God!

11.8.3 Rick: Motivations

A major factor in the closure of health facilities in Monrovia during the Ebola crisis was the interactions staff were having with their families. Many of our staff told me that their spouses had told them "*If you go to the hospital to work today, don't come back here tonight to sleep.*" Fear of Ebola was a major problem, and not without basis. Many of ELWA's staff affirmed that the only reason they were coming to work was their own personal faith in God and the knowledge that it was the right thing to do. A theme that was raised frequently by the local leadership in our meetings was that, if we close the hospital, then if your spouse, or your sibling, or your child got sick, where would they go to get care? This seemed the most cogent and convincing reason that people were using to get themselves and their fellow health care workers to come to work—if we don't come, and one of our loved ones gets sick, what then?

11.8.4 Rick: Parables

Several weeks after reopening, the staff requested a meeting with the hospital leadership. The meeting was led by Dr. Jerry Brown, our Medical Director. The staff expressed their concern that they were not receiving any extra pay to account for the risk they were taking in coming to work. Staff members who had been hired to work in ETUs were receiving much higher "hazard" pay rates than our regular nurses, midwives, and aides. They felt the situation was unfair, since it had become clear that working in the hospital carried as much risk as work in the ETU. The President of Liberia, Ellen Johnson-Sirleaf, had met with health care workers and promised that something would be done, but there was no specific timeline given.

Dr. Brown answered the staff with a parable. Liberians often illustrate their points with a "parable" or little story. This is viewed as the most compelling form of argument—more salient than data or scientific proof. In fact, in many West African societies, an apt parable can put an end to a dispute (Penfield 1981). Dr. Brown remarked, "*When the blind man says he will chunk you, you know the rock already in the man's hand*." Or in more standard English, "*When the blind man says he will throw a rock at you, you know that he already has the rock in his hand*." I had to ask several people for help in grasping the meaning of this parable. What Dr. Brown was saying was that he did not want to make an empty promise. If he was going to promise some sort of extra allowance or "hazard" pay, he wanted to know that he had the money to provide it, and wanted to know how it would be paid for. Since he did not have the resources now, the staff would have to wait.

11.8.5 John: Drama

One of the largest hospitals in Monrovia was hit hard by Ebola. After losing several staff members to the disease, the hospital closed down. In late October, the midwives there were interested in resuming duties. They had been watching our hospital and were aware that ELWA was providing maternity care without any of our midwives becoming infected. They requested a meeting in which our midwives shared their insight and procedures for maternity care during such a risky time. Twenty midwives and physicians came to ELWA Hospital where our midwives put on a drama, reenacting the arrival of a sick patient in labor. There was mutual respect, thoughtful instruction, encouragement, and laughter as our midwives proved to be quite melodramatic. The following week, all of these midwives returned to work at this local hospital, providing safe and much needed obstetrical care. For me, it was a time of reflection and admiration as I saw these women bravely embracing the responsibility of their profession at a time when the risk was acute and ever apparent (See Fig. 11.13).

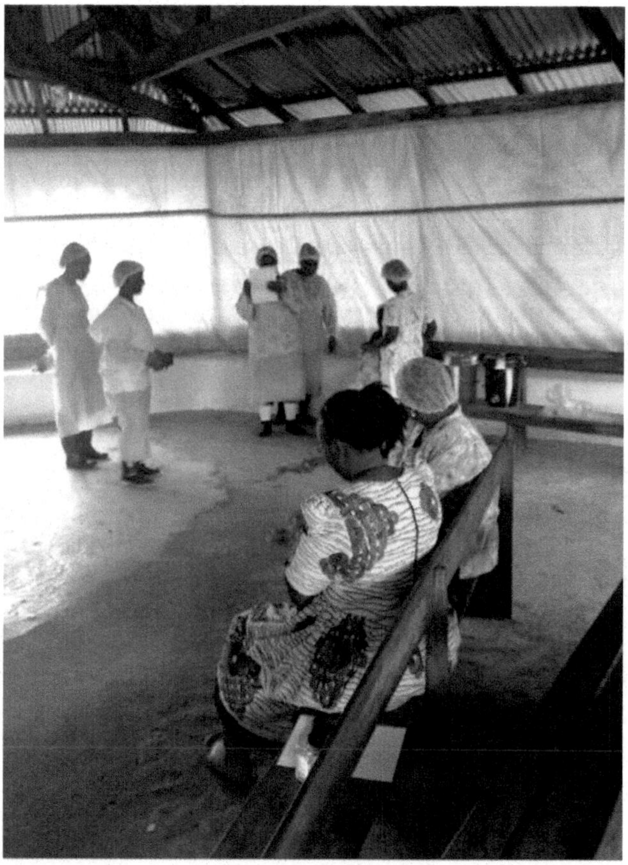

Fig. 11.13 ELWA midwives dramatize safe delivery practices at a session with midwives from another facility. Photo courtesy of John Fankhauser

11.9 Discussion

Caring for pregnant women during the Ebola crisis was like trying to walk across a bridge in a hurricane. The stress and pressure of the situation (both internationally and locally), the illnesses of colleagues getting sick with EVD despite use of PPE, the loss of many patients' lives, shortages of equipment and supplies, and staff absences all served to disorient and buffet health care providers. Thus, thrown off balance, and suddenly concerned about their own safety, doctors found it difficult to maintain their commitment to patients. As we reflect on doctor-patient relationships during the Ebola crisis, we gain insight from three perspectives: the professional, the spiritual, and the philosophical. Each of these perspectives or lenses highlights an aspect or commitment in the doctor-patient relationship which was specifically challenged and damaged by Ebola. And, each provides a portion of the antidote to that challenge.

Looking through the professional lens, we are family physicians, and we embrace that identity wholeheartedly. The concepts that undergird Family Medicine are as old as medicine itself, but the field began to be defined as a specialty in the 1940s, culminating in the formation of the American Board of Family Medicine (USA) in 1969. Globally, Family Medicine is promoted by WONCA, the World Organization of National Colleges, Academies and Academic Associations of General Practitioners/Family Physicians. James E. Bryan explains the perspective of the Family Medicine specialist: "*The ultimate distinction between the new family physician and members of the existing clinical specialties will be the former's ability to relate the parts to the whole, the machinery to the purpose, the special talent to the basic task... His view of patient care is oriented to the patient rather than the disease, and his concern is the continuing welfare of the patient in the full context of his life situation rather than the episodic care of a presenting complaint...*" (Bryan 1968). This orientation to the patient rather than the disease was challenged during Ebola. One way that physicians support the integrity of the patient, and let them know of our concern for their entire person, is by performing a complete history and physical examination. For instance, in a patient with a headache, we inquire about symptoms from a variety of organ systems and physically examine the heart, lungs, and abdomen, which lets them know we view their whole person as important. During the Ebola crisis, the complete intake history was replaced by an interview entirely dedicated to assessing the patient's risk of EVD. On the hospital wards, the physical examination was largely eliminated. As explained by a colleague who served in Liberia during the crisis, the first priority became keeping staff safe, rather than caring for the patient (Personal communication, Dr. Steven Hatch, 2015). These constraints made it very difficult to maintain a holistic view of the patient.

The story of Fatu illustrates the difficulty in establishing a strong, multidimensional relationship with patients during the Ebola crisis. When she first arrived, it was unclear why she delayed 4 days in presenting for definitive care, despite knowing that she needed a cesarean section and living just a mile from the hospital. But in the operating room, while she was under spinal anesthetic, our conversation allowed us to establish rapport and get past the impediments to communication. I developed a more holistic view of Fatu and her family and community and began to understand the issues that were keeping people from seeking timely care. A holistic view, understanding our patients not only as people with a medical condition needing treatment, but as members of a family, community, and society, informed our ability to be effective caregivers during the Ebola crisis.

Looking at our work through the spiritual lens yields further insights. As Christian physicians, our view of patients is informed by what the Bible says about them. Genesis 1:27 says, "*God created man in his own image, in the image of God he created him; male and female he created them*" (Genesis 1:27, English Standard Version). We view the patients we treat as bearers of the image of God. In the health care setting, we often lose sight of the uniqueness, the sacredness of the human being we are caring for. Sometimes our inability to recognize the divine in our patients is caused by the stress, over-commitment, and fatigue we are experiencing. Sometimes it is a result of our own biases, even unconscious ones. During the Ebola crisis, often the condition of the patients themselves made them appear less than human. We saw many patients who arrived at the hospital incapacitated, carried by a family member in a wheelbarrow or in their arms. Maintaining the dignity of these patients was a major challenge. Near the hospital, we could see very ill individuals with Ebola waiting on the ground outside packed ETUs hoping that a bed would become available. The constraints of the local situation and the lack of provisions needed to care for these patients left them helpless and dehumanized.

Keeping a focus on the spiritual became a way of maintaining perspective at ELWA, with daily times of Scripture reading, singing, and prayer. As we looked to God together, health care workers sought divine strength to keep serving and dignifying our patients in the midst of a situation which threatened to dehumanize both provider and patient. Miatta is an example of such a patient, who by the time she arrived at ELWA was unable to walk, feverish, with an open draining episiotomy wound, and a long-dead fetus causing a very unpleasant odor. It is easy to see how health facility staff, fearing for their own safety, could respond to such a patient with rejection, tempted to send her away to seek care elsewhere. Even though in the end an overwhelming bacterial infection claimed Miatta's life, the perspective that she is a child of God, deserving of love and dignity, helped us to provide care to her that was compassionate and responsive.

Looking at the doctor-patient relationship through the philosophical lens, the ideas of philosopher Martin Buber are salient. Buber defines genuine "dialogue" as "*where each of the participants really has in mind the other or others in their present and particular being and turns to them with the intention of establishing a living, mutual relationship between himself and them.*" (Buber 2007). When we consider interactions with a patient in the health care setting, it is useful to ask, "*Am I fully recognizing my fellow human being*?" Or, in the terms utilized by Buber, am I merely pursuing a "technical dialogue" in which the transfer of information is the objective? There were both physical and interpersonal barriers to establishing genuine dialogue with patients during the Ebola crisis. Health care providers performing any sort of procedure, like delivering a baby, were completely shrouded in protective gear, even their eyes covered by plastic goggles. With these physical constraints, it is especially challenging for providers to effectively communicate care and comfort to the patient. Buber speaks of "turning to" another, implicitly communicating love and recognition. During the Ebola crisis, many of the nonverbal cues we use to communicate care to the patient were lost—physical touch, unimpeded eye contact. Even tone of voice was muffled by a protective mask. Other authors who provided care in ETUs have shared how they attempted to make these connections with patients through gestures, tone of voice, and mutual gaze (Mobula 2014).

One of the major interpersonal barriers to dialogue was the lack of trust between patient and health care provider, manifested by dishonesty, often built on fear. Prior to a patient encounter at ELWA Hospital in June 2014, a referring doctor passed along the information that the patient had recently attended the funeral of a relative who had died. However, on arrival at our facility, the patient answered "*no*" to all the routine questions about attending a funeral or having anyone in the family who recently passed. The patient was only willing to own up to these facts when directly confronted by the provider with the clear-cut information provided by the referring doctor. This type of lying during patient intake, a privacy management strategy, is very common in the health care setting in Liberia and was more noticeable during Ebola. The lack of truth-telling formed a major barrier to the successful establishment of dialogue. Data from a study of Africans in the United States showed that privacy concerns were especially high in the African population (Sriphanlop et al. 2014). We found that, in order to successfully provide compassionate care, our approach had to be unconditional. The health care provider "turning to" the patient with care, concern, and compassion had to do so in a manner that was not contingent on the actions of the patient, specifically truth-telling. Especially in a desperate situation like the Ebola crisis, the establishment of the health care relationship depends primarily on the posture of the health care provider.

11.10 Conclusions

We have seen how the Ebola epidemic in 2014 made the most routine of human interactions—shaking a hand, getting into a taxi together, visiting the hospital for evaluation of a fever—into fraught experiences. The tiniest details, the splash of a droplet of body fluid or the application of a few ounces of bleach, could make the difference between normal life and a fatal outcome. Usually the relationship between health care provider and patient is one in which a professional, acting within their field of expertise, and at minimal risk to themselves, evaluates a patient who is ill and experiencing the fear of loss of life or limb. The patient hopes to get help and sound advice to guide them to health and recovery. During the Ebola crisis, however, we saw the relationship between health care provider and patient transformed into one in which both parties had to assess and evaluate risk. Many simply decided to avoid health care interactions altogether. Some health care workers in West Africa chose to stay away from work, resulting in the closure of health facilities. Most patients also chose to avoid health care facilities, eschewing routine care, vaccinations for children, and follow-up for chronic diseases like diabetes or HIV/AIDS. This mutual avoidance led to outbreaks of vaccine-preventable illnesses like measles among children in post-Ebola West Africa (Suk et al. 2016), as well as increases in mortality from stroke and AIDS (Ekyinabah et al. 2016). Hard national or regional data on increases in mortality is scant; studies using computer models provide estimates (Parpia et al. 2016). In addition, women who experienced normal labors stayed away from health facilities in Liberia (Ly et al. 2016; Strong and Schwartz 2016). However, women with complicated labors were unable to avoid health care, caught in situations without clear choices about where to go or what to do to navigate their way to a successful outcome.

Practicing medicine in Monrovia in 2014 was a humbling experience, in which professional expectations of what was "normal" and what was "acceptable risk" were overcome by the reality. Colleagues, even those who seemed to be carefully following guidelines for prevention of EVD, were getting sick with the disease. Initially told by health authorities that screening for fever would identify all Ebola cases, we later learned that pregnant women with EVD could present without a fever (Akerlund et al. 2015; Oduyebo et al. 2015). Changes in protocols and procedures had to be made on the fly, depending on the latest surprising events.

Health care providers during the Ebola crisis so often felt powerless to bring about a good outcome (Mobula 2014; Spencer 2015). It is especially distressing to know you have done your best and feel that things are going to work out, only to find out later that the patient has died. Health care providers entering a crisis situation should prepare themselves for the strong emotions created by both witnessing trauma and loss, and by the experience of impotency to impact it.

We have discussed factors which allowed ELWA Hospital, and specifically the authors of this chapter, to continue to provide health care to pregnant women during the Ebola crisis, a challenging environment that led other health facilities to close. In this chapter, we used several lenses to give perspective and clarity to our relationships with patients. Each health care provider approaches their work with their own identities, their own professional, spiritual, and philosophical resources. When a health care provider enters into a crisis situation, it is important to develop the self-awareness to identify and tap into those foundational attitudes and commitments. Such awareness will both provide a measure of resilience to health care providers and improve the quality of their therapeutic relationships.

Fig. 11.14 The authors, 2015 (Rick Sacra, left, John Fankhauser, right)

We are very grateful that ELWA Hospital was able to help many women during the Ebola crisis. We deeply regret the loss of several of our coworkers to EVD. In the end, we found ourselves practicing medicine outside of our comfort zones and could only trust God that we were doing the best we could with what we had, and that He would take that effort and bless it (Fig. 11.14).

References

Akerlund, E., Prescott, J., & Tampellini, L. (2015). Shedding of Ebola virus in an asymptomatic pregnant woman. *New England Journal of Medicine, 372*(25), 2467–2469. Retrieved December 14, 2017, from http://www.nejm.org/doi/10.1056/NEJMc1503275.

Bryan, J. E. (1968). *The role of the family physician in America's developing medical care program: A report and commentary*. St. Louis: Warren H. Green, Inc.

Buber, M. (2007). Dialogue. In R. Craig & H. Muller (Eds.), *Theorizing communication readings across traditions* (pp. 225–237). Los Angeles: Sage.

Centers for Disease Control and Prevention and World Health Organization. (1998). *Infection control for viral hemorrhagic fevers in the African health care setting*. Retrieved December 1, 2017, from https://www.cdc.gov/vhf/abroad/vhf-manual.html.

Ekyinabah, E. K., Okiror, D., Ssentamu, J. V., Babua, C., & Njoh, J. N. (2016). *Conditions associated with mortality among hospitalized patients in the medical ward of the John F Kennedy Medical Center-Monrovia, Liberia*. Monrovia: Liberia College of Physicians and Surgeons Scientific Session.

Fadiman, A. (1998). *The spirit catches you and you fall down: A Hmong child, her American doctors, and the collision of two cultures*. New York: Noonday Press.

Lopez, K. A., & Willis, D. G. (2004). Descriptive versus interpretive phenomenology: Their contributions to nursing knowledge. *Qualitative Health Research, 14*(5), 726–735.

Ly, J., Sthananthan, V., Griffiths, T., Kanjee, Z., Kenny, A., Gordon, N., et al. (2016). Facility-based delivery during the Ebola virus disease epidemic in rural Liberia: Analysis from a cross-sectional, population-based household survey. *PLoS Medicine*. https://doi.org/10.1371/journal.pmed.100209. Retrieved November 15, 2017, from http://journals.plos.org/plosmedicine/article?id=10.1371/journal.pmed.1002096.

Medecins Sans Frontieres. (2014). *Field notes: The largest Ebola management centre ever built*. MSF's ELWA 3, Monrovia, Liberia. Retrieved December 1, 2017, from https://www.youtube.com/watch?v=loUQWgf00Uc.

Mobula, L. (2014). Courage is not the absence of fear: Responding to the Ebola outbreak in Liberia. *Global Health: Science & Practice, 2*(4), 487–489. Retrieved February 25, 2018, from http://www.ghspjournal.org/content/2/4/487.

Nyenswah, T.G., Westercamp, M., Kamali, A.A., Qin, J., Zielinski-Gutierrez, E., et al. (2014). Evidence for declining numbers of Ebola cases-Montserrado County, Liberia, June–October 2014. *Morbidity and Mortality Weekly Report, 63*, 1–5n. Retrieved December 12, 2017, from https://www.cdc.gov/mmwr/preview/mmwrhtml/mm63e1114a2.htm.

Oduyebo, T., Pineda, D., Lamin, M., Leung, A., Corbett, C., & Jamieson, D. (2015). A pregnant patient with Ebola virus disease. *Obstetrics and Gynecology, 126*(6), 1273–1275.

Parpia, A., Ndeffo-Mbah, M., Wenzel, N., & Galvani, A. (2016). Effects of response to 2014-2015 Ebola outbreak on deaths from malaria, HIV/AIDS, and tuberculosis, West Africa. *Emerging Infectious Diseases, 22*(3), 433–441 Retrieved November 18, 2017, from https://wwwnc.cdc.gov/eid/article/22/3/15-0977_article.

Penfield, J. A. (1981). Quoting behavior in Igbo society. *Research in African Literatures, 12*(3), 309–337.

Smith, J. A. (1996). Beyond the divide between cognition and discourse: Using interpretive phenomenological analysis in health psychology. *Psychology & Health, 11*(2), 261–271. https://doi.org/10.1080/08870449608400256.

Solomon, M. (1985). The rhetoric of dehumanization: An analysis of medical reports of the Tuskegee syphilis project. *Western Journal of Speech Communication, 49*(4), 233–247.

Spencer, C. (2015). Having and fighting Ebola — Public health lessons from a clinician turned patient. *New England Journal of Medicine, 372*(12), 1089–1091 Retrieved December 14, 2017, http://www.nejm.org/doi/10.1056/NEJMp1501355.

Sriphanlop, P., Jandorf, L., Kairouz, C., Thelemague, L., Shankar, H., & Perumalswami, P. (2014). Factors related to hepatitis B screening among Africans in New York City. *American Journal of Health Behavior, 38*(5), 745–754. https://doi.org/10.5993/AJHB.38.5.12.

Strong, A., & Schwartz, D. A. (2016). Sociocultural aspects of risk to pregnant women during the 2013-2015 multinational Ebola virus outbreak in West Africa. *Health Care for Women International, 37*(8), 922–942.

Suk, J., Jimenez, A., Kourouma, M., Derrough, T., Balde, M., Honomou, P., et al. (2016). Post-Ebola measles outbreak in Lola, Guinea, January-June 2015. *Emerging Infectious Diseases, 22*(6), 1106–1108 Retrieved November 13, 2017, from https://www.ncbi.nlm.nih.gov/pmc/articles/PMC4880080/.

12 Risk and Recognition: The Traditional Midwives Who Filled the Gap in the Time of Ebola

Theresa Jones

12.1 Introduction

Long before Western health systems were established in Liberia, women delivered their babies alone, with a family member, or with a traditional birth attendant (TBA). Lori (2009) recounts the practices, beliefs, and spirituality of traditional approaches to delivery and the trust placed in these community-based sources of care. In the 1990s, a number of TBAs were formally linked to the increasingly established health system after receiving training from the Liberian Ministry of Health on safe home deliveries. They were given the name Traditional Trained Midwives (TTM).

In 2010, the maternal mortality ratio (MMR) in lower- and middle-income countries was calculated as 15 times higher than that of high-income countries, with Liberia the seventh highest at 770 maternal deaths per 100,000 live births (WHO 2010). Among the various strategies to address this disparity, skilled attendance at all births was recommended to curb maternal and neonatal mortality and morbidity (WHO 2005; Campbell and Graham 2006).

The Liberian National Reproductive and Sexual Health Policy (2010) outlined engaging traditional midwives as birth supporters *rather than* as birth attendants. New roles and responsibilities included counseling on essential newborn care, breastfeeding, and nutrition; encouraging women to seek care at health facilities; recognizing and referring pregnant women with danger signs to health facilities; and assisting women to develop birth preparedness plans. With income from actually carrying out deliveries potentially lost, the policy recognized the need to compensate traditional midwives for their new role. This was supported by a number of aid agency programs, yet there was no formal government structure established for this.

In March 2014, the most widespread Ebola epidemic ever recorded entered Liberia. Without the capacity to contain or withstand the virus, numerous health care facilities across the country restricted services or, in many cases, shut down entirely. This included a mass withdrawal of facility-based basic and comprehensive emergency obstetric care.

One key example was the Redemption Hospital, the only free-of-charge government hospital in Liberia's capital city Monrovia, located in the slum community of New Kru Town. The hospital site, opened in 1982, was never intended to be a hospital with the emergency room previously a market hall. Prior to the Ebola outbreak, there were 400–500 staff and only 206 beds at Redemption, serving a city of 1.5 million.

T. Jones (✉)
International Rescue Committee, Monrovia, Liberia

D. A. Schwartz et al. (eds.), *Pregnant in the Time of Ebola*, Global Maternal and Child Health,
https://doi.org/10.1007/978-3-319-97637-2_12

Denial of the virus' existence and rumors of it being introduced and transmitted by health workers led to violent clashes between Redemption hospital staff and the surrounding community. At the peak of Ebola in July 2014, the in-patient services closed ending the availability of obstetric and neonatal care for a period of approximately 6 months. Hospital management reports show few health workers remained during its temporary status as a holding center for suspected Ebola cases, and many found work in Ebola Treatment Units (ETUs) (see Miller (2016) for full accounts). Anecdotal evidence suggests that traditional midwives (including both TTMs and TBAs) from communities around Redemption mobilized to fill the resulting gap in care (S. Saytue, personal communication, July 18, 2015).

This chapter narrates the accounts of three such women. Specifically, it relates what they did, why they had to, how they were able to, and what the global health world can learn for the future. We finish by hearing one attempt to rebuild a maternal health system by recognizing the needs and resources of the community it serves.

12.2 Under The Bullets—Theresa Jayennah

Theresa Jayennah is a 60-year-old resident of Funday Community in New Kru Town.[1] Her mother was a professional nurse at Catholic Hospital, Monrovia, and she started by practicing deliveries with her. On May 19th, 1984, Theresa graduated from traditional midwife training in JFK Hospital, Monrovia, and was sent to Niklay Clinic to support the maternity clinic there. Liberia entered its first civil conflict in 1989. She relates:

> *In 1990 the world fell down and there was nowhere to go. No hospitals. That time I was fresh and strong, so I collected people to deliver them. We were doing this under the bullets because if the rebel caught a pregnant woman they will say "we will open your stomach". Men were begging me to help their wives, but we were all under curfew. If someone went into labour after 5pm, you would be surrounded by bullets to get there.*
> *1999, World War 2 was the same thing again. Everyone ran away and no-one was at hospital. If someone comes to you crying you cannot turn them away. World War 1 everything fell down. World War 2 everything fell down again. So Ebola was nothing new to me.*
> *I got enough chlorine, wearing two socks and two gloves. Afterwards I take a bath and spray all over the house. Sometimes I would have four or five women everyday coming to me. The community around you are angry at you for delivering, but what can you say?*
> *I was happy because of the lives saved. If women had Ebola I sent them to the centre. If not, I kept them in my house and kept them safe. Every night I would pray. I asked the pregnant women questions. One lady said she was vomiting. This is normal but she would be refused at hospital or sent to the Ebola centre. I told her it was normal, she was relieved.*
> *During the wars I can do all this with my bare hands. Ebola could not allow us that. But people are holding me, they are not satisfied until you touch or pet them. People would say "Old Ma is touching you too much".*
> *Our big big people say we should carry to Redemption Hospital again now.*

Theresa's story touches on why such gaps in medical services became a reality during the Ebola crisis. The Ebola epidemic revealed health system weaknesses in terms of numbers of qualified health workers, infrastructure, logistics, health information, surveillance, governance, drug supply systems, and management of health services (Keiny et al. 2014). A reliance on care outside of the formal health system during crises points in Liberia was already well-documented. Lori (2009) has discussed the confidence that community-based and traditional systems earned because they stayed available during the wars, while many health facilities closed.

[1]New Kru Town is a town and northwestern coastal suburb of Monrovia, Liberia. It is located on the north end of Bushrod Island, was officially founded by the Liberian National Legislature as the Borough of New Kru Town in 1916, and initially grew as a planned "transplant" town of Old Kru Town. Old Kru Town had been evacuated for the development of a new breakwater for the new port. Being located on the corner of the Atlantic Ocean and the Saint Paul River estuary, fishing is an important source of income. New Kru Town, considered to be a slum area, and has been subject to ethnic tension and contains the Redemption Hospital.

Acceptance of traditional methods, belief, and comfort in community, as well as poor understanding and negative rumors about medical interventions, was also well-documented before the Ebola epidemic (Lori 2009). Ebola exacerbated these, with trust and fear of health facilities recognized as key to the fall in facility deliveries across Liberia—even in counties where facilities remained open, facility-based deliveries dropped by 30% (Ly et al. 2016).

Despite the policy of the World Health Organization (WHO) for facility births, even at the height of health system strength in 2013, 39% of women were delivering outside of health facilities (LISGIS et al. 2014). Theresa's account suggests that community systems of care were still very much alive, and thus available for women to rely on. This brings to light a resilience to operate during times of crisis and to remain a source of competition even in good times for the formal health system.

12.3 A Gift from God—Susie Saytue

Susie Saytue is many things to her community of St Paul Bridge. She is a local leader, she heads a women's group, and she is a traditional trained midwife (TTM). Everyone knows her as "*Mama Gee*" ("Gee" for General), a name given to powerful women inspired by the Nigerian movie scene. During the Ebola epidemic, she was a member of her community's Ebola Task Force, a group of volunteers working to identify cases, support quarantined families, and educate people of preventative methods.

> *This issue of Ebola we never knew, it was our first experience. As soon as they told us I took precaution. We had no PPE, no protective gear. I used to go to market and buy a raincoat, use gloves and double gloves. Rubber bands around our wrist.*
> *There was nowhere for women to go. All the people trained I called together. Let us save people's lives I said. Let us pick the girls up who are in pain and were rejected by the facility and bring them here by car. Even get the ambulances to bring them here. Let us not let them die in front of Redemption in a wheelbarrow.*
> *For some women the baby died in their stomach and was rotting. When women came they had fear I would not accept them either. When they came I encourage them. I talk to them. They relax. I was doing those things free.*
> *I delivered several Ebola patients. They had red eyes, stomach running, vomiting. After delivery I told them to find an ETU. I stayed in communication with them if they went inside.*
> *One girl. Her whole family died of Ebola. All the signs and symptoms showing. The girl's stomach was running all the way inside my house.*
> *"Mama can you help me?"*
> *"Yes, I can help you"*
> *The baby had died inside 2-3 days before. The head was soft and rotting hair was coming off. I told the family and they said the girl could smell a rotting thing. I told the family to take her to an ETU when we were through and gave them a wheelbarrow. They took her to a church instead. Two others then died. I put myself on 21 days every day. I kept away from my family.*
> *Even though fear came to me. My skin would get hot, my head would hurt. But it was fear. God was on my side. I told them my work is life-saving work.*
> *My community was too happy with me. I was not picking people out. No problem, she saved our women's lives.*
> *After Ebola subsided we had a small church gathering. A government Minister was there. I asked her:*
> *"You know you have TTMs and TBAs in this district?"*
> *"Yes, before."*
> *"No, now. July to September I did 150 deliveries in the presence of Ebola"*
> *"Wow"*
> *"So, minister. Where was your degree when Ebola entered this country?"*
> *She said she would make sure we were recognised.*
> *But Liberians believe in theory, not practical. A book is one thing. Practice is another.*
> *Nothing else is as important as this work. It is a gift from God. Anything given to you by God is hard to leave you.*

Traditionally, the birthing process is more than a medical procedure for Liberian women; it is a deeply cultural and spiritual experience. Western medicine is still unfamiliar to many, and problems during childbirth are largely explained by spiritual or supernatural causes (Lori 2009). Susie explains how her belief in a divine purpose and a higher protection helped her decision to face risk and offer care for her community.

Theresa's and Susie's account both show a reshaping of their traditional models of care to include more biomedical aspects. Susie described augmenting the birthing skills learned from her mother with the "*qui side*" or "*the civilised*" medical theories. During Ebola, perhaps it was making these conflicting philosophies fit that kept many protected, alive, and able to offer care. Kleinman (1980) argued the flipside, that to improve uptake, effectiveness, and resilience of western medicine, we need to "*reshape the biomedical model*" and include more social and cultural aspects of care.

In a post-Ebola survey conducted in New Kru Town, pregnant women defined "*good quality care*" during the birthing process as being about more than drugs and medicine. In fact, the harsh way health workers talk to them was the major deterrent to returning to reopened health facilities (Jones et al. 2018). Reassurance and good communication is mentioned in both Theresa's and Susie's stories. Models of care that fit the beliefs and expectations of those seeking it are more sought after (Kleinman 1980) and may explain how traditional midwives have been able to retain resilience in good times for the Liberian health system and at crisis points for Liberia.

12.4 When the Big Big Doctors Run Away, We Will Still Be Here—Finda Halay

Finda Halay is a 32-year-old woman from Popo Beach, New Kru Town. She has one son. Finda is a registered nurse and is soon to be a qualified registered midwife (RM). Although she has had medical training, she considers herself first as a traditional midwife.

> *I have worked in counties with no doctors, only nurses. Most die from the placenta not coming out. In the counties it is almost like war - woman are in labour for and have to walk 12 hours to a health facility, it is like fighting a war.*
> *Throughout the country this community had the highest mortality rate for Ebola. In Popo Beach they were collecting 7 to 8 people every day to take them to the Ebola hospital. All the NGOS came here.*
> *I excommunicated myself from my family and was engaging in the work in my house. I have a clinic I made in the back here. I would make chlorine water. I got sacks, shoes, gloves. I bought my PPE from Redemption. Usually when the next person came it was still wet from washing it*
> *I was not only doing delivery, but treatment too. It was more than only Big Bellies. Everyone was coming to me. One lady she lied and said she was pregnant to see me. I did a pregnancy test and touched the urine. I took risks and 2 to 3 days later that person had died.*
> *I was afraid small when a friend of mine died. He was not careful, he was not protecting himself. He believed it was all witchcraft. I was too confident, no fear came in me. The only fear was when my friend died.*
> *I have applied as a RM. It is only the paper now really. I will remember where I came from. Realistically speaking, TTMs and TBAs are for people to hold. It is not good to push them aside. Make them feel they are part of something. When the Big Big doctors run away, we will still be here.*
> *When the last cases of Ebola came to surprise everyone the chairman at our community level gave us chlorine. Even at the Ministry level they gave first priority to us. If anything pulls back again the Ministry of Health told us to be careful. To be careful, not to not do it. They still depend on us. That is why they should not push us aside.*

Finda's account looks a little more to the future. World Health Organization (2015) reported that the Ebola epidemic had *"exacerbated the pre-existing shortage of health workers, high rates of attrition, uneven distribution, poor employment conditions and gaps in OHS*[2] *in the three countries."* A total of 8.07% of health workers in Liberia died from Ebola virus disease. Other ongoing impacts of the Ebola outbreak include new Infection Prevention Control (IPC) measures limiting bed space at the largest free government health facility (Jones et al. 2018). The resources lost due to Ebola have been predicted to risk bringing Liberia back to its 1995 rates of maternal mortality (Evans et al. 2015).

[2] Occupational Health and Safety.

Costello et al. (2006) identified that every woman deserves a skilled attendant during her childbirth, but in many developing countries the infrastructure and resources in human capital to support this strategy do not exist. Also in the context of Ebola virus infection, anthropological perspectives have highlighted the harm that promoting a purely biomedical approach can do by breaking down established social structures (Minor Peters 2014; Chandler et al. 2014). With new crises emerging across the world and post-Ebola funding to support health systems unlikely to last, Finda's account gives a stark warning not to undermine the supports that you "*still depend on*".

In a literature review of the role of traditional midwives, Piper (1997) suggests that their role should be recognized as long as pregnant women continue to seek their services, and that any training should be a two-way process with both parties learning from one another. Redemption Hospital in its reopening phase offers one example of a recognizing and learning process.

12.5 Participating in Health Service Reopening

After Redemption Hospital was reopened in January 2015, in comparison to the other departments, the Delivery Service uptake was slow to improve. The reasons for this were unknown by the hospital management and the International Rescue Committee (IRC) who were supporting the reopening process. In early 2015, the New Kru Council, Redemption Hospital, IRC, and TARSC (Training and Research Support Centre) initiated participatory action research (PAR).

The methodology aims to establish processes that address power dynamics and build relationships between stakeholder groups and service providers. PAR facilitates those affected by health circumstances to be researchers themselves and to generate and use their own knowledge of the local context to improve access to and quality of maternal health services.

The key questions for the project were how Ebola had weakened the system and what solutions could sustainably address these weaknesses and build on system strengths. The aims were to improve and sustain communication and participation from different levels of the maternal health system, to identify shared points of action, and to learn from these actions. A baseline and follow-up survey, qualitative data from the PAR meetings, and a number of facility indicators helped to answer these questions and to measure impact.[3]

The process involved pregnant women, community traditional midwives, local leadership, and health workers from antenatal care and maternity services. The group included Theresa, Susie, and Finda. Meetings were held with each group on maternal health needs and services, the impact of the Ebola epidemic, and the priority issues for service improvement. A joint meeting brought all groups together to identify common priorities, a shared action plan to make things better, and a coordination committee to review actions.

Each group had their own motivations to take part. Broadly, the traditional midwives wanted to be recognized for their efforts during Ebola and to remain a valid part of the health system, the health workers wanted to build bridges with the local community after the tense Ebola conflicts, and pregnant women wanted to secure a safe and respectful place to deliver.

The process highlighted the disconnection between community supports, primary health services, and the hospital, which Ebola had worsened. There were key social and cultural barriers to facility delivery. As already mentioned, "good quality care" for pregnant women in New Kru Town meant more

[3]The experiences, methods and tools of the Regional Network for Equity in Health in East and Southern Africa (EQUINET) and particularly its learning network on PAR coordinated by Training and Research Support Centre (TARSC) were used (Mbwili-Muleya et al. <CitationRef CitationID="CR15" >2008</Citation Ref>; Loewenson et al. <CitationRef CitationID="CR12" >2014</Citation Ref>; Loewenson et al. <CitationRef CitationID="CR11" >2006</ Citation Ref>). The full report can be found Jones et al. (in press).

than the availability of drugs and required more of the human side of care their traditional midwives provided. These factors were less well-recognized and underestimated by hospital personnel. The process revealed that traditional midwives did not feel recognized for risks taken during Ebola. They also felt they had not yet received enough training, had their roles as "*supporters*" formalized or received an ID badge as outlined in the pre-Ebola Liberian National RSH policy (Ministry of Health and Social Welfare, 2010). This resentment at the community level created another barrier for pregnant women to access reopened health facilities.

The actions selected by the group included better collaboration of traditional midwives and hospital staff through training, monthly meetings, and being allowed to accompany women into the delivery room; improving capacity of local clinics through training and equipment donation to relieve pressure on Redemption Hospital; and formation of drama groups of pregnant women to enhance community awareness of hospital procedures and good communication skills.

A review meeting and survey readministration after 4 months found high agreement that the communication and interaction between communities and health services improved, as did the perceived quality of hospital services. However, the primary care services remained weak and underutilized, and the number of hospital beds was inadequate. Through the PAR process in New Kru Town, the traditional midwives and birth attendants and hospital staff felt they had learned to work together better. The approach took the spirit of learning from one another and accepting that "*there are things you know that I do not*".

By March 2016 with the deadly epidemic firmly in the past, the Liberian health system was still in a changing climate. The traditional midwives were navigating whether patients' families will reimburse them for their role as "*accompaniers*" to hospital, or whether they will lose their livelihoods as people take themselves straight to hospital or the clinics. One traditional midwife has plans to build a community clinic to relieve pressure on the Redemption Hospital and to offer safe and respectful care which she hopes will be validated by the Ministry of Health. Again, an action that highlights resilience from a readiness to adapt with the needs of a context and a changing system.

12.6 Conclusion

The three accounts shone light on what these women did during the Ebola epidemic. They operated in a hazardous environment with little equipment for infection control, often using plastic bags or rewashing black-market PPEs, in an environment of great fear, rumor, and uncertainty, especially regarding the health status of delivering pregnant women. Hayden (2015) argues that conditions naturally arising during deliver can be difficult to distinguish from Ebola symptoms, meaning these women were a highly feared but extremely vulnerable social group.

The reasons why they had to do this included the structural weaknesses and under-resourcing of the formal health system and ongoing preference of traditional methods for delivery rather than medical approaches. This preference to be attended to by those who understand customs and community norms is universal (Adams et al. 2005; Berry 2006; Chapman 2006).

For traditional midwives to operate during times of crisis and be a genuine source of competition even in good times can be linked to their role being more a vocation than a profession, and their definition of good care being more in-line with that of the pregnant women they support. The ability to unite traditional and biomedical philosophy undoubtedly helped keep them alive and relevant during the outbreak.

In terms of what we can learn as a global health body, we saw how the system of facility deliveries crumbled during Ebola, and people reverted to traditional ways to deliver. We reviewed some of the factors that make these traditional supports able to operate during crisis points and even in good times for the Liberian formal health system. A key takeaway is to ensure meaningful two-way learning between different approaches to maternal health, especially as the residual effects of crises can make the need to work with community-based supporters greater than ever.

The post-Ebola epidemic climate presented an opportunity for both the New Kru Town community and Redemption Hospital to recognize, rather than to condemn, risks taken by all groups during the crisis and a chance to build more shared definitions of a safe and respectful birthing process. Whether this can translate into a more resilient and acceptable health system is not yet clear; however, having a negotiated process rather than a prescribed one may help ensure this community does not suffer such gaps in health care in the future.

As a final word, this chapter has not focused significantly on the experiences of pregnant women themselves. In common with many resource-poor countries, Lori (2009) described how pregnant women in Liberia tend to be the least powerful person in their birth process in Liberia. This highlights the risk of their voice being lost in power struggles between formal and informal systems. Methods for learning must give pregnant women a voice and allow them to be directly involved in their obstetrical care.

12.7 Photo Gallery (Figs. 12.1, 12.2, 12.3, 12.4, 12.5, 12.6, 12.7, 12.8, and 12.9)

Fig. 12.1 The New Kru Town community. Photo by the author

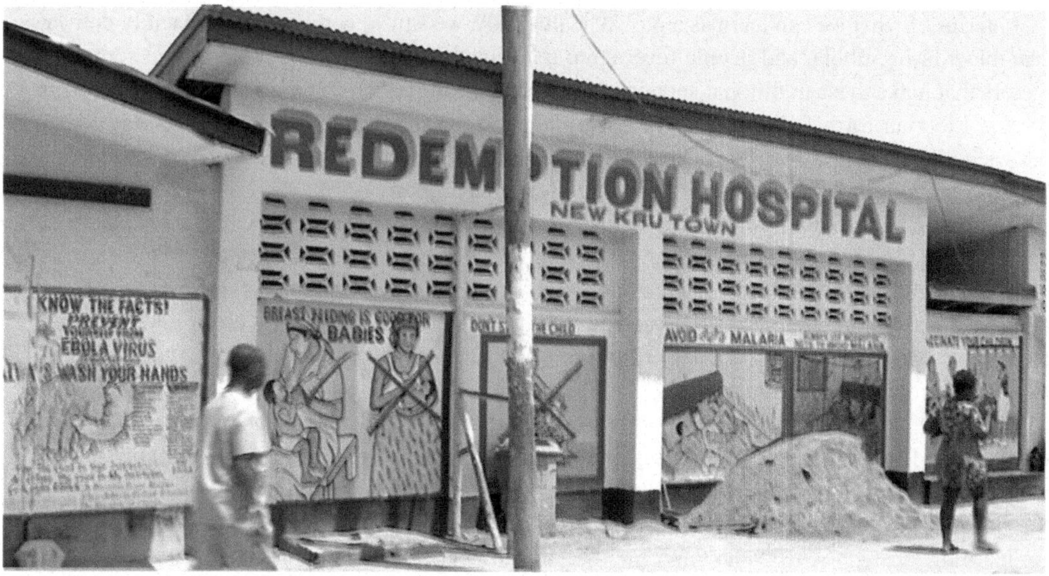

Fig. 12.2 The Redemption Hospital in Monrovia. In August 2014, the hospital was at the center of the Ebola epidemic in Liberia. It quickly became overwhelmed and, for a time, had to close its in-patient department. Photograph courtesy of the World Health Organization

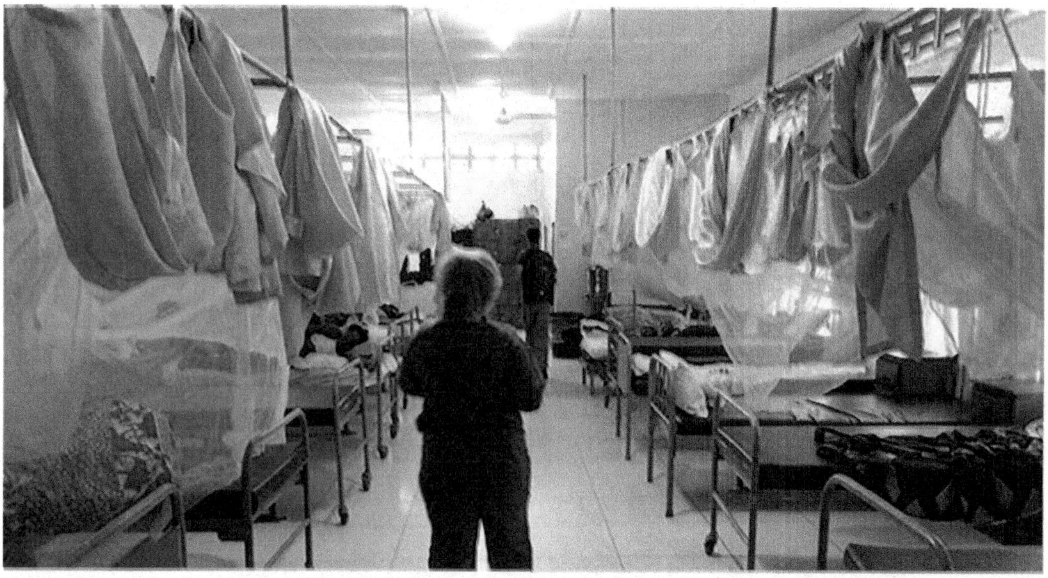

Fig. 12.3 Before infection precaution policies (IPC) were in place at Redemption Hospital, the beds were very close to each other. Patients often doubled up together on a single bed, or even slept on the floor due to overcrowding, supporting the potential for disease spread. Now, the beds are spaced safely 1 m apart, the ward is clean and uncrowded, and the mosquito netting protects against insects carrying malaria, which causes fever and can be mistaken for Ebola. These measures are now part of Liberia's minimum standards for safe care provision. Photograph courtesy of the World Health Organization

Fig. 12.4 Mural painted outside of the Redemption Hospital upon reopening in January 2015 to prepare community members on the new hygiene measures to expect when entering the hospital. Photo from the author

Fig. 12.5 The exterior of the Redemption Hospital is covered with murals with public health messages painted by local artists. This mural illustrates what the community can do daily to create a more hygienic environment that can help stop the spread of disease. Photograph courtesy of the World Health Organization

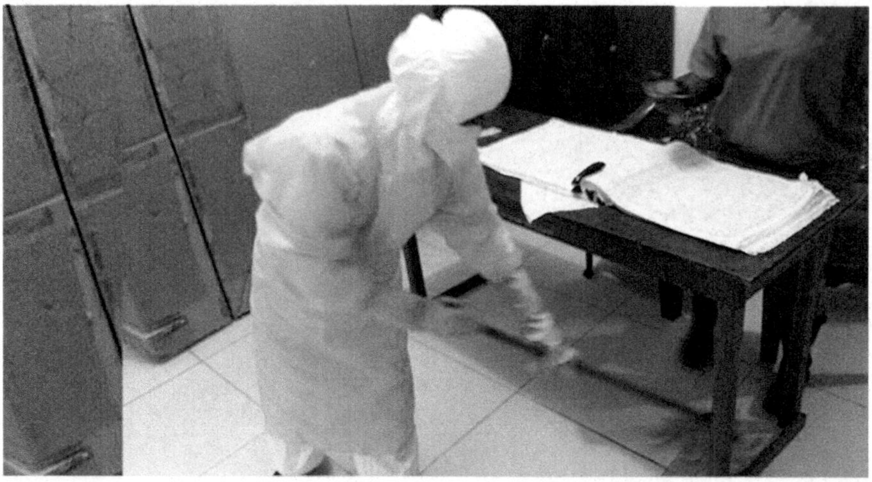

Fig. 12.6 Nonclinical health care workers, such as the maintenance woman in this photograph at the Redemption Hospital, were essential to breaking the infection transmission chain at hospitals; however, they were also at risk of becoming infected. Proper IPC training, personal protection equipment, and staff payment systems that account for hazardous conditions are key to motivating the workforce, from doctors to janitors, and keeping them on the job. Photograph courtesy of the World Health Organization

Fig. 12.7 A billboard being prepared to show New Kru Town community members what was happening to disinfect and renovate the hospital in January 2015. Photo by the author

Fig. 12.8 Community meeting in New Kru Town during the reopening process in January 2015. Photo from the author

Fig. 12.9 A Participatory Action Research meeting between Redemption health workers, pregnant women of New Kru Town, and New Kru Town TBAs and TTMs. Photo by the author

References

Adams, V., Miller, S., Chertow, J., Craig, S., Samen, A., & Varner, M. (2005). Having a "safe delivery": Conflicting views from Tibet. *Health Care for Women International, 26*, 821–851.

Berry, N. (2006). Kaqchikel midwives, home births, and emergency obstetric referrals in Guatemala: Contextualizing the choice to stay home. *Social Science & Medicine, 62*, 1958–1969.

Campbell, O. M., & Graham, W. J. (2006). Strategies for reducing maternal mortality. Getting on with what works. *Lancet, 368*, 1284–1299.

Chandler, C., Fairhead, J., Kelly, A., Leach, M., Martineau, F., Mokuwa, E., et al. (2014). Ebola: Limitations of correcting misinformation. *Lancet, 385*, 1275–1277.

Chapman, R. (2006). Chikotsa—Secrets, silence, and hiding: Social risk and reproductive vulnerability in central Mozambique. *Medical Anthropology Quarterly, 20*(4), 487–515.

Costello, A., Azad, K., & Barnett, S. (2006). An alternative strategy to reduce maternal mortality. *Lancet, 368*, 1477–1479.

Evans, D. K., Goldstein, M. & Popova, A. (2015). Health-care worker mortality and the legacy of the Ebola epidemic. *The Lancet Global Health, 3*(8), E439–E440.

Hayden, E. C. (2015). Maternal health: Ebola's lasting legacy. *Nature, 519*, 24–26.

Keiny, M., Evans, D., Schmets, G., & Kadandale, S. (2014). Health-system resilience: Reflections of the Ebola crisis in western Africa. *Bulletin of the World Health Organization, 92*, 850. Retrieved November 12, 2017, from http://www.who.int/bulletin/volumes/92/12/14-149278/en/.

Jones, T., Ho, L., Kun, K., Shakpeh, J. & Loewenson, R. (2018). Rebuilding people-centred maternal health services in post-Ebola Liberia through participatory action research. *Global Public Health, 13*(11), 1650–1669. https://doi.org/10.1080/17441692.2018.1427772.

Kleinman, A. (1980). *Patients and healers in the context of culture*. Los Angeles: University of California Press.

LISGIS [Liberia Institute of Statistics and Geo-Information Services], Ministry of Health and Social, Welfare [Liberia], National AIDS Control Program [Liberia], ICF International. (2014). *Liberia Demographic and Health Survey 2013*. Monrovia, Liberia.

Loewenson, R., Kaim, B., Chikomo, F., Mbuyita, S., & Makemba, A. (2006) *Organizing people's power for health: Participatory methods for a People Centred Health System*. PRA toolkit. Ideas Studio, South Africa.

Loewenson, R., Laurell, A. C., Hogstedt, C., D'Ambruoso, L., & Shroff, Z. (2014). *Participatory action research in health systems: A methods reader*. Harare: IDRC/CRDI Canada and World Health Organization, EQUINET.

Lori, J. R. (2009). *Cultural childbirth practices, beliefs and traditions in Liberia* (Doctoral dissertation, College of Nursing, University of Arizona).

Ly, J., Sathananthan, V., Griffiths, T., Kanjee, Z., Kenny, A., Gordon, N., et al. (2016). Facility-based delivery during the Ebola virus disease epidemic in rural Liberia: Analysis from a cross-sectional, population-based household survey. *PLoS Medicine, 13*(8), e1002096. https://doi.org/10.1371/journal.pmed.1002096. Retrieved November 12, 2017, from http://journals.plos.org/plosmedicine/article?id=10.1371/journal.pmed.1002096.

Mbwili-Muleya, C., Lungu, M., Kabuba, I., Zulu Lishandu, I., & Loewenson, R. (2008). *EQUINET Participatory Research Report. An EQUINET PRA project report*. Harare: EQUINET. Retrieved November 12, 2017, from www.equinetafrica.org/sites/default/files/uploads/documents/PRAequitygauge2008.pdf.

Miller, L. (2016). *Whenever light enters darkness, the places becomes bright. Evaluation of IRC support of the Restoration of Health Services at Redemption Hospital*. Liberia: International Rescue Committee.

Ministry of Health & Social Welfare. (2010). *National reproductive & sexual health policy*. Liberia: Ministry of Health & Social Welfare. Retrieved November 12, 2017, from www.liberiamohsw.org/Policies%20&%20Plans/National%20Sexual%20&%20Reproductive%20Health%20Policy.pdf.

Minor Peters, M. (2014). *Community perceptions of Ebola response efforts in Liberia: Montserrado and Nimba counties*. Ebola Response Anthropology Platform, 2015. Retrieved November 12, 2017, from http://www.ebola-anthropology.net/case_studies/community-perceptions-of-ebola-response-efforts-in-liberia-montserrado-and-nimba-counties/.

Piper, C. J. (1997). Is there a place for traditional midwives in the provision of community health services? *Annals of Tropical Medicine and Parasitology, 91*(3), 237–245.

World Health Organization. (2005). *The world health report (2005) —make every mother and child count*. Geneva: World Health Organization. Retrieved November 10, 2017, from http://www.who.int/entity/whr/2005/whr2005_en.pdf?ua=1.

World Health Organization. (2010). *Working with individuals, families and communities to improve maternal and newborn health*. Retrieved October 4, 2017, from http://whqlibdoc.who.int/hq/2010/WHO_MPS_09.04_eng.pdf.

World Health Organization. (2015). *Health worker Ebola infections in Guinea, Liberia and Sierra Leone*. A preliminary report 21 May 2015. Retrieved November 18, 2018, from http://www.who.int/hrh/documents/21may2015_web_final.pdf

13 Having Belly During Ebola

Janice L. Cooper and Meekie J. Glayweon

13.1 Introduction

In 2014, the Liberia declared a public health emergency because of an outbreak of Ebola Virus Disease (EVD), an emergency that lasted nearly 2 years. The West African Ebola epidemic was by far the largest ever recorded, impacting women who were pregnant in a disproportionally severe manner. Research has shown that pregnant women are more susceptible to infectious diseases and are at higher risks for poor clinical outcomes when compared with nonpregnant women (Beigi 2017). In Liberia, over 4800 people are estimated to have lost their lives during the Ebola outbreak. An estimated 10,700 were infected, including approximately 550 children. Nearly 8000 children lost one or both of their parents as a result of the epidemic (Ministry of Health 2016a). At some points during the outbreak in Liberia, all health facilities were closed. In some places where health facilities were open or reopened, the staff refused to accept or treat pregnant women (Hessou 2014). There were drastic declines in the use of maternal and child health services. Iyengar and colleagues reported that access to formal primary health services, especially maternal and child health care, was so severely compromised that prenatal care access declined drastically reaching only 9–14% of peak utilization rates, with facility-based deliveries reduced to less than 10% (Iyengar et al. 2015). Iyengar et al. also found that 65% of health facilities shut down during the peak of the EVD epidemic. The World Bank predicted over 110% increase in maternal mortality and a 20% increase in infant mortality (Evans et al. 2015).

For women who were pregnant during the EVD epidemic in Liberia, closure of health facilities, the fact that some health facilities were often sources of infections, and the occurrence of infections and deaths from EVD at health facilities all contributed to the belief that healthy pregnant women should avoid these facilities. Even in places where the prevalence of EVD was relatively low, utilization of

J. L. Cooper (✉)
The Carter Center Mental Health Program, Monrovia, Liberia

Incident Management Team and National Ebola Survivor Network, Monrovia, Liberia
e-mail: janice.cooper@cartercenter.org

M. J. Glayweon
Incident Management Team and National Ebola Survivor Network, Monrovia, Liberia

PREVAIL-III Natural History Study, J.F.K. Hospital, Monrovia, Liberia

D. A. Schwartz et al. (eds.), *Pregnant in the Time of Ebola*, Global Maternal and Child Health,
https://doi.org/10.1007/978-3-319-97637-2_13

pregnancy-related services declined by 30% (Ly et al. 2016). In addition, women who were pregnant faced anxiety at the impending delivery of their children. Early post-EVD research suggested that the closure of hospitals, combined with fear of infections on the part of patients and health care workers, led to declines in accessing prenatal care that ranged from 9% to 14% of peak utilization during months with the highest rates of EVD cases (Iyengar et al. 2015).

We document experiences of women of reproductive age who were pregnant during the EVD crisis. This chapter also outlines some of the factors that impacted their prenatal health-seeking behaviors and the factors that propelled their choice of providers and the outcomes of their pregnancies. Linking the crisis in maternal health in Liberia during Ebola with the fears that were associated with EVD, this study focuses on what women and girls who were pregnant experienced during the epidemic, how their experiences were mitigated, crafted, or influenced by their mental health status, and relates these factors to access to mental health services and supports. It also focuses on health care providers and their reactions to the crisis and how they were prepared for working with pregnant women during an infectious disease outbreak.

Information gathered for this chapter came from two sources—interviews and focus-group discussions with individuals who survived Ebola and were pregnant, health care providers who attended pregnant women, and secondary data sources that included public information sources. Three in-depth interviews were conducted with women who were infected with the Ebola virus and who were pregnant during the outbreak, and one woman whose partner survived EVD. The service providers interviewed were all nurses, midwives, or administrators. Two of the service providers interviewed were nurses licensed to practice as mental health providers. Collectively, the women who were interviewed had lost four pregnancies during and immediately after having been infected with the Ebola virus, and often in very graphic ways. They represent a microcosm of women in Liberia exposed to Ebola infection. The health care workers we interviewed were all Liberian women—females dominate the health care work force. Indeed, one of the most famous stories of pregnancy, Ebola, and the intersection between being a health care provider and a pregnant woman was the death of a pregnant health care worker who had survived Ebola (Mukpo 2015).

Universally, the women interviewed reported on a service delivery system that was largely chaotic, engendered fear, and fostered stigma. As the outbreak increased and resources were brought into the system of care, that fear dissipated. Respondents differed on whether an outbreak of this scale of EVD could happen again, but not on its devastating consequences for women.

By most accounts, the first casualty of the EVD outbreak was a 2-year-old baby in Guinea (Baize et al. 2014). By August 2014, President Ellen Johnson-Sirleaf, Liberia's first elected female President, had declared a national emergency (Executive Mansion 2014). Less than 1 month into the outbreak, the presentation of pregnant women with Ebola began to raise alarms. All prior public health advice regarding pregnant women appeared to be discarded. Despite millions of dollars of investment in messages that declared women should only go to health facilities for care these were irrelevant as, in many cases, available care was largely in the community among traditional midwives (Hessou 2014). Prior to the emergency, Liberia had the sixth highest maternal mortality rate (MMR) in the world at 994/100,000 live births (Index Mundi 2015).

One study of pregnancy during Ebola epidemic reported that pregnant women only accessed care in emergencies, and that this care was primarily from traditional midwives (Moddares and Berg 2016). Moreover, accessing pregnancy-related services entailed both financial and social barriers. The lack of maternity care service was compounded by the reports of poor infant viability and prognosis for the mother in cases where the mother was infected with Ebola virus (Baggi et al. 2014). Other studies suggested that either the data on pregnancy-related mortality showed no differences in mortality rates or that there were insufficient data (Henwood et al. 2017).

Exacerbating this confusion and limited access was stigma and discrimination that all individuals with EVD experienced, but that was most pronounced with pregnant women with EVD (Doucleff 2014). Even women who were not exposed to Ebola infections but were pregnant were considered at increased risk and were treated with suspicion and could not access care (Strong and Schwartz 2016, 2019).

This chapter documents the stories of women who were with "belly," a Liberian colloquial word for pregnancy, and their caregivers during the EVD outbreak in Liberia. The first part focuses on health care for pregnant women in Liberia before the country declared the Ebola outbreak a national public health emergency. In the second part, we describe what services were available to pregnant women during Ebola from the perspectives of health care workers and female survivors of EVD who were pregnant during the outbreak. Part 3 explores pregnancy outcomes from women who were pregnant during EVD based on interviews and a focus-group discussion. Among the respondents were women who experienced pregnancy during EVD and service providers working with providers. Part 4 focuses on barriers to care and specific experiences of women who were pregnant. Part 5 describes the stigma that was pervasive during the outbreak. In part 6, we discuss how women and their caregivers view the future of the health care system and its capacity to withstand another outbreak.

13.2 Maternal Health in Liberia Before the Ebola Outbreak

Despite investments in maternal health in Liberia, the country languishes as the sixth worst in the world, remaining without progress on this Sustainable Development Goal (SDG) indicator. The maternal mortality ratio (MMR) in Liberia rose from 578/100,000 live births to 1072/100,000 live births during the interval 2005–2013 (Index Mundi 2015). It continues to rise. The major reasons for this high level of maternity-related deaths include a lack of emergency obstetric services and high volumes of home births delivered by unskilled birth attendants. Additionally, among adolescents, high rates of pregnancy and poor access to prenatal care exacerbate the problem. Similar to other countries in the developing world, hemorrhage, sepsis, abortion, and the hypertensive disorders of pregnancy represent the leading causes of maternal deaths (Ministry of Health 2015a; Schwartz 2015a, b). Maternal health workers largely lack essential emergency obstetric skills, equipment, drugs, and basic sanitation and infrastructure (Otolorin et al. 2015). Poor access to health facilities also poses a significant challenge for pregnant women as they near delivery. Abortions are illegal in Liberia, yet research suggests approximately one third of women have had an abortion, and unsafe, or "back-room," abortions carried by unskilled personnel also contribute to the mortality rates (Schwartz 2015c). Reports also suggest that, while there were reductions in access to health facilities for prenatal care, deliveries at health facilities, and those assisted by skilled birth attendants (SBAs), declines were also noted in community births (Shannon et al. 2017).

Liberia is among the 15 countries globally that have the highest rates of infant mortality, thus impeding the country's development (World Health Organization 2015). High rates of teenage pregnancy abound in Liberia, where 30% of adolescents and girls having begun childbearing during these ages (Ministry of Health 2015a). As a result, pregnancy and childbirth take a toll on Liberian women. Maternal mortality here is among the highest in the world, even surpassing the levels in their impoverished neighboring countries (Ministry of Health 2015a). Nearly one third of female deaths in Liberia were maternal deaths, with births to adolescents being a high-risk factor for both the mother and the child (Ministry of Health 2015a). Deliveries to adolescents by unskilled birth attendants also were high at 30% (Ministry of Health 2015a). The mother's age was a significant predictor of avoidable risk in child mortality in Liberia, with children of adolescents under 19 years of age having highest child mortality rates (Ministry of Health 2015a). Other predictors for improved maternal and child health outcomes such as prenatal care, access to skilled attendants at delivery, access to postdelivery care, and nutritional status according to the Infant and Young Feeding (IYCF) standards developed by the WHO and UNICEF were also poor for Liberian mothers (Ministry of Health 2015a). Rural versus urban settings also influenced these outcomes. Access to prenatal care for women and adolescent girls can also be hampered by lack of permission to seek health care—8% of Liberian women reported that they need permission to seek health care (Liberia Institute of Statistics and Geo-Information Services (LISGIS) et al. 2008). Additional risk factors for poor outcomes include low utilization of pregnancy prevention strategies (20% nationally) and low use of child spacing and contraception (Ministry of Health 2015a).

The constant stock out of drugs for maternal health contributed to poor maternal health outcomes (Ministry of Health 2015b). Laboratory capacity was inadequate (Ministry of Health 2015a). The "stuff" (supplies, medications, fluids, etc.) to provide emergency obstetrical care as well as basic skilled deliveries were often out of stock or at low levels. Additionally, conditions including basic electricity, water, and availability of skilled attendants also often hamper safe deliveries. In 2015, only 55% of health facilities in Liberia had water and electricity (Ministry of Health 2016b). Indeed, Ebola surfaced amidst a national health care worker strike provoked by low pay. A demoralized workforce, including maternal health care staff, often needed to take money out of their own pockets to purchase supplies for delivery (Worzi 2014).

13.3 Service Availability During the Ebola Virus Epidemic

In August 2014, the President of Liberia declared a public health emergency and ordered the closure of public and private institutions that might contribute to further spread of the Ebola virus (Executive Mansion 2014). Businesses, factories, and markets were closed, and Liberia declared that nonessential government workers should stay at home. The President cited the "immense strain" and the "chilling effects of Ebola" on the country's health system. Major health facilities were closed for between 1 to 7 months (Muchler 2014). Some health care facilities were simply abandoned (Arwady et al. 2015). In some health facilities, Ebola virus infections among health care workers had already led to closure of the facilities (Dawson 2014). There were reports of health care workers performing prenatal care and deliveries in communities during Ebola, but there was evidence of an overall reduction in community-based deliveries (Shannon et al. 2017; Gizelis et al. 2017). To make matters worse, there were nearly 400 reported EVD infections that occurred among health care workers, of which nearly 200 died (Ministry of Health 2016a). In some facilities, the prevalence of Ebola virus infection of health care workers was so high that it spurred closure of the facilities. For example, in a county close to the capital of Monrovia, over 50 health care workers lost their lives as the county struggled with bringing the epidemic under control (Iyengar et al. 2015). In one study, 42% of people reported that closure of hospitals and the health personnel's refusal to provide patient care were the main reasons they didn't seek basic primary care (Gizelis et al. 2017). Even where facilities reported that they remained open, facility-based obstetric delivery services declined by 77% (Lori et al. 2015). However, according to one clinician interviewed, a positive side of the closure of health facilities was that it enabled the training of health care workers in Safe and Quality Services, so that the facility was not a source for Ebola virus infection.

A Ministry of Health (MOH) report indicates that skilled attendants-assisted prenatal care and deliveries dropped dramatically. Skilled delivery coverage dropped from 52% to 40% (Ministry of Health 2015a). The lack of access to health care facilities resulted in some pregnant women accessing traditional birth attendants, many of whom lacked skills, particularly in emergency and complicated prenatal and delivery care. Additionally, the rebuilding of the entire post-civil war system of health care, in particular maternity and obstetrical care, had been based on using skilled health workers, and not using traditional birth attendants. The MOH maternal health policy, as stated in the national 2011–2021 strategic plan, called for increasing the competencies and number of providers to care for pregnant girls and women (Ministry of Health and Social Welfare 2011). Among the top priorities were facility-based delivery targets of 80%, increase in the proportion of births by skilled attendants (50%), establishing comprehensive emergency obstetrics and newborn care (CEmONC) at all county health hospitals, and basic emergency obstetrics and newborn care (BEmONC) based on population density. Yet, according to one health care worker from a rural county interviewed for this study, shortly following the announcement of the national EVD emergency, 100% of women did not have facility-based delivery because facilities had closed; thus only home-based services were available (Shannon et al. 2017; Gizelis et al. 2017). She went on to state that after being trained in Safe and Quality Services (SQS), health care workers started performing deliveries. During that interval, women were delivered by traditional midwives, which the interviewee considered 'unsafe and unskilled" deliveries. The situation was desperate, explained another health care worker from a referral hospital outside of the capital, who admitted that:

> "*As it relates to pregnant women during Ebola I had a friend who was pregnant at the heat of Ebola, she was afraid, crying I used to talk with her, council her give her other pills to take...but when she got into labor she is somebody that can bleed, because it was not her first child and at that time hospital was closed, everybody was afraid, everybody was staying home. The hospital was closed, I took her to B facility because she was bleeding and nobody could take her. Even me, myself I was afraid, but she was a close friend. I wanted to help, I was keeping distant but while trying to help we had to end up to a traditional midwife area and she gave birth. She gave birth safely. It was time the facility was not open, the breakout had come in X facility.*"

13.4 What Ebola Looked Like to Pregnant Women and Health Care Workers

During this study, we spoke to women who experienced pregnancy during the Ebola epidemic. One woman lost a pregnancy when *"Ebola was getting hot"*. She is not sure how she lost the baby. She explained that she alternated between a government facility and a private facility in seeking care. She further recalled that 1 month after being told she was pregnant by the private facility which she indicated she preferred because they *"didn't give you paper"* to get medicine but rather provided the drugs. She said,

> "*I started bleeding small-small, that time Ebola was coming small, small, Ebola was getting hot. I went to my friend place to use the bathroom then everything just came down, I said to myself wow, if it was going to come down when I was on the street, people were going to call ambulance for me.*"

By this time she said, she was 3–4 months into the pregnancy.

> "*The last one came down, then I knew it was gone. They gave me medicine to stop the bleeding it couldn't stop.*"

This mother would later contract Ebola virus infection after visiting a friend in the hospital. Her next experience with pregnancy would be as an Ebola survivor. She details this experience meticulously.

> "*My second baby after Ebola I went through surgery. I took surgery... like in X hospital when I started vomiting, then I went through the surgery, I got sick. When I went home they brought me back to X hospital. Ambulance went for me, I started convulsing, went back X hospital, I started vomiting, I started throwing up, the nurses were afraid. No one could come and clean me up, no one could come and help me. That when Dr. W, when he came, I called his attention. At that time I was weak. At that time, I had just left from W hospital. I called Dr. W, I said "Dr. W, I am dying oh." He said, "what happened, but S you just left from Z hospital, I thought you just gave birth this month, but what happened?" That when I told him that since I started vomiting no one came come around me. Then I said but since I came to Y hospital, all the nurses ran away from Y hospital. That how he called everybody to come and clean up and that is how he started working on me that night. That how he order for some*

things to bring for me, pass certain tube through my nose because the water was too much in me, started burning my chest that night but when he passed the tube throughout my nose the water started running out. Two-time operation. The water bag burst, cervix was not open. I don't know how they can call it self. They get the terms for it. They had to do the second operation and just left the place open, they started cleaning it like a sore. Almost one month plus I spent to the hospital. It was not easy."

Another woman who was the wife of a survivor explained her experience, after acquiring Ebola virus infection, when she was expecting her fifth child. She said,

"My experience was very bad because, number one, when I was having the pregnancy, I started using the chlorine the people gave me. When I went to the hospital after they discharged my ma, they gave me chlorine. They gave me chlorine but they never teach me how to use the chlorine. I started using in the bathroom, not measuring it. I use to use the chlorine every Saturday. So one Saturday after I finish cleaning the bathroom, I started bleeding. I just felt that I was seeing my period but after I looked when I went I saw it, I say 'ah I pregnant but how come I bleeding'. So I thought that was normal thing, I start working, I started working but when I look the bleeding started coming fast, fast, when I lay down it just coming, it just running so I say but 'ah what thing happening to me? For then my one was in the room that how I sent somebody, to go call my sister. I not able to walk, so your come. That how they came, they met the blood in the bucket, it was full. So they came. They met the blood in the bucket. It was full. When they came around they rushed me [to the] clinic to the community. While they were about to tend to me first that how my uncle said, 'your call Y oh, he not around, your call Y. That is when the woman around said: "oh that Ebola survivor wife there, that how the woman that was treating me she put stop to me."

The community clinic health care worker stopped serving her.

"That how all the nurses started getting afraid of me. They say they can't touch me. So they must hurry up and come take me from their hospital. That is how X (my husband) called for ambulance, the ambulance came and they put me in the ambulance."

The ambulance took her to the national referral hospital. But her ordeal was far from over. There, the health care workers refused to see her without "a paper" (a certificate or medical order stating EVD-free status). From there, she was transferred to the Ebola Treatment Unit (ETU) in Monrovia.

One study (from the same counties from which we drew participants) of survivors who were pregnant or became pregnant after developing EVD suggests that rates of adverse pregnancy outcomes averaged 28%, ranging from 11.5% to 56% in areas from which women for this study were enlisted.

Stillbirths that were associated with proximity to being discharged from an Ebola treatment unit (ETUs) occurred an average of 2.5 months after leaving the units. However, miscarriages appeared to be more closely related to pregnancies that were at least 4 months post-discharge from an ETU (Fallah et al. 2016).

An EVD survivor, Ms. F delivered after being discharged from the ETU. Ms. F is a 32-year-old woman from northern Liberia who became pregnant during the Ebola outbreak, developed EVD and lost the pregnancy—she vividly described her experience. Like other Liberian women, Ms. F attended a renowned facility operated by a well-known obstetrician. She reported that she went there because of media reports that emphasized the excellence of care. She also attended a faith-based facility that was well-known for its expatriate physicians. But when all the services closed, she was not able to get routine maternity care. She recalled:

"Because I remember once I was sick, and before I even came in contact (with Ebola), I went to X facility and they told me that because of the situation they are not really seeing patients and all the people were leaving, I was not seeking any treatment."

She explained her ordeal when she gave birth to a dead child on the street.

"Actually when I lost the pregnancy, I felt, I felt a little bit relieved. Because the tension I was under, it was very high at that time. But one thing there that not made me feel really good was doing it in the street, that was the only major problem that I really had with it."

In describing the circumstances that led her to delivery on the side of a major street, she said,

> *"After I got discharged from the ETU, and the doctor visited me one time."*

She explained that at the ETU they knew she was pregnant. She further stated,

> *"When I got discharged and I was going home with the pregnancy, the doctor came and he said, "X, you will have to be taken to the hospital on Monday, so you will come back to the ETU on Monday where will we will try to do ultrasound to test. "Then they said they want to test the placenta if the placenta was positive, then they will have to do an operation to get the child out, but if is negative then they will have to keep me to study it because during delivery I may also someone may get infected and they will not want me to leave me in the community to do such. So I agreed. But then the very day they left, by 1 o'clock that night I started to feel pain... the pain started to intensify. I started seeing fluid, from there the water bag all ruptured and I was still home. Because there was no car, and besides it was in the night and it was Ebola time. No one could take someone associated with Ebola in the car or so forth. So I myself went was on the road trying stop car to see how best I can go back to the ETU and I called to the ETU the people said the ambulance that there they get body in the ambulance people got to come to take down this body and disinfect before they could come to take me back to the ETU. So I was on all that process and, as God could have it as for that child came out easy but right on the street where people were looking at me, helping to find car. That was the only bad part. But as for that child that coming down dead, actually, I never have problem with it. What I had problem with was how it happened in the street and people were standing by. The women had to go make a circle around to let say cover me small. That being the first time, God really gave me the strength when it happen end it happen everything came out at once and the people asked me because they couldn't touch. No one touched me that day and they asked me "oh you got strength?" I said, "yeah, I got strength." someone in the group gave me towel and I took that dead child, I wrapped it. I myself, I put it under my arm, that how I went home. That's how I went home and I buried the everything."*

That same day she went to the ETU, where they said they heard the call and that the body was still in the ambulance. She reported that, the same doctor, who had visited before she miscarried in the street, wearing his PPE (personal protective equipment), checked her and gave her chlorine and antibiotics. No referrals were made. Ms. F. went home.

> *"Nobody carry me hospital, nobody carry me for further treatment anywhere."* She further stated, after the miscarriage, *"nobody counseled me oh, after they gave me that medicine, I went home."*

From the perspective of providers in the system, the lack of access to equipment and supplies together with the inability to maintain their own personal safety contributed to the closure of hospitals. Nationally, 163 health care workers were infected, of which 54% were female (Personal Communication, Luke Bawo, Ministry of Health, 2017). At one county referral hospital that was closed subsequent to the infection of 14 health care workers, the providers explained that they were frequently out of infection control supplies well into the outbreak, and as a result, the director of the hospital and the county health team had to make frequent trips to the capital to press for supplies to be provided (Table 13.1).

The fear and suspicion of health care workers by prospective patients arose from their potential feelings of vulnerability. In some cases, health care workers who were exposed to Ebola virus infection, fearing being admitted to ETUs, would use medications to mask their symptoms. This created much fear and suspicion among health care workers. One worker stated:

> *Like for me, I was like, because I was in the status of being in the mentor position but even though I was not on the OB ward, but, like the 12 or 14 persons,*[1] *that you are talking about and I dealt with all of them. So, during*

Table 13.1 Cohort of births to Ebola survivors being followed by the Liberian PREVAIL Study

Pregnancies	Live births	Deaths	Stillbirths	Miscarriages
120	110	3	2	5

From the Prevail Study, Birth Cohort, 2017. Source: Glayweon, M. Co-author.

[1] Referring to the hospital staff that contracted Ebola.

the two days, I don't know, I kind of because of humanity, with all the rumors that we had about Ebola, I did not feel Ebola was so harmful at the time when I met my colleagues who I had been working with for the past years came down with fever and lot of other symptoms. We all stayed in the ER for like 2 days like for 11 of them I established IV lines, but there were no PPEs. But two days later, when I was reminded back that Ebola was so desperate, that Ebola really exists, there was an orange tree before the ER (that) in the evening by 6 o'clock I sat down there and I was like meditating, "What have I done, am I going to die like these other people?" Because the rumors was if you go to the ETU you were not going to come back. I was like selfless that night, that evening, and while I was sitting meditating, another nurse aide of ours that died, the late XX, they brought him. The wife brought him, and he was on his son's back. And they entered the ER then he called me and said, "My daughter, I dying, help me." I actually said "no", because all that I did came back to my memory, that I was not even safe any longer that I could die any moment. So I told him, I said, "XX I can't just help. Because they had just brought the late X and I established her line and I admitted her. So I said him, "XX I can't help, I am just tired, I am beat up". Because I was just confused. Later, I got up called the PA and I told him, well, "I am leaving for home", I am going to town today because I feel like I can die any moment...Sister X if you go I am also going to leave the hospital. So I left with that mind. As soon as I got home, my daughter she had the water in the bucket, she is putting enough ...we never knew about the harms of chlorine at that time.... Those things when it come to my memory sometimes for me to sleep it was hard. Then we had people coming from the ETU talking about mental health, talking about coping mechanisms, I started getting over.... Sometimes when I enter the hospital I feel so hurt that my colleagues died, their parents are left abandoned. They serve humanity and nothing is happening for them."

This facility, as well as many other major public and private facilities, were subsequently closed for between 1 and 7 months.

However, not all experiences with the health care system were negative ones. A pregnant woman explained her positive experience while at an Ebola Treatment Unit when she was pregnant:

"I was, let's say cared for, I must admit. Because, Ebola is very bad on a pregnant woman. What I see there you will bleed until the child can come out and the bleeding will not stop and you will die in there. I saw it plenty, plenty, plenty. So, I was I was cared for, because I was still walking, and I was still strong. I was still eating. Yes, so I was cared for because I was counseled, people were talking to me people talking to me because my chances of survival was very slim at the time, so lot of people were concerned. They were talking to me."

The fear of pregnant women with Ebola virus infection was particularly deep-seated. This may have come from some of the messages that were disseminated, or from the knowledge of the profuse amount of bleeding that was associated with Ebola infection. One instance that service providers recalled may illustrate this point:

"At X clinic (an ETU), when they came, huh that is no small bleeding....even the child self it can't be easy, the bleeding can come from all over from the head, ear, whatsoever field opening, that it can't be easy even that child self in the womb, when that child start coming out, so so bleeding. That child can't survive. We had midwives among us that used to do the delivery." Another asserted, "The encounter came with severe weakness, bleeding and they delivered. When they delivered the child came dead, the babies came dead already ...but you could see from every opening of the baby there was blood flowing out. Yes, ok. One of the babies, the first woman came dead, the second the baby came alive the mother died. Soon as she gave birth she died. And the baby too, there was no breathing, the baby was not breathing so we said the baby is dead, we laid the baby down by the mother but after some hours we heard the baby crying, I don't know if it was you, somebody went to the window saying the baby is crying inside there but we could not enter... the blood ...the blood. Every one was afraid... The mother bled. We could not walk in to go to rescue the baby. So the baby just cried, cried. After some hours, the baby too died. We were just waiting for surveillance. We were alike all afraid."

They explained that the profuse amount of blood deterred them from entering. They recalled that even establishing intravenous lines, a procedure that they were constantly performing, especially for their co-workers, put them at the forefront of danger of becoming infected themselves.

13.5 Stigma During the Ebola Virus Outbreak for Pregnant Women

Stigma against pregnant women during and after the Ebola outbreak was pervasive (Strong and Schwartz 2016, 2019). One pregnant survivor described a particular time that has stuck with her,

> *"Like, when I was in labor pain, my stomach was not really hurting, only my thigh. I was feeling pain in my thigh and later on when they checked me. My water bag burst and my cervix was not open. The other woman made the statement* (referring to a nurse) *said: " 'eh you know survivors them, it hard for their children to live, their children can't make it.... The doctor said the child heart beat was going, that was when the other woman made the statement. I told her, "I rebuke you: my child will live, nothing will happen to that child, how will you use that statement... As God would have it, I came home with my baby."*

Another survivor explained the incident that had a profound impact of stigma on her.

> *"After the EVD thing subsided small, I think went to the clinic and I was throwing up. But, then I started to have my medical. I have my certificate with me and I wanted them to know that I am a survivor and so that they can know how to treatment me. So when they got to know, and they saw the certificate with me, they were even telling me say, "stand aside". And I went standing, so the lady left me and she went to the man and they "hang head,"*[2] *they talked a bit and from there he came back. So I asked him, I say "but I feeling pain, I throwing up, I nah feeling too fine, so, can you people tend to me?" And he left there and he said, "I will give you medication". So when he took the medicine to give to me, when I hold (out) my hand he said, "No, just stand" and he put the medicine down. And that how I take it from there. And besides that I already had prescription in my hand from the doctor office. So when I gave him the prescription too. He didn't hold it from my hand. He tell me to put it down the same way. I took it from him because, I was carrying the certificate and I heard them talking "that survivor, that survivor".*

A health care worker affirmed this stigma, she noted:

> *When they found out the level of EVD exposure or EVD status the health worker would violate confidentiality, call other colleagues, point at the person and make everyone to know their status, the woman would become embarrassed.*

Much of the stigma and discrimination against pregnant women during Ebola was a result of fear and lack of knowledge about the disease (Strong and Schwartz 2016, 2019). Even though health care workers had studied hemorrhagic fevers including Ebola virus infection during their training, they were still caught unprepared. Ironically, less than a year before the outbreak, a Harvard-trained Liberian epidemiologist returned home and started a course to train health care workers on infectious diseases (Fallah et al. 2015). One of his students was from a facility that had shut down because it was completely overwhelmed by the epidemic. Despite this handful of students, Liberia was ill-prepared for the convergence of events that led to the maelstrom that was the Ebola virus epidemic.

Eventually this lead to the closure of many facilities—even in counties where there were relatively few cases of infection, maternity care provision was only at 30% of normal levels (Ly et al. 2016). In one county referral hospital, a high-ranking member of the obstetrics team explained that:

> *"Knowing very well [that] they told us you can get Ebola from the body fluid, the pregnant woman they come in labor, you encounter the blood, the urine, the whatever fluid. So I was wondering, how will I help my daughter? I was wondering, it played on my mind and at the time, we never had the correct equipment to handle them....so it played on my mind..., the only thing we have at the time was this gloves and the masks so I just coached my girls that we have to be properly dressed to go to these people and whenever we get to them, right after handling them we wash our hands vigorously before touching our papers and what have you. Wash our hands so we can at least survive to be of help to the others that are coming, we shouldn't just do things carelessly."*

The fear and suspicion was not limited to their prospective patients, but was also directed at fellow health care workers. One nurse-midwife noted:

[2]Liberian-English term that means to confer.

"Another fear was even though we as health workers were taught to wash our hands but how many of us were washing our hands? So people was [were] not used to washing hand or wearing gloves because sometimes gloves used to be out of stock in the hospital. So my fear was, if some of us don't know how to wash hands from one patient to others I was thinking people were really going to be infected with this virus. Many of us were very careful, for us to come on the Ob ward, it was difficult."

A phrase that our interviewees commonly used was, it is "*playing on my mind*." Besides being worried, they also frequently expressed being afraid using terms such as "*it was a great fear*" and "*frightened*". Recognizing the fear and stigma created by the Ebola virus epidemic and its effects on service delivery, individuals we interviewed discussed how the health care system could be strengthened to serve as a buffer against a whole system collapse as happened in Ebola. One interviewee offered:

> *"Even normally besides Ebola, women die from let's say complications. Becomes sometimes, women go to give birth and they are not properly cared for. One thing government needs to do is to put an end totally to let's say women doing that home deliveries or so. They need to stop it. They need to enforce it. Once you are pregnant go to the hospital to seek proper care.... (The) road leading to hospital need to be opened. We need to have ambulances, because sometimes there is an emergency and there is no ambulance. I have been around several times when you see pregnant woman sitting on motorbike to get to far distances to hospital, its not healthy for them."*

Could the confluence of factors that propelled EVD happen again? Among the health care workers interviewed, the vast majority believed that they would be able to handle another disease outbreak like Ebola, based on their current knowledge. However, they struggle with the reality that they don't have same level of funding or procurement as was available during the epidemic. Then, an estimated $1.62b was spent in the three countries, with $385 m spent in Liberia alone (World Bank 2016). Within 1 month of the Ebola outbreak, USAID released USD $12 million to UN agencies for the purchase of infection control supplies such as gloves and masks (Paterson and Widner 2017). Today, they bemoan the current lack of supplies. There are few surgical gloves, often not the right sizes, no boots, nor other supplies they need to be safe.

> *"When it comes to materials' availability, that is left with our government. It's a challenge to us. For instance, I needed gloves, I can wear size 8 you can wear size 7 another person can wear size 6 and materials came only in size 6. For me, that using size 8 gloves it becomes difficult for me to attend to my patients. Because I can't wear the size 6 gloves. Even the PPEs, the gowns it was like that.....For our side, for me I can say with the knowledge gained we can stand but when it comes to logistics we need more awareness on that."*

Another provider asserted that government's failure to provide supplies and drugs undermines the trust in hospital.

> *"The community tends to have loss of confidence when it comes to the health workers. Now the government our patient come with the expectation at the end of the day we give prescription. That is another "loss of confidence in us" when it comes to the community and the hospital."*

It is a haunting reminder that undermining the trust began with the Ebola epidemic (Mukpo 2015).

Since February 2015, the Prevail Birth Cohort Study has provided health and mental health services for survivors of EVD at three sites—John F. Kennedy Medical Center and Duport Road Health Center in Monrovia, and at CH Rennie, the county referral hospital in Kakata, Margibi County. Some of the women in this study regularly access this care. Others EVD survivors have died from what appear to be complications from Ebola infection that are still not fully understood. In one case, the daughter of a health care worker who was an EVD survivor described her mother's persistent ill-health that eventually led to her death. The daughter related a post-Ebola life for her mother marred by pain and illness 1 year after her discharge from the Ebola Treatment Unit (ETU), ultimately leading to paralysis and death. She tells of her mother's pain and inability of the post-Ebola health care delivery system to care for her, as she took her from facility to facility to receive treatment. Plans to evacuate her had not materialized when her mother died.

For the vast majority of women in Liberia who are not survivors of Ebola infection, the factors that propelled the epidemic remain. One health care provider confidently asserted that it could happen again. She explained that, at a major facility in Monrovia that cares for pregnant women, she asked a provider there about how they were coping with the lack of supplies. They told her that pregnant women purchased and brought all the medical supplies they needed for delivery.

This chapter documents the experiences of women of reproductive age who were pregnant during the EVD crisis. We discuss stories of health care workers who attended to women during EVD. We report on those factors and circumstances that women we interviewed indicated impacted their prenatal health-seeking behaviors, the factors that propelled their choice of providers, outcomes of their pregnancies, and their experiences with their health care providers. We also report on the experiences of health care workers, as providers, as women and as mothers themselves.

An important lesson learned as a result of the epidemic includes the need for resources to provide quality maternal health services during "normal times." Factors that exacerbated the effects of the Ebola epidemic on women, children, and their health caregivers included the lack of continuous training of health care workers, failure to provide the basic tools of work for providers including regular pay and supplies to facilitate infection control, and ultimately an overall lack of trust in the health care delivery system. Veteran nurses talked about not knowing about the "Five Moments for Hand Hygiene", a commonly taught and used approach to hygiene that was used during the Ebola outbreak (World Health Organization 2018).

Several years after the cessation of the West African Ebola virus epidemic, Liberia has a system of care which, while much improved, does not seem to have learned the lessons of the past. The Ministry of Health's 2016 Health Review noted that skilled health care workers attend fewer than 50% of the births (Ministry of Health 2016c). Targets for prenatal care remain exceedingly low in some counties, with some counties in rural parts of the country having only met 40% of coverage targets.[3] Even with substantial improvements in the health care system, there continues to be significant gaps and challenges.

Acknowledgments Janice L. Cooper, PhD, and Meekie Glayweon, MA, worked on the Incident Management System for the Liberia Government's response to Ebola. Dr. Cooper headed the Psychosocial Pillar and Rev. Glayweon was coordinator of the National Ebola Survivor Network. Reverend Glayweon is with the PREVAIL III Natural History Study, JFK Hospital, Monrovia, Liberia. The University of Liberia PIRE's IRB approved the research study upon which this article is based. The research for this chapter was approved by the University of Liberia, Pacific Institute For Research and Evaluation Insitution Review Board, UL-PIRE. We thank the participants who were willing to discuss their experiences in Ebola.

References

Arwady, M. A., Bawo, L., Hunter, J. C., Massaquoi, M., Matanock, A., Dahn, B., et al. (2015). Evolution of Ebola virus disease from exotic infection to global health priority, Liberia, mid-2014. *Emerging Infectious Diseases, 21*(4), 578–584. https://doi.org/10.3201/eid2104.141940. Retrieved March 28, 2018, from https://wwwnc.cdc.gov/eid/article/21/4/14-1940_article.

Baggi, F. M., Taybi, A., Van Herp, M., Di Caro, A., Wofel, R., Gunther, S., et al. (2014). Management of pregnant women affected with Ebola virus in a treatment center in Guinea. *EuroSurveillance, 19*(49). pii: 20983. Retrieved March 27, 2018, from http://eurosurveillance.org/content/10.2807/1560-7917.ES2014.19.49.20983.

Baize, S., Pannetier, D., Osetereich, L., Rieger, T., Koivogul, L., Magassouba, N., et al. (2014). Emergence of Zaire Ebola virus disease in Guinea. *New England Journal of Medicine, 371*(15), 1418–1425. Retrieved March 26, 2018, from http://www.nejm.org/doi/full/10.1056/NEJMoa1404505.

[3]Ministry of Health Op. cit.

Beigi, R. H. (2017). Emerging infectious diseases in pregnancy. *Obstetrics & Gynecology, 129*(5), 896–906.

Dawson, S. (2014). Exclusive: Liberia health system collapsing as Ebola spreads. *Reuters News*. Retrieved March 26, 2018, from https://www.reuters.com/article/us-health-ebola-liberia/exclusive-liberia-health-system-collapsing-as-ebola-spreads-idUSKBN0G72FC20140807.

Doucleff, M. (2014). Dangerous deliveries: Ebola leaves moms and babies without care. *NPR*. Retrieved March 26, 2018, from https://www.npr.org/sections/goatsandsoda/2014/11/18/364179795/dangerous-deliveries-ebola-devastates-womens-health-in-liberia.

Evans, D., Goldstein, L., & Popova, A. (2015). Correspondence: Health care worker mortality and the legacy of the Ebola epidemic. *The Lancet Global Health, 3*, e439–e440. https://doi.org/10.1016/S2214-109X(15)00065-0. Retrieved March 26, 2018, from http://www.thelancet.com/journals/langlo/article/PIIS2214-109X(15)00065-0/fulltext.

Executive Mansion. (2014). *President Sirleaf declares 90-day state of emergency, as government steps up the fight against the spread of the Ebola virus disease*. Monrovia: Executive Mansion. Retrieved March 27, 2018, from http://www.emansion.gov.lr/2press.php?news_id=3053&related=7&pg=sp.

Fallah, M., Nyenswah, T., Wiles, W., Baawo, S., Tarpeh, M., Kollie, S., et al. (2015). Communication as the key to guide workforce development in the health sector in public stakeholder partnerships: A case study in Liberia. *The Lancet, 2*(Special Issue), S43. Retrieved March 28, 2018, from http://www.thelancet.com/journals/langlo/article/PIIS2214-109X(15)70065-3/fulltext.

Fallah, M. P., Skrip, L. A., Dahn, B., Nyenswah, T., Flumo, H., Glayweon, M., et al. (2016). Pregnancy outcomes in Liberian women who conceived after recovery from Ebola virus disease. *The Lancet: Global Health, 4*(10), e678–e679. Retrieved March 28, 2018, from http://www.thelancet.com/journals/langlo/article/PIIS2214-109X(16)30147-4/fulltext.

Gizelis, T., Karim, S., Østby, G., & Urdal, H. (2017). Maternal health care in the time of Ebola: A mixed method exploration of the impact of the epidemic on delivery services in Monrovia. *World Development, 98*, 169–178. https://doi.org/10.1016/j.worlddev.2017.04.027. Retrieved March 28, 2018, from https://www.sciencedirect.com/science/article/pii/S0305750X17301377.

Henwood, P. C., Bebell, L. M., Roshania, R., Wolfman, V., Mallow, M., Kalyanpur, A., et al. (2017). Ebola virus disease and pregnancy: A retrospective cohort study of patients managed at 5 Ebola treatment units in West Africa. *Clinical Infectious Diseases, 65*(2), 292–299. Retrieved March 28, 2018, from https://www.ncbi.nlm.nih.gov/pmc/articles/PMC5850452/.

Hessou, C. (2014). Pregnant in the shadow of Ebola: Deteriorating health systems endanger women. *UNFPA*. Retrieved March 28, 2018, from https://www.unfpa.org/news/pregnant-shadow-ebola-deteriorating-health-systems-endanger-women.

Index Mundi. (2015). Country comparison: Maternal mortality rates.

Iyengar, P., Kerber, K., Howe, C. J., & Dahn, B. (2015). Services for mothers and newborns during the Ebola outbreak in Liberia: The need for improvement in emergencies. *PLOS Currents*. https://doi.org/10.1371/currents.outbreaks.4ba318308719ac86fbef91f8e56cb66f. Retrieved March 28, 2018, from https://www.ncbi.nlm.nih.gov/pmc/articles/PMC4404271/.

Liberia Institute of Statistics and Geo-Information Services (LISGIS), Ministry of Health and Social Welfare, National AIDS Control Program, & Macro International Inc. (2008). *Demographic and Health Survey 2007*. Monrovia: Liberia Institute of Statistics and Geo-Information Services (LISGIS) and Macro International Inc.

Lori, J. R., Rominski, S. D., Perosky, J. E., Munro, M. L., Williams, G., Bell, S. A., et al. (2015). A case series study of the effect of Ebola on facility-based deliveries in rural Liberia. *BMC Pregnancy and Childbirth, 15*, 254. https://doi.org/10.1186/s12884-015-0694-x. Retrieved March 27, 2018, from https://bmcpregnancychildbirth.biomedcentral.com/articles/10.1186/s12884-015-0694-x.

Ly, J., Sathananthan, V., Griffiths, T., Kanjee, Z., Kenny, A., Gordon, N., et al. (2016). Facility-based delivery during the Ebola virus disease epidemic in rural Liberia: Analysis from a cross-sectional population-based household survey. *PLoS Medicine, 13*(8), 1–17. https://doi.org/10.1371/journal.pmed.1002096. Retrieved March 29, 2018, from https://www.ncbi.nlm.nih.gov/pmc/articles/PMC4970816/.

Ministry of Health. (2015a). *Investment case for reproductive, maternal, newborn, child and adolescent health 2016–2020*. Monrovia: Ministry of Health, Republic of Liberia.

Ministry of Health. (2015b). *Investment plan for building a resilient health system 2015–2021*. Monrovia: Ministry of Health.

Ministry of Health. (2016a). *The Republic of Liberia Ebola survivors care and support policy*. Monrovia: Ministry of Health, Republic of Liberia.

Ministry of Health. (2016b). *Joint annual health sector review report. National Health Sector Investment Plan*. Monrovia: Ministry of Health.

Ministry of Health. (2016c). *Liberia Service Availability and Readiness Assessment (SARA) and quality of care report*. Monrovia: Ministry of Health.

Ministry of Health and Social Welfare. (2011). National Health and Social Welfare Policy and Plan: Final Draft for Validation. Monrovia.

Moddares, N., & Berg, K. (2016). *Qualitative assessment on health system trust and health service utilization in Liberia*. Baltimore: Johns Hopkins Bloomberg School of Public Health, Center for Communications Programs.
Muchler, B. (2014). Liberia's JFK hospital reopens after temporary Ebola closure. *VOA*. Retrieved Match 29, 2018, from https://www.voanews.com/a/liberias-jfk-hospital-reopens-after-temporary-closure/2483132.html.
Mukpo, A. (2015). Surviving Ebola: Public perceptions of governance and the outbreak response in Liberia. *International Alert*. Retrieved Match 29, 2018, from https://reliefweb.int/sites/reliefweb.int/files/resources/Liberia_SurvivingEbola_EN_2015.pdf.
Otolorin, E., Gomez, P., Currie, S., Thapa, K., & Dao, B. (2015). Essential basic and emergency obstetric and newborn care: From education and training to service delivery and quality of care. *International Journal of Gynaecology and Obstetrics, 130*, 546–553. Retrieved March 28, 2018, from https://www.sciencedirect.com/science/article/pii/S0020729215001368.
Paterson, D., & Widner, S. (2017). Offering a lifeline: Delivering critical supplies to Ebola-affected communities in Liberia, 2014–2015. *Princeton University*. Retrieved March 28, 2018, from https://successfulsocieties.princeton.edu/publications/ebola-delivering-critical-supplies-liberia.
Schwartz, D. A. (2015a). Interface of epidemiology, anthropology and health care in maternal death prevention in resource-poor nations. In D. A. Schwartz (Ed.), *Maternal mortality: Risk factors, anthropological perspectives, prevalence in developing countries and preventive strategies for pregnancy-related death* (pp. ix–xiv). New York: Nova Science.
Schwartz, D. A. (2015b). The pathology of maternal death—The importance of accurate autopsy diagnosis for epidemiologic surveillance and prevention of maternal mortality in developing countries. In D. A. Schwartz (Ed.), *Maternal mortality: Risk factors, anthropological perspectives, prevalence in developing countries and preventive strategies for pregnancy-related death* (pp. 215–253). New York: Nova Science.
Schwartz, D. A. (2015c). Unsafe abortion: A persistent cause of maternal death and reproductive morbidity in resource-poor nations. In D. A. Schwartz (Ed.), *Maternal mortality: Risk factors, anthropological perspectives, prevalence in developing countries and preventive strategies for pregnancy-related death* (pp. 425–439). New York: Nova Science.
Shannon, R. Q., Horace-Kwemi, E., Najjemba, R., Owiti, P., Edwards, J., Shringarpure, K., et al. (2017). Effects of the 2014 Ebola outbreak on antenatal and delivery outcomes in Liberia: A nationwide analysis. *Public Health Action, 7*(S1), S88–S93. Retrieved March 27, 2018, from https://www.ncbi.nlm.nih.gov/pmc/articles/PMC5515570/.
Strong, A., & Schwartz, D. A. (2016). Sociocultural aspects of risk to pregnant women during the 2013–2015 multinational Ebola virus outbreak in West Africa. *Health Care for Women International, 37*(8), 922–942. https://doi.org/10.1080/07399332.2016.1167896.
Strong, A. E., & Schwartz, D. A. (2019). Effects of the West African Ebola epidemic on health care of pregnant women—Stigmatization with and without infection. In D. A. Schwartz, J. N. Anoko, & S.A. Abramowitz (Eds.), *Pregnant in the time of Ebola: Women and their children in the 2013-2015 West African epidemic*. New York: Springer.
World Bank. (2016). *World Bank group Ebola response fact sheet*. Retrieved March 29, 2018, from http://www.worldbank.org/en/topic/health/brief/world-bank-group-ebola-fact-sheet.
World Health Organization. (2015). *Liberia: Country cooperation strategy at a glance*. Retrieved March 28, 2018, from http://apps.who.int/iris/bitstream/10665/136911/1/ccsbrief_lbr_en.pdf.
World Health Organization. (2018). *About save lives: Clean your hands*. Retrieved March 28, 2018, from http://www.who.int/gpsc/5may/background/5moments/en/.
Worzi, A. (2014). Redemption hospital worker protest. *The Daily Observer*. Retrieved March 29, 2018, from https://www.liberianobserver.com/news/redemption-hospital-workers-protest/.

14 Health Workers, Children, and Families: Child Protection and Communication Challenges in the Context of the Ebola Virus Epidemic in Liberia

Dominique de Juriew

14.1 Communication During Emergency Interventions: General Considerations

In most of the emergency interventions in which I participated as a Child Protection Consultant for the United Nations Children's Fund (UNICEF), communication has represented a significant challenge that has had to be understood and overcome. This was largely due to the presence in these crisis situations of multiple stakeholders and a significant need for effective communication with the affected population on the situation and its consequences, the resources available to respond materially and psychologically to the crisis, and the perception from the rest of the world and displays of solidarity. Clear communication in emergency contexts needs to ensure continuity of connections between affected people and their families as well as their caregivers in order to prevent anxiety and fear. It also helps to prevent or neutralize rumors. The capacity to communicate clearly and precisely is not always easily accomplished, or even comfortable, in an unfamiliar sociocultural context. This is especially true in an emergency situation, such as the West African Ebola epidemic, and requires knowledge, competencies, and sensitivities that are not always available when needed.

During the Ebola virus outbreak in Liberia, there were intensive efforts to produce ethnographic material to assist in understanding the epidemiology and social aspects of the Ebola virus and its effects on the affected populations, including the occurrence of death and performance of funeral rites. One of the major goals of this massive informational effort was to ensure that epidemiological measures were culturally adequate, acceptable, and effective. There existed questions that had to do both with the content of the communications, as well as the most culturally sensitive and effective methods that could be used to communicate with the communities at risk.

The partners in this effort, including government, civil society, nongovernmental organizations, communities and parents, national and international humanitarian actors, all rarely addressed the communication needs of children on issues such as separation with parents, siblings and relatives, illness, the risk of and occurrence of death, and relations between children and adults. Effective communication with children was not a high priority of the overall Ebola response; this responsibility was left to a group of psychologists who were insufficiently prepared to respond to their young patient's needs.

D. de Juriew (✉)
UNICEF and UNHCR, Montreal, QC, Canada

D. A. Schwartz et al. (eds.), *Pregnant in the Time of Ebola*, Global Maternal and Child Health,
https://doi.org/10.1007/978-3-319-97637-2_14

This report highlights challenges we faced of caring for children, highlighting a psychologist's efforts to respond to an urgent and clearly demonstrable demand from children immersed in an anxiety-producing and unimaginably stressful situation. This report also suggests solutions for improving communication with children in the context of an emergent and life-threatening infectious disease epidemic as occurred during the Ebola virus outbreak. The information shared in this report are notes and observations from the author, collected during her direct working experience as child protection specialist in Monrovia during the period October 2014 to December 2015.

14.2 The Liberian Context of Caring for Children Exposed to the Ebola Virus Disease

14.2.1 Children Were Placed in Observation Centers[1]

Some children who had been in contact with persons infected with the Ebola virus were placed in Observation Centers for 3 weeks. This period corresponded to the incubation period of the disease. However, depending upon the results of case investigation, this confinement in an Observation Center could be extended if new risks of exposure to the infection were identified.

From the point of view of child protection, these centers were considered as a final option, or a "last resort". Family placement was preferred for two main reasons: to avoid both psychological distress and difficult separations, as well as the threat of child abandonments[2] as an unintended consequence of separations.

From a pediatric point of view, babies, infants, and young children need specific and efficient follow-up delivered by trained and experienced professionals. As the Ebola virus disease (EVD) symptomatology of these groups is somewhat different from adults, it is highly beneficial to thoroughly know the child in order to establish an early diagnosis. These perspectives on children confinement were discussed extensively and measures for mitigatng their impact on the children were identfied.

The main challenge was on maintaining the lines of communication between the children, their parents, and the extended family. This could be problematic, as in some cases one or even both parents as well as relatives were under care in Ebola Treatment Units (ETUs). Both children and parents faced many difficulties in staying informed of one another's situation, because the sharing of information was neither easy nor systematic. It was unfortunate that, in many cases, the only information children received was related to the death of one or both parents.

Although numerous efforts were made to avoid the separation of siblings, it was not always possible to place all of the children in the same care centers. Sometimes, siblings were separated following the development of Ebola virus disease symptoms and were then taken to Ebola treatment centers. During these separations, children were not given their relatives' telephone contacts, nor were they provided with updated news in order to stay in contact with one another. An additional challenge that affected the well-being of these children was the lack of communication with their extended relatives in order

[1]In Liberia these centers were known as Interim Care Centers. We avoided using the word "quarantine" so as to not frighten children and families and cause stigmatization.

[2]One phenomenon encouraged us to remain vigilant regarding the separation and placement of children in observation centers. Prior to the Ebola virus outbreak, and because parents wanted to send children to school and/or provide them with better health care, some children were moved from their villages of origin to live in urban settings in orphanages/residential care. However, the main goal was to enroll them very often in illegal international adoption process. With the close support provided by UNICEF to its national counterpart, some orphanages were closed. Therefore, we directed our efforts to ensure that such placements would not increase the risk of child trafficking and abandonment.

to prepare for family and community reintegration after being released from the Observation Centers or ETUs.

Unfortunately, there were few initiatives taken to psychologically prepare children for the possible death of their parent(s), brother(s), and/or sister(s), although they were old enough to understand the concept of death. When a family death did occur, a psychologist was in charge of informing the child, and then scheduling a psychological follow-up following a certain period of time.

14.2.2 Children in Treatment Centers

On too many occasions, the hospitalization of children in treatment centers resulted in the loss of family ties and the total absence of communication with relatives during their confinement. The adequacy of information that was received on the disease and its evolution seems to have been inconsistent from one treatment center to another.

The information supplied about surveillance measures and available resources for community reintegration of the survivors also varied, not only from one treatment to another, but also from one organization to the other.

14.2.3 Children in the Community

In some cases, contact with those children confined in Observation Centers or in ETUs was maintained by members of their extended family. In other cases, following their discharge from care, children were reintegrated into their extended families, and subsequently were dependent on those adults to receive updated news of their parents and their health status. A system for follow-up on these children was established; however, its effectiveness was suboptimal because of logistical difficulties.

14.3 Communication with Children: The Left Behind

14.3.1 The Absence of a Specific Assessment of Children

During meetings held by anthropologists and ethnologists in Monrovia, child and infant issues were rarely addressed. Additionally, there was not much attention to addressing the perceptions and representations of such issues such as age groups, the stages of child development, and the age-dependent appropriateness of what should (and should not) be told to children.

As partners, we did not try to familiarize ourselves with the children, teenagers, and young people of the Liberian society. There was little information gathered relating to their specific needs in terms of communication (content and methods) and how to expedite the participation of adults from appropriate and available family and social networks to assume this responsibility. This absence of critical information resulted in a deficiency in the orientation of workers at the ETUs and health care centers to the needs of children. As a result, in many cases communication with children was dependent on interventions from qualified psychologists.

14.3.2 The Absence of Specific Messages Addressed to Children

Apart from the few traditional messages and communication tools that promoted personal hygiene and prevention of EVD transmission that had been developed for children, the majority of messaging during the outbreak was mainly oriented and directed to adults. Children who were old enough to understand and analyze the informational content of these messages received no specific information outside the scope of child protection projects. We identified a significant gap between child protection practitioners and the medical team when addressing caring for children. This gap left children, and those issues concerning them, to child protection and social workers; other stakeholders, believing they were ignorant in this area, generally did not wish to be involved with it. An additional difficulty was due to the fact that only a few attempts of coordinating these different services were successful. This was at least partially due to the large number of organizations involved in the Ebola epidemic response and the difficulty of reaching consensus and identifying interests and priorities in caring for children. Because of the lack of consideration for children's communication needs, they were not protected against the many rumors that were in circulation during the epidemic.

14.4 Failure of Adults to Communicate with Children

14.4.1 The Lack of Material Resources

We were quickly able to observe the lack of information and methods for effective communication to maintain the ties between children and adults that had resulted from the myriad of adverse situations occurring during epidemic. During a meeting with the nursing staff from the treatment centers, we were not only able to observe their exhaustion, but also that they were disorientated and confused about their roles when they had to deal with children.

Nursing staff were not supplied with training or informational materials on techniques for preserving links with family members, the type of information on the Ebola outbreak to share with patients, methods for communicating the death of a relative, or the mechanism for the community reintegration process after being discharged from the treatment center. Some nursing staff were uninformed about children's admissions' operational procedures in Ebola treatment centers.

14.4.2 The Challenge of Communicating the Risk of Death and the Demise of Parents

The most alarming question for nursing staff was how to effectively and compassionately communicate the risk of death from Ebola disease for parents and other family members, as well as the occurrence of the death of a family member, to children during their stay in a treatment center. Approaching this sensitive question seemed difficult for nursing staff and, in certain cases, impossible to address because of cultural, social, and psychological reasons.

Nursing staff also faced difficulties resulting from limitations in communication during their duty of caring for patients. In order to prevent contamination, they had to significantly limit the amount of time spent with patients, prioritizing the time spent with providing treatment, and thus, limiting communication with their patients. The effectiveness of psychologist therapy sessions with children also had to be limited. This was because some psychologists who regularly visited the children were not specially trained in addressing psychological distress in pediatric patients; as a result, children did not always feel at ease during these sessions.

In some cases, patients succeeded in solving communication issues and coped with the situation. Some adult patients and health care workers were able to build one-on-one relationships with children in treatment centers and allowed them to use their personal mobile phones to talk with their parents during the several hours of inactivity in the treatment centers.[3] Caregivers reported that these informal connections and communications between adults and children in the treatment centers often produced a positive psychological effect on both parties by reducing feelings of isolation, loneliness, and distress caused by the illness.

14.5 Late Solutions for Improved Communication with Children

The workers in the treatment centers were mostly caregivers, not communication specialists. Because of this, some people in key positions had to ensure maintaining communication links between parents and children. While this had the advantage of having established lists of easily identifiable stakeholders, it also had disadvantages—there was no systematic approach to collecting personal information from or about children upon their entrance to the treatment center, and it was challenging to provide follow-up of patients due to a high patient/caregiver ratio and the absence of established standards of sharing personal information.

The issue of sharing information between child protection partners and health partners was not considered until late in the outbreak response. Although the technical coordination between the child protection services and the health sector was established through working groups, allowing the development of such tools as manuals and standard operating procedures, the practical aspects of coordination were not devised until very late in the outbreak. This point was highlighted during meetings with the nursing staff working in the treatment centers. The need for training in communicating with children was identified and placed at the top of the list of priorities. Unfortunately, the implementation of this training was delayed for budgetary reasons.

14.6 Conclusion: Questions and Ideas for Reflection

14.6.1 Should We Consider Communicating with Children on Difficult and Complex Topics such as the Illness and Death of a Parent as Reserved only for Psychologists?

In many cases during the EVD outbreak in Liberia, the burden of communication with children, and especially when parents or siblings had died, was left to the psychologists. This situation presented as a temporary solution at the beginning, but had become permanent by the close of the epidemic. While providing a mechanism for children to be informed of the loss of their parent(s), it did not prepare them for the eventuality of this loss, nor to have the capabilities (when they were old enough) to deal with their new situation. This leads us to question the role(s) that the various actors must play regarding children and the chain of communication implemented during emergency contexts to enable children to remain informed and to participate and communicate with their family members (both nuclear and extended). This is all the more relevant and important in the African and European contexts, where elder brothers and sisters very often play a very important role with the younger siblings.

[3]It was difficult to find appropriate activities for children as medical teams were concerned that the availability of too many objects in the treatment units would lead to re-infection or relapse of patients during their recovery.

14.6.2 How to Make Communication Between Patients and Caregivers More Effective?

In the Liberian context, the terror generated by the Ebola epidemic and the lack of a definitive knowledge about the disease made it a particularly difficult situation for health care workers. Health center staff had to communicate with frightened people about a life-threatening situation that they did not understand. Unfortunately, the intricacies of the situation did not encourage them to communicate with the children. It should also be noted that the family role of some individuals (as parent, aunt, elder brother, etc.) took precedence over their socio-professional function in their vision of children. In practical terms, this meant that many health workers saw themselves as parents rather than as caregivers in their relationship with sick children, and therefore, communicated with them according to Liberian codes and practices.

This form of communication had the advantage of giving children a sense of security and familiarity, but did not always correspond to the informational needs that were beneficial for the children. In addition, it remained anchored in the customs of the various Liberian ethnic groups regarding what could be said to children and what could not. We interviewed several people who told us that talking about death, explaining illness, or similar difficult situations to children was not done in many families. In some of them, the main interlocutors of children are women, especially mothers, grandmothers, and aunts. Fathers do not start communicating with children until they are teenagers. Most of the children's requests go through the mother who acts as an intermediary between fathers and children. Grandparents can also be a privileged interlocutor of children. It would be interesting to open a dialogue with parents in order to gather their visions and practices of communicating with children and to reflect with them on the tools that could be useful to them to interact with their children on issues related to epidemics, diseases, death, etc. Working with parents would create spaces open to address, understand, and share children's issues and their need. The use of children's loved stories could be of great help in creating this space for dialogue and exchange.

14.6.3 What to Do with So Many Emotions? The Challenge of Communication in a Traumatic Context

Caregivers and other humanitarian stakeholders had another challenge during the Ebola epidemic. The rumors that circulated during the Ebola outbreak exacerbated strong emotions and feelings among health workers and patients, and in the worst case, could lead to conflicts or tragic incidents both for adults and children. We have not found any literature on the perception and treatment of emotions in the Liberian society that would have allowed us to refine our understanding of this dimension. The collection of information on this dimension should be included in the list of data to be collected for future interventions in similar contexts. This issue is directly related to the issue of psychosocial support for parents to support their role as communicators and protectors of children and as intermediaries between their children and staff in health centers or hospitals. A better understanding of their vision and perception of their emotions and feelings and the emotions of their children would enrich the practice and promote dialogue between generations.

14.6.4 Maintaining the Links Between Families, Family Members, and Stakeholders

The Ebola virus epidemic has demonstrated the importance of having a consensual approach to the issue of maintaining effective communication links between parents and children, between families, and with stakeholders in order to avoid tragic separations and to enhance the development of resilience. New technologies can be used for this purpose, allowing families to contact each other by cell phone or messaging and permitting health workers to quickly locate patients belonging to the same family in order to keep them in contact. Effective communication systems also make it easier to coordinate and streamline interventions, as well as to monitor children who are transferred to health centers or hospitals.

14.6.5 Child Protection and Humanitarian Work

The Ebola virus outbreak has shown that child protection is a field that is little appreciated, or even known, by humanitarian actors as a whole, even among those agencies and organizations with specialized departments in the field. As a result, all child-related issues are often left to specialized organizations or departments, by humanitarians, nongovernmental agencies, governments, and civil society organizations. This lack of knowledge and integration of the field of child protection with the practices of other fields is an important factor resulting in increased vulnerability for children in the context of a humanitarian crisis such as that occurred in the Ebola outbreak. Indeed, it took valuable time for all stakeholders to be aware of the resources available for discussions on appropriate practices for caring for children, as well as for the training needs of medical practitioners, to be identified. Some children ended up facing adults who were unable to understand their needs and to effectively communicate with them. It was very difficult to follow-up on some of these cases because there was no information on the services to which they were entitled. Furthermore, no chain of communication between them and the stakeholders had been clearly established.

We have also found that child participation is a little-known principle of the Convention for the Rights of the Child.[4] It is difficult to apply because it involves dialogue with children and educational practices that are not always compatible with the sociocultural context and the educational system of the countries involved. In order to ensure that the guiding principles of the Convention are taken into account, it is important to understand what child participation means within the local context, as well as which local or cultural practices or values it challenges or threatens in order to encourage adults (parents and other stakeholders) to integrate it into their communication with children.

During the Ebola virus epidemic, we found that the health system in Liberia was disorganized and ill-prepared for the psychosocially appropriate care of children. Stakeholders were neither trained in child protection nor effectively communicating with children and adolescents. Health centers and hospitals were not equipped with dedicated spaces reserved only for children. In

[4]The United Nations Convention on the Rights of the Child (UNCRC) is a human rights treaty which sets out the civil, political, economic, social, health, and cultural rights of children, with a child defined being under the age of 18. The United Nations General Assembly adopted the Convention, opening it for signature on November 20th, 1989, and coming into force on September 2nd, 1990.

general, the specific needs of children over 5 years-of-age were not identified and considered. It is important that, when restoring health systems in countries that have been affected by the Ebola epidemic, provisions are made that will provide child protection stakeholders with the opportunity to work hand-in-hand with health partners to mainstream child protection throughout all levels of the medical system.

The issue of communication between children and health center staff is of critical importance and forms a basis for the relationship between the caregivers and care receivers in health systems throughout West Africa. Surveys conducted by teams of anthropologists and ethnographers have shown that these relationships were sometimes difficult, based on inappropriate practices, and could lead to the abuse of power. These data support reinforces the stories of caregivers who wanted to be trained in communicating with children during the Ebola outbreak. It should be considered that this type of training be integrated into a curriculum that includes the appropriate approaches and practices for the development of healthy and equitable caring relationships (Jaffré and Olivier de Sardan 2003).

Reference

Jaffré, Y., & Olivier de Sardan, J.-P. (2003). *Une médecine inhospitalière. Les difficiles relations entre soignants et soignés dans cinq capitales d'Afrique de l'Ouest*. Paris: APAD.

Maternal and Reproductive Rights: Ebola and the Law in Liberia

15

Veronica Fynn Bruey

15.1 Introduction

> *The health care system in the county is now under immense strain and the Ebola epidemic is having a chilling effect on the overall health care delivery. Out of fear of being infected with the disease, health care practitioners are afraid to accept new patients, especially in community clinics across the country. Consequently, many common diseases which are especially prevalent during the rainy season, such as malaria, typhoid and common cold, are going untreated and may lead to unnecessary and preventable deaths. The virus currently has no cure and has a fatality rate of up to 90 percent. (…) I instructed all non-essential government staff to stay home for 30 days, ordered the closure of schools, and authorized the fumigation of all public buildings. We have shut down markets in affected areas and have restricted movement in others. We have improved our response time and contact tracking and have begun coordinating with regional and international partners. Despite these and other continuing efforts, the threat continues to grow. The actions allowed by statues under the Public Health Law [1976] are no longer adequate to deal with the Ebola epidemic in as comprehensive and holistic as the outbreak requires. The scope and scale of the epidemic, the virulence and deadliness of the virus now exceed the capacity and statutory responsibility of any one government agency or ministry. The Government and people of Liberia require extraordinary measures for the very survival of our state and for the protection of the lives of our people. Therefore, and by the virtue of the powers vested in me as President of the Republic of Liberia, I, Ellen Johnson Sirleaf, President of the Republic of Liberia, and in keeping with Article 86(a) (b) of the Constitution of the Republic of Liberia, hereby declare a State of Emergency throughout the Republic of Liberia effective as of August 6, 2014 for a period of 90 days. Under this State of Emergency, the Government will institute extraordinary measures, including, if need be, the suspensions of certain rights and privileges* (Sirleaf 2014).

In 2005, Liberia democratically elected the first female president in Africa (Encyclopaedia Britannica 2017; NBC News 2006), yet it is a struggling nation recovering from 14 years of civil war. Liberia ranks 177 of 188 countries on the Human Development Index (UNDP 2015, p. 50) and 83 of 168 countries on the Corruption Perceptions Index (Beddow 2016, p. 22). Between 2002 and 2012, 83.3% of Liberians survived on US $1.25 per day (UNDP 2015, p. 228). The illiteracy rate for adults 15 years and over is 52.3% for years 2005 and 2010 (UNESCO 2016). Even though the unemployment rate is not a good indicator of the state of the labor market due to the relatively high informal employment rate (56.6%), Liberia's labor force participation rate is 48.5% at a base rate of 1,133,000 employable persons, aged 15 years and over (LISGIS 2011, p. xii, xiii, 26). Although global statistics should always be interpreted through a critical lens (Merry 2016), it is no surprise that Liberia was the hardest

V. F. Bruey (✉)
Seattle University School of Law, Seattle, WA, USA
e-mail: fynnbruv@seattleu.edu

D. A. Schwartz et al. (eds.), *Pregnant in the Time of Ebola*, Global Maternal and Child Health,
https://doi.org/10.1007/978-3-319-97637-2_15

hit country during the Ebola virus disease (EVD) epidemic that occurred in West Africa. Liberia experienced 10,666 cases and 4806 deaths as a direct result of EVD; these figures not including indirect deaths. Nevertheless, the World Health Organization (WHO) was able to pronounce Liberia Ebola-free on 14 January, 2016 (WHO 2016a).

In writing this chapter, I reflected on the warm, beautiful, summer evening of 22 June, 2015. I sat in the back row of a University of British Columbia lecture theater listening to Dr. Francis Kateh, then Chief Medical Officer of Liberia (Kateh 2015). Doctor Kateh was invited by the Korle-Bu Neuroscience Foundation to share Liberia's "success story" in fighting the Ebola virus outbreak (Nyenswah et al. 2016). I remember thinking to myself, how could such an extremely poor country curb the Ebola epidemic? Can Liberia be put on the same footing with Cuba, which is critically acclaimed as becoming a world-class medical power house with limited resources (Fitz 2013)? Far from it. Liberia is no Cuba.

While Liberia's success quelling the Ebola outbreak is primarily owed to a combination of international support (e.g., the United States military) (Department of Defense 2014) and an innovative model called the "sector approach"—a quality management system in public health (Cordier-Lasalle 2015), coupled with the strong political will of President Ellen Johnson Sirleaf and daily sacrifices made by health workers and ordinary citizens, Liberia's health system cannot be given much credit. In fact, the Ebola crisis exposed the many preexisting weaknesses in Liberia's dilapidated and civil war-torn health care system. During the Ebola crisis, the health care system in Liberia was overwhelmed with a panic that hindered delivery of essential health care and mired in ineffective public health law and policy. Unfortunately, the negative effects of the inadequate system extended well beyond EVD-infected individuals. As the health care system buckled under the pressure of the Ebola crisis, the most vulnerable of Liberians—women, mothers, and children—bore the greatest burden. The efficacy of applying law and policy to protect girls' and women's maternal and reproductive rights is paramount to ensuring the highest attainable health outcomes in Liberia. This chapter examines the extent to which international, regional, and national law and policy impacted upon reproductive and maternal health outcomes of girls and women during the EVD outbreak in Liberia.

15.2 Liberia and the Ebola Crisis

Long before West Africa's 2014 Ebola epidemic, Liberia was already stricken by poverty, corruption, hunger, and civil war. Stunted by a 14-year civil war that killed 250,000 people and displaced another 850,000 from 1989 to 2003, Liberia is classified as a fragile state by the Organization for Economic Cooperation and Development and ranked "highly vulnerable" on the Global Vulnerability and Crisis Assessment Index (Parshley 2016). Before the outbreak, the country's population of 4.1 million was served by 50 licensed Liberian-trained doctors, now upgraded to 298 for 4.5 million people in 2016 (Ballah 2016a). As of 2008, Liberia only had 437 health facilities (Black et al. 2016, p. 292). For many Westerners, Liberia is seen not as one of two African countries that pride themselves on never having been colonized, a haven for freed black slaves sent to the continent by the American Colonization Society, or the first African country to democratically elect a female president, but "as a weak and fragmented" (Parshley 2016) country that "sits prominently on various unfortunate lists" (Parshley 2016). The irony is how a low-income, underdeveloped country could successfully manage the impossible, i.e., to control the Ebola virus in record time (Cordier-Lasalle 2015; World Health Organization 2015c).

The U.S. Centers for Disease Control and Prevention (CDC) described the 2014 EVD crisis as the largest Ebola epidemic in history (CDC 2016). Affecting multiple countries, WHO reports that, as of 27 March, 2016, a total of 11,323 deaths and 28,646 cases[1] occurred in 10 countries. There were

[1] Noteworthy is the Center for Disease Control and Prevention's estimate of 250,000 confirmed, probably, and suspected

10,666 cases and 4806 deaths in Liberia alone (World Health Organization 2016). In response to a case fatality rate of 72.3% for persons between the ages of 15 and 44 years with a known clinical outcome of the infection, the WHO Ebola Response Team warned the number of cases and deaths from EVD were expected to increase from hundreds to thousands per week, in the absence of drastic control measures (WHO Ebola Response Team 2014). On 6 August, 2014, President Ellen Johnson Sirleaf declared a state of emergency, citing serious public health risks to the safety, security, welfare, and economic instability posed by the Ebola virus on the Liberian nation (Government of Liberia 2014b). The state of emergency permitted the implementation of many intensive and sometimes controversial intervention measures (Meltzer et al. 2014). The state of emergency declared by President Sirleaf saw a 9:00 PM–6:00 AM curfew and quarantine order imposed on one of the poorest communities in the country (Aizenman 2014; Los Angeles Times 2014; Onishi 2014a). Extraordinary measures used by the government included arresting and prosecuting journalists who failed to obtain written permission from the Ministry of Health and Social Welfare before contacting Ebola patients for interviews, filming, or photographing (Felix 2014). The controversial quarantine of Monrovia's West Point neighborhood garnered international attention when Shakie Kamara was shot by soldiers of the Liberia Armed Forces while attempting to enforce the presidential order (Onishi 2014b, 2014c).

The enormity of the EVD outbreak in Liberia drew the attention of an international community perceiving a global security threat. First, the world reacted to the sensation surrounding the death of Patrick Sawyer, who died in Nigeria (Odunayo 2014). Between 2 August and 19 September, the next three cases in the United States received much attention from the international media. Lawyers, attorneys, and legal practitioners in the United States discussed the role of law in public health emergency preparedness and the legality (i.e., duties and powers of personnel) associated with infected persons entering the US (Penn et al. 2014). Ebola-infected persons entering the United States included Dr. Kent Brantly (BBC 2014b), Nancy Writebol (BBC 2014a), Dr. Rick Sacra (Fox 2014), and Liberia's own Thomas Eric Duncan (Voorhees and Mathis-Lilley 2014). Duncan, who was likely infected while assisting a pregnant friend in Monrovia who had little or no chance of survival (Sepkowitz and Haglage 2014), was the first and only Ebola death in the United States (Botelho and Wilson 2014). While high profile cases of Americans captured the attention of international media, girls and women at the lowest end of the economic gradient in Liberia had little chance of survival during the Ebola crisis.

The vulnerable position of Liberian women and girls exposed them to a disproportionately negative impact of the Ebola outbreak. The Liberia Demographic Health Survey reported in 2013 that 35% of all households in Liberia were headed by women (Government of Liberia 2014a, pp. 19–20), with 69% of rural women employed in the agricultural sector (Government of Liberia 2014a, pp. 45–47). As of 15 April, 2015, 83% of women were more likely to be out of work compared to 27% of men during the EVD crisis (Himelein 2015, p. 7). Liberia is also among the developing countries that have the highest rates of adolescent pregnancy in the world. For young women, early pregnancy and childcare responsibilities cut short their education, increase the occurrence of gender violence and abuse, and heighten their risk of illness and death. According to the Global Fund for Women, 75% of those who died from Ebola were women (Akanni 2014; Global Fund For Women 2014; UN Women 2014). With particular regard to pregnant women, a Khaki and Bayatmakoo study of 109 women revealed that the mortality rate among pregnant women could be higher (93.3%) than the overall mortality rate of the epidemic (77%) in Ebola-affected areas (Khaki and Bayatmakoo 2015). The above statistics suggest that the EVD epidemic probably reversed any miniscule progress in reducing the maternal

cases of Ebola virus in Guinea, Liberia and Sierra Leone in 2015 obscures the actual 28, 646 cases occurring over the period of 3 years (see the Center for Disease Control in Liberia (July 2015) available at: https://www.cdc.gov/globalhealth/countries/liberia/).

mortality rate that Liberia showed during its relatively short postwar reconstruction. Fear of health workers combined with shortage of midwives and extortionate fees charged by doctors to women to give birth led to the deaths of many pregnant women (IRIN News 2014).

High death rates of women and girls resulting from the Ebola virus epidemic is partly explained by Liberian cultural norms. Generally, women and girls are charged with providing food, cooking, cleaning, washing, and caring for ill family members, including those infected with the Ebola virus. Women are also instrumental in burial rites, which require the handling of infected bodies (Global Fund For Women 2014). Although contact with infected bodily fluids is the primary transmission mechanism of the disease, the painful suffering associated with the progress of the disease adjures family members to comfort patients by touching. The impact of the EVD on girls and women was even more profound in rural communities where access to institutional medical care is absent. Janet Kincaid's story (Kincaid 2015) of what happened to the women in the village of Joeblow (Schwartz 2019a) is a case in point:

> The village of Joeblow, Liberia, is like so many outposts in West Africa – a cluster of mud brick homes situated off of a dirt road that cuts through dense, forested expanses. What makes this village different, though, is there are no women. Last summer, the Ebola virus arrived in the village, carried by the husband of one family. In a twist of events none foresaw, the wife cared for her sick husband. When she became ill, all of the women of the village shared the task of caring for her. Then, one by one, each of them fell sick. All of the women died, leaving husbands widowed, children without mothers, and an entire village devoid of the females in their lives.

Further impacting children, stigmatization often made it difficult to find someone to provide basic medical care for non-Ebola-infected children with infected mothers, even if they were orphaned (Brewinski Isaacs 2016; Kolleh and Doucleff 2014). Some 3700 orphan children, whose parents died from EVD, were abandoned and outcast for fear that they would pass on the disease (Global Fund For Women 2014). Unfortunately, the imperative of caring for Ebola-stricken community members was forced on local women by a failing public health system.

The Ebola crisis was further exacerbated by a complete breakdown of Liberia's health care system, which exposed medical care workers and patients to the highest risks. A lack of basic personal protective equipment (PPE) left health care workers in untenable position of providing care only by putting their own lives at risk. Desperation compelled some health care providers with meager resources to introduce drastic measures. For example, it is reported that health care providers set up triage areas (restricted units for patients suspected of having Ebola infections) and utterly avoided body contact with all sick individuals (Fleischman 2014), or in some cases, completely abandoned patient care. Many abandoned patients, including pregnant mothers, both Ebola virus-infected and uninfected, died because of the lack of care (Fink 2014a; Strong and Schwartz 2016). During the Ebola outbreak, 184 health workers died from infection. Of the fewer than 300 medical doctors available for the entire population of 4.1 million people, 117, including five obstetrician gynaecologists, were claimed by the virus (Boaden 2016). Nurses and nursing aids (mostly females) accounted for the highest proportion (35%) of the 810 Ebola health worker infections reported by mid-August 2014 (Nyenswah et al. 2016). Even when health care practitioners were available, the strained health infrastructure in Liberia failed patients. For example, according to the WHO Ebola Health Team, some 995 patients needed clinical care in the week of September 8 through 14 alone, which far exceeded the bed capacity of Liberia at the time (i.e., without accounting for underreporting) (WHO Ebola Response Team 2014). Other public institutions in Liberia suffered under the Ebola epidemic, resulting in indirect negative health outcomes for women and girls.

The crippled education system in Liberia predisposed communities to a lack of basic education and knowledge of the disease and promoted a reluctance to seek treatment and upheld traditional beliefs (e.g., cause of illness is due to witchcraft). A Save the Children evaluation study demonstrated that the Ebola epidemic placed increased pressure on a fragile educational system that was already struggling to meet the scholastic needs of the population. In July 2014, schools were closed for 7 months. During that period, local communities used and, unfortunately, damaged 82% of 213 school facilities for water, sanitation, and hygiene, as well as such items as tables, chairs, cabinets, and other resources (Save the Children 2015, pp. 29–30). Schools in Liberia provide a relatively safe haven for girls, helping prevent childhood pregnancy and marriage. However, with the closure of schools due to Ebola, those risks to girl's and women's health were increased (Fleischman 2014).

15.3 The State of the Liberia Health System: International, Regional, and National Law

15.3.1 International Law Instruments

Liberia's legal record regarding international and regional health and human rights instruments deserves nothing short of applause. Liberia is party to the three International Bills of Rights: the Universal Declaration of Human Rights (UDHR), 1948; the International Covenant on Civil and Political Rights (ICCPR) 1966; and the International Covenant on Economic, Social, and Cultural Rights (ICESCR), 1966.[2] On 10 December 1948, Liberia, one of 48 countries of the United National General Assembly, voted in favor of the UDHR. Under customary international law, Liberia is obliged to refrain from acts which would defeat the object and purpose"[3] of the UDHR. Thus, Liberia agrees to uphold Article 25(1) and (2) of the UDHR, which states:

> (1) Everyone has the right to a standard of living adequate for the health and well-being of himself and of his family, including food, clothing, housing and medical care and necessary social services, and the right to security in the event of unemployment, sickness, disability, widowhood, old age or other lack of livelihood in circumstances beyond his control.
> (2) Motherhood and childhood are entitled to special care and assistance. All children, whether born in or out of wedlock, shall enjoy the same social protection.

Article 12 of the ICCPR states that "*everyone has the right to liberty of movement, except when it is necessary to protect national security or public health, in accordance with the law, shall such provision be subject to restrictions.*" Article 12 of the ICESCR requires States Party to "*recognize the right of everyone to the enjoyment of the highest attainable standard of physical and mental health*" including providing for "*the reduction of stillbirth-rate and of infant mortality and for healthy development of the child.*"

Dr. Joseph N. Togba represented Liberia at the International Health Conference, held in New York from 19 June to 22 July 1946.[4] As a result, Liberia is among the first 61 countries to adopt the WHO Constitution, which entered into force on 7 April 1948. Pursuant to the principles of the WHO Constitution, Liberia consented to ensuring that all Liberians enjoy the highest attainable standard of health as a fundamental right as well as taking necessary steps to combat and eradicate epidemics and other diseases.[5]

Liberia acceded to the International Convention on the Elimination of All Forms of Discrimination against Women (hereafter CEDAW), 1979, and signed its Optional Protocol, 2000 on 17 July 1984

[2] Liberia signed and ratified both the International Covenant on Civil and Political Rights and the International Covenant on Economic, Social and Cultural Rights on the same dates, i.e., 18 April 1967 and 22 September 2004.

[3] *See*, Article 18 of the *Vienna Convention on the Law of Treaties*, 1969, which Liberia signed and ratified on 23 May 1969 and 29 August 1985, respectively.

[4] Official Records of WHO, no. 2, p. 9, available at: http://apps.who.int/iris/bitstream/10665/85573/1/Official_record2_eng.pdf.

[5] *See*, Articles 2(g) and 28(i) of the World Health Organisation Constitution, 1946.

and 22 September 2004, respectively. Article 2(f) obliges Liberia to take all appropriate measures, including legislation, to modify or abolish existing laws, regulations, customs, and practices which constitute discrimination against women. Liberia is also required to protect women's right to health, including safeguarding the function of reproduction (article 11(f)). Partly due to the 14-year civil war, Liberia combined and submitted its initial, second, third, fourth, fifth, and sixth periodic reports of States parties to CEDAW on 13 October, 2008. Liberia provided no information on maternal health regarding the role of rural women in the economic survival of their families, justifying that, "[t]he only maternal health information available at the time of this report is the national average and information provided in the health section of the report."[6]

The United Nations Convention on the Rights of the Child (CRC), 1989, is the most widely and rapidly ratified international human rights instrument in history.[7] Liberia signed and ratified the CRC on 26 April 1990 and 4 June 1993, respectively. Once again, Article 24 of the CRC provides for the enjoyment of the highest attainable standard of health so that no child is deprived of his or her right to access such health care services. Article 19 of the CRC urges States Party to take all appropriate legislative, administrative, social, and educational measures to protect the (disabled) child (Article 23) from all forms of physical or mental violence including sexual abuse and exploitation. Befittingly, Liberia also signed the CRC's Optional Protocol on the involvement of children in armed conflict. To this effect, Liberia's Ministry of Defence recently collaborated with the United Nations Children's Fund (UNICEF) to enshrine the CRC in its military code (UNICEF 2009).

Liberia signed and ratified the United Nations Convention on the Rights of Persons with Disabilities (CRPD), 2008, on 30 March 2007 and 26 July 2012, respectively. Liberia also signed the Optional Protocol of the CRPD on the same date as it did the actual Convention. Article 25 of CRPD lays out a string of provisions for the enjoyment of the highest attainable health without discrimination on the basis of disability, including providing the same quality of care in the area of sexual and reproductive health as it does for other persons.[8]

15.3.2 Regional Legal Frameworks

Regionally, Liberia has ratified and/or signed key instruments earmarked for promoting and protecting children's and women's health and human rights. Of the 55[9] countries on the continent, only one (South Sudan) has not signed the African Charter on Human and Peoples' Rights, 1981 (hereafter the Banjul Charter).[10] Article 16(1) and (2) of the Charter enshrine the right to health, instructing that every individual has the right to enjoy the best attainable state of physical and mental health. States parties are urged to take necessary measures to ensure that medical care is provided for those that are

[6] *See*, section 16.13, p. 73 of the *Consideration of reports submitted by States parties under article 18 of the Convention on the Elimination of All Forms of Discrimination against Women*, 13 October 2008, CEDAW/C/LBR/6 available at http://www.refworld.org/country,,CEDAW,,LBR,,49672b4f2,0.html.

[7] The United States of America and Somalia are the only two countries that have not signed the Convention on the Rights of the Child.

[8] Article 25(a) of the Convention on the Rights of Persons with Disabilities provides that States Parties shall: Provide persons with disabilities with the same range, quality and standard of free or affordable health care and programmes as provided to other persons, including in the area of sexual and reproductive health and population-based public health programmes.

[9] Morocco was not a member of the African Union in the last 30 years until on 30 January 2017, when African Union readmitted it as a member state despite on-going dispute in Western Sahara. See: http://www.bbc.com/news/world-africa-38795676.

[10] Liberia signed, ratified and deposited with the Banjul Charter on 31 January 1983, 04 August 1982, and 29 December 1982 respectively.

sick. Article 18 specifically provides for the protection of the rights of children, women, the disabled, and the elderly.

The African Charter on the Rights and Welfare of the Child, 1990 (hereafter the African Children Charter), is hailed as one of the most progressive international human rights instruments to secure rights peculiar to the African child. Thomas Kaime affirms that:

> ...CRC addressed every aspect of children's lives, specific provisions on aspects peculiar to Africa fell victim to the overriding aim reaching a compromise and the African Children's Charter was intended to fill that void in terms of African concerns. The Children's Charter incorporates the universalist outlook of CRC, but at the same time clothes its conceptions within the 'African cultural context.' It is, therefore, a document with a cultural-universalist outlook and a perfect starting point for the consideration and elucidation of children's rights in Africa.

Like its "sister" CRC, the African Children Charter is also widely signed and/or ratified. Only four African countries (Democratic Republic of Congo, Sao Tome and Principe, South Sudan, and Sudan)[11] have not yet signed and ratified the African Children Charter. While the African Children Charter encapsulates all aspects of African children's rights, Article 14 is dedicated to protecting their health rights. Precisely, Article 14(1) provides that every child shall have the right to enjoy the best attainable state of physical, mental, and spiritual health, admonishing state parties "to ensure appropriate health care for expectant and nursing mothers" (Article 14(2)(e)).

Considering the inseparable bond between a mother and a child, the Second Session of the Assembly of the Africa Union adopted the Protocol to the African Charter on Human and Peoples' Rights on the Rights of Women in Africa, 2003 (hereafter the Maputo Protocol), exactly 13 years after the African Children Charter was established. To date, only three African States have not signed or ratified the Maputo Protocol. Subject to Article 28 of the Maputo Protocol, Liberia signed, ratified, and deposited its instrument with the Chairperson of the Commission of the African Union on 16 December 2003, 14 December 2007, and 15 July 2008, respectively. As State Party to the Maputo Protocol, Liberia agrees to uphold its provisions, particularly Article 14, which instructs States Parties to ensure women's right to sexual and reproductive health including strengthening and establishing pre- and postnatal health services for women.

Unfortunately, violation of the Banjul Charter, The African Children Charter, and the Maputo Protocol does not lead to legal redress, remedy, or relief via the regional courts (i.e., the Court of Justice of the African Union, 2003, and the African Human Rights Courts, 1998). The Court of Justice of the African Union and the African Court on Human and Peoples' Rights, which are the "principal judicial organ" with jurisdiction to interpret, apply, and validate the African Union's treaties as well as rule on all disputes,[12] were both dysfunctional. The two courts are now merged into one African Court of Justice and Human Rights,[13] but are yet to see the adjudication of cases pertinent to African women's reproductive and health rights. On a brighter note, recent activities with the Economic Community of West African States (ECOWAS) Court of Justice, 2005, offer hope for trying human rights violations in subregional West Africa (Alter et al. 2013). Health is a fundamental right, and nation states are required by international, regional, and national instruments to take steps toward ensuring its access, affordability, and availability.

---

[11] Liberia signed the African Charter on the Rights and Welfare of the Child on 14 May 1992.

[12] *See,* Articles 2(2) and 19 of the *Protocol of the Court of Justice of the African Union*, 2003 and Article 3 of the *Protocol to the African Charter on Human and Peoples' Rights on the Establishment of an African Court of Human and Peoples' Rights*, 1998.

[13] See Chapter 1: Merger of the African Court on Human and Peoples' Rights and the Court of Justice of the African Union of the *Protocol on the Statute of the African Court of Justice and Human Rights*, 2014. As at 1 April 2016, only nine (Liberia is not included) out of the 54 countries have signed the Statute of the African Court. There is zero ratification and deposit on its instrument.

15.3.3 National Law and Policy

In Liberia, a catalogue of legal instruments, policies, and institutions embodies a theoretical perfection for protecting children and women's rights. Public, private, international (including United Nations agencies and nongovernmental organizations (NGOs)), local, community-based, and ad hoc organizations encompass the nation's health implementation and intervention scheme. Six key public health institutions, laws, and policies in Liberia are worth mentioning, with respect to educating health workers and service delivery in health and social welfare. They are: the Tubman National Institute of Medical Arts (TNIMA), the Achille Mario Dogliotti College of Medicine and Health Sciences (hereafter Dogliotti College of Medicine), the Ministry of Health and Social Welfare (hereafter the Ministry of Health), the Public Health Act, the Ministry of Gender, Children and Social Protection (hereafter the Ministry of Gender), and the National Commission on Disabilities.

15.3.3.1 Tubman National Institute of Medical Arts

The Tubman National Institute of Medical Arts (TNIMA) furthers educational, health, and social development in Liberia by training health workers for the rural areas. The Tubman National Institute of Medical Arts was established in 1945 in collaboration with the Liberian National Public Health Services and the United States Mission in Liberia. To date, TNIMA operates four programs: School of Environmental Health, School of Midwifery, School of Physician Assistant, and School of Professional Nursing (Kennedy 2011). It has graduated over 3000 paramedical health workers as of 2013 (Kennedy 2011), although not without challenges (All Africa 2013; Wolokolie 2013). However, recent "investment into trained midwives" across the country brings hope for improvement (WHO 2016b).

15.3.3.2 Achille Mario Dogliotti College of Medicine and Health Sciences

In 1966, then President William V. S. Tubman, with assistance from the Italian Government, the Vatican, and the Achille Mario Dogliotti Foundation, established the Monrovia-Torino Medical School in hopes of training medical doctors for Liberia as well as Africa at large (AM Dogliotti College of Medicine 2010, p. 3). Initially affiliated with the Faculty of Medicine and Surgery of the University of Turin in Italy, in 1970 the Monrovia-Torino Medical School merged with the University of Liberia and was renamed Achille Mario Dogliotti College of Medicine, becoming the second professional school in Liberia (AM Dogliotti College of Medicine 2010, p. 3). After the inauguration of the John F. Kennedy Memorial Hospital (hereafter JFK Hospital) in 1971, it became the teaching hospital of Dogliotti College of Medicine. From its opening in 1966 until 2009, the Dogliotti College of Medicine graduated 273 doctors (Executive Mansion 2009). In 2016, enrollment at the AM College of Medicine included 231 medical students, 79 pharmacy students, and 20 students in nursing and midwifery (Ballah 2016b). In 2006, a situational and observational report of Doglotti College of Medicine highlighted its rickety and neglectful state, noting broken down structures with no access to running water, electricity, or internet (Nyan 2006). As not much has changed today (Ballah 2016b), the question remains whether Liberia will ever be prepared for a future epidemic in the face of the lingering possibility of the resurgence of the EVD.

15.3.3.3 Ministry of Health and Social Welfare

Amended Chapter 30 of the Executive Law, 1972,[14] obliges the Ministry of Health and Social Welfare to administer government activities regarding protection and improvement of public health and social

[14] Act to Amend Chapter 30 of the Executive Law of 1972. 30.2 Duties of the Minister of Health and Social Welfare. Sub-section (b).

welfare, including provision of medical care and treatment through public hospitals, health centers, and clinics across the country (Ministry of Health and Social Welfare 2008, p. 3). On 8 December, 2016, the House of Representatives agreed with the Senate (from 21 September, 2016) to establish the new Ministry of Health and National Public Health Institute of Liberia Acts. The new Acts amended the Ministry of Health and Social Welfare Act (1972) and the National Public Health Institute of Liberia (hereafter National Public Health Institute), formerly but now defunct Liberian Institute for Biomedical Research (1975) (Bolay 2016), respectively (Carter 2016; Front Page Africa 2016a). The mandate of the newly formed National Public Health Institute is to create a center of excellence in preventing and controlling public health threats. The immediate mandate of the National Public Health Institute is to "*enhance and expand surveillance; develop and build the public health workforce and improve epidemic preparedness, laboratory capacity and research.*" The new Act is the Legislature's and the Government of Liberia's direct response to enforcing a law that would effectively address future outbreaks (Front Page Africa 2016a).

Eager to delve into post-conflict reconstruction, the Ministry of Health developed a five-year Emergency Human Resources for Health Plan in 2007 (Government of Liberia 2007a, p. 3) based on rapid assessment of health workers, facilities, and community access. By 2010, the Ministry of Health had increased its clinical workforce from 1396 in 1998 to 4653, including 3394 midwives and nurses (Varpilah et al. 2011). Despite adopting strategies to augment human resources in health, Tornorlah Varpilah et al. argue that payroll management, equitable distribution, retention of health workers in remote communities, and improving performance for high quality service remain outstanding challenges for Liberia's health system (Varpilah et al. 2011).

In 2012, the Government of Liberia launched the National Vision 2030—aspiring to become a middle-income country by 2030 (Government of Liberia 2012). Closely linked to the National Vision *2030* are Liberian citizens' health and social welfare. In achieving the National Vision's health and social welfare goal, the National Health and Social Welfare Policy: 2011–2021 (hereafter National Health Policy Plan) aims to improve equitable access to health and social welfare (Government of Liberia 2011, pp. 1–2) by: (1) increasing access to and utilization of a comprehensive package of quality health; (2) making health more responsive to people's needs, demands, and expectation; and (3) making health and social protection available to every Liberia at an affordable cost regardless of their social position in society.

Predating the National Health Policy Plan was the Government of Liberia's Basic Package of Health and Social Welfare Services, 2008 (hereafter, Basic Package of Health). Based on the fundamental principles of primary health care and decentralization of health services management, the Basic Package of Health targets maternal, child, adolescent, reproductive, mental health, communicable disease control, and emergency care. Thus, the main purposes of the Basic Package of Health are to: (1) standardize the service delivery package at the five levels of the Liberian health system; and (2) redistribute essential health services to ensure universal access throughout Liberia (Government of Liberia 2008, pp. 1–3).

In spite of the Basic Package of Health's strong emphasis on sexual and reproductive health, Bayard Roberts et al. observe major challenges for sexual and reproductive health services in post-conflict countries such as Liberia. According to Roberts et al., disruptions in care and fragmented health service delivery caused by damaged infrastructure, limited human resources, weak stewardship, and proliferation of NGOs are common symptoms of countries emerging from violent conflict (Roberts et al. 2008). Dörte Petit et al. conducted interviews with 63 sexual and reproductive health service providers and affirmed that post-conflict governments, such as Liberia, tend to subcontract NGOs to deliver the Basic Package Health Service (Partners in Health 2015, 2016, 2017; Petit et al. 2013). Findings from the Petit et al. study also showed that,

> "While health workers attributed low uptake of [sexual and reproductive health] services to low demand arising from socio-cultural factors and the role of trained traditional midwives, policy makers attributed it largely to a supply-side problem because of poor quality of services and poor attitudes towards patients particularly to women delivering at health facilities. (…) This limited engagement by policy makers with the perspectives of health workers is also reflected by the fact that policy makers focused more on the need to expand the list of [Basic Package Health Service] services, and rather less on improving the working conditions and constraints to this expansion."

Accreditation is one way to conduct continuous assessment of the Basic Package Health Service program as it does not only improve quality of care, but also facilitates monitoring of trends and evaluation of health systems. While successful accreditation systems have existed in high- and middle-income countries, many low-income countries in Africa have fallen behind in establishing such health facilities programs. In 2007, the Ministry of Health and Social Welfare did not have basic information about health care in Liberia. With assistance from the Clinton Foundation, a Basic Package of Health Service accreditation program was initiated amidst extreme logistical and implementation challenges. In particular, the Basic Package Health Service program was designed to track quality improvement or measure of quality of care. Nevertheless, the accreditation program allowed the Ministry of Health to gather unprecedented amounts of information about the current status of service provision across the country, well beyond directly tracking and measuring quality of care (Cleveland et al. 2011).

As a result of the Ebola epidemic, the Government of Liberia mobilized to develop yet another policy to address the suffering, economic ruin, and deaths of health workers. Emphasizing the fact that health workers experienced a 30 times higher risk of infection compared as to the general population, (i.e., as of 8 April 2015, 372 health care workers were infected and 184 had died), *the Investment Plan for Building a Resilient Health System* emerged in 2015. Like the National Health Policy and Plan, the Investment Plan for Building a Resilient Health System (Government of Liberia 2015, p. iii and 1) aims to provide universal access to safe and quality services through improved capacity of the health network for provision of safe, quality Essential Packages of Health Services, as well as a robust Health Emergency Risk Management System through building public health capacity for prevention, preparedness, alert, and response for disease outbreaks and other health threats.

The policies listed above promise to create an enabling environment that restores trust in the health authorities' ability to provide services through community engagement in service delivery and utilization, improved leadership, governance, and accountability at all levels.

In 2014, the Global Fund for Women reported that only 7% (or US$10 per person) of Liberia's national budget is allocated to its fragile health system (Global Fund For Women 2014). However, the National Budget for fiscal year 2016/17 approved on 20 September, 2016 (Brooks 2016; Yates 2016) reveals a stunning 5% increase in health allocation, i.e., 77.4 million of its 600 million projected revenue (Government of Liberia 2016a, pp. 203–240). In spite of the health budget increase, Martin Kollie contests, "*…even though Liberia's national budget of US$600 million is far less than the United States (US$4.147 trillion), the world's economic and military superpower, [Liberia] speaker and senate pro tempore receive more money annually those of the U.S.*" (Kollie 2017). According to Kollie, the Liberian Legislature spends US$40.6 million over a 12-month period with the Speaker of the House receiving US$445,000 and the Senate Pro Tempore's US$482,000 (Kollie 2017). Together, the Speaker and the Senate Pro Tempore amass receive US$927,148, "*while over 800 Liberian children die every year as a result of unsafe drinking water…,*" Kollie laments (Kollie 2017).

15.3.3.4 The Public Health Law

The Liberia Public Health Law, 1976 (hereafter the PHL), was first established in 1956. In 1977, an act of legislation set up the Liberian Association of Public Health Inspectors (now the Liberian Association of Public Health Technicians) to protect Liberians' health through preventive health measures (Front Page Africa 2017). Chapter 14.2 of the PHL invests power in the Minister of Health to make rules regarding a cadre of matters, including but not limited to, (1) conducting house visitation (Chapter 14.2(b)); (2) preventing any person from leaving an infected area without undergoing medical examination (Chapter 14.2(d)); (3) for establishing hospital and observation camps or stations (Chapter 14.2(e)); (4) for removal of person with the disease and those in contact with such persons (Chapter 14.2 (g)); and (5) for removal of corpses (Chapter 14.2(h)). However, as stated in President Sirleaf's above declaration, provisions of the PHL are no longer effective to address a contemporary epidemic such as the EVD. In its current state, the PHL makes no provision for controlling the epidemic using quarantine or other isolation measures.[15]

Regardless, authority for enforcing the PHL is entrusted to the Minister of Health. But, the President of Liberia, affirming the outdated nature of the Law, appointed herself the "*chair along with the Minister of Internal Affairs in his role as Chairman of the National Disaster Relief Commission*" (Johnson Sirleaf 2014a) of a National Task Force with support from a technical team headed jointly by the Minister of Health and Social Welfare and the Country Representative of the World Health Organization (Johnson Sirleaf 2014b; WHO 2014a). Understandably, a desperate situation (the EVD outbreak) called for desperate measures, thus President Sirleaf exercised strong political will and commitment in taking charge of security and health safety of her people. Nevertheless, such haphazard measures do not always amount to positive outcomes. So was the experience of 15-year old Shakie Kamara who was fatally shot when the Liberian Armed Forces tried to enforced a lockdown of one of the most impoverished communities in the country (Onishi 2014b; Shute 2014). Criminalizing disease in Liberia partly explains young Kamara's death, as it relates vulnerable populations (Diaz 2016).

As with the amendment providing for the control of HIV/AIDS (All Africa 2014a, 2014b; The Center for HIV Law and Policy 2016), the PHL criminalizes "risky" behaviors and practices. For instance, Chapter 14.6 of the Public Health Act states: "*Any person other than a public officer who violates any of the provisions of this chapter or any of the rules made thereunder shall, upon conviction, be liable to a fine not exceeding two hundred dollars or to imprisonment not exceeding thirty days, or to both such fine and imprisonment.*" The Public Health Law is further critiqued for being very general and incongruent to the health realities at the time (World Health Organization 2005), yet heavily focused on regulating drugs, standards for health workers, and licensing of health professionals. Much needed revision of the PHL started in 2011 with assistance from Jeannette Austin (another individual foreign to the Liberian health system), serving with a team of in-country lawyers in developing a new public health law (Austin 2011).

15.3.3.5 Ministry of Gender, Children, and Social Protection

An Act to Amend the Executive Law Title 12 (The Gender and Development Act), 2001, established the Ministry of Gender. Mandated to effectively coordinate mainstreaming government-wide efforts to ensure gender equality and empowerment of women and development of children, inter alia, the Ministry of Gender is the driving force behind the enforcement of the Universal Declaration of Human Rights, CEDAW, CRC, the Maputo Protocol, the African Children Charter, and the United Nations Security Council 1325 on Women Peace and Security. The roles of both the Ministries of Health and Gender are significantly increasing in the lives of children and women in postwar Liberia (Fuest 2008).

[15]Chapter 11(i) and (k) of the Public Health Law, defines isolation and quarantine as it pertains "Control of acute communicable diseases and conditions". These terms are not used anywhere in Chapter 14: "Formidable Epidemic, Endemic or Infectious Diseases."

15.3.3.6 National Commission on Disabilities

In 2005, 7 years after the Liberian Legislature enacted the National Commission on Disabilities Act 2005, the Government ratified the UN Convention on the Rights of Persons with Disabilities 2006 on 26 July, 2012 (All Africa 2012). The National Commission on Disabilities is an autonomous agency having *"jurisdiction over all matters involving and appertaining to general welfare, wellbeing, education of all disabled persons within the Republic of Liberia"*.[16] Section 7 of the Act authorized the Government of Liberia to make budgetary appropriation for the education, social, and economic development of all disabled persons within minutes of passing the Act. Despite its overzealous aspiration, the National Commission on Disabilities continues to struggle. In an address by President Johnson Sirleaf to the National Union of Organisations of the Disabled, the National President, Reverend Fallah S. Boima, discloses: "…the [National Union of Organisations of the Disabled], which is the mother organization of all disabled persons in Liberia, does not have a stable source of funding to support its activities and programs… That the subsidy provided by government through the National Commission on Disabilities cannot cover the monthly rent of the Union's office and it stands to be evicted from its office shortly" (Executive Mansion 2015). One can only imagine the situation of people with disabilities living in remote Liberia when its centralized location in Monrovia is at risk for being "homeless" due to lack of funding. Nevertheless, the country deserves some recognition for ongoing progress made in health and education budgets.

15.4 Liberia's Public Health Emergency and Response

On the 8th of August, 2014, the Ebola outbreak in West Africa was declared an international public health emergency by WHO. The epidemic provided an opportunity to canvass the many fields of law responsive to the threat of the epidemic disease, with respect to maternal and reproductive health care in Liberia. When WHO declares a "public health emergency of international concern," the language for application of the International Health Regulations is triggered in effort to provide a framework for global response designed to strengthen international public health security (Price 2015; 2014, p. 4).

Article 13(a) of the *Constitution of Liberia*, 1986 (hereafter the Liberian Constitution), instructs that "[e]very person lawfully within the Republic shall have the right to move freely throughout Liberia, to reside in any part thereof and to leave therefrom subject however to the safeguarding of public security, public order, public health or morals or the rights and freedoms of others." It is this provision of the Liberian Constitution that supports President Sirleaf's compulsion to isolate and quarantine segments of the country as above. The Public Health Law that puts the Ministry of Health in charge of isolation and quarantine in case of controlling acute communicable diseases and conditions fails to include Ebola on the list of "notifiable disease or condition" that should be reported during a public health emergency. Albeit, Markey argues that contemporary views regarding individual liberties sometime make public health authority reluctant to use force to restrict movement (Markey et al. 2015). At the very onset of the EVD crisis, the Human Rights Watch admonished West African governments of the immediacy of addressing the gender dimensions of the outbreak, protecting health workers from infection, limiting the use of quarantines, and ensuring that security forces respond to the crisis respect basic rights (Human Rights Watch 2014, p. 3). For example, in Sierra Leone, prior to the President Ernest Bai Koroma's proclamation of a state of emergency on 7 August, 2015, the Parliament of Sierra Leone moved to enact a new law, which imposed a maximum jail term of 2 years for anyone caught sheltering an Ebola patient (BBC 2014c).[17]

---

[16] Section 2, National Commission on Disabilities Act.

[17] See Article 8 of *The Public Emergency Regulations*, 2014.

Standard public health actions critical to arresting the Ebola outbreak included early identification of cases, setting up isolation care facilities (Yamin et al. 2015), and treatment, among others (WHO 2014b, p. 1). The World Health Organization established the community care centers (hereafter CCC) (WHO 2014c, p. 1). The CCCs were run by community health workers with the goal of isolating Ebola patients so as to prevent further transmission of the virus while providing "rapid and high coverage" (Kucharski et al. 2015), curative, and palliative care (WHO 2015, pp. 1–2, 12). Mathematical modelling analysis predicted that allocation of 4800 additional beds at the CCC would have averted 77,312 cases by 15 December, 2014 (Lewnard et al. 2014). Evidently, only 28,616 cases and 11,310 cumulative cases occurred in all countries as of the 13th of May, 2016 (WHO 2016c).

Interestingly, some perceived the CCC as "Ebola death camps" where relocation to the centers tended not to be voluntary (Adams 2014), thus raising legal and human rights concerns. For example, a risk and benefit research analysis conducted in Sierra Leone found, even though CCCs somehow shifted the EVD transmission pattern, infection control could worsen should Ebola virus-negative patients have a high risk of exposure in the community (Kucharski et al. 2015). West Point, an impoverished slum in Monrovia, Liberia, epitomizes a site hard hit and tested for human rights violations during the EVD crisis (Crowe 2014; Liljas 2014; Onishi 2014b; Parshley 2016; Shute 2014). In August 2015, 15-year-old Shakie Kamara died from indiscriminate stray bullets shot by soldiers of the Liberia Armed Forces on residents of West Point who were protesting a blanket quarantine imposed on their community by the Government of Liberia (Onishi 2014b, c).

The regional response to the Ebola epidemic also created border tensions. For 2 years (i.e., between 23 August and 9 September 2016), La Cote d'Ivoire closed its land border with Liberia to prevent spread of the EVD (BBC 2014d). Bruno Kone, the Government of La Cote d'Ivoire spokesperson, asserts that "[w]e had to take these measures to protect our country. And the fact we didn't have a single case must be considered a real success," (Reuters 2016).

Overall, Liberia did not seem prepared for response to the Ebola virus epidemic, but succeeded with managing the outbreak through a coordinated, united, and immediate response from the international community including WHO (World Health Organization 2015a), US military (du Lac 2014), CDC (Centre for Disease Control 2014), local health workers (Evans et al. 2015), and individual sacrifices (Moore 2014) and acts of compassion (Nyenswah et al. 2016).

15.5 Maternal Health, Reproductive Rights, and Protection

As noted above, the protection of women's and children's rights in Liberia under international, regional, and national law is well-documented. In addition to these instruments, global conferences such as the International Conference on Population and Development (ICPD), 1994, followed by the Beijing Conference (1994), popularized the global mantra: "women's [reproductive] rights are human rights" (Office of the United Nations High Commissioner for Human Rights 2014, p. 1). Also, the Commission on Human Rights Resolution 2003/68 affirmed: *"[s]exual and reproductive health are integral elements of the right of everyone to the enjoyment of the highest attainable standard of physical and mental health*."[18] Despite these historical trademarks designed to protect women's sexual and reproductive rights, this section shows that Liberian women have not yet attained an adequate standard of maternal and reproductive health.

The state of maternal and reproductive health of girls and women in Liberia is troubling. For 2006–2015, Liberia ranks fifth among developing countries with the highest adolescent birth rate per 1000 women age 15–19 with 149 births per 1000 (UNFPA 2016a, pp. 94–98). Thirty-one percent of

[18] Commission on Human Rights resolution 2003/28, preamble and para. 6.

women in Liberia begin childbearing between the ages of 15 and 19 years (Government of Liberia 2014a, p. 79). In 2015 alone, some 221,000 pregnancies were expected in Liberia (UNFPA et al. 2014, p. 129). Research shows that over 55% of neonatal mortality occurs among adolescent mothers under 15 years, whereas 6% is observed among women over 19 years of age (Government of Liberia 2016b, p. 14), an indication that delaying childbearing until after 20 years results in fewer infant deaths. Liberia's under-five mortality rate is 248 per 1000 live births and the maternal mortality ratio is 1200—increased from 598 in 2007 (per 100,000 live births) (Front Page Africa 2016b; USAID Liberia Mission 2016, pp. 4–5; World Health Organization 2005, 2015b).

The major causes of maternal death are either almost entirely preventable or treatable, provided skilled birth attendants or emergency obstetric care is present (Meyers 2013, p. 5; Schwartz 2015). The Liberia Demographic Health Survey of 2013 found that the maternal mortality ratio (MMR) was 1072 deaths per 100,000 live births, one of the highest in the world. Sixty-one percent of births take place at home, assisted by traditional birth attendants (Government of Liberia 2014a, pp. 162, 285). Doctor Walter T. Gwenigale, a former Minister of Health, suggests that the appalling situation of mothers and their newborns is partly due to an acute shortage of skilled health workers, inadequacy of reproductive health supplies, widespread cultural practices, and lack of public transportation to facilitate transfer of obstetric emergencies (Government of Liberia 2007b, p. 5).

Additional barriers to effective maternal and reproductive health in Liberia include (but are not limited to) poverty, harmful traditional practices (e.g., female genital cutting), social myths, and long distances between health facilities (Government of Liberia 2016b, p. 13; Meyers 2013, p. 6; Subah et al. 2014, p. xi). Of 550 women interviewed for the Liberia Demographic Health Survey 2013, 50% between ages 15 and 49 years said they were members of the Sande Institution (traditional schools for girls). Membership of the Sande institution is marked by female genital cutting, where parts of the clitoris or labia are removed during ceremonious rites of passage into adolescence (Government of Liberia 2014a, pp. 275–276; Schwartz 2019b).

The Ebola crisis in Liberia compounded health risks for children and pregnant women. The EVD crisis diverted resources away from pregnant women. For example, in Bong County, one of the most populous counties, the ambulance normally used for obstetric emergencies was transferred to the Ebola response unit. In the capital city of Monrovia, the surgical and emergency departments of the JFK Hospital, the country's largest referral hospital, were closed (Hessou 2014a). This diversion of resources and attention resulted in a direct increase of pregnant women dying from preventable causes (Hessou 2014a). The leading cause of maternal mortality was postpartum haemorrhage, accounting for 34% of all deaths (Subah et al. 2014, p. ix). Other major causes of maternal mortality included anaemia, ruptured uterus, sepsis, unsafe abortions, and eclampsia. Teen pregnancy surged (Werber 2015), and as health facilities closed, pregnant women had very few places to deliver their babies safely (Crowe 2014). Facility-based deliveries also dwindled during the EVD crisis due to the common belief that Ebola was spread by health workers (USAID Liberia Mission 2016, p. 23). Furthermore, children died from vaccine preventable diseases (e.g., measles) (Crowe 2014). Childhood vaccination rates declined during the Ebola outbreak mainly due to the same parental mistrust in the health system.

One example of how social myths impacted maternal mortality was when a pregnant woman bled to death after refusing an oxytocin injection due to her fear of contracting Ebola from the injection. Seemingly justified, "*the myth that health workers were the reservoirs for Ebola is now being diminished via radio and trained traditional midwives who have been effectively disseminating this information to their communities*" (USAID Liberia Mission 2016, p. 23). Tragically, the prevalence of Ebola social myths predisposed pregnant women and babies to even more health risks as some health workers were not willing to touch patients, especially in rural areas. For instance, in Nimba County, it was reported that pregnant patients were forced to leave the local clinic to deliver themselves on the bare ground of the street (Global Fund For Women 2014).

Horrendous stories of women and girls gripped international media, especially stories of women hemorrhaging or bleeding due to pregnancy complications (Lori and Starke 2012) who were turned away by overwhelmed health workers (Fleischman 2014), or women simply refusing medical care for fear of being infected with the disease by health workers. Fear of the "general collapse of the health care system" compelled pregnant women to navigate curfew on foot or travel the city in taxis for untold hours searching for care (Couch 2014). Relief Web reported on Comfort Fayiah, a 36-year old mother who gave birth to twins on the side of the road amidst a downpour after struggling to find a health center. No amount of begging, blood, sweat, or tears would suffice. Four different clinics and hospitals "*bluntly refused*" to admit Mrs. Fayiah (Hessou 2014b). Mrs. Fayiah's welfare was left to the hands of passersby, "*including women who formed a human chain barrier using their clothes,*" (Hessou 2014b) to help protect Mrs. Fayiah's privacy and dignity. Sheri Fink of the New York Times tells of a young woman bleeding profusely who prematurely gave birth. Told that her boyfriend had recently received treatment for Ebola, a health worker asked her to remain outside the locked gate while they dressed up with masks and a chlorine sprayer. Unfortunately, by the time the doctor reached out to the young woman she was already dead (Fink 2014b). As terrible as these conditions were for healthy mothers, infected mothers suffered even more.

Yana Richens and Jacque Gerrard suggest that not only do pregnant women with the Ebola virus present clinically more severe symptoms and complications, but also have increased rates of spontaneous abortion, other hemorrhagic infections, and mortality when compared to nonpregnant patients with the EVD (Yana and Jacque 2014). If an infant born of an EVD-infected mother does survive, direct contact with the mother should be avoided to prevent transmission of the disease (Sepkowitz and Haglage 2014; Yana and Jacque 2014). Noteworthy, there has only been one case of a baby surviving when born to a mother with EVD—Baby Nubia in Guinea (Farge 2015; Sepkowitz and Haglage 2014). Sadly, above and beyond clinical medical care, the Ebola crisis also disadvantaged women and girls with respect to other social determinants of health.

Pregnancy can have devastating effects on women and girls, especially teenage girls, who tend to have an increased risk of contracting the virus (Menéndez et al. 2015). The Ebola virus is proven to be spread through semen from infected men who have recovered, and sexual and gender-based violence, precisely rape, is pervasive in Liberia (Deen et al. 2015). Raped and abandoned in September 2014, 14-year old Tina Williams exemplifies the many traumatic woes of women and young girls (Yasmin 2016). Without a doubt, the risks of Ebola infection for women and girls is exacerbated by gender discrimination, inequalities, and disparity (Fleischman 2014). A systematic review conducted by Meghan Bohren et al. affirms the mistreatment of women during childbirth globally (Bohren et al. 2015). According to Rebecca Cook, neglect and discrimination against women can sometimes be perpetuated by laws that obstruct women's access to reproductive health (Cook et al. 2003). For example, the ad hoc imposition of curfew and quarantine noted above prevented girls and women from accessing health care and other social services, especially outside congested and confined space, such as in West Point.

Women's and girls' increased vulnerability to maternal mortality and gender-based violence during the Ebola crisis subsequently reduced their access to education, economic empowerment, and political participation. Being tasked with domestic caregiving, most girls and women have less opportunity to engage politically with issues that affect their health and well-being. In Liberia, severe disruption of health services as a result of the Ebola outbreak did not only wreck the already broken health services, but also caused other social ills, as it made thousands of children orphans. In addition, school closures left many children without access to education. Even if schools remained open, teenage girls who become pregnant are generally forced to leave school, thereby jeopardizing any of her chances for future economic prosperity (UNFPA 2016b). The intersection of losing husbands and parents, school closures, and teenage pregnancy affected many young women in Liberia during the outbreak. As Fleischman posits: "*With the schools closed, girls may now be at higher risk for teenage pregnancy and child marriage. In addition, property and inheritance laws may discriminate against women and girls affected by Ebola, with some reports of Ebola widows being shunned by their families and denied the ability to inherit their husband's property*" (Fleischman 2014). Unfortunately, the social impact of EVD on girls and women did not stop with the end of the outbreak.

Society and the media disproportionately and unfairly linked the outbreak to Liberian women. Even after Liberia was pronounced Ebola-free by WHO in May 2015, the news heralded headlines such as "Female survivor may be cause of Ebola flare-up in Liberia" (Farge and Giahyue 2015), stigmatizing and stereotyping women as the "cause" of an EVD resurgence without realizing that women who suffer stigma and discrimination are more vulnerable to sexual abuse (Nyenswah et al. 2016). According to one news report, researchers expressed concerns about any possibility of controlling the epidemic, when at least some 17,000 survivors, many of whom are girls and women, could be potential human reservoir of the virus (Farge and Giahyue 2015). Media releases with a taint of gender bias undermine the goal of safeguarding women's health as they can also be used for justifying violence, exclusion, and denial of access to care for women and girls. Hence, ostracism of survivors (especially women and girls), Nyenswah suggests, "*would be an unacceptable conclusion to this unique event in global health*" (Nyenswah et al. 2016). Nonetheless, the Government of Liberia's restriction on media coverage should not have only concerned itself with preventing reporters from coming in direct contact with Ebola virus-inflicted persons; it should have also tasked media personnel with the responsibility of reporting bias-free news during and after the Ebola outbreak, when everyone seemed irrationally suspicious of others and nursed their fears with deliberation over who should face sanctions.

15.6 The Way Forward: Challenges and Opportunities

This chapter demonstrates over and again how women as mothers, primary caregivers, and health workers were at higher risk during the Ebola outbreak in Liberia. However, safeguarding girls' and women's reproductive health rights is often not prioritized by public institutions. The Government of Liberia's persistent lack of protection of women's reproductive health is a clear reflection of the extremely poor health system, whereby medical personnel did not only turn away and refused to treat pregnant patients for fear of contracting the disease (Couch 2014), but also neglected to address the disproportionate impact of the Ebola outbreak on girls and women. Though at high risk, midwives-turned-surgeons worked overtime, only accepting limited cases involving emergency pregnancy complications (Al Jazeera 2016).

As the impact of the Ebola virus disease stretches well beyond a public health epidemic towards social, political, and economic development, there is more reason for intervention programs to adopt gender equality and nondiscriminatory approaches. Hessou laments that the "*lack of access by women, especially pregnant women, to reproductive health services is a major disaster in waiting*" (Hessou 2014a). Meanwhile, Redd reasons that finding external partners who are willing to invest in long-term recovery and development efforts is often far more challenging than seeking emergency funds for immediate crises such as the Ebola epidemic (Redd 2015, p. 12).

Fleischman suggests two strategies for responding to Hessou's concern: (1) collection of sex-disaggregated data about infection and deaths that are publicly available, reliable, and unequivocal; and (2) investigation of abuse against women and girls, in order to enhance monitoring and protection of human rights (Fleischman 2014). Sophie Harman's feminist research further corroborates the stark, conspicuous invisibility of women and gender in international response plans of global health governance during the Ebola outbreak (Harman 2016):

> "Women are conspicuously invisible in global health governance: everyone knows they are there and that they do the majority of the care work, but they remain invisible in global health policy. The 2014 Ebola outbreak provides an acute case study on conspicuous invisibility, where issues of women and gender have been invisible in both the emergency response and in long-term planning on health system resilience. The short- and long-term responses to Ebola show that the male bias is very much present in thinking about disease outbreaks: there is little to no discussion about gendered impacts of the disease in framing the crisis, data disaggregated by sex were late in coming, and no strategy includes gender indicators."

As the UN Secretary General and WHO expressed grave concern for the EVD epidemic as an international public health crisis requiring unprecedented emergency action, Sarah Roache et al. warned that moving forward was not only about scaling up a proper health system, but also about, inter alia, (1) investing into a chronically unfunded, (2) building public distrust in the government and health workers; (3) promoting public education regarding harmful traditional practices; (4) developing drugs and vaccines; and (5) planning for the future by establishing a robust response to public health emergency (Roache et al. 2014, pp. 3, 4, 10–11). Doctor Margaret Brewinski, a medical officer in the National Institute of Child Health and Human Development Office of Global Health who also participated in the Ebola response to Liberia, identifies three research priority questions regarding care for pregnant women during the Ebola outbreak: (1) interweaving the needs of mothers and children into research protocols; (2) achieving the best possible outcomes for the Ebola-infected mother and her child, while also minimizing the risk of infection to health workers; and (3) assisting Ebola survivors care for children who have no other caregiver (Brewinski Isaacs 2016). Paul Hunt and Judith Bueno De Mesquita recommend that devoting increased attention to properly understand reproductive health and reproductive rights is not only crucial, but also necessary in the "struggle against intolerance, gender inequality and global poverty" (Hunt and De Mesquita 2006, p. 11).

15.7 Conclusions

The Ebola crisis heightened Liberia's capacity to establish health care law and policies as well as identify long-term commitments to significantly reduce the impact of maternal deaths, amidst doubts and fear to halt the disease (Rivers et al. 2014). However, as with many law and policy initiatives, the challenge for postwar Liberia is conducting effective implementation and intervention programs, especially in remote rural communities where girls and women, including those who are disabled, are gravely marginalized. Meyers observes that Liberia's reliance on third party institutions (precisely international not-for-profit organizations) to design policy documents alongside local administrations and public institutions' persistent lack of capacity, power, and resource to implement long-term health intervention programs offer a unique conundrum (Meyers 2013, p. 30) for the postwar recovery of the country.

Strengthening data capturing processes, oversight, and management at the municipal and county level is crucial to rebuilding and improving maternal and child health care outcomes. Although efforts to engage pregnant women in the formal health sector are challenging among populations of women who have not employed primary health care measures throughout their life, without a doubt, the Ebola crisis presents Liberia with a new window of opportunity to establish such care for girls and women. Introducing women to the formal health sector in preparation for their first pregnancy requires a community-based approach facilitated by training of traditional midwives to build trust and educate women about the value of skilled births to their community, their family, and their overall health (USAID Liberia Mission 2016, p. 24). Thus, the recent introduction of the new Bachelor of Science in Midwifery and Nursing at the University of Liberia is optimistic and timely (All Africa 2015; WHO 2016b).

While no single law or policy explains Liberia's successful management of the Ebola epidemic, Nyenswah et al. identify six issues worth mentioning: (1) government leadership and sense of urgency; (2) coordinated international assistance; (3) sound technical work; (4) flexibility guided by epidemiologic data; (5) transparency and communication; and (6) efforts made by the local community (Nyenswah et al. 2016). In closing, consider President Sirleaf's response to an interview by the National Public Radio, which aptly positions Liberia as a global entity capable of drawing from adverse experience to improve its growth and development. When asked, whether she thinks the image of Liberia has changed through the Ebola crisis, President Sirleaf responds:

> *"We were the poster child of everything that could go wrong: disaster, death, destruction all over the place... [but] [t]here's a lot to celebrate in Liberia: The number of new Ebola cases have been declining, kids are going back to school and life is returning to some semblance of normalcy... I think we all finally realized that all of our lives were at stake. Everything we had worked for was at stake. That brought the coming together... Yes, we have problems, we don't have perfection. Yes, we do have corruption in the country. It's been there a long time. The deprivation that people suffered led them to a place of survival. And survival meant they had to do anything they could to feed their children and to live... [so I say] [g]o fix it. That's a call to everybody. Go fix it. In other words, do it yourself. Take charge. Empower yourself."*—Ellen Johnson Sirleaf, President of Liberia 2015 (Zambelich 2015).

References

Adams, M. (2014). Ebola death camps in Liberia: Disease victims to be rounded up and removed from their own homes by force. *Global Research*. Retrieved November 17, 2016, from http://www.globalresearch.ca/ebola-death-camps-unveiled-in-liberia-disease-victims-to-be-rounded-up-and-removed-from-their-own-homes-by-force/5403765.

Aizenman, N. (2014). An Ebola quarantine triggers A riot in a Liberian slum. *NPR*. Retrieved February 16, 2017, from http://www.npr.org/sections/goatsandsoda/2014/08/20/341958704/in-liberia-an-ebola-quarantine-descends-into-riots.

Akanni, T. (2014). Confronting Ebola in Liberia: The gendered realities. *openDemocracy*. Retrieved February 12, 2017, from http://www.opendemocracy.net/5050/tooni-akanni/confronting-ebola-in-liberia-gendered-realities-0.

Al Jazeera. (2016). Liberia after Ebola: Turning midwives into surgeons. *Al Jazeera*. Retrieved July 26, 2016, from http://www.aljazeera.com/programmes/thecure/2016/06/liberia-ebola-turning-midwives-surgeons-160606141205329.html.

All Africa. (2012). Liberia: Government finally ratifies UN Convention On Disability—Disable community lauds Liberian government. *The Informer (Monrovia)*. Retrieved January 10, 2017, from http://allafrica.com/stories/201208071099.html.

All Africa. (2013). Liberia: TNIMA administrator under pressure. *New Democrat (Monrovia)*. Retrieved February 12, 2017, from http://allafrica.com/stories/201304152366.html.

All Africa. (2014a). Liberia: House reviews bill to criminalize the concealing of information of persons with communicable or contiguous infectious diseases. *All Africa*. Retrieved December 27, 2017, from http://allafrica.com/stories/201408261562.html.

All Africa. (2014b). Liberia: House passes law to criminalize the concealing of information of persons with communicable or contiguous infectious diseases. *All Africa*. Retrieved January 27, 2017, from http://allafrica.com/stories/201410021746.html.

All Africa. (2015, October 1). Liberia: UL introduces Bachelor's in Nursing, Midwifery. *Liberia News Agency (Monrovia)*. Retrieved February 12, 2017, from http://allafrica.com/stories/201510011471.html.

Alter, K. J., Helfer, L., & McAllister, J. R. (2013). A new international human rights court for West Africa: The ECOWAS community court of justice. *The American Journal of International Law, 107*(4), 737–779. https://doi.org/10.5305/amerjintelaw.107.4.0737.

AM Dogliotti College of Medicine. (2010). New curriculum (Revised) A. M. Dogliotti College of Medicine, University of Liberia. *University of Liberia*. Retrieved February 12, 2017, from http://www.iss.it/binary/ures/cont/Curriculum_AMDCM_Part_1.pdf.

Austin, J. (2011). *Lessons from the field: Revising the public health laws in Liberia, a post-conflict state*. Lecture presented at the O'Neill Institute for National and Global Health Law Talk. Washington, DC. Retrieved from http://www.law.georgetown.edu/oneillinstitute/documents/2011-06-29_JeannieAustin.pdf.

Ballah, Z. (2016a). Liberia's 4.5 million population has only 298 medical doctors. *The Bush Chicken*. Retrieved February 11, 2017, from http://www.bushchicken.com/liberias-4-5-million-population-has-only-298-medical-doctors/.

Ballah, Z. (2016b, September 3). Medical students find that A. M. Dogliotti not as free as advertised. *The Bush Chicken*. Retrieved February 12, 2017, from http://www.bushchicken.com/medical-students-find-that-a-m-dogliotti-not-as-free-as-advertised/.

BBC. (2014a). Ebola patient Nancy Writebol arrives in Atlanta. *BBC Studio*. Retrieved January 24, 2017, from http://www.bbc.com/news/world-us-canada-28663449.

BBC. (2014b). Ebola crisis: Infected doctor Kent Brantly lands in the US. *BBC News*. Retrieved from http://www.bbc.com/news/world-us-canada-28596416.

BBC. (2014c). Ebola crisis: Sierra Leone law makes hiding patients illegal. *BBC News*. Retrieved November 17, 2016, from http://www.bbc.com/news/world-africa-28914791.

BBC. (2014d). Ebola crisis: Ivory Coast closes land borders. *BBC News*. Retrieved November 28, 2016, from http://www.bbc.com/news/world-africa-28913253.

Beddow, R. (2016). *Corruption Perceptions Index 2015* (p. 20). Berlin, Germany: Transparency International.

Black, R., Laxminarayan, R., Temmerman, M., & Walker, N. (2016). *Reproductive, maternal, newborn, and child health (Private Report)* (p. 419). Washington. DC: Disease Control Priorities, World Bank Group. Retrieved from https://openknowledge.worldbank.org/bitstream/handle/10986/23833/9781464803482.pdf?sequence=3.

Boaden, D. (2016). The life-saving work of one Liberian midwife. *Aljazeera Features Health*. Retrieved November 12, 2016, from http://www.aljazeera.com/indepth/features/2016/07/life-saving-work-liberian-midwife-160713153725961.html.

Bohren, M. A., Vogel, J. P., Hunter, E. C., Lutsiv, O., Makh, S. K., Souza, J. P., et al. (2015). The mistreatment of women during childbirth in health facilities globally: A mixed-methods systematic review. *PLOS Medicine, 12*(6), e1001847. https://doi.org/10.1371/journal.pmed.1001847.

Bolay, F. (2016). History of the Liberian Institute of Biomedical Research. *Liberian Institute for Biomedical Research*. Retrieved from http://libresearch.org/contacts.html.

Botelho, G., & Wilson, J. (2014). Thomas Eric Duncan: First Ebola death in U.S. *CNN*. Retrieved January 24, 2017, from http://www.cnn.com/2014/10/08/health/thomas-eric-duncan-ebola/index.html.

Brewinski Isaacs, M. (2016). Ebola reveals needs of women, children as research priority. *National Institute of Health*. Retrieved November 15, 2016, from https://www.nichd.nih.gov/news/resources/spotlight/Pages/040416-brewinski-global-health.aspx.

Brooks, C. (2016). Liberia Legislature finally approve U.S.$600 Million -. *Global News Network*. Retrieved January 5, 2017, from http://gnnliberia.com/2016/09/20/liberia-legislature-finally-approve-u-s-556-million/.

Carter, J. B. (2016). Liberia: Senate passes MoH, National Public Health Institute Act. *Liberian Observer*. Retrieved January 5, 2017, from http://allafrica.com/stories/201609211149.html.

CDC. (2016). *State Ebola protocols*. Retrieved from http://www.cdc.gov/phlp/publications/topic/ebola.html.

Centre for Disease Control. (2014). *CDC supplements Ebola assistance to Liberia, Sierra Leone, and Guinea by preparing neighboring countries to rapidly detect and contain Ebola*. Retrieved February 12, 2017, from https://www.cdc.gov/globalhealth/stories/ebola_assistance.htm.

Cleveland, E. C., Dahn, B. T., Lincoln, T. M., Safer, M., Podesta, M., & Bradley, E. (2011). Introducing health facility accreditation in Liberia. *Global Public Health, 6*(3), 271–282. https://doi.org/10.1080/17441692.2010.489052.

Cook, R. J., Dickens, B. M., & Fathalla, M. F. (2003). *Reproductive health and human rights: Integrating medicine, ethics, and law*. Oxon, Oxford: Clarendon Press.

Cordier-Lasalle, T. (2015). Liberia succeeds in fighting Ebola with local, sector response. *WHO*. Retrieved January 26, 2017, from http://www.who.int/features/2015/ebola-sector-approach/en/.

Couch, R. (2014). How pregnant women are becoming victims of the Ebola outbreak. *Huffington Post*. Retrieved November 17, 2016, from http://www.huffingtonpost.com/2014/09/30/ebola-outbreak-pregnant-women_n_5901816.html.

Crowe, S. (2014). Ebola crisis in Liberia hits child health and well-being. *UNICEF—Liberia*. Retrieved November 15, 2016, from https://www.unicef.org/media/media_75860.html.

Deen, G. F., Knust, B., Broutet, N., Sesay, F. R., Formenty, P., Ross, C., et al. (2015). Ebola RNA persistence in semen of Ebola virus disease survivors Preliminary report. *New England Journal of Medicine*. https://doi.org/10.1056/NEJMoa1511410.

Department of Defense. (2014). *DoD helps fight Ebola in Liberia and West Africa*. Retrieved February 11, 2017, from http://archive.defense.gov/home/features/2014/1014_ebola/.

Diaz, D. (2016). *When Ebola hit Liberia, refugees took frontline health role*. Retrieved November 23, 2016, from http://www.unhcr.org/news/latest/2016/6/5750093e4/ebola-hit-liberia-refugees-took-frontline-health-role.html.

du Lac, J. F. (2014). The U.S. military's new enemy: Ebola. Operation United Assistance is now underway. *Washington Post*. Retrieved February 12, 2017, from https://www.washingtonpost.com/news/to-your-health/wp/2014/09/30/the-u-s-military-forces-fighting-the-war-on-ebola/.

Encyclopaedia Britannica. (2017). *Ellen Johnson Sirleaf*. Retrieved February 10, 2017, from https://www.britannica.com/biography/Ellen-Johnson-Sirleaf.

Evans, D. K., Goldstein, M., & Popova, A. (2015). Health-care worker mortality and the legacy of the Ebola epidemic. *The Lancet Global Health, 3*(8), e439–e440. https://doi.org/10.1016/S2214-109X(15)00065-0.

Executive Mansion. (2009). A. M. Dogliotti Medical College at 40—President recommits government support. *Executive Mansion News*. Retrieved February 12, 2017, from http://www.emansion.gov.lr/2press.php?news_id=1110&related=7&pg=sp.

Executive Mansion. (2015). President Sirleaf addresses the National Union of Organizations of the Disabled; assures them of government's continued support. *Executive Mansion News*. Retrieved from http://www.emansion.gov.lr/2press.php?news_id=3250&related=7&pg=sp.

Farge, E. (2015). *Guinea's Last Ebola case, a Baby Girl, leaves hospital Reuters (London, UK)*. Retrieved December 27, 2017, from https://www.reuters.com/article/us-health-ebola-guinea/guineas-last-ebola-case-a-baby-girl-leaves-hospital-idUSKBN0TH0PB20151128.

Farge, E., & Giahyue, J. H. (2015). Female survivor may be cause of Ebola flare-up in Liberia. *Reuters*. Retrieved November 17, 2016, from http://www.reuters.com/article/us-health-ebola-liberia-idUSKBN0U02EJ20151217.

Felix, B. (2014). Liberia imposes media restrictions on "invasive" Ebola coverage. *Reuters*. Retrieved November 17, 2016, http://www.reuters.com/article/us-health-ebola-liberia-idUSKCN0HS15Q20141003.

Fink, S. (2014a). Life, death and grim routine fill the day at a Liberian Ebola clinic. *The New York Times*. Retrieved November 15, 2016, from http://www.nytimes.com/2014/10/08/world/life-death-and-careful-routine-fill-the-day-at-a-liberian-ebola-clinic.html.

Fink, S. (2014b). Heart-rending test in Ebola zone: A baby. *The New York Times*. Retrieved November 15, 2016, from http://www.nytimes.com/2014/10/10/world/africa/heart-rending-test-in-ebola-zone-a-baby.html.

Fitz, D. (2013, April 25). Why is Cuba's health care system the best model for poor countries? *Realcuba's Blog*. Retrieved February 11, 2017, from https://realcuba.wordpress.com/2013/04/25/why-is-cubas-health-care-system-the-best-model-for-poor-countries/.

Fleischman, J. (2014). US Ebola response: Strategies for women and girls. *Center for Strategic & International Studies*. Retrieved from https://www.csis.org/blogs/smart-global-health/us-ebola-response-strategies-women-and-girls.

Fox, M. (2014). Ebola patient Dr. Rick Sacra gets "everything we had." *NBC News*. Retrieved January 24, 2017, from http://www.nbcnews.com/storyline/ebola-virus-outbreak/ebola-patient-dr-rick-sacra-gets-everything-we-had-n201426.

Front Page Africa. (2016a). *Liberian legislature passes National Public Health Institute of Liberia*. Retrieved January 5, 2017, from http://frontpageafricaonline.com/index.php/news/2776-liberian-legislature-passes-national-public-health-institute-of-liberia.

Front Page Africa. (2016b). *Health team progresses in reducing maternal death in Lofa County*. Retrieved from http://www.frontpageafricaonline.com/index.php/county-news/2528-health-team-progresses-in-reducing-maternal-death-in-lofa-county.

Front Page Africa. (2017). *Association of Public Health Technicians inducts new corps of officers*. Retrieved January 4, 2017, from http://frontpageafricaonline.com/index.php/news/2929-association-of-public-health-technicians-inducts-new-corps-of-officers.

Fuest, V. (2008). "This is the time to get in front": Changing roles and opportunities for women in Liberia. *African Affairs, 107*(427), 201–224. https://doi.org/10.1093/afraf/adn003.

Global Fund For Women. (2014). *Women in Liberia and Ebola*. Retrieved November 17, 2016, from https://www.globalfundforwomen.org/liberia-ebola-crisis/.

Government of Liberia. (2007a). *Emergency human resources for health plan, Ministry of Health and Social Welfare 2007–2011 (Government Policy)* (p. 34). Monrovia, Liberia: Ministry of Health and Social Welfare. Retrieved from http://apps.who.int/medicinedocs/documents/s18060en/s18060en.pdf.

Government of Liberia. (2007b). *Road map for accelerating the reduction of maternal and newborn morbidity and mortality in Liberia (Government Report)* (p. 51). Monrovia, Liberia: Ministry of Health and Social Welfare.

Government of Liberia. (2008). *Basic package of health and social welfare services for Liberia (Government Policy)* (p. 55). Monrovia, Liberia: Ministry of Health and Social Welfare. Retrieved from http://www.basics.org/documents/Basic-Package-of-Health-and-Social-Welfare-Services_Liberia.pdf.

Government of Liberia. (2011). *National Health and Social Welfare Policy: 2011–2021 (Government Policy)* (p. 114). Monrovia, Liberia: Ministry of Health and Social Welfare. Retrieved from http://www.nationalplanningcycles.org/sites/default/files/country_docs/Liberia/ndp_liberia.pdf.

Government of Liberia. (2012). *National Vision 2030 (Government Policy)* (p. 26). Monrovia, Liberia: Governance Commission. Retrieved from http://governancecommissionlr.org/doc_download/VISION%202030%20%20%20summary%20for%20the%20conference%20(25%20pgs)%20for%20GC%20%20Website.pdf?a4705305cd27e04fb1f66830e7e0ef9d=NjQ%3D.

Government of Liberia. (2014a). *Liberia Demographic and Health Survey 2013 (Government Report)* (p. 480). Monrovia, Liberia: Liberia Institute of Statistics and Geo-Information Services. Retrieved from http://dhsprogram.com/what-we-do/survey/survey-display-435.cfm.

Government of Liberia. (2014b, August 6). President Sirleaf declares 90-day state of emergency, as governments steps up the fight against the spread of the Ebola virus disease. *Executive Mansion Press Release*. Retrieved from http://www.emansion.gov.lr/2press.php?news_id=3053&related=7&pg=sp.

Government of Liberia. (2015). *Investment plan for building a resilient health system in Liberia (Government Report)* (p. 71). Monrovia, Liberia: Ministry of Health and Social Welfare. Retrieved from http://pages.au.int/sites/default/files/LIBERIA-%20Investment%20Plan%20for%20Building%20a%20Resilient%20Health%20System.pdf.

Government of Liberia. (2016a). *National budget fiscal year 2016–2017 (Government Report)* (p. 526). Monrovia, Liberia: Ministry of Finance and Development Planning.

Government of Liberia. (2016b). *Investment case for reproductive, maternal, newborn, child and adolescent health (Draft Report) (Government Report)* (p. 60). Monrovia, Liberia: Ministry of Health and Social Welfare.

Harman, S. (2016). Ebola, gender and conspicuously invisible women in global health governance. *Third World Quarterly, 37*(3), 524–541. https://doi.org/10.1080/01436597.2015.1108827.

Hessou, C. (2014a). *Liberia's Ebola outbreak leaves pregnant women stranded*. Retrieved November 17, 2016, from http://www.unfpa.org/news/liberias-ebola-outbreak-leaves-pregnant-women-stranded.

Hessou, C. (2014b). Pregnant in the shadow of Ebola: Deteriorating health systems endanger women. *Relief Web*. Retrieved from http://www.unfpa.org/news/pregnant-shadow-ebola-deteriorating-health-systems-endanger-women.

Himelein, K. (2015). *The socio-economic impacts of Ebola in Liberia* (p. 19). Monrovia, Liberia: World Bank. Retrieved from http://www.worldbank.org/content/dam/Worldbank/document/Poverty%20documents/Socio-Economic%20Impacts%20of%20Ebola%20in%20Liberia,%20April%2015%20(final).pdf.

Human Rights Watch. (2014). *West Africa: Respect rights in Ebola response*. Retrieved November 17, 2016, from https://www.hrw.org/news/2014/09/15/west-africa-respect-rights-ebola-response.

Hunt, P., & De Mesquita, J. B. (2006). *The rights to sexual and reproductive health (University Report)* (p. 12). Essex, England: University of Essex Human Rights Centre. Retrieved from http://repository.essex.ac.uk/9718/1/right-sexual-reproductive-health.pdf.

IRIN News. (2014). *Ebola effect reverses gains in maternal, child mortality*. Retrieved November 15, 2016, from http://www.irinnews.org/feature/2014/10/08.

Johnson Sirleaf, E. (2014a). A special statement by the Chair on the National Task Force on Ebola, President Ellen Johnson Sirleaf. *Executive Mansion Press release*. Retrieved December 1, 2016, from http://www.emansion.gov.lr/doc/Special%20Statement%20by%20President%20Ellen%20Johnson%20Sirleaf%20-1_1.pdf.

Johnson Sirleaf, E. (2014b). Special statement by President Ellen Johnson Sirleaf on additional measures in the fight against the Ebola viral disease. *Executive Mansion Press Release*. Retrieved January 27, 2017, from http://www.emansion.gov.lr/doc/Special_State_Delivered_July%2030.pdf.

Kateh, F. N. (2015). *Dr. Francis Kateh: Post Ebola Liberia*. Retrieved February 9, 2017, from http://kbnf.org/post_ebola/.

Kennedy, J. (2011). *About the school (TNIMA)*. Retrieved from http://www.tnimaa.org/contactus.html.

Khaki, A., & Bayatmakoo, Z. (2015). Ebola virus disease (EVD) and women's health care in pregnancy. *International Journal of Women's Health and Reproduction Sciences, 3*(1), 1. https://doi.org/10.15296/ijwhr.2015.01.

Kincaid, J. (2015). International women's day: The repercussions of Ebola on women's health. *UN Foundation Blog Global Connections—Special Edition: Ebola*. Retrieved November 15, 2016, from http://unfoundationblog.org/ebola/international-womens-day-empower-the-women-battling-ebola/.

Kolleh, E., & Doucleff, M. (2014). Dangerous deliveries: Ebola leaves moms and babies without care. *NPR*. Retrieved November 17, 2016, from http://www.npr.org/sections/goatsandsoda/2014/11/18/364179795/dangerous-deliveries-ebola-devastates-womens-health-in-liberia.

Kollie, M. K. N. (2017). *OP-ED: The Liberian legislature—An unholy political theater of unrepented crooks*. Retrieved February 16, 2017, from http://www.bushchicken.com/op-ed-the-liberian-legislature-an-unholy-political-theater-of-unrepented-crooks/.

Kucharski, A. J., Camacho, A., Checchi, F., Waldman, R., Grais, R. F., Cabrol, J.-C., et al. (2015). Evaluation of the benefits and risks of introducing Ebola Community Care Centers, Sierra Leone. *Emerging Infectious Diseases, 21*(3), 393–399. https://doi.org/10.3201/eid2103.141892.

Lewnard, J. A., Ndeffo Mbah, M. L., Alfaro-Murillo, J. A., Altice, F. L., Bawo, L., Nyenswah, T. G., et al. (2014). Dynamics and control of Ebola virus transmission in Montserrado, Liberia: A mathematical modelling analysis. *The Lancet Infectious Diseases, 14*(12), 1189–1195. https://doi.org/10.1016/S1473-3099(14)70995-8.

Liljas, P. (2014). Liberia's West Point slum reels from the nightmare of Ebola. *Time*. Retrieved from http://time.com/3158244/liberia-west-point-slum-ebola-disease-quarantine/.

LISGIS. (2011). *Report on the Liberia Labour Force Survey 2010* (p. 168). Monrovia, Liberia: Liberia Institute of Statistics and Geo-Information Services. Retrieved February 9, 2017, from http://www.ilo.org/wcmsp5/groups/public/%2D%2D-dgreports/%2D%2D-stat/documents/presentation/wcms_156366.pdf.

Lori, J. R., & Starke, A. E. (2012). A critical analysis of maternal morbidity and mortality in Liberia, West Africa. *Midwifery, 28*(1), 67–72. https://doi.org/10.1016/j.midw.2010.12.001.

Los Angeles Times. (2014, August 19). Liberia imposes Ebola quarantine and curfew in a Monrovia slum. *Los Angeles Times*. Retrieved February 16, 2017, from http://www.latimes.com/world/africa/la-fg-africa-ebola-liberia-curfew-20140819-story.html.

Markey, M., Ranson, M. M., & Sunshine, H. (2015). Ebola: A public health and legal perspective. *American Bar Association Health eSource, 11*(5). Retrieved from http://www.americanbar.org/publications/aba_health_esource/2014-2015/january/ebola.html.

Meltzer, M. I., Atkins, C. Y., Santibanez, S., Knust, B., Petersen, B. W., Ervin, E. D., et al. (2014). Estimating the future number of cases in the Ebola epidemic—Liberia and Sierra Leone, 2014–2015. *MMWR Morbidity and Mortality Weekly Report, S63*(3), 1–14.

Menéndez, C., Lucas, A., Munguambe, K., & Langer, A. (2015). Ebola crisis: The unequal impact on women and children's health. *The Lancet Global Health, 3*(3), e130. https://doi.org/10.1016/S2214-109X(15)70009-4.

Merry, S. E. (2016). *The seductions of quantification: Measuring human rights, gender violence, and sex trafficking*. Chicago: The University of Chicago Press.

Meyers, N. (2013). *Assessing the implementation barriers to the "Accelerated Action Plan to Reduce Maternal and Newborn Mortality" in Liberia (Masters of Public Health)*. Boston, MA: Harvard University. Retrieved from https://cdn1.sph.harvard.edu/wp-content/uploads/sites/112/2012/09/Final_Draft.pdf.

Ministry of Health and Social Welfare. (2008). National health policy on contracting 2008–2011. *Government of Liberia*. Retrieved February 7, 2017, from http://liberiamohsw.org/Policies%20&%20Plans/Contracting%20policy%20doc_MOHSW_%20Oct.%2014%20-%2008.doc.

Moore, J. (2014). *The hidden heroes of Liberia's Ebola crisis*. Retrieved February 12, 2017, from http://www.buzzfeed.com/jinamoore/meet-some-of-the-hidden-heroes-of-liberias-ebola-crisis.

NBC News. (2006). Liberian becomes Africa's first female president. *NBC*. Retrieved February 10, 2017, from http://www.nbcnews.com/id/10865705/ns/world_news-africa/t/liberian-becomes-africas-first-female-president/.

Nyan, D. C. (2006). Observational and situational report: Liberia Medical and Dental Association of the USA (LMDA-USA) visiting professorship program at The A. M. Dogliotti College of Medicine, University of Liberia. *The Perspective*. Retrieved February 12, 2016, from http://www.theperspective.org/documents/nyan.htm.

Nyenswah, T. G., Kateh, F. N., Bawo, L., Massaquoi, M., Gbanyan, M., Fallah, M., et al. (2016). Ebola and its control in Liberia, 2014–2015. *Emerging Infectious Diseases, 22*(2), 1–9.

Odunayo, A. (2014). EBOLA: Why Patrick Sawyer deliberately travelled to Nigeria - wife. *Naij.com—Nigeria news*. Retrieved January 24, 2017, from https://www.naij.com/276765-ebola-patrick-sawyer-deliberately-travelled-nigeria-wife.html.

Office of the United Nations High Commissioner for Human Rights. (2014). *Women's rights are human rights* (UN Report No. HR/PUB/14/2) (p. 123). UN High Commission for Human Rights. Retrieved from http://www.ohchr.org/Documents/Events/WHRD/WomenRightsAreHR.pdf.

Onishi, N. (2014a). Clashes erupt as Liberia sets an Ebola quarantine. *The New York Times*. Retrieved February 16, 2017, from https://www.nytimes.com/2014/08/21/world/africa/ebola-outbreak-liberia-quarantine.html.

Onishi, N. (2014b). Liberian boy dies after being shot during clash over Ebola quarantine. *The New York Times*. Retrieved January 14, 2017, from https://www.nytimes.com/2014/08/22/world/africa/liberian-boy-dies-after-being-shot-during-clash-over-ebola-quarantine.html.

Onishi, N. (2014c). Inquiry faults Liberia force that fired on protesters. *The New York Times*. Retrieved February 12, 2017, from https://www.nytimes.com/2014/11/04/world/africa/soldiers-faulted-in-deadly-crackdown-during-ebola-protests-in-liberia.html.

Parshley, L. (2016). After Ebola: The disease has left a terrible legacy—And another outbreak is likely. *The Atlantic*. Retrieved November 17, 2016, from http://www.theatlantic.com/magazine/archive/2016/07/after-ebola/485609/.

Partners in Health. (2015). *Liberia: Building a health system*. Retrieved January 26, 2017, from http://www.pih.org/pages/pleebo.

Partners in Health. (2016). *New health center opens in Liberia*. Retrieved January 26, 2017, from http://www.pih.org/blog/new-health-center-opens-in-liberia.

Partners in Health. (2017). *Liberia*. Retrieved January 26, 2017, from http://www.pih.org/country/liberia.

Penn, M., Bird, B., Jordan, J., & Hodge, J. (2014). Ebola and the law: What you need to know. Retrieved November 16, 2016, from https://www.networkforphl.org/webinars/2014/08/11/481/ebola_and_the_law_what_you_need_to_know.

Petit, D., Sondorp, E., Mayhew, S., Roura, M., & Roberts, B. (2013). Implementing a basic package of health services in post-conflict Liberia: Perceptions of key stakeholders. *Social Science & Medicine, 78*, 42–49. https://doi.org/10.1016/j.socscimed.2012.11.026.

Price, P. J. (2015). *Ebola and the Law in the United States: A short guide to public health authority and practical limits*. Emory University School of Law.

Redd, S. (2015). *The aftermath of Ebola: Strengthening health systems in Liberia*. Retrieved January 26, 2017, from http://www.vanderbilt.edu/vigh-sac/case/2016.GHCC.Case.Prompt.pdf.

Reuters. (2016). Ivory Coast re-opens western borders closed during Ebola epidemic. *Thomson Reuters Foundation*. Retrieved November 28, 2016, from http://news.trust.org/item/20160909103921-1jmqw/.

Rivers, C. M., Lofgren, E. T., Marathe, M., Eubank, S., & Lewis, B. L. (2014). Modeling the impact of interventions on an epidemic of Ebola in Sierra Leone and Liberia. *PLoS Currents*. https://doi.org/10.1371/currents.outbreaks.4d41fe5d6c05e9df30ddce33c66d084c.

Roache, S., Gostin, L. O., Hougendobler, D., & Friedman, E. (2014). *Lessons from the West African Ebola epidemic: Towards a legacy of strong health systems*. O'Neill Institute for National and Global Health Law, Georgetown Law. Retrieved from http://www.law.georgetown.edu/oneillinstitute/resources/documents/Briefing10Ebola2inTemplate.pdf.

Roberts, B., Guy, S., Sondorp, E., & Lee-Jones, L. (2008). A basic package of health services for post-conflict countries: Implications for sexual and reproductive health services. *Reproductive Health Matters, 16*(31), 57–64. https://doi.org/10.1016/S0968-8080(08)31347-0.

Save the Children. (2015). *Assessment of the effect of Ebola on education in Liberia* (p. 47). Monrovia, Liberia: United Nations Children's Fund. Retrieved February 11, 2017, from http://educationcluster.net/?get=002241|2015/02/Liberia_Education_Cluster_Ebola_Assessment_Report_FINAL.pdf.

Schwartz, D. A. (2015). Interface of epidemiology, anthropology and health care in maternal death prevention in resource-poor nations. In D. A. Schwartz (Ed.), *Maternal mortality: Risk factors, anthropological perspectives, prevalence in developing countries and preventive strategies for pregnancy-related death* (pp. ix–xiv). New York: Nova Science Publishers.

Schwartz, D. A. (2019a). All the mothers are dead – Ebola's chilling effects on the young women of one Liberian town named Joe Blow. In D. A. Schwartz, J. N. Anoko, & S.A. Abramowitz (Eds.), *Pregnant in the time of Ebola: Women and their children in the 2013-2015 West African epidemic*. New York: Springer.

Schwartz, D. A. (2019b). The Ebola epidemic halted female genital cutting in Sierra Leone – Temporarily. In D. A. Schwartz, J. N. Anoko, & S.A. Abramowitz (Eds.), *Pregnant in the time of Ebola: Women and their children in the 2013-2015 West African epidemic*. New York: Springer.

Sepkowitz, K., & Haglage, A. (2014). The only thing more terrifying than Ebola is being pregnant with Ebola. *The Daily Beast*. Retrieved November 17, 2016, from http://www.thedailybeast.com/articles/2014/10/02/america-s-patient-zero-was-trying-to-save-a-pregnant-ebola-victim.html.

Shute, J. (2014). Ebola: Inside Liberia's West Point slum. *Telegraph*. Retrieved January 27, 2017, from http://www.telegraph.co.uk/news/worldnews/ebola/11295271/Ebola-inside-Liberias-West-Point-slum.html.

Sirleaf, E. J. (2014). Statement on the declaration of a state of emergency. *Ministry of Foreign Affairs*. Retrieved from http://www.mofa.gov.lr/public2/2press.php?news_id=1223&related=7&pg=sp&sub=44.

Strong, A., & Schwartz, D. A. (2016). Sociocultural aspects of risk to pregnant women during the 2013–2015 multinational Ebola virus outbreak in West Africa. *Health Care for Women International, 37*(8), 922–942. https://doi.org/10.1080/07399332.2016.1167896.

Subah, M., Gebeh, C., Taylor, N., Gbassie, V., & Gavin, A. (2014). *Maternal and Child Health Integrated Program Liberia: End of Project Report (Private Report)* (p. 59). Monrovia, Liberia: USAID.

The Center for HIV Law and Policy. (2016). *Criminalizing disease in Liberia, Gambia and Uganda*. Retrieved from http://www.hivlawandpolicy.org/fine-print-blog/criminalizing-disease-liberia-gambia-and-uganda.

UN Women. (2014). *Ebola outbreak takes its toll on women*. Retrieved November 15, 2016, from http://www.unwomen.org/en/news/stories/2014/9/ebola-outbreak-takes-its-toll-on-women.

UNDP (Ed.). (2015). *Work for human development*. New York, NY: United Nations Development Programme. Retrieved from http://hdr.undp.org/sites/default/files/2015_human_development_report.pdf.

UNESCO. (2016). *Effective literacy programmes*. Retrieved February 10, 2017, from http://www.unesco.org/uil/litbase/?menu=4&programme=217.

UNFPA. (2016a). *The state of the world population 2016 (UN Report)* (p. 116). New York, NY: United Nations Populations Fund. Retrieved from http://www.unfpa.org/sites/default/files/sowp/downloads/The_State_of_World_Population_2016_-_English.pdf.

UNFPA. (2016b). *Adolescent pregnancy, UNFPA—United Nations Population Fund*. Retrieved January 20, 2017, from http://www.unfpa.org/adolescent-pregnancy.

UNFPA, WHO, & International Confederation of Midwives. (2014). *The state of the world's midwifery 2014: A universal pathway. A woman's right to health* (p. 228). New York, NY: United Nations Populations Fund. Retrieved from http://www.unfpa.org/sowmy.

UNICEF. (2009). *Liberia enshrines the convention on the rights of the child in military code*. Retrieved January 28, 2017, from https://www.unicef.org/infobycountry/liberia_51970.html.

USAID Liberia Mission. (2016). *Assessment of the accelerated action plan (AAP) for the reduction of maternal and neonatal morbidity and mortality in Liberia* (p. 29). Monrovia, Liberia: USAID.

Varpilah, S. T., Safer, M., Frenkel, E., Baba, D., Massaquoi, M., & Barrow, G. (2011). Rebuilding human resources for health: A case study from Liberia. *Human Resources for Health, 9*(1), 11. https://doi.org/10.1186/1478-4491-9-11.

Voorhees, J., & Mathis-Lilley, B. (2014). Everything that went wrong in Dallas. *Slate*. Retrieved January 24, 2017, from http://www.slate.com/articles/health_and_science/medical_examiner/2014/10/dallas_ebola_timeline_the_many_medical_missteps_at_texas_health_presbyterian.html.

Werber, C. (2015). How Ebola Led to more teenage pregnancy in West Africa. *Quartz*. Retrieved November 17, 2016, from http://qz.com/543354/how-ebola-led-to-more-teenage-pregnancy-in-west-africa/.

WHO. (2014a). *The president of Liberia supports Ebola outbreak response*. Retrieved November 22, 2016, from http://www.afro.who.int/liberia/press-materials/item/6463-the-president-of-liberia-supports-ebola-outbreak-response.html?lang=en.

WHO. (2014b). *Key considerations for the implementation of an Ebola care unit at community level*. Retrieved from https://assets.documentcloud.org/documents/1303518/0920-key-considerations-ecu-rev-ipc-5.pdf.

WHO. (2014c). *Key considerations for the implementation of community care centres*. Retrieved from https://extranet.who.int/emt/sites/default/files/Key%20considerations%20for%20the%20implementation%20of%20Community%20Care%20Centres.pdf.

WHO. (2015). *Manual for the care and management of patients in Ebola care units/community care centres: Interim emergency guidance*. Retrieved from http://apps.who.int/iris/bitstream/10665/149781/1/WHO_EVD_Manual_ECU_15.1_eng.pdf.

WHO. (2016a). *Latest Ebola outbreak over in Liberia; West Africa is at zero, but new flare-ups are likely to occur*. Retrieved February 10, 2017, from http://www.who.int/mediacentre/news/releases/2016/ebola-zero-liberia/en/.

WHO. (2016b). *Investing in trained midwives across Liberia*. Retrieved February 12, 2017, from http://www.who.int/features/2016/liberia-training-midwives/en/.

WHO. (2016c). *Ebola data and statistics: Situation summary data published 11 May 2016*. Retrieved February 12, 2017, from http://apps.who.int/gho/data/view.ebola-sitrep.ebola-summary-20160511?lang=en.

WHO Ebola Response Team. (2014). Ebola virus disease in West Africa—The first 9 months of the epidemic and forward projections. *New England Journal of Medicine, 371*(16), 1481–1495. https://doi.org/10.1056/NEJMoa1411100.

Wolokolie, A. M. (2013). TNIMA students protest. *The Inquirer*. Retrieved February 12, 2017, from http://theinquirer.com.lr/content1.php?main=news&news_id=1949.

World Health Organization. (2005). *Liberia health system profile (UN Report)* (p. 18). Brazzaville, Congo: Regional Office for Africa.

World Health Organization. (2015a). *Liberia: Country health profile*. Geneva: World Health Organization.

World Health Organization. (2015b). *How Liberia got to zero cases of Ebola*. Retrieved January 26, 2017, from http://www.who.int/features/2015/liberia-ends-ebola/en/.

World Health Organization. (2015c). *2015 WHO strategic response plan: West Africa Ebola outbreak*. Geneva, Switzerland: World Health Organization. Retrieved February 12, 2017, from http://apps.who.int/iris/bitstream/10665/163360/1/9789241508698_eng.pdf.

World Health Organization. (2016). *Ebola data and statistics*. Retrieved December 17, 2017, from http://apps.who.int/gho/data/view.ebola-sitrep.ebola-summary-latest?lang=en.

Yamin, D., Gertler, S., Ndeffo-Mbah, M. L., Skrip, L. A., Fallah, M., Nyenswah, T. G., et al. (2015). Effect of Ebola progression on transmission and control in Liberia. *Annals of Internal Medicine, 162*(1), 11. https://doi.org/10.7326/M14-2255.

Yana, R., & Jacque, G. (2014). Ebola in pregnancy. *British Journal of Midwifery, 22*(11), 771–775.

Yasmin, S. (2016). The Ebola rape epidemic no one's talking about. *Foreign Policy*. Retrieved November 12, 2016, from https://foreignpolicy.com/2016/02/02/the-ebola-rape-epidemic-west-africa-teenage-pregnancy/.

Yates, D. A. (2016). MFDP presents Legislature FY 2016/17 Draft Budget. *Daily Observer*. Monrovia, Liberia.

Zambelich, A. (2015). Liberia's President: Ebola re-energized her downtrodden country. *NPR*. Retrieved November 21, 2016, from http://www.npr.org/sections/goatsandsoda/2015/03/02/389478897/liberias-president-ebola-re-energized-her-downtrodden-country.

Uncovering More Questions: Salome Karwah and the Lingering Impact of Ebola Virus Disease on the Reproductive Health of Survivors

16

Christine L. Godwin and David A. Schwartz

16.1 Salome Karwah: A Person of the Year

In 2014, Salome Karwah contracted Ebola virus disease (EVD) in Libera, recovered, and then returned to the Ebola Treatment Unit (ETU) to care for other victims of the disease as a nursing assistant. For her heroism, Salome was recognized on the cover of Time Magazine as a Person of the Year for her work in combating the Ebola epidemic (Time 2014). In 2017, Time wrote about Salome again, but this time in a very different light. Salome had died just days after giving birth, a tragic victim of maternal death that is so prevalent in Liberia and other African countries before, during, and after the Ebola virus epidemic.

Salome fell ill with Ebola in 2014 after an uncle unknowingly passed the virus to her and other relatives when he sought treatment at the family's medical clinic (Von Drehle 2014). She stated:

> *"It all started with a severe headache and a fever. Then, later, I began to vomit and I got diarrhoea. My father was sick and my mother too. My niece, my fiancé and my sister had all fallen sick. We all felt helpless. It was my uncle who first got the virus in our family. He contracted it from a woman he helped bring to hospital. He got sick and called our father for help, and our father went to him to bring him to a hospital for treatment.*
> *A few days after our father came back, he too got sick. We all cared for him and got infected too"* (Medecins Sans Frontieres 2017).

Salome and her family, including her physician father, were treated at the Ebola Management Centre operated by Medicins Sans Frontieres (MSF) in Monrovia. In an interview, she describes their initial visit to the Centre:

> "*On 21 August, my whole family and I made our way to MSF's Ebola management centre in Monrovia. When we arrived at the treatment unit, the nurses took my mother and me to the same tent. My fiancé, my sister, my father and my niece were taken to separate tents. My sister was pregnant and had a miscarriage. They took our blood and we waited for them to announce the results. After the lab test, I was confirmed positive. I thought that was the end of my world. I was afraid, because we had heard people say that if you catch Ebola, you die. The rest of my family also tested positive for the virus*" (Medecins Sans Frontieres 2017).

C. L. Godwin
Department of Maternal and Child Health, Gillings School of Global Public Health, University of North Carolina at Chapel Hill, Chapel Hill, NC, USA

D. A. Schwartz (✉)
Department of Pathology, Medical College of Georgia, Augusta University, Augusta, GA, USA

D. A. Schwartz et al. (eds.), *Pregnant in the Time of Ebola*, Global Maternal and Child Health,
https://doi.org/10.1007/978-3-319-97637-2_16

The pain that Salome was subjected to by her Ebola virus disease was unbearable—she described it to an MSF interviewer:

> *"After a few days in the isolation ward, my condition became worse. My mother was also fighting for her life. She was in a terrible state. At that point, the nurses made the decision to move me to another tent. By then, I barely understood what was going on around me. I was unconscious. I was helpless. The nurses had to bathe me, change my clothes and feed me. I was vomiting constantly and I was very weak. I was feeling severe pains inside my body. The feeling was overpowering. Ebola is like a sickness from a different planet. It comes with so much pain. It causes so much pain that you can feel it in your bones. I'd never felt pain like this in my lifetime. My mother and father died while I was battling for my life. I didn't know they were dead. It was only one week later, when I had started recovering, that the nurses told me that they had passed away. I was sad, but I had to accept that it had happened. I was shocked that I had lost both my parents. But God spared my life from the disease, as well as the lives of my sister, my niece and my fiancé. Though I am sad at the death of my parents, I'm happy to be alive. God could not have allowed the entire family to perish. He kept us alive for a purpose."* (Medecins Sans Frontieres 2017).

While the Ebola virus had killed her parents and many other family members, Salome survived, together with her sister Josephine Manley. She was released from the Ebola Unit on August 28th, 2014. While she was hospitalized, the MSF staff had noticed that Salome and her boyfriend James Harris had been active in voluntarily caring for other victims of Ebola infection, regardless of risks to their own health. Consequently, MSF hired Salome and Joseph as mental health counselors to provide care on the Ebola units—Salome returned to the unit 1 month after her release as a nurse and counselor (Mukpo 2017). Being a survivor, Salome had some immunity to the virus and was thus able to care for other victims without the same crippling fear of infection experienced by other healthcare workers. In discussing her new role as an Ebola survivor and helping her infected countrymen in a report from October 2014, Salome related:

> *"Now I am back at the treatment centre, helping people who are suffering from the virus to recover, working as a mental health counsellor. Helping people brings pleasure, and that is what brought me here. My efforts here may help other people to survive. When I am on a shift, I counsel my patients. If a patient doesn't want to eat, I encourage them to eat. If they are weak and are unable to bathe on their own, I help to bathe them. I help them with all my might because I understand the experience – I've been through the very same thing.*
> *I feel happy in my new role. I treat my patients as if they are my children. I talk to them about my own experiences. I tell them my story to inspire them and to let them know that they too can survive. My elder brother and my sister are happy for me to work here. They support me in this 100%. Even though our parents didn't survive the virus, we can help other people to recover."* (Karwah 2014).

Regarding her work in the Ebola unit, her husband recalled:

> *"She was so caring…They told us we should only spend 30 minutes in [protective gear], but sometimes she would stay in the ward for 2 or 3 hours, just talking to patients and telling them to have hope."*

Facing down the virus that destroyed her family and nearly took her life, in the spirit of helping others in need, made Salome a 2014 Time Magazine Person of the Year (Von Drehle 2014). However, she faced stigma from neighbors and friends after returning to her home, cured of Ebola disease (Fig. 16.1):

> *"I arrived back home feeling happy, but my neighbours were still afraid of me. Few of them welcomed me back; others are still afraid to be around me – they say that I still have Ebola. There was a particular group that kept calling our house 'Ebola home'. But, to my surprise, I saw one of the ladies in the group come to my house to ask me to take her mother to the Ebola centre because she was sick with Ebola. I did it, and I felt happy that at least she knows now that someone cannot go to a supermarket to buy Ebola. It's a disease that anyone, any family, can get. If someone has Ebola, it isn't good to stigmatise them, because you don't know who is next in line to contract the virus."* (Medecins Sans Frontieres 2017).

After the West African Ebola outbreak ended, Liberia was declared Ebola Free four different times as sporadic cases continued to be identified, with the final declaration on September 1st, 2016 (Akwei

Fig. 16.1 Photo of Salome Karwah outside of the Ebola Treatment Unit administered by Medecins Sans Frontieres in Monrovia, Liberia. Photo by John W. Poola/NPR

2016). During that time, Salome married her longtime fiancée and fellow Ebola survivor, James Harris. Salome, together with her fiancée and sister, also reopened the clinic that her physician father had to close before he died from Ebola virus disease. When interviewed in November 2014, Salome said that she envisioned a kind of super-clinic, whose survivor nurses would able to go where other medical personnel feared to tread because of their immunity. "I can do things that other people can't," she said then. "If an Ebola patient is in his house, and his immediate relative cannot go to him, I can go to him. I can take [care of] him." (Baker 2017). In the months following their wedding, Salome gave birth to a healthy baby girl, named Destiny. Salome became pregnant again and, in February 2017, delivered a baby boy via cesarean section at a hospital on the outskirts of Monrovia. Salome and her baby were discharged from the hospital 3 days later and returned to their home with James. This pregnancy had been complicated by a dangerously high spike of her blood pressure, termed pre-eclampsia, which occurred just prior to delivery and had reoccurred in the days following her cesarean section.

Only hours after returning home, Salome began having seizures. Her husband and sister rushed her back to the same hospital. Salome was admitted to the hospital and, tragically, died the following day, February 21st, 2017, just 4 days after giving birth (Baker 2017; Mukpo 2017).

The cause of Salome's seizures and ultimate death remains unknown, although her symptoms were consistent with a form of severe preeclampsia that affects the central nervous system, termed eclampsia. The hypertensive diseases of pregnancy—preeclampsia and eclampsia—complicate less than 5% of pregnancies in developed countries. But these conditions are more severe in developing countries, affecting from a low of 4% to as many as 18% of all pregnancies in parts of Africa (Preeclampsia Foundation 2013; Schwartz 2015, 2018; Villar et al. 2003). However, central nervous system abnormalities have been identified in EVD survivors (Scott et al. 2016; Wong et al. 2016), and there remains the possibility, albeit a small one, that her death may have been a result of her prior Ebola infection.

In October 2017, the Ministry of Health of Liberia announced that they would create a Foundation to honor Salome. The chief medical officer of the country, Dr. Francis Kateh, stated "What is important here is we lost an energetic female who had a desire to become a nurse in order to save the lives of others." The mission of the Foundation is to provide for children who lost their parents from Ebola and children who also lost their mothers during childbirth (All Africa 2017).

After defeating the physiological effects of EVD and being celebrated internationally as a hero of the epidemic, Salome could not escape the high prevalence of maternal death that has remained a public health problem throughout much of West Africa both before and after the epidemic. Following the cessation of the outbreak, and with a maternal mortality ratio of 1072 maternal deaths for every

1000 live births (UN Women 2017), Liberia remains one of the ten most dangerous countries in the world to give birth.

16.2 Men with Ebola and Their Partners

The plight of female survivors, such as Salome, has received precious little attention in the aftermath of the largest Ebola outbreak in history. Due to the occurrence and persistence of the Ebola virus in the semen of male Ebola survivors and, therefore, potential risk that male survivors might unknowingly pass the virus on to sexual partners months after recovery (Deen et al. 2015; Thorson et al. 2016), a tremendous amount of well-warranted research has been focused on understanding and mitigating that risk. Understanding the potential for Ebola's reemergence via sexual transmission is necessary to avert additional flare ups of Ebola in West Africa. Because this was the largest epidemic of Ebola virus to have occurred, large numbers of infected persons (almost 29,000) and EVD survivors available for study have shed light on many factors that had previously not been thoroughly evaluated.

The occurrence of Ebola virus in semen is one of these important factors. In March 2015, a 44-year-old female from Monrovia, Liberia, contracted Ebola virus disease, subsequently dying while in an Ebola treatment unit (ETU). An epidemiological investigation into her case revealed an unexpected connection to Ebola virus exposure—she had been engaged in unprotected sexual intercourse with a male Ebola virus disease survivor (Christie et al. 2015). A sample of semen obtained from the Ebola virus disease survivor was found to be positive for Ebola virus RNA by real-time RT-PCR (rRT-PCR) 199 days after he first developed symptoms of Ebola virus disease. The Men's Health Screening Program was instituted on July 7, 2015 in Liberia by the Liberian Ministry of Health, in collaboration with WHO, the Academic Consortium Combating Ebola in Liberia, and the US Centers for Disease Control and Prevention (CDC). This program had several goals—to provide male survivors of EVD with semen testing for persistence of Ebola virus RNA by rRT-PCR, behavioral counselling on safe sex practices, provision of condoms and instructions on condom use, and referrals for health care services. (Purpura et al. 2016; Soka et al. 2016). Operating from three locations—Redemption Hospital in Montserrado County, Phebe Hospital in Bong County, and Tellewoyan Hospital in Lofa County—as of May 2016, the program had enrolled 466 male survivors aged 15 years and older. Of these, approximately 63% of survivor-provided semen samples had fragments of Ebola virus in them 1 year after recovery from EVD (Soka et al. 2016). One man who was HIV-positive had remnants of the Ebola virus in his semen more than 18 months (565 days) after he recovered from Ebola virus disease (Purpura et al. 2017; Soka et al. 2016). More recently, Daouda Sissoko and colleagues have provided important new data on the clearance of Ebola virus from semen (Sissoko et al. 2017). They evaluated a group of 26 Ebola survivors from Guinea for a median 197 days following inclusion in the study (255 days after disease onset) and performed laboratory testing of 130 semen specimens by real-time reverse transcriptase-polymerase chain reaction (RT-PCR). In addition, they experimentally inoculated severe combined immune deficiency (SCID) mice with aliquots of seminal fluid from eight participants to test for infectivity of the semen. Sissoko et al. identified Ebola ribonucleic acid (RNA) in the semen of 19 of the 26 convalescent participants (73%). They also isolated infectious Ebola virus in the inoculated SCID mice from semen of seven of eight participants. The median duration of Ebola virus RNA in semen was 158 days, with a maximum duration extending up to the close of the study at 407 days. Sissoko and colleagues utilized biostatistical modeling to demonstrate that the mean clearance rate of Ebola virus RNA in semen is −0·58 log units per month. Their investigation also demonstrated the occurrence of significant variability between survivors in the rate of clearance of Ebola RNA from seminal fluid. In another investigation from Liberia conducted on male survivors of Ebola infection from the ELWA Hospital outside of Monrovia, Fischer and colleagues found prolonged persistence of Ebola virus in the semen (Fischer et al. 2017). In this cohort of 149 male survivors who donated semen from 260 to 1016 days following the onset of EVD, Ebola virus RNA was identified in the semen of 13 men (9%). Of 137 men who had donated their semen 2 years after the onset of EVD, 11 (8%) had an Ebola virus RNA-positive specimen. In the results of a study performed in Sierra Leone among 220 male survivors of EVD, Ebola virus RNA was detected by RT-PCR in the semen of all 7 men having a specimen collected within 3 months after their discharge from an ETU. Additional data demonstrated Ebola virus RNA in the semen of 26 of 42 (62%) men with a specimen obtained at 4–6 months, in 15 of 60 (25%) with a specimen obtained at 7–9 months, in 4 of 26 (15%) with a specimen obtained at 10–12 months, in 4 of 38 (11%) with a specimen obtained at 13–15 months, and in 1 of 25 (4%) with a specimen obtained at 16–18 months. No viral RNA was identified in men with a specimen obtained at 19 months or later after discharge from the ETU (Deen et al. 2017).

There have been at least a handful of cases where a female sexual partner of a male survivor has become infected with Ebola (Christie et al. 2015; Mate et al. 2015; Thorson et al. 2016).

The clinical significance of persistence of the Ebola virus in semen occurring remotely from active EVD was evident from a common source outbreak that occurred in the Koropara subprefecture of Guinea in 2016. Following the declaration by the World Health Organization of the termination of Ebola virus (EBOV) transmission in the Republic of Guinea on 29 December 2015, a new cluster of EVD-infected persons, consisting of three probable and seven confirmed cases, were identified between 27 February and 15 March 2016. Of these, six were admitted to an Ebola treatment unit—four patients died in the community and four patients in the ETU, for an overall case fatality rate of 80% (Diallo et al. 2016). Subsequent epidemiological investigation found that this outbreak cluster had its origin from a male EVD survivor who had been engaged in intercourse with one of the infected female patients. Analysis of the semen from this male EVD survivor, who had been treated in the ETU of Guéckédou, Guinea, from 3 to 14 November 2014, found Ebola virus's persistence for 531 days after the onset of his symptoms. The host factors that aid in persistence of Ebola virus in the semen of some men remain unknown. However, a recent study has provided evidence that seminal amyloid fibrils, a normal component of semen in healthy men, enhance the in vitro infectivity of Ebola virus by protecting the virions from stresses encountered during transmission including dessication and degradation, and stabilize viral infectivity (Bart et al. 2018).

Nonetheless, the critical importance of understanding the impact of Ebola virus on its male survivors and their sexual partners should not detract from the need to understand the serious, lingering effects of Ebola on female survivors. A stark contrast exists between the growing body of knowledge around the reproductive health of male survivors and the relative dearth of research regarding female survivors' reproductive health. This vacuum of research on female survivors exists despite evidence that these women continue to be uniquely affected by Ebola, both in their reproductive health and social interactions.

16.3 Reproductive Health of Female Survivors

The World Health Organization's (WHO) Clinical Guidelines on Care of Ebola Survivors, released in 2016, indicated that female survivors of Ebola suffer from menstrual issues, as well as a greater risk of adverse pregnancy outcomes in subsequent pregnancies (World Health Organization 2016). With regard to these issues, the report states:

> "*Menorrhagia/metrorrhagia and amenorrhea are all frequently reported, although the causal link of these conditions with EVD remains to be determined... Numerous anecdotal reports exist of stillbirths in women who have conceived after recovering from EVD but it is still uncertain whether the rate of stillbirth in EVD survivors is higher than that of the general population.*"

These warnings regarding pregnancies conceived after Ebola recovery were confirmed in a study by Fallah et al. (2016) that was released later that year. The study, which included 70 female survivors from Montserrado ($n = 54$) and Margibi Counties ($n + 16$) in Liberia who had become pregnant after recovering from Ebola, found that the women were at a statistically significant increased risk of stillbirth and miscarriage. Of the 70 pregnancies evaluated, 15 resulted in miscarriage, 4 ended in stillbirth ≥28 weeks gestation, 2 pregnancies were terminated by the mother, and 49 pregnancies resulted in a normal birth. Notably, half of the pregnancies conceived within 2 months of discharge from the ETU resulted in stillbirth. There were no significant differences observed between the groups of 49 women whose pregnancies resulted in normal births and the 19 women who experienced stillbirth or miscarriage either in terms of the mean age or number of antenatal care visits per month of pregnancy. In addition, of six pregnancies that were conceived within 2 months following release of the mother from the hospital, one half (3) resulted in stillbirths. All of the 15 miscarriages in this cohort occurred in women who became pregnant 4 months or longer after their hospital discharge. While the authors do not speculate on the potential mechanisms through which Ebola survivorship can affect pregnancies, they do raise the concerns that the reduced immune response that accompanies pregnancy could potentially allow for reactivation of any Ebola virus that might persist in immune-privileged sites of the body (Fallah et al. 2016). However, there have been no reports so far of sexual transmission of the Ebola virus from a woman (Soka et al. 2017). Although Ebola virus RNA has been identified in vaginal fluid 33 days after initial infection, the presence of RNA may or may not indicate the existence of viable infectious virions (Vetter et al. 2016). In a recent study, Ebola virus secretion was evaluated by RT-PCR among 112 male and female survivors of EVD in Sierra Leone at an average of 142 days following discharge from the Kerry Town ETU (Green et al. 2016). Of the 21 swabs of vaginal secretions that were tested, no positive results for the Ebola virus were found.

An analysis of a family cluster of Ebola virus infections occurring in Liberia has provided evidence for the long-term persistence of the virus in some infected women (Dokubo et al. 2018). Following the November 2015 infection of a 15-year-old boy in Monrovia who subsequently died of EVD, evaluation of the other family members revealed an 8-year-old brother with Ebola virus RNA in his blood; a 5-year-old brother with no evidence of infection; a 2-month-old brother born in September 2015 with IgG antibodies to Ebola virus which were attributable to maternal transfer; and the father with Ebola virus RNA in the blood and an antibody profile positive for Ebola-specific IgM and IgG antibodies suggesting previous Ebola infection. The mother of the children had provided care for her adult brother in July 2014 - he had previous exposure as a nurse's aide to persons with EVD and subsequently died of presumptive EVD. Soon after her brother's death, she developed an illness compatible with EVD, but did not seek care for her condition, had a miscarriage in August 2014, and was found to have high titers to IgG and low titers to IgM anti-Ebola antibodies. These findings, together with results of genomic analysis, indicated that the most plausible explanation for this family cluster was that the woman had survived EVD in 2014 having acquired it while taking care of her brother, developed persistent Ebola infection, and transmitted the virus to her three family members one year later.

A separate study of female survivors across three counties in Liberia, described in another chapter in this book,[1] found that 46% of the 69 women interviewed reported some form of menstrual irregularity since being released from the ETU, ranging from amenorrhea to black, painful menses and suspected infection. In the same study, 13 pregnancies conceived after Ebola were carried to terminus, resulting in 10 live births and 3 miscarriages or stillbirths. Yet, we do not know or understand the mechanisms through which Ebola has caused these outcomes.

[1] C.L. Godwin, A. Buller, M. Bentley, & K. Singh. Understanding the personal relationships and reproductive health changes of female survivors of Ebola infection in Liberia.

We have only scratched the surface in discovering how Ebola affects its female survivors—ostensibly the lucky ones. Rather than providing concrete answers, early research has uncovered more questions.

The tragic case of Salome Karwah cast new light on Ebola's deadly impact, even after the outbreak. Yet Salome's passing, as well as continuing research, is helping to illuminate just how much we do *not* know about the lingering effects of Ebola—both physical and social. Sadly, it is exactly this *lack* of information that perpetuates the stigma still shrouding some of the survivors today. The only way to free survivors of Ebola—for them to truly defeat the virus—is for us to gain thorough understanding of and then demystify its lingering effects.

References

Akwei, L. (2016). For the fourth time, Liberia declared Ebola-free as surveillance continues. *Africa News*. Retrieved November 7, 2017, from http://www.africanews.com/2016/06/09/for-the-fourth-time-liberia-declared-ebola-free-as-surveillance-continues//.

All Africa. (2017). Liberia: MOH to establish Salome Karwah's Foundation. *All Africa*. Retrieved February 24, 2018, from http://allafrica.com/stories/201710030765.html.

Baker, A. (2017). Liberian Ebola fighter, a TIME person of the year, dies in childbirth. *Time Magazine*. Retrieved January 19, 2018, from http://time.com/4683873/ebola-fighter-time-person-of-the-year-salome-karwah/.

Bart, S. M., Cohen, C., Dye, J. M., Shorter, J., & Bates, P. (2018). Enhancement of Ebola virus infection by seminal amyloid fibrils. *Proceedings of the National Academy of Sciences, 115*(28), 7410–7415. Retrieved September 22, 2018, from http://www.pnas.org/content/early/2018/06/20/1721646115

Christie, A., Davies-Wayne, G. J., Cordier-Lassalle, T., Blackley, D. J., Scott Laney, A., Williams, D. E., et al. (2015). Possible sexual transmission of Ebola virus-Liberia, 2015. *MMWR Morbidity and Mortality Weekly Report, 64*, 479–481. Retrieved January 19, 2018, from https://www.cdc.gov/mmwr/preview/mmwrhtml/mm6417a6.htm.

Deen, G. F., Broutet, N., Xu, W., Knust, B., Sesay, F. R., McDonald, S. L. R., et al. (2017). Ebola RNA persistence in semen of Ebola virus disease survivors—Final report. *New England Journal of Medicine, 377*(15), 1428–1437. Retrieved February 22, 2018, from https://www.ncbi.nlm.nih.gov/pmc/articles/PMC5798881/.

Deen, G. F., Knust, B., Broutet, N., Sesay, F. R., Formenty, P., Ross, C., et al. (2015) Ebola RNA persistence in semen of Ebola virus disease survivors—Preliminary report. *New England Journal of Medicine*. https://doi.org/10.1056/NEJMoa1511410. Retrieved January 18, 2018, from http://www.nejm.org/doi/suppl/10.1056/NEJMoa1511410/suppl_file/nejmoa1511410_preliminaryreport.pdf.

Diallo, B., Sissoko, D., Loman, N. J., Bah, H. A., Bah, H., Worrell, M. C., et al. (2016). Resurgence of Ebola virus disease in Guinea linked to a survivor with virus persistence in seminal fluid for more than 500 days. *Clinical Infectious Diseases, 63*(10), 1353–1356. Retrieved January 18, 2018, from https://www.ncbi.nlm.nih.gov/pmc/articles/PMC5091350/.

Dokubo, E. K., Wendland, A., Mate, S. E., Ladner, J. T., Hamblion, E. L., Raftery, P., et al. (2018). Persistence of Ebola virus after the end of widespread transmission in Liberia: An outbreak report. *The Lancet Infectious Diseases, 18*(9), 1015–1024. Retrieved September 22, 2018, from: https://www.thelancet.com/action/showPdf?pii=S1473-3099%2818%2930417-1

Fallah, M. P., Skrip, L. A., Dahn, B. T., Nyenswah, T. G., Flumo, H., Glayweon, M., et al. (2016). Pregnancy outcomes in Liberian women who conceived after recovery from Ebola virus disease. *Lancet Global Health, 4*, e678–e679. Retrieved January 18, 2018, from http://www.thelancet.com/journals/langlo/article/PIIS2214-109X(16)30147-4/fulltext.

Fischer, W. A., Brown, J., Wohl, D. A., Loftis, A. J., Tozay, S., Reeves, E., et al. (2017). Ebola virus ribonucleic acid detection in semen more than two years after resolution of acute Ebola virus infection. *Open Forum Infectious Diseases, 4*(3), ofx155. Retrieved January 20, 2018, from https://academic.oup.com/ofid/article/4/3/ofx155/4004818.

Green, E., Hunt, L., Ross, J. C. G., Nissen, N. M., Curran, T., Badhan, A., et al. (2016). Viraemia and Ebola virus secretion in survivors of Ebola virus disease in Sierra Leone: A cross-sectional cohort study. *Lancet Infectious Diseases, 16*(9), 1052–1056. Retrieved February 21, 2018, from https://www.sciencedirect.com/science/article/pii/S1473309916300603#!.

Karwah, S. (2014). I survived Ebola for a reason—To help others recover. *The Guardian*. Retrieved January 19, 2018, from https://www.theguardian.com/commentisfree/2014/oct/12/ebola-survivor-now-i-am-helping-others-fight-the-disease.

Mate, S. E., Kugelman, J. R., Nyenswah, T. G., Ladner, J. T., Wiley, M. R., Cordier-Lassalle, T., et al. (2015). Molecular evidence of sexual transmission of Ebola virus. *New England Journal of Medicine, 373*, 2448–2454. Retrieved January 17, 2018, from http://www.nejm.org/doi/full/10.1056/NEJMoa1509773#t=article.

Medecins Sans Frontieres. (2017). A tribute to Salome Karwah: "I survived Ebola to help others." Retrieved January 18, 2018, from https://www.msf.org.uk/article/tribute-salome-karwah-%E2%80%9Ci-survived-ebola-help-others.

Mukpo, A. (2017). A husband loses his 'best friend'—Salome Karwah, Ebola hero. *NPR*. Retrieved January 14, 2018, from https://www.npr.org/sections/goatsandsoda/2017/03/01/515822228/a-husband-loses-his-best-friend-salome-karwah-ebola-hero.

Preeclampsia Foundation. (2013). Preeclampsia and maternal mortality: A global burden. Retrieved February 23, 2018, from https://www.preeclampsia.org/health-information/149-advocacy-awareness/332-preeclampsia-and-maternal-mortality-a-global-burden.

Purpura, L. J., Rogers, E., Baller, A., White, S., Soka, M., Choi, M. J., et al. (2017). Ebola virus RNA in semen from an HIV-positive survivor of Ebola. *Emerging Infectious Diseases, 23*(4), 714–715. Retrieved January 20, 2018, from https://www.ncbi.nlm.nih.gov/pmc/articles/PMC5367423/.

Purpura, L. J., Soka, M., Baller, A., White, S., Rogers, E., Choi, M. J., et al. (2016). Implementation of a national semen testing and counseling program for male Ebola survivors—Liberia, 2015–2016. *MMWR Morbidity and Mortality Weekly Review, 65*(36), 963–966. Retrieved from https://www.cdc.gov/mmwr/volumes/65/wr/mm6536a5.htm.

Schwartz, D. A. (2015). Pathology of maternal death—The importance of accurate autopsy diagnosis for epidemiologic surveillance and prevention of maternal mortality in developing countries. In D. A. Schwartz (Ed.), *Maternal mortality: Risk factors, anthropological perspectives, prevalence in developing countries and preventative strategies for pregnancy-related death* (pp. 215–253). New York: Nova Scientific Publishing.

Schwartz, D. A. (2018). Hypertensive mothers, obstetric hemorrhage and infections: Biomedical aspects of maternal death among indigenous women in Mexico and Central America. In D. A. Schwartz (Ed.), *Maternal death and pregnancy-related morbidity among indigenous women of Mexico and Central America: An anthropological, epidemiological and biomedical approach* (pp. 35–50). New York: Springer.

Scott, J. T., Sesay, F. R., Massaquoi, T. A., Idriss, B. R., Sahr, F., & Semple, M. G. (2016). Post-Ebola syndrome, Sierra Leone. *Emerging Infectious Diseases, 22*(4), 641–646. Retrieved February 21, 2018, from https://www.ncbi.nlm.nih.gov/pmc/articles/PMC4806950/.

Sissoko, D., Duraffour, S., Kerber, R., Kolie, J. S., Beavogul, A. H., Camara, A.-M., et al. (2017). Persistence and clearance of Ebola virus RNA from seminal fluid of Ebola virus disease survivors: A longitudinal analysis and modelling study. *Lancet Global Health, 5*(1), 80–88. Retrieved January 19, 2018, from http://www.thelancet.com/journals/langlo/article/PIIS2214-109X(16)30243-1/fulltext.

Soka, M. J., Choi, M. J., Baller, A., White, S., Rogers, E., Purpura, L. J., et al. (2016). Prevention of sexual transmission of Ebola in Liberia through a national semen testing and counselling programme for survivors: An analysis of Ebola virus RNA results and behavioural data. *Lancet Global Health, 4*, e736–e743. Retrieved January 19, 2018, from http://www.thelancet.com/journals/langlo/article/PIIS2214-109X(16)30175-9/fulltext.

Soka, M. J., Choi, M. J., Purpura, L. J., Ströher, U., Knust, B., & Rollin, P. (2017). Screening of genital fluid for Ebola virus—Authors' reply. *Lancet Global Health, 5*(1), e33. Retrieved February 24, 2018, from http://www.thelancet.com/journals/langlo/article/PIIS2214-109X(16)30290-X/fulltext.

Thorson, A., Formenty, P., Lofthouse, C., & Broutet, N. (2016). Systematic review of the literature on viral persistence and sexual transmission from recovered Ebola survivors: Evidence and recommendations. *BMJ Open, 6*, e008859. Retrieved January 19, 2018, from http://bmjopen.bmj.com/content/6/1/e008859.

Time. (2014). The Ebola fighters. *Time magazine*. Retrieved January 19, 2018, from http://time.com/4683873/ebola-fighter-time-person-of-the-year-salome-karwah/.

UN Women. (2017). *Maternal health gets a new boost in Liberia*. Retrieved February 21, 2018, from http://www.unwomen.org/en/news/stories/2017/7/feature-maternal-health-gets-a-new-boost-in-liberia.

Vetter, P., Fischer 2nd., W. A., Schibler, M., Jacobs, M., Bausch, D. G., & Kaiser, L. (2016). Ebola virus shedding and transmission: Review of current evidence. *Journal of Infectious Diseases, 214*(suppl 3), S177–S184. Retrieved February 20, 2018, from https://academic.oup.com/jid/article/214/suppl_3/S177/2388183.

Villar, J., Say, L., Gulmezoglu, A. M., Meraldi, M., Lindheimer, M. D., Betran, A. P., et al. (2003). Eclampsia and pre-eclampsia: A health problem for 2000 years. In H. Critchly, A. MacLean, L. Poston, & J. Walker (Eds.), *Pre-eclampsia* (pp. 189–207). London: RCOG Press.

Von Drehle, D. (2014). The Ebola fighters. *Time Magazine*. Retrieved January 19, 2018, from http://time.com/time-person-of-the-year-ebola-fighters/.

Wong, G., Qiu, X., Bi, Y., Formenty, P., Sprecher, A., Jacobs, M., et al. (2016). More challenges from Ebola: Infection of the central nervous system. *Journal of infectious Diseases, 214*(suppl 3), S294–S296. Retrieved February 21, 2018, from https://www.ncbi.nlm.nih.gov/pmc/articles/PMC5050477/.

World Health Organization. (2016). *Clinical care for survivors of Ebola virus disease: Interim guidance*. Geneva: World Health Organization. Retrieved January 19, 2018, from http://apps.who.int/iris/bitstream/10665/204235/1/WHO_EVD_OHE_PED_16.1_eng.pdf.

17 All the Mothers Are Dead: Ebola's Chilling Effects on the Young Women of One Liberian Town Named Joe Blow

David A. Schwartz

17.1 The Liberian Epidemic

The Ebola virus epidemic in Liberia was catastrophic to the people of this impoverished country that was still recovering from many years of civil strife and armed conflict. The first cases of Ebola infection were reported at the close of March 2014, in Foya, Lofa County (WHO 2015). Just one week later, there were 21 confirmed, suspected, or probable infections in the country and 10 deaths. All five of the initial laboratory-confirmed cases of Ebola virus disease (EVD) died, including three health care workers. There followed a lull in new cases, mostly in Lofa County, with only 51 cases reported up to the end of June. But then things changed—new cases of Ebola infection began occurring in Monrovia, the nation's capital. This was a unique epidemiological occurrence for Ebola virus infection, as it had previously been concentrated in rural areas of Africa and never become epidemic in a major city since its initial recognition in 1976.

Redemption Hospital in Monrovia (Fig. 17.1) was ground zero at the beginning of the epidemic in Liberia. A government-run hospital that provided care free of charge, the first Ebola victim in Monrovia was treated there in June 2014, after which a number of staff members became ill,resulting in the deaths of 12 doctors and nurses (Aizenman 2014; Guilbert 2016). On July 2nd, the Chief Surgeon of Redemption Hospital died from the infection (Soy 2015). Following his death, persons became frightened to visit the hospital, it was shuttered (Fig. 17.2), and patients were transferred or referred to other facilities.

Beginning in June, Bong County (Fig. 17.3), a predominantly rural region on the border with Guinea, became involved in the Ebola epidemic, resulting in the closing of its three hospitals and most of the private clinics, and causing the deaths of hundreds of individuals, hospital staff members, and doctors (Levine 2014).

The situation became so dire that, on August 6th, 2014 the president of Liberia, Ellen Johnson Sirleaf, declared a 3-month national state of emergency. This was followed by an investigation of the status of the epidemic by a surveillance team from the Ministry of Health (MOH) that included the World Health Organization (WHO) and Centers for Disease Control and Prevention (CDC) led by Dr. Hans Rosling, which revealed that 14 of the country's 15 counties had Ebola virus cases. It also found

D. A. Schwartz (✉)
Department of Pathology, Medical College of Georgia, Augusta University, Augusta, GA, USA

D. A. Schwartz et al. (eds.), *Pregnant in the Time of Ebola*, Global Maternal and Child Health,
https://doi.org/10.1007/978-3-319-97637-2_17

Fig. 17.1 Redemption Hospital, one of Liberia's largest health care facilities, became ground zero for the country's Ebola epidemic. Photograph courtesy of USAID

Fig. 17.2 The Emergency Department at Redemption Hospital in Monrovia after being closed to the public as a result of the Ebola epidemic. Photograph courtesy of USAID

that the epidemic was taking a high toll on the handful of health care workers including doctors and nurses—152 had become infected, and 79 had died. Transmission of the virus was relentless, and by September 8th, Liberia had the highest cumulative number of EVD cases among the three involved countries—almost 2000 infected persons and greater than 1000 fatalities. The growth of the epidemic had completely overwhelmed the limited number of surviving Liberian doctors and nurses, available hospital beds, and ambulance transport to provide any semblance of care for the exponentially increasing number of suspected cases. By the 23rd September 2014, Liberia had seen 3458 total cases of infection, 1830 deaths, and 914 laboratory-confirmed cases. The epidemic was having an especially devastating effect on reproductive-aged women, and in particular, those who were pregnant—high fatality rates, lack of availability and access to medical care for obstetric complications, fear of conta-

Fig. 17.3 Ebola survivors leave their handprints on a wall of the Bong County Ebola Treatment Unit—the facility that saved their lives. Photograph courtesy of USAID

gion from the body fluids occurring during delivery, and stigmatization. Dr. Rick Brennan, the director of a WHO emergency assessment team, stated that delivering a baby in Liberia was "one of the most dangerous jobs in the world" (WHO 2014a). With the assistance of outside agencies and organizations, new treatment centers were constructed, hospital beds were increased, and there was an influx of skilled medical assistance to help the beleaguered country respond to the epidemic. Despite this, in Monrovia, the ill waited in holding centers for a hospital bed to become available, and many died without having received medical care.

Burials of the Ebola virus infection became problematic, which was compounded by the high viral loads present in the recently dead, making their bodies highly infectious (Figs. 17.4 and 17.5). The number of deaths prevented safe burial of human remains, and according to one report from the WHO (WHO 2014a; Snyder et al. 2014), bodies of EVD victims from the West Point slum were dumped into nearby rivers in order to help control the outbreak and deal with the high death rate. In a burial area hastily set up in a swamp near the town of Johnsonville, the body bags containing EVD victims were thrown into graves—during the night the water table rose, floating the bodies up to the surfaces of their graves, resulting in the complete abandonment of the burial site (Silver 2014). There were also reports of families dragging the dead bodies of Ebola victims from inside their houses and leaving them in the streets in order to evade quarantine regulations (MacDougall and Flynn 2014). Liberian Information Minister Lewis Brown said "*They are therefore removing the bodies from their homes and are putting them out in the street. They're exposing themselves to the risk of being contaminated,*" "*We're asking people to please leave the bodies in their homes and we'll pick them up*" (MacDougall and Flynn 2014).

At the end of September, Liberia's chief medical official, Dr. Bernice Dahn, went on a self-enforced 21-day quarantine because of exposure to the virus from her assistant, who had developed EVD and died (Capelouto and Smith 2014)—she was later found to be uninfected. Safe collection and processing of bodies of Ebola virus victims for burial or cremation was an important part of the preventive measures for interrupting viral transmission, as the traditional Liberian funeral arrangements, usually supervised by women, were an important risk factor in the spread of Ebola. Because the dead body contains high levels of virus, 12 teams of "body collectors," protected by personal protective equipment (PPE) or

Fig. 17.4 A team of Ebola response workers in Monrovia, Liberia, carries the body of an Ebola victim. Photograph courtesy of USAID

Fig. 17.5 Members of the Liberian Red Cross and Global Communities burial team remove the body of a 6-year-old girl who died of EVD from her family home in Arthington, Montserrado County, Liberia. Photograph courtesy of USAID

hazmat suits, collected and disposed of the bodies of victims from the dwellings and streets in Monrovia throughout the height of the epidemic (Chhabra 2015). Unfortunately, because of the stigma of Ebola infection, as well as refusing to believe that Ebola infection was contagious, there were families that did not disclose that their relative died from Ebola, and the virus-laden body of the deceased would be handled by relatives and given a traditional funeral, contributing to the spread of the disease (McConnell 2014). Cokie van der Velde, a sanitation specialist with Médecins Sans Frontières (MSF), observed "*Very, very few of those dying in the community are being brought forward," and "That means they're being kept* hidden and buried in secret," (McConnell 2014). In some rural Liberian villages, the bodies of EVD victims were buried without notifying local health officials, and consequently, with no investigation of the cause of death. In order to approximate the number of EVD deaths, some epidemiologists resorted to counting the number of fresh graves that were dug in villages as a crude indicator of suspected cases (WHO 2014b). As of the 5th of November 2014, Liberia reported 6525 cases (including 1627 probable and 2447 suspected cases) and 2697 deaths from Ebola infection (WHO 2014c). As increases in human resources, infrastructure, hospital beds, safe burial practices, case finding and contact tracing, and additional public health measures began to take effect into November and December, a decline of new cases was noted in Monrovia as well as in Lofa County, but infections in other regions of the country remained active (Sharma et al. 2014). The Liberian president announced the lifting of the state of emergency on November 13th as a result of the decrease in new infections, although new cases were occurring, and by the middle of January 2015, there were only two counties—Montsterrado and Grand Cape Mount counties—where Ebola transmission continued (Scientific American 2015). Liberia reopened its land borders on February 20th, 2015, and by the first week of March 2015, the WHO announced that the last Ebola patient was released from a treatment center following one week without any new EVD cases occurring (Fallah 2015). Following this, sporadic cases occurred throughout the country—one man tested positive on March 20th and died on March 27th; on June 29th, the body of a boy with malaria tested positive for Ebola virus in Margibi County; and on July 1st and 2nd, two additional persons were found to be infected. WHO declared Liberia to be Ebola-free on September 3rd, 2015. On November 20th, Ebola virus returned to Liberia when a 15-year-old boy and his two family members were all diagnosed with the infection (Fox 2015)—the boy subsequently died. Finally, Liberia was declared free from the virus once again on January 14th, 2016, following a 42-day mandatory period of no new cases. Unexpectedly, a new and fatal Ebola case in a young woman occurred in Liberia on April 1st, 2016 (Reuters 2016), to be followed on April 3rd by another infected person on Monrovia. Three new positive Ebola cases were identified on April 7th, leading to the monitoring of 97 contacts including 15 health care workers. On June 9th, 2016 and for the fourth and last time, Liberia was declared by the WHO to be free of Ebola virus transmission, thus ending the outbreak that had begun in Guinea two years before. The final toll of the Ebola outbreak in Liberia was 10,666 cases and 4806 deaths, including 378 infections among health care workers resulting in 192 deaths (Statista 2015).

17.2 Joe Blow Town

Prior to the Ebola virus epidemic in Liberia, Joe Blow was a just another village, one so small that it didn't appear on most maps (Fig. 17.6). At that time, fewer than 150 persons lived there (ZOA n.d.). But the Ebola virus changed that.

A traditional funeral was held in this tiny Liberian village, situated at the end of a long narrow road in Margibi County, following the death of a prominent woman resident of the village. Her death was the result of Ebola virus infection. She had acquired the infection from her husband, who became infected while outside of the village and brought it home to his wife (Knapton 2015). During the course of her Ebola virus infection, she was cared for by 14 mothers living in the village, in accordance

Fig. 17.6 The road leading into Joe Blow Town. Photograph courtesy of ZOA/Sandra Vogd

with Liberian custom, who then laid out her body at the time of her death. According to Tom Dannatt, founder of the nongovernmental organization Street Child, all of the women in the village bathed in the water that was used to clean the dead body, as tradition dictated (Brekke 2015). According to Dannatt, "*They got Ebola, one by one … and literally the entire [population of] young women of that village were wiped out.*" He also stated "*When my colleague arrived in that village in December, what she saw was a couple of old ladies, a couple of grandmas, just looking after a village full of children. The men were in the field working, and, you know, this is the sort of extraordinary sight that Ebola is creating.*" Chloe Brett, who was working with Street Child in the village following the death of the mothers, observed "*When we visited Joeblow, it seemed normal at first, with children in the street, men, a couple of old women. But then we realized there were no other women anywhere.*" She went on to say "*We talked to a man who had survived Ebola and he told us what had happened. All of the women had caught the disease. It's now a village of no mothers and very confused children with blank looks on their faces*" (Delay 2015).

The virus spread through the village, taking additional victims. It took 4 h to bury one of the dead village mothers, Jartu Karkula, in the rain (BBC 2015; Street Child 2015). Gravediggers prefer burying victims in the sun, as they believe that the sunshine kills the virus. As the gravediggers left the house with the bag containing the body of Jartu, her niece, Jackie Smith, approached the bag and attempted to take pictures with her cellphone, but the leader of the team ordered her to withdraw. "*The family is not burying her, if only to see her one last time, and I feel really bad,*" said Smith. The family could not participate in her burial due to fear of contagion. According to her brother, the Reverend John Singbae. "*You have to think about children*" and "*They do not know what's going on. So that they really accept the reality that their mother is dead, they will have to live with this nightmare ... They did not follow the burial of their mother*" (VOA 2014). Jartu was a Christian and had been a pillar of her parish in Margibi County. Following her death, her four children were interned in the center of nearby Dolo Town, where they remained under observation for possible Ebola infection (VOA 2014).

The village outbreak left numerous children with no mothers. For one motherless boy, 11-year-old Montgomery Phillip, the loss of his own mother as well as all the other mothers in his village has forced him to grow up quickly. He had to provide care for his 10-month-old baby brother Jenkie (Knapton 2015). "*Seeing Montgomery struggle to change the baby's nappy without any guidance is something that made me realize just how devastating this disease can be on those left behind,*" said Chloe Brett. "*He was a helpless 11-year-old having to become a man well before his time*" (Delay 2015). Discussing the orphaned children, she made several observations "*I saw Montgomery carrying his 10-month-old brother—that is life for him now. He won't be able to go back to school if he is looking after his brother.*" "*All the children wear rags because all their clothes and possessions have had to be burnt as a precaution because of the disease.*" "*We try to find relatives or neighbors to take the children in, but the community is scared*" (Knapton 2015).

Following the Ebola virus outbreak and death of all mothers in the village, Joe Blow Town was shut off from the outside world for some time to prevent further Ebola infection. Human contact was not possible, buckets of chlorine were distributed, the small school was closed, and trading with other villages was not permitted. A nongovernmental organization, ZOA from the Netherlands, worked to help keep the villagers maintain contact with the outside world by distributing prepaid telephone cards (ZOA n.d.). One of the villagers remarked to a ZOA staff member, "*Only two cars arrived here, those of the Ebola funeral team and that of your organization.*"

17.3 Traditional Health Care and Burial Practices

There is no evidence that women are more biologically susceptible to Ebola virus infection. However, the contribution of traditional health care and burial practices to the increased risk for acquiring Ebola virus infection, in particular by the girls and women who perform most of these tasks, has been well-documented. It must be remembered that there exists great variation in these cultural practices—in Liberia alone, there are 16 major ethnic and cultural groups, each having its own language and associated customs and traditions.

Ebola virus transmission occurs when an individual had direct contact with the viral-infected body fluids, including urine, blood, amniotic fluid, sweat, saliva, vomit, and other fluids from an Ebola-virus-infected individual, or with objects that have been contaminated with viral-infected tissues or fluids (CDC 2015). Because of this, there is a heightened likelihood for Ebola virus transmission among caregivers, during preparation of dead bodies for a funeral, and unsafe burial practices (Tiffany et al. 2017). Women and girls are the primary caregivers in their homes, communities, and health facilities in most West African cultures—they care for, bathe, feed, clean up after, and otherwise assist most infected individuals, placing themselves at an increased risk for contracting the virus. Women fulfill their gender roles by being the primary caregivers when their children, husbands, brothers and sisters, and other relatives are ill; they bathe the infirm, nurse their children, and perform other tasks that place them in direct contact with potentially infected individuals (Nkangu et al. 2017). Thus, they take on the role of "nurse" to their family and networks. Unlike skilled nurses, however, they are not trained in hygienic methods for caring for the sick and infection control practices.

Traditional funerary and burial practices, which are typically performed by women, also places them at higher risk for acquiring infection (Brainard et al. 2016; Hewlett and Amola 2003; Menéndez et al. 2015). In Liberia and the other countries of West Africa, traditional mourning and burial practices in most ethnic groups incorporate preparation of the body of the deceased, resulting in an anthropological process in which death is viewed as a continual process as well as a rite of passage. The mortuary preparations that occur are consistent with their religious philosophy, ensuring that the soul of the deceased makes its way to the ancestral world safely, and once there, does not return to the land of the living. If the traditional mortuary preparations remain incomplete, the body remains in a liminal state in which it is no longer part of the living world, and not in the afterlife. Thus, the importance of preparing the body for burial cannot be overemphasized, but it brings these rites into direct conflict with the recommendations of Ebola health care responders to avoid touching, washing, or kissing the deceased prior to burial. Generally, the bodies of dead women are washed by women, and dead men are washed by other men (Richards et al. 2015). Another mortuary practice that creates risk for Ebola virus transmission is that wives are required to shave the heads of their dead husbands, after which they are smeared with mud made from their husband's body's washings (Shah 2015).

Overwhelmed with the numbers of deaths from Ebola infection and with fear of transmission of the infection from unburied or unsafely handled corpses, the government of Liberia declared in August of 2014 that bodies of Ebola victims were to be cremated instead of being buried (BBC 2014). The bodies of the deceased were collected by government-organized Dead Body Management teams, who were charged with the safe collection and transport of the corpses for cremation. In many cases the dead bodies could not be tested for confirmation of Ebola infection prior to their mass cremations, which were usually conducted at night. "*We take every body, and burn it,*" said Nelson Sayon, a member of one of the Dead Body management teams (Baker 2014). This policy resulted in hostility from communities in which funeral practices were an important part of traditional respect and mourning for the dead. Kenneth Martu, a community organizer from the Westpoint neighborhood of Monrovia, confirmed that the mass cremations have caused a deep rift in a country where there was already distrust of the government. He said "*In West Africa we don't cremate bodies at all. So when the government takes away our bodies, and can't even tell us if they died of Ebola or not, it breeds resentment*" (Baker 2014). This order led to the practice of "secret burials" by families who kept their ill at home to die in order to avoid cremation (Pellecchia et al. 2015; Ryeng 2015; The Guardian 2014). This resulted in additional unsafe burials. "*For fear of cremation, do not stay home to die,*" Tolbert Nyenswah, Assistant Health Minister and director of Liberia's Ebola response, requested of the Liberian community during a news conference. "*We know cremation is not our culture in our country,*" Nyenswah said. "*But now we have disease, so we have to change the way we used to do business*" (The Guardian 2014).

Interviews were conducted with Liberians in both urban and peri-urban areas to understand their attitudes toward mass burials, cremation, memorialization, and remembrance ceremonies in regard to the Ebola epidemic and preventive practices that had been instituted to prevent transmission of the virus following death. These interviews revealed that although there was great variation in the attitudes of people towards these practices, there was widespread concern for the correct handling of the bodies of the deceased (Abramowitz and Omidian 2014). Some members of the community were worried that their family members with Ebola infection might simply "disappear," and there was concern that "health workers were injecting people to death and then selling body parts, so cremation concealed the theft of body parts," and that the body parts of the deceased were being circulated through a global network of trade. Among interviewees in Monrovia, there was a recurring concern that centered on the apparent disappearance of both sick people and corpses following their removal by health teams and burial teams (Abramowitz and Omidian 2014).

Ebola virus disease victims are the most infectious after they die. By the time of death, the viral load is at its height, infectious bodily secretions are released, and the washing, touching, and kissing of the deceased place persons at great risk for contracting the infection. These were the risk factors occurring in Joe Blow Town, which resulted in the tragic deaths of all the mothers in the village.

References

Abramowitz, S., & Omidian, P. (2014). Attitudes towards Ebola related funerary practices and memorialization in urban Liberia. *Ebola Response Anthropology Platform*. Retrieved May 5, 2018, from http://www.ebola-anthropology.net/key_messages/attitudes-towards-ebola-related-funerary-practices-and-memorialization-in-urban-liberia/.

Aizenman, N. (2014). Why patients aren't coming to Liberia's Redemption Hospital. *NPR*. Retrieved May 1, 2018, from https://www.npr.org/sections/goatsandsoda/2014/08/27/343494521/why-patients-arent-coming-to-liberias-redemption-hospital.

Baker, A. (2014). Liberia burns its bodies as Ebola fears run rampant. *Time Magazine*. Retrieved May 8, 2018, from http://time.com/3478238/ebola-liberia-burials-cremation-burned/.

BBC. (2014). Liberia orders Ebola victims' bodies to be cremated. *BBC News*. Retrieved May 2, 2018, from http://www.bbc.com/news/world-africa-28640745.

BBC. (2015). Liberia's Ebola orphans. Interview with Chloe Brett. *BBC*. Retrieved April 3, 2018, from https://soundcloud.com/bbc-world-service/liberias-ebola-orphans.

Brainard, J., Hooper, L., Pond, K., Edmunds, K., & Hunter, P. R. (2016). Risk factors for transmission of Ebola or Marburg virus disease: A systematic review and meta-analysis. *International Journal of Epidemiology, 45*(1), 102–116. Retrieved January 25, 2018, from https://www.ncbi.nlm.nih.gov/pubmed/26589246.

Brekke, K. (2015). Ebola killed every mother in this Liberian village. *Huffington Post*. Retrieved January 10, 2018, from http://www.businessinsider.com/ebola-killed-all-mothers-in-joeblow-liberia-2015-1.

Capelouto, S., & Smith, M. (2014). Liberia's top medical officer is in Ebola quarantine. *CNN*. Retrieved February 7, 2018, from http://www.cnn.com/2014/09/27/world/africa/ebola-liberia/index.html.

CDC. (2015). Ebola (Ebola virus disease). *Transmission*. Retrieved February 1, 2018, from https://www.cdc.gov/vhf/ebola/transmission/index.html.

Chhabra, E. (2015). Ebola's body collectors. *The Atlantic*. Retrieved February 5, 2018, from https://www.theatlantic.com/health/archive/2015/09/ebolas-body-collectors/407965/.

Delay, J. (2015). In one Liberian village, the Ebola outbreak did not spare a single mother. *National Post*. Retrieved April 4, 2018, from http://nationalpost.com/news/in-one-liberian-village-the-ebola-outbreak-did-not-spare-a-single-mother.

Fallah, S. (2015). Liberia's last Ebola patient discharged from center. *USA Today*. Retrieved January 20, 2018, from https://www.usatoday.com/story/news/world/2015/03/05/last-ebola-patient-released-liberia/24421305/.

Fox, M. (2015). Ebola returns to Liberia, again. *NBC News*. Retrieved February 7, 2018, from https://www.nbcnews.com/storyline/ebola-virus-outbreak/ebola-returns-liberia-again-n467146.

Guilbert, K. (2016). Redemption in Liberia: A hospital's painful recovery from Ebola. *Reuters*. Retrieved May 1, 2018, from https://www.reuters.com/article/us-liberia-ebola-health-idUSKCN0V40IA.

Hewlett, B. S., & Amola, R. P. (2003). Cultural contexts of Ebola in northern Uganda. *Emerging Infectious Diseases, 9*(10), 1242–1248.

Knapton, S. (2015). Ebola wipes out every mother in Liberian village. *The Telegraph*. Retrieved January 10, 2018, from http://www.telegraph.co.uk/news/worldnews/ebola/11304584/Ebola-wipes-out-every-mother-in-Liberian-village.html.

Levine, A. (2014). One woman walked in, and the Ebola nightmare began. *CNN*. Retrieved April 30, 2018, from https://www.cnn.com/2014/09/24/health/ebola-epidemic-liberia/index.html.

MacDougall, C., & Flynn, D. (2014). Bodies dumped in streets as West Africa struggles to curb Ebola. *Reuters*. Retrieved April 3, 2018, from https://www.reuters.com/article/us-health-ebola-africa/bodies-dumped-in-streets-as-west-africa-struggles-to-curb-ebola-idUSKBN0G51VF20140805.

McConnell, T. (2014). Some people would rather die of Ebola than stop hugging sick loved ones. *Public Radio International*. Retrieved February 1, 2018, from https://www.pri.org/stories/2014-10-10/some-people-would-rather-die-ebola-stop-hugging-sick-loved-ones.

Menéndez, C., Lucas, A., Munguambe, K., & Langer, A. (2015). Ebola crisis: The unequal impact on women and children's health. *Lancet Global Health, 3*(3), e130. Retrieved January 25, 2018, from http://www.thelancet.com/journals/langlo/article/PIIS2214-109X(15)70009-4/fulltext.

Nkangu, M. N., Olatunde, O. A., & Yaya, S. (2017). The perspective of gender on the Ebola virus using a risk management and population health framework: A scoping review. *Infectious Diseases of Poverty, 6*(1), 135. Retrieved February 3, 2018, from https://www.ncbi.nlm.nih.gov/pmc/articles/PMC5635524/.

Pellecchia, U., Crerstani, R., Decroo, T., Van den Bergh, R., & Al-Koirdi, Y. (2015). Social consequences of Ebola containment measures in Liberia. *PLOS One, 10*(12), e0143036. https://doi.org/10.1371/journal.pone.0143036.

Reuters. (2016). Liberia records new Ebola death, months after end of its outbreak. *Thomson Reuters Foundation News*. Retrieved January 18, 2018, from http://news.trust.org/item/20160401113911-cthc6.

Richards, P., Amara, J., Ferme, M., & Kamara, P. (2015). Social pathways for Ebola virus disease in rural Sierra Leone, and some implications for containment. *PLOS Neglected Tropical Diseases, 9*(4), e0003567. https://doi.org/10.1371/journal.pntd.0003567.

Ryeng, H. S. (2015). Ebola in Liberia: From secret burials to safe burials. *UNICEF*. Retrieved May 10, 2018, from https://blogs.unicef.org/blog/ebola-in-liberia-from-secret-burials-to-safe-burials/.

Scientific American. (2015). *Ebola spread now limited to 2 counties in Liberia—Official*. Retrieved February 5, 2018, from https://www.scientificamerican.com/article/ebola-spread-now-limited-to-2-counties-in-liberia-official/.

Shah, J. A. (2015). The dead bodies of the West African Ebola epidemic: Understanding the importance of traditional burial practices. *Inquiries, 7*(11), 1–4. Retrieved April 4, 2018, from http://www.inquiriesjournal.com/articles/1300/3/the-dead-bodies-of-the-west-african-ebola-epidemic-understanding-the-importance-of-traditional-burial-practices.

Sharma, A., Heijenberg, N., Clement, P., Bolongei, J., Reeder, B., Alpha, T., et al. (2014). Evidence for a decrease in transmission of Ebola virus—Lofa County, Liberia, June 8–November 1, 2014. *MMWR Morbidity and Mortality Weekly Report, 63*(46), 1067–1071. Retrieved February 5, 2018, from https://www.cdc.gov/mmwr/preview/mmwrhtml/mm63e1114a1.htm?s_cid=mm63e1114a1_w.

Silver, M. (2014). A fiasco at the burial ground, a prank at the shop: Covering Ebola. *NPR*. Retrieved April 2, 2018, from https://www.npr.org/sections/goatsandsoda/2014/08/14/340153946/a-fiasco-at-the-burial-ground-a-prank-at-the-shop-covering-ebola.

Snyder, R. E., Marlow, M. A., & Riley, L. W. (2014). Ebola in urban slums: The elephant in the room. *Lancet Global Health, 2*(12), e685. https://doi.org/10.1016/S2214-109X(14)70339-0. Retrieved April 2, 2018, from https://www.ncbi.nlm.nih.gov/pmc/articles/PMC5004591/#R9.

Soy, A. (2015). Redemption of an Ebola-hit hospital in Monrovia. *BBC News*. Retrieved January 27, 2018, from http://www.bbc.com/news/world-africa-33465459.

Statista. (2015). *Ebola cases and deaths among health care workers due to the outbreaks in West African countries as of November 4, 2015*. Retrieved January 20, 2018, from https://www.statista.com/statistics/325347/west-africa-ebola-cases-and-deaths-among-health-care-workers/.

Street Child. (2015). The Liberian village where ebola has taken all the mothers—S. Telegraph, BBC World Service, BBC News 24. *Street Child*. Retrieved April 3, 2018, from https://www.street-child.co.uk/ebola-crisis-update/2015/1/7/the-liberian-village-where-ebola-has-taken-all-the-mothers-s-telegraph-bbc-world-service-bbc-news-24.

The Guardian. (2014). Ebola cremation ruling prompts secret burials in Liberia. *The Guardian*. Retrieved May 5, 2018, from https://www.theguardian.com/world/2014/oct/24/ebola-cremation-ruling-secret-burials-liberia.

Tiffany, A., Dalziel, B. D., Kagume Njenge, H., Johnson, G., Nugba Ballah, R., James, D., et al. (2017). Estimating the number of secondary Ebola cases resulting from an unsafe burial and risk factors for transmission during the West Africa Ebola epidemic. *PLoS Neglected Tropical Diseases, 11*(6), e0005491. https://doi.org/10.1371/journal.pntd.0005491.

VOA. (2014). Ebola: la suppression des rites funéraires exacerbe la douleur des survivants. *Voice of America Africa*. Retrieved February 1, 2018, from https://www.voaafrique.com/a/ebola-la-suppression-des-rites-funeraires-exacerbe-la-douleur-des-survivants/2477044.html.

WHO. (2014a). *Ebola in Liberia: Misery and despair tempered by some good reasons for hope*. Retrieved February 1, 2018, from http://www.who.int/csr/disease/ebola/ebola-6-months/liberia/en/.

WHO. (2014b). *Why the Ebola outbreak has been underestimated*. 22 August 2014. Retrieved April 2, 2018, from http://www.who.int/mediacentre/news/ebola/22-august-2014/en/.

WHO. (2014c). *Ebola response roadmap. Situation report update*. 14 November 2014. Retrieved February 7, 2018, from http://apps.who.int/iris/bitstream/10665/143216/1/roadmapsitrep_14Nov2014_eng.pdf?ua=1.

WHO. (2015). *Liberia: a country—And its capital—Are overwhelmed with Ebola cases*. Retrieved January 15, 2018, from http://www.who.int/csr/disease/ebola/one-year-report/liberia/en/.

ZOA. (n.d.). *Ebola. Noodhulp. Een jaar na de uitbrak van ebola* [Ebola. Emergency aid. One year after the outbreak of Ebola]. Retrieved November 24, 2018, from https://www.google.com/url?sa=t&rct=j&q=&esrc=s&source=web&cd=1&ved=0ahUKEwjEgrzTgZnZAhXFtVMKHaTJAW4QFggrMAA&url=https%3A%2F%2Fwww.zoa.nl%2Fcontent%2Fuploads%2FRapportage-1-jaar-noodhulp-Ebola-low-res.pdf&usg=AOvVaw08q7QLnv2IeXYO35uA1T2W.

Part III

Guinea

Removing a Community Curse Resulting from the Burial of a Pregnant Woman with a Fetus in Her Womb. An Anthropological Approach Conducted During the Ebola Virus Epidemic in Guinea

18

Julienne Ngoundoung Anoko and Doug Henry

18.1 Introduction

In June 2014, in the "Forestière" region of Guinea, a 24-year-old pregnant woman from the Kissi[1] ethnic group presented to the Prefectural hospital in Guéckédou, Guinea, with fever and hemorrhaging. She was admitted to the emergency services ward with the diagnosis, "severe infection due to tearing of the amniotic sac in a post-term pregnancy of 10 months." Despite attempts to save her, death came quickly, only 4 hours later. Because of the timing, symptoms, context, and location of her death, Ebola was clinically suspected and tested for.[2] Unfortunately, the test sample taken by the Mobile European Laboratory was poorly handled and became damaged. The result, therefore, was filed as "inconclusive." The medical staff, however, under extreme clinical suspicion of EVD, decided to act on the side of caution and recommended a "safe" burial. This meant placing the body whole in an impermeable vinyl body bag, to be transported to a secured cemetery for burial. Significantly, it also meant that the traditional burial normally expected to be associated with the death of a pregnant woman would be forbidden by the district health authorities (Fig. 18.1).

When the family became informed of the clinical mandate, it set off a dangerous chain of events and anxious confrontations between medical staff, the district health officers, the surviving family, and community leaders. The surviving family and community demanded release of the body so that they could

[1]The Kissi people are an ethnic group living in Guinea, Liberia and Sierra Leone. Although in Guinea they are a minority group, constituting less than 5% of the total population (CIA (Central Intelligence Agency) 2017), the Kissi are the largest ethnic group of the "Forestiere" region of Guinea and in the Makona River triangle between Sierra Leone, Liberia and Guinea. Mostly working as agriculturalists, they speak the Kissi language—a Mel language of West Africa belonging to the Niger-Congo language family. They still practice their traditional ethnic beliefs despite many conversions amongst them to Christianity.

[2]The index case for Ebola came from Meliandou Village, only about 8 km away from Guéckédou, 6 months prior to this event.

J. N. Anoko
University of Rene Descartes Paris V La Sorbonne, Paris, France

D. Henry (✉)
Department of Anthropology, University of North Texas, Denton, TX, USA
e-mail: doug.henry@unt.edu

D. A. Schwartz et al. (eds.), *Pregnant in the Time of Ebola*, Global Maternal and Child Health,
https://doi.org/10.1007/978-3-319-97637-2_18

Fig. 18.1 The village of Yeredou. Photograph from Julienne Anoko

follow custom and perform a traditional "washing" of the body, which included a postmortem cesarean section. They feared a different kind of contagion would result should they not be allowed to follow tradition—that a widely feared curse upon the woman's home village would result, with disastrous implications for the broader community's reproductive health. Indeed, there was alarm that two curses could result from the death—one caused by the death of the pregnant woman herself, the other caused by burying her body with the also-deceased baby remaining in her womb. Out of their own concern with community health, both the family and community adamantly insisted on the traditional postmortem operation and ritual washing before burial.

The district medical authorities, however, charged with responding to the expanding Ebola epidemic, became resolute that the operation be prohibited, even in the relatively secure confines of the medical center, on the grounds that performing surgery on an Ebola-infected body so soon after death would put too many people at risk of contamination; virus-laden blood is judged far too contagious (World Health Organization 2014).[3] The family present at the time (the husband, a brother, and the mother of the dead woman) strongly disputed the Ebola diagnosis and denied that the deceased had participated in any funeral where she could have become infected. They categorically refused the secured burial with the baby remaining in the womb, and instead, demanded an immediate release of the body for the customary cesarean section and communal burial. They also refused to provide the district medical officers with a list of the deceased mother's contacts for tracing.[4]

[3]The World Health Organization estimates that at least 20% of new Ebola infections occur during burials of deceased Ebola patients.

[4]Contact tracing, along with surveillance, education, active case finding, safe burials, and safe transportation, have long been key components to breaking the chain of Ebola transmission. Contact tracing is particularly labor intensive, involving interviewing patients (or surviving family members) to find out all of a patient's contacts within the past 21 days, and sending outreach workers to contact them, and if need be—quarantine them to watch for the potential development of symptoms.

Fig. 18.2 Map of Guinea, showing the location of the Forestière region in the southeastern part of the country. Available from: https://en.wikipedia.org/wiki/File:Un-guinea.png

These events set the stage for a precarious conflict between the mandates of medical/humanitarian relief workers attempting to respond to Ebola virus disease (EVD) and the demands of local custom. Alerted to the potential for confrontation, the World Health Organization (WHO) called the first author (JNA), who was coincidentally on-scene as a consultant and had training in both applied anthropology and global health, to assess available options and to search for a potential solution that could respect epidemiological safety (i.e., excluding a cesarean section) while also respecting the concerns of the community.

18.2 Ebola and Confrontation in Guinea

By 2014, the Forestière region (Fig. 18.2) was becoming a well-documented locus of conflict between local communities and Ebola responders, with educators, clinic staff, and burial teams sometimes encountering distrust, resistance, and even open hostility to their actions. Communities began suspecting that Ebola responders were actually spreading the virus and reacted by cutting off roads to ambulances, stoning vehicles, and attacking and even killing intruding Red Cross workers (Anoko et al. 2014; Buchanan 2015; (Izadi 2014).

18.3 Pregnancy and the Meanings of Traditional Funerary Practice

During the EVD epidemic in Guinea, Liberia, and Sierra Leone, as well as during past breakouts of Ebola in Central Africa, funerals have become recognized as potential sites for increased risk of emergence, exposure, transmission, and spread of Ebola infection (World Health Organization 2014). Given the tensions mentioned above, first understanding how a community perceived the entirety of risks associated with death became crucial to understanding how to begin sensitively talking with them about possible flexibilities within funeral arrangements. Therefore, the beginning of the anthropological investigation in Guinea became centered around the goal of conceptually situating and understanding this current death within the broader meanings of traditional regional funerary practices.

In Africa, the events surrounding death are one of the key cultural events that families and communities face, involving burial, but also community discussions of status, identity, family bonds, social structure, moral standing, and more. During a funeral rite, it is customary for family and community members to pay last homage to the deceased through sometimes complex ceremonies and rituals which require direct contact with the body. These include ritual washing, dressing up, the application of make-up, caressing, kissing, transport, placing ritual objects into the coffin, and burial (Anoko 2014; Anoko et al. 2012, 2014; Epelboin 2009, 2014; Epelboin and Formenty 2011; Epelboin et al. 2008; Fairhead 2015). Funerals are spaces for negotiating family and social challenges and resolving tensions through the exchange of gifts and counter-gifts. They express grief and give homage to the deceased, but also celebrate the dead person's passage into the afterlife, and/or their return to their origins, as the deceased becomes reborn into the world of the ancestors. The body must be shown respect, so that the departed soul will feel at ease during this transition. This passage and rebirth also function as an opportunity for the living to make amends with the ancestors from any past grievances—this is to protect the living from potential punishment or rebuke from the ancestors (Van Gennep 1960; Malinowski 1922; Mauss 1926; Paulme 1954; Thomas 1982; Thomas 1985; Pradelles de la Tour 1996; Bloch 2005; Anoko et al. 2014; Godelier 2014).

For the researcher, it was crucial to spend initial time conducting a rapid assessment in and around the research area in order to understand death, particularly untimely death, from the point of view of the community where the pregnant woman lived. Several key informants knowledgeable of Kissi deaths were interviewed (elderly men and women, chiefs of the sacred forest, excision practitioners, traditional midwives, traditional healers, national authorities, etc.).

Among the Kissi ethnic group, every person who dies in the world of the living undertakes their last and long journey to the afterlife where they start a new life with the ancestors. During this journey, the deceased has to pay what are considered "border fees." Gifts and cash money from surviving family members are, thus, often put in the coffin to allow the dead to pay these fees while on this voyage. If the deceased are not properly sent off by the living, there will be concern that they will not reach the world of the ancestors and their spirit could come haunting those still alive.

Importantly for this research, though funeral ceremonies are often heavily scripted and complex, there can be flexibility within traditions. For example, if a body becomes badly burned during a farm-clearing accident, the family may decide to forgo the traditional washing and touching of the corpse, and instead, enlist relatives to pray over the body. Water can be sprinkled over the corpse without fully washing and touching it. If the deceased is male, dirt can be taken from nearby the corpse and put on the widow's head, instead of taking the "wash-water" and making mud of it (which is more typical). Local religious leaders can be enlisted for prayer. Money given to the deceased's family can be immensely appreciated, in that it can allow them to throw a "proper" large funeral, being able to feed many friends and relatives over separate 7- and 40-days ceremonies, thereby showing with much respect, even though actual touching may not be part.

As with other ethnic groups in Africa, the Kissi also have special elaborations in the case of a pregnant woman. In particular, the untimeliness of her death will arouse suspicions as to the ultimate cause, and the malevolent motivations of witchcraft will be suspected first. As one elder informant told the first author:

> *Here in the land of the Kissi, normal women don't typically die while pregnant- it is rare. If it does happen, people will suspect that the woman who died was a witch, who was trying to steal all the riches of the village, or of a few rich people. If a witch becomes pregnant, it becomes very difficult for them to give birth, because the riches they have swallowed through their witchcraft will weaken them during childbirth. That is what causes their death. Of course, other pregnant women also die from common diseases during childbirth.*

In cases where the deceased pregnant woman does become suspected of witchcraft, a family will feel shame by the sudden public scrutiny and will become anxious to prove her innocent. The same postmortem cesarean section that separates that fetus from the womb can also be used as part of a religious ceremony to assess potential involvement in witchcraft. So in addition to the motivations of relieving the woman of an eternal burden, of preserving the purity of the world of the ancestors, and of protecting the community from all bad repercussions, the Kissi have an additional motivation for the postmortem cesarean: to potentially relieve the family and community from the specter of witchcraft.

There are additional concerns. Upon the death of a pregnant woman, a community may fear additional violence from two fronts. From one, the death itself presents a kind of "pollution," resulting from the duality that the deceased is at once both one and two persons in the same body; family survivors, thus, have to deal with the symbolic discord where the burial of one corpse could contain another corpse (the fetus). On the second front, a fetus having not yet been born and properly socialized into the larger community is itself a problematic abnormality. Having something unnamed, unsocialized, and unbaptized moving into the world of the ancestors via the deceased woman's belly would represent significant symbolic violence, in that the fetus' lack of socialization would represent a special burden that the mother would have to carry for eternity, throughout the afterlife.

In the specific case in Guéckédou, these abnormalities together were functioning to provoke significant collective anxiety about the possible repercussions and misfortunes that the ancestors (now including the deceased woman) could bring. Most feared of all was that all women of childbearing age could be condemned to die while pregnant or during delivery, thus putting several generations of women in childbearing age at risk (see also Fairhead 2015). Women from the three villages associated with the deceased woman were asked to leave the area immediately. As the widow of the deceased woman said:

> *Our traditional religious leaders and elders women have asked girls and women of childbearing age to leave their homes to escape the ancestors' anger. They don't want them to be "contaminated" by the misfortune that my wife's death has brought to us... Both my family and family-in-law are responsible of this critical situation. We have to respect our traditions and bury her.*

It was now understood by the researcher that the community felt an indispensable need to "wash" the pollution away from the area, through a specific, complex ritual performed to avoid angering the ancestors which could reestablish the disturbed order, and above all to clean the community to prevent wider contamination. Yet before generating and assessing possible solutions, it was important to understand the steps and intentions of the specific, complex ritual performed that could reestablish order and "clean" the community to prevent wider contamination (Douglas 1971; Epelboin et al. 2008).

18.4 Photo Journal: Removing a Community Curse in Guéckédou

18.4.1 The Preparation of the Ritual

See Figs. 18.3, 18.4, 18.5, 18.6, 18.7, and 18.8.

Fig. 18.3 Individuals buying the ingredients for the ritual at a Guéckédou Marcket (l. to r. WHO staff member, Ministry of Health staff member and sages). Photograph from Julienne Anoko

Fig. 18.4 The family of the deceased and the village elder (prefectural director of education). Photograph from Julienne Anoko

Fig. 18.5 The subprefect, the mayor, and the prefectural director of education (dean of the family of the deceased). Photograph from Julienne Anoko

Fig. 18.6 The representative of the Ministry of Health and the zone's administrative manager. Photograph from Julienne Anoko

Fig. 18.7 The presentation of gifts to the widower by the Subprefect (in hat). Photograph from Julienne Anoko

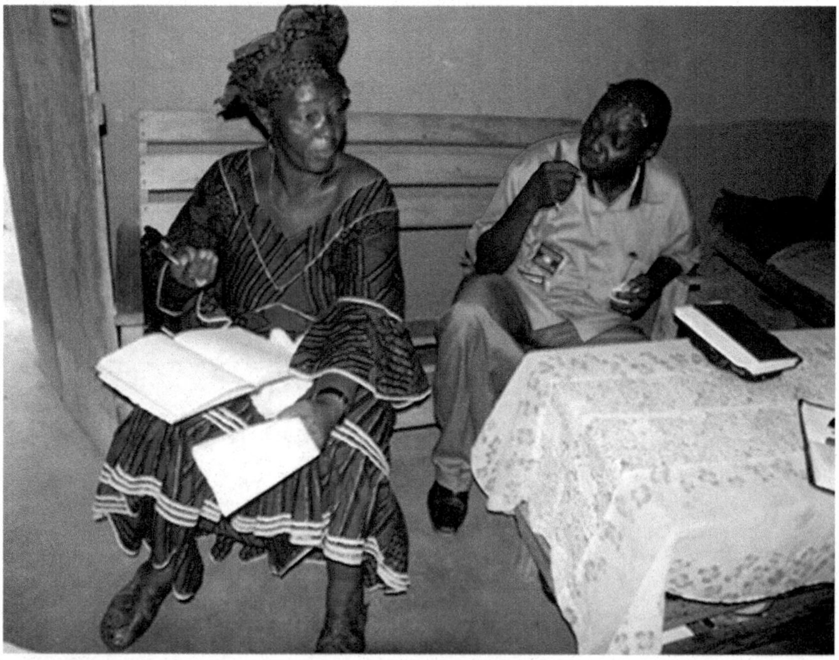

Fig. 18.8 The establishment of a list of the 81 contacts of the deceased between the wise and the Ministry of Public Health representative. Photograph from Julienne Anoko

18.4.2 Setting Up Groups for the Ritual

See Figs. 18.9, 18.10, and 18.11.

Fig. 18.9 The women's group. Photograph from Julienne Anoko

Fig. 18.10 The men's group. Photograph from Julienne Anoko

Fig. 18.11 The young male space. Photograph from Julienne Anoko

18.5 The Ritual of a "Sinister" Death

In Guinea, as in much of West Africa, the required involves the separation of the fetus from the mother's womb, through a surgery performed by either a society of initiated men or women, or in a medical facility. There are variations, even within the local "Forestière" region of Guinea. Sometimes only postmenopausal women, infertile women, or those who have witnessed the death of all their children may approach the corpse. Depending on the culture group, the mother and the fetus may be buried in completely separate graves and locations, or in separate graves in the same location (e.g., placed at the feet of the mother before being laid in a grave). Women may announce that a "sinister" death has occurred, but the men will immediately take the body away, saying that it is crucial to "forget" this death as soon as possible. The logic is that misfortune and curses may be chased away from the village by making the body seem to "disappear." The mother's body may be buried by being "thrown" roughly into a grave dug beside a termite hill (on the assumption that the corpse will become decomposed quickly by the termites), while the fetus will be buried on the bottom of a pool of still water, covered by protective plants carried by ritual specialists. In some Muslim communities in same region, the child must be extracted from the mother's womb through a surgery performed on the left side of the body, conducted by the either the husband, or by another person whom the mother had the right to marry (Al-Bostani 2010).

In a Kissi village, shortly after being informed of a pregnant woman's death, postmenopausal women in the village will begin howling, as one informant described it, "*like a hunted chimpanzee's cry*." Upon hearing this sound, men, children, and young women shut themselves in their homes. All women of childbearing age must leave the area immediately, to avoid being contaminated by the curse. They are only allowed back after the burial takes place and the contamination has been washed away. The village appears deserted except for the elderly women, who roam around the streets both singing and wailing.

Next, a female messenger carries the news of the death to neighboring villages, particularly those associated with the deceased woman's family. She is naked except for a cloth around her waist, her body smeared with dirt streaks, red and white. Once she arrives at a village, she will first circle around it three times without speaking to anyone, and then run screaming into people's houses, calling out the grievous news, knocking over baskets, and tearing at mats and clothes. Women surround her, to calm her by trying to get her to accept small gifts, such as handfuls of rice or salt. When calmed, the messenger responds to questioning from the elders and announces the deceased woman's name and village. When she is questioned further, she will add the detail, "*-and a child came out, with its feet forward.*" This will clue the elders in to the circumstances surrounding the situation. They may nod and say among themselves, "*Surely while the woman was pregnant she wished for the child to die out of hate for her husband; what else should we expect in such a situation? The omniscient ancestors have punished her by killing both her and the child.*"

Back in the original village, the older women begin preparation for the postmortem cesarean section. From the widower, they require a large sum of money, as well as chickens, palm oil, white cloth, rice, salt, kola nuts, and a goat. These goods will partly be used to pay the women, but also allow the preparation of a large communal feast to be held after the ritual, which then completes the "washing" of danger away from the village. The widower is dressed in dry leaves. The surgical procedure is led by a group of postmenopausal excision practitioners called locally "*Sokonö bémbélo*" (*Sokonö* refers to "circumcisers" and *Bémbélo* "a deceased pregnant woman") as well as by postmenopausal women alternately called *Touboulnö*, *Gbeguelnö,* or *Salilnö*, initiated into the sacred society (in general, the *Salilnö* are responsible for corpses from what are considered "sinister" deaths, such as a death in pregnancy, or a suicide). While the circumcisers can only be women, others involved in the ritual can be of both genders and operate according to the gender of the victim.

The older women will circle around the deathbed of the corpse, ritually sprinkling it with a decoction made of leaves. Then they move on to sprinkle the corpse. They then pick up the corpse and carry it quickly to a sacred area by the river, a place used by the women's society for ritual activities. The eldest woman, who is also the guardian of this sacred place, cuts open the belly with a knife to retrieve the fetus. The separation of the mother's fetus is carried out on the riverbank, and the mother's organs are examined by the specialists who are able to assess by their appearance if witchcraft was indeed involved. In the case that the deceased is found guilty of witchcraft, she will be said to have carried "wealth" in her womb along with the baby, and the women will return the money received from the widower without explanation. This act will cast shame upon the family of the diseased.

The corpse is then washed, stuffed with strips of cloth, and sewn back up. Some informants reported that the fetus should be buried on the right bank of the river, while the mother on the left bank. However, others reported that the fetus should be buried in the streambed, while the mother's corpse would be brought to the village and buried according to normal, traditional rituals.

Once the burial procedures begin, the widower is allowed to have his mourning clothes removed and will have his contamination washed away, using a decoction similar to that of the older women. The foliage that he wore is thrown away on the side of the village nearest the burial of the deceased woman. Upon return from the burial, the communal feast is prepared. All the villagers line up, and the *Touboulnö* serve food with a horn to each villager twice with the right hand and twice with the left hand. After the meal, a traditional specialist will purify the place where the woman died, as well as her husband and all the villagers gathered at this occasion, saying: "*Let such misfortune never be repeated again, may the crops yield good harvest, may women give birth to many children....*"

This ceremony lifts the curse, and life will return back to normal in the village. The husband will observe a mourning period for 40 days.

18.6 Rapid Community Participatory Involvement

This preliminary rapid ethnographic work achieved several results: (1) it introduced the lead author to important community leaders and began the process of involving the community in the search for mediation, (2) it contextualized the community's responses, so that their actions did not appear irresponsible or irrational, (3) it built trust between the anthropologist and the community under study, which enabled mediation to take place, and (4) it uncovered a probable transmission chain of EVD, despite the previously strong denials from the widower. Although the pregnant woman had not herself participated in a funeral, her mother-in-law did attend a communal burial of her best friend, who had died of unexplained causes 3 weeks before in the nearby village of Houndoni,[5] one of the villages most affected by EVD. When the mother-in-law returned, she brought with her the best friend's two children and entrusted them to her pregnant daughter-in-law for care. When the daughter-in-law later began showing EVD symptoms, she first consulted a traditional midwife, then a traditional healer in a neighboring village who, after several days of unsuccessful treatment, recommended a transfer to the Guéckédou Hospital.

When the death of the pregnant woman was announced, all the women in childbearing age followed their own concerns with community health and tradition and left the three villages to protect themselves from the curse. A group of initiated old women immediately gathered to prepare for the funeral rite, including the separation of the woman and the fetus. An order by the village elders was given to the family of the woman to remove the pollution caused by the curse according to the tradition and to reestablish the disturbed order. The widower and the family of the deceased became threatened with expulsion by some of the villagers, notably by the women and the youth. Facing such pressure from their own community, the family of the deceased became increasingly agitated and demanded that the hospital either release the body for a proper village burial, or that the cesarean section be conducted by the medical team there on site in the clinic. At the time of the anthropologist's entrance on-scene, the two sides had reached a tense and threatening impasse, as the hospital medical staff refused to conduct the cesarean section themselves, and the community as a whole rejected the option of the clinically safe burial in a vinyl body bag.

18.7 Mediating the Curse-Removal Ritual

The next task for the anthropologist was to search for flexibility inherent within the funeral traditions. At that time of increasing numbers of cases, while attempting to respond to local, national, and international pressure, the clinical staff offered none and were insistent that the secured burial protocol be followed. The anthropologist then organized several meetings with traditional practitioners, including circumcisers and society initiates of both genders, as well as older chiefs of the forest. Importantly, all shared memories of a specific reparation ritual is called the "*Wolilé*," or "appeasement," which could be done in cases where the circumstances surrounding a sinister death prevented the traditional burials. This reparation ceremony was held up by the community as a possibility to move forward, given the hospital's refusal of a postmortem cesarean section.

The *Wolilé* ritual "washes" the village from curses in seeking to obtain forgiveness from the ancestors for the disgrace committed. The deceased woman becomes buried outside the village, but is symbolically replaced by a dog. Taken into the forest, the animal is sacrificed, and the blood collected in a gourd. The remains of the animal are buried on the spot. A large amount of edible plants known to have purifying

[5]Houndoni is a small village that is 6 km from Gueckedou, an administrative center where the Ebola virus Outbreak was initially declared. The village was devastated by the epidemic beginning in December 2013—of a population of approximately 300 persons, 27 died after becoming infected with EVD.

properties are collected in the forest; a decoction is made from them and mixed with the blood of the sacrificed animal. Each villager is given this to drink with a wild animal horn. The remainder of the mixture is taken and sprayed on all the houses and their inhabitants. The *Salilnö* recites the following words: "*May God, the ancestors and the witches spare us from this evil and may all the village women be promptly released from the curse in a good condition. May the ancestors forgive us for having committed a disgrace, and protect us from all the evil and bad omens.*"

Most importantly to come from these meetings, an elder *Salilnö* who was familiar with the *Wolilé* ritual was located living in the village of Wonde, some 50 km away. WHO agreed to underwrite the costs associated with the repair, on the condition that the family agreed to follow the recommended clinical protocol for secured burial (with no cesarean section), and additionally, make an exhaustive list of contacts of the deceased. Suddenly, the *Wolilé* became a possibility, and the family of the deceased contacted the *Salilnö* to begin leading the repair ritual.

A special mission comprising the anthropologist, village elders, an elder of the dead woman's family (her uncle), a sociology student, and a WHO volunteer met with both the family and the religious authorities to agree on the specifics of the reparation ritual, under the condition that the deceased woman would still be buried with the baby in her womb. The agreement was settled after negotiating for a day, despite reluctance and protests from the women and the village youth. Given the exceptional nature of the situation, the ritual was organized on the seventh day after the death of the pregnant woman, thereby also functioning as the traditional "7 days" funeral ceremony.

The anthropologist asked the deceased family's formal written consent in order for the burial to take place. The consent below was drafted jointly by the deceased woman's husband and her oldest brother and reads:

> *The husband, Mr. [NAME] and the parents [NAMES] have made the decision that the deceased [NAME] be buried in her pregnant state (with the baby in the womb) at Nongoa, in a secured burial overseen by the Red Cross.*
>
> Signed: [Husband] [Mother] [Father]

According to tradition, the *Salilnö* had to get on his way (walking) immediately after receiving the request and immediately start the repair ritual as soon as he left his house. This phase is marked by the payment of "transition costs" to all rivers crossed on its way. These fees are to ask forgiveness and protection of the ancestors, and on the return route, the fees are aimed to thank them for their indulgence and magnanimity. Since it was the rainy season, he had to cross 75 rivers and backwaters. This required WHO to pay 1,500,000 GNF (US $215) at a rate of 20,000 GNF per passage. It also requires 450,000 GNF (US $64) in labor fees. The World Health Organization gave the anthropologist the task of making purchases for the preparation of the ritual. A goat (instead of a dog), four chickens, 11 m of white cloth, a traditional woven fiber mat, 70 kg traditional white rice, 10 L palm oil, a 25 kg bag of salt, and cola nuts are purchased. The widower, on his side, had to borrow a large amount of traditional rice flour. The total cost of the ritual was 2,652,500 GNF (US $379).

The viewing of the corpse happened on the third day after the death, overseen by two logisticians of the WHO. Burial teams from the Guinean Red Cross were supposed to have overseen this, but members stated that their own traditions and beliefs forbade them to see the body of a pregnant woman; they also feared risk of contagion of bad omens on women within their own families. Two cars were made available to the family to carry the body to the village. Finally, on the seventh day at night (because the body was now in an advanced state of decomposition, the family did not want the entire village to observe it), the funeral was held, under the discrete surveillance of police officers. It is the anthropologist's belief that family and community participation in planning the process, along with the signed consent of the family, plus very visible official concern about the community's health, helped to calm any residual reluctance or protests. Following improvised custom, the village both mourned and celebrated, in the presence of administrative authorities, the WHO team, and the anthropologist. The community thanked everyone involved with traditional songs of peace.

18.8 Conclusion

This case illustrates the value and potential role of anthropological methods as part of health teams responding to epidemiological crises. In Guéckédou, rapid, community-based ethnography allowed the World Health Organization to understand, respect, and respond to the local lay epidemiology, thereby averting a dangerous confrontation between a group of Guinean villages and a regional Ebola response team. Involving the community in the search for a solution helped identify sources of flexibility within tradition and deescalate violence, while empowering participants, establishing trust between communities and responders, and stemming further Ebola transmission within this chain of infection.

To our knowledge, this kind of utilization of a social scientist within a crisis situation seems to be unusual and not often replicated elsewhere. This oversight is unfortunate, we feel, and can (and in the case of the 2014 epidemic, did) have calamitous consequences. The skill set of an anthropologist to rapidly uncover so-called "lay epidemiology," how health risks and situations are perceived and appraised by lay people, and how those local perceptions trigger public responses has been well-documented and shown to be a crucial part of managing epidemics (Mesquita and Haire 2004; Smith et al. 2006). Lay epidemiology has also been shown to have a significant bearing on the public's evaluation of the plausibility of health information and health promotion messages (Davison et al. 1991) and on maladaptive responses like avoidance, impotence, confusion, anger, blame, paranoia, denial, and even the kind of violence experienced in Guinea, Liberia, and Sierra Leone (Hastings et al. 2004; Rains and Turner 2007). Unfortunately, lay epidemiology is often been considered irrational by health officials, as a barrier to public health, or something to blame for when the public disbelieves or fails to act on a "more rational" public health mandate (Allmark and Tod 2006; Smith et al. 1999). We hope we have shown that lay epidemiology is neither irrational nor irrelevant and that the same models of risk that shape community lay response to epidemic threat can be harnessed, if done correctly, to match the goals of public health. "Traditional songs of peace" dedicated to community health and order should be sung by everyone.

Acknowledgments The authors wish to express their gratitude to Heejin Ahn, Research/Administrative Assistant to Dr. Abramowitz (Center for African Studies), for translating the original French version of this paper written by the first author into English and Marie Ouendeno, translator and assistant to Dr. Anoko during the field mission in Guinea, and a WHO Volunteer.

References

Al-Bostani, A. A. (2010). *Le Guide du Musulman: Abrégé des principaux décrets religieux des juristes musulmans contemporains et notamment de l'Ayatollâh A.Q. Al-Khoî.* Retrieved October 20, 2017, from http://www.bostani.com/livre/le-guide-du-musulman.htm.

Allmark, P., & Tod, A. (2006). How should public health professionals engage with lay epidemiology? *Journal of Medical Ethics, 32*(8), 460–463.

Anoko, J. (2014). *Communication with rebellious communities during an outbreak of Ebola virus disease in Guinea: An anthropological approach.* Retrieved September 10, 2017, from http://www.ebolaanthropology.net/case_studies/communication-with-rebellious-communities-during-anoutbreak-of-ebola-virus-disease-in-guinea-an-anthropological-approach/.

Anoko, J., Epelboin, A., & Formenty, P. (2012). *Humanisation de la réponse aux fièvres hémorragique à virus de ebola et Marburg en RDC, Congo, Gabon, Angola 2003-2012: une approche anthropologique.* Retrieved December 11,

2017, from https://hal.archives-ouvertes.fr/hal-01090299/file/2014_07_30%20Anoko%20Epelboin%20Guinee%20 Rapport%20socioanthropo%20Ebola.pdf.

Anoko, J., Epelboin, A., & Formenty, P. (2014). *Humanisation de la réponse à la fièvre hémorragique Ebola: Une approche anthropologique en République de Guinée*. Rapports de mission mars-juillet 2014, Organisation mondiale de la santé (OMS).

Bloch. (2005). Where did anthropology go?: Or the need for 'human nature'. In M. Bloch (Ed.), *Essays on cultural transmission. LSE monographs on social anthropology* (pp. 1–20). Oxford, UK: Berg. ISBN 9781845202866.

Buchanan, E. (2015). *Ebola crisis: Red Cross workers attacked as virus conspiracies create panic in Guinea*. Retrieved September 20, 2017, from http://www.ibtimes.co.uk/ebola-crisis-red-cross-workersattacked-virus-conspiracies-create-panic-guinea-1488865.

CIA (Central Intelligence Agency). (2017). *The World Factbook*. Retrieved from https://www.cia.gov/library/publications/the-world-factbook/geos/gv.html.

Davison, C., Smith, G., & Frankel, S. (1991). Lay epidemiology and the prevention paradox—The implications of coronary candidacy for health education. *Sociology of Health and Illness, 13*(1), 1–19.

Douglas, M. (1971). *De la souillure. Essai sur les notions de pollution et de tabou* (A. Guérin, Trans.). Paris: Maspero.

Epelboin, A. (2009). L'anthropologue dans la réponse aux épidémies: science, savoir-faire ou placebo? *Bulletin Amades*, 78. Retrieved October 20, 2017, from http://amades.revues.org/index1060.html.

Epelboin, A. (2014). *Approche anthropologique de l'épidémie de FHV Ebola 2014 en Guinée Conakry*. [Rapport de recherche] OMS. p. 34.

Epelboin, A., & Formenty, P. (2011). Anthropologie sociale et culturelle et lutte contre les épidémies de fièvre Ebola et Marburgin. In L. Catherine & J. Guégan (Eds.), *Les maladies infectieuses émergentes: État de la situation et perspectives* (pp. 111–113). Haut Conseil de la santé publique La Documentation française, Collection Avis et rapports.

Epelboin, A., Formenty, P., Anoko, J., & Allarangar, Y. (2008). Humanisations et consentements éclairés des personnes et des populations lors des réponses aux épidémies de FHV en Afrique centrale (2003-2008). In *Humanitarian Stakes N°1. MSF Switzerland's Review on Humanitarian Stakes and Practices* (pp. 25–37). September 2008. Retrieved November 4, 2017, from www.msf-ureph.ch/sites/default/files/fichiers/humanitarian_stakes_no1.pdf.

Fairhead, J. (2015). *The significance of death, funerals, and the afterlife in Ebola-hit Sierra Leone, Guinea, and Liberia: Anthropological insights into infection and social resistance*. Retrieved September 7, 2017, from http://www.ebola-anthropology.net/key_messages/the-significance-of-death-funerals-and-the-after-life-in-ebola-hit-sierra-leone-guinea-and-liberia-anthropological-insights-into-infection-and-social-resistance/.

Godelier, M. (2014). *La mort et ses au-delà (Collectif)*. Paris: CNRS Éditions.

Hastings, G., Stead, M., & Webb, J. (2004). Fear appeals in social marketing: Strategic and ethical reasons for concern. *Psychology & Marketing, 21*(11), 961–986.

Izadi, E. (2014). Red Cross volunteers attacked in Guinea while trying to bury an Ebola victim. *Washington Post*. 24 September 2014. Retrieved October 11, 2017, from https://www.washingtonpost.com/news/to-your-health/wp/2014/09/24/red-cross-volunteers-attacked-in-guinea-while-trying-to-bury-an-ebola-victim/?utm_term=.b8658baa1056.

Malinowski. (1922). *Argonauts of the Western Pacific*. London/New York: G Routledge/E.P. Dutton.

Mauss, M. (1926). Rites funéraires en Chine in Marcel Mauss. *Extrait de la revue Année Sociologique, 1899*(2), 221–226. Retrieved March 1, 2016, from http://classiques.uqac.ca/classiques/mauss_marcel/oeuvres_2/oeuvres_2_14b/rites_funeraires_chine.html.

Mesquita, B., & Haire, A. (2004). Emotion and culture. In C. Spielberger (Ed.), *Encyclopedia of applied anthropology* (Vol. 1, pp. 731–737). Waltham: Academic.

Paulme, D. (1954). *Les Gens du Riz: Les Kissi de Haute-Guinée*. Paris: Librairie Plon.

Pradelles de la Tour, C.-H. (1996). Les morts et leurs rites en Afrique: À propos de Systèmes de pensée en Afrique noire 9, 11 & 13: Le deuil et ses rites. *L'Homme, 138*, 137–142.

Rains, S., & Turner, M. (2007). Psychological reactance and persuasive health communication: A test and extension of the intertwined model. *Human Communication Research, 33*(2), 241–269.

Smith, B., Sullivan, E., Bauman, A., Powell-Davies, G., & Mitchell, J. (1999). Lay beliefs about the preventability of major health conditions. *Health Education Research, 14*(3), 315–325.

Smith, R., Drager, N., & Hardimann, M. (2006). *The rapid assessment of the economic impact of public health emergencies of international concern*. Oxford: Oxford University Press.

Thomas, L. (1982). *La mort africaine. Idéologie funéraire en Afrique noire*. Paris: Payot.

Thomas, L. (1985). *Rites de mort, pour la paix des vivants*. Paris: Fayard.

Van Gennep, A. (1960). *The rites of passage* (M. B. Vizedon & G. L. Caffee, Trans.). Chicago: University of Chicago Press.

World Health Organization. (2014). *How to conduct safe and dignified burial of a patient who has died from suspected or confirmed Ebola or Marburg virus disease*. Retrieved September 10, 2017, from http://www.who.int/csr/resources/publications/ebola/safe-burial-protocol/en/.

19 Ebola-Related Complications for Maternal, Newborn, and Child Health Service Delivery and Utilization in Guinea

Janine Barden-O'Fallon, Paul Henry Brodish, and Mamadou Alimou Barry

19.1 Introduction

Guinea was one of the three most affected countries in the West African outbreak of the Ebola Virus Disease ("Ebola" hereafter). During the largest, longest, and deadliest outbreak of Ebola, Guinea recorded a total of 3814 Ebola cases and 2544 deaths between March 2014 and the declared end of the outbreak in Guinea on December 29, 2015 (CDC 2016). While the number of Ebola cases in Guinea was much lower than in the other most affected countries of Liberia (10,678) and Sierra Leone (14,124), the mortality rate was much higher, at 67%, compared to 45% and 28%, respectively (CDC 2016). The weekly caseload of Ebola in Liberia peaked at 442 cases in early October 2014 and in Sierra Leone at 570 cases in early December 2014 (WHO 2015a). In Guinea, it fluctuated throughout the epidemic, peaking at 171 cases during the final week of 2014 (WHO 2015a). Ebola in Guinea disproportionally affected adults (persons aged 15–44 years and 45+ years were three and five times more likely to be infected, respectively, than were children and infants aged <14 years) and individuals living in certain regions of the country, such as Conakry, Guéckédou, Macenta, and N'Zérékoré (WHO 2015a).

Notwithstanding the severity of the direct effects of Ebola on the health of the population, experts were also concerned with the indirect effects of the disease on the access to health care services and potential for mortality from other health conditions. Diseases such as malaria, pneumonia, and typhoid in many cases went untreated due to Ebola-related effects on health systems such as closures of clinics, patients who were afraid to visit facilities for fear of contracting Ebola, or because patients with Ebola-like symptoms were turned away from care (Delamou et al. 2014; Paye-Layleh and DiLorenzo

---

J. Barden-O'Fallon (✉)
Carolina Population Center, University of North Carolina at Chapel Hill, Chapel Hill, NC, USA

Department of Maternal and Child Health, Gillings School of Global Public Health, University of North Carolina at Chapel Hill, Chapel Hill, NC, USA
e-mail: bardenof@email.unc.edu

P. H. Brodish
Carolina Population Center, University of North Carolina at Chapel Hill, Chapel Hill, NC, USA

M. A. Barry
John Snow, Inc., Chapel Hill, NC, USA

D. A. Schwartz et al. (eds.), *Pregnant in the Time of Ebola*, Global Maternal and Child Health,
https://doi.org/10.1007/978-3-319-97637-2_19

2014; Strong and Schwartz 2016, 2019). In some areas, certain important services, such as vaccinations, were suspended due to the lack of personal safety equipment, the inability to conduct real-time tests for Ebola, or insufficient staff to meet the additional health burdens caused by the disease. Projections of measles outbreaks occurring 6–18 months after vaccination disruptions from Ebola were estimated to result in 2000 to 16,000 additional deaths in the three most affected countries (Takahashi et al. 2015). The exacerbation of weaknesses in the health system led to calls for more investment in general system strengthening and the development of "resilient" health systems (CBC News 2014; Barbiero 2014; Menendez et al. 2015; Schlein 2014).

At the time, information on the "indirect" effects of Ebola at the health-facility level was spotty across the three most affected countries. One study using data on inpatient admissions rates and surgery in the neighboring Sierra Leone found a 70% drop in the median number of admissions between May and October 2014 among 40 surveyed facilities (Bolkan et al. 2014). The authors also found a similar 50% drop in the median number of surgeries during the same time period and estimated that 35,000 sick Sierra Leoneans would be excluded from inpatient care from the onset of the Ebola epidemic through the end of 2014 if the low admissions rates continued (Bolkan et al. 2014).

In Guinea, declines in service utilization in some areas were also being reported. For example, the number of women giving birth in facilities with a skilled birth attendant (SBA) in the hardest-hit prefectures of N'Zérékoré and Conakry fell by 87% from the period of October through December 2013 to the period July through September 2014 (Jhpiego 2015). It was also reported that the hospital in Kissidougou was seeing only 12–15 patients a day at the end of September 2014, compared to a typical pre-epidemic intake of 200–250 patients per day (Paye-Layleh and DiLorenzo 2014). There were indications that women were avoiding services due to stigma and fear. Statistics from Matam Maternity Hospital in Conakry showed a decrease in attendance during the advanced stage of the epidemic, with 123 patients for the July through September 2014 quarter, compared to 760 patients for the same quarter in 2013 (Delamou et al. 2014).

Guinean women and children may have been especially vulnerable to worsening health care conditions. Prior to the Ebola epidemic, a key indicator of maternal health, the maternal mortality ratio or MMR was 650 per 100,000 live births (2012) (WHO 2015b), one of the highest in the world. Facility-based deliveries were occurring for only 41% of births, and the contraceptive prevalence rate was 6%. In addition, child health indicators also ranked low: the under-five mortality rate was 101 per 1000 live births, with malaria as the top cause of death, and full immunizations were received by only 36.5% of children (WHO 2015b; ICF International 2015). Furthermore, the health system lacked the necessary elements to control an epidemic, including a strong health care workforce. Prior to the epidemic, health workforce density in Guinea (numbers of physicians, nurses, midwives, dentists, pharmacists, and psychiatrists) was less than 1.5 per 10,000 population, and per capita government expenditure on health was 9 US$ per year (WHO 2016). Also, there were only three hospital beds per 10,000 population (WHO 2014). The health systems infrastructure was similar in the other two affected countries, Liberia and Sierra Leone. Thus, Guinea could not afford declines in maternal, newborn, and child health (MNCH) care service delivery or utilization, nor in its already meager health care workforce and infrastructure.

Based on early reports, it was assumed that levels of service delivery for routine MNCH care had fallen precipitously over the course of the Ebola outbreak in Guinea. However, no data were available to assess the extent to which this may have been true across geographic zones and different prefectures, or to inform planning and resource allocation in response to the challenge created by the sudden and severe Ebola outbreak.

The authors were involved with a rapid assessment effort requested and funded by the United States Agency for International Development (USAID)/Guinea to better understand how the delivery and utilization of routine MNCH services may have been affected by the extraordinary strain placed on the health system and its client population by the Ebola outbreak in Guinea. At the time the assessment was undertaken, no such information was available to guide the response of local officials and donors concerned about MNCH services. The assessment was commissioned as a way to provide a quick yet systematic look at the status of MNCH service delivery and utilization in selected facilities during the period immediately before recognition of the Ebola outbreak, compared to the conditions in Guinea approximately 1 year later. The following section describes the assessment—its design, methodology, and selected results. Full details of the assessment and its comprehensive results can be found at Barry et al. (2015) and Barden-O'Fallon et al. (2015).

19.2 Rapid Assessment: Study Design

A retrospective facility records review combined with structured interviews with health personnel was conducted from January to February 2015. Twelve prefectures and three city districts in the Conakry region were selected for the assessment and included all four geographic zones (Upper, Lower, Middle, and Forest) of Guinea. The Ebola status of prefectures over the 9-month period from March through December 2014 was reviewed and three categories were constructed: "Active" (prefectures that reported Ebola cases throughout the time period beginning in March 2014); "Inactive" (no diagnosed Ebola cases reported during the surveyed time period); and "Changing" (classification of prefecture varied—status classified as active, changing, and/or inactive during the surveyed period). Active and changing zones were oversampled compared to inactive zones. The 12 prefectures and three city districts were categorized as follows: Active (Guéckédou, and Conakry City Districts of Dixinn, Ratoma, and Matam), Changing (Boffa, Coyah, Dabola, Dalaba, Faranah, Fria, Kissidougou, N'Zérékoré, and Siguiri), and Inactive (Mamou and Mandiana) (Fig. 19.1).

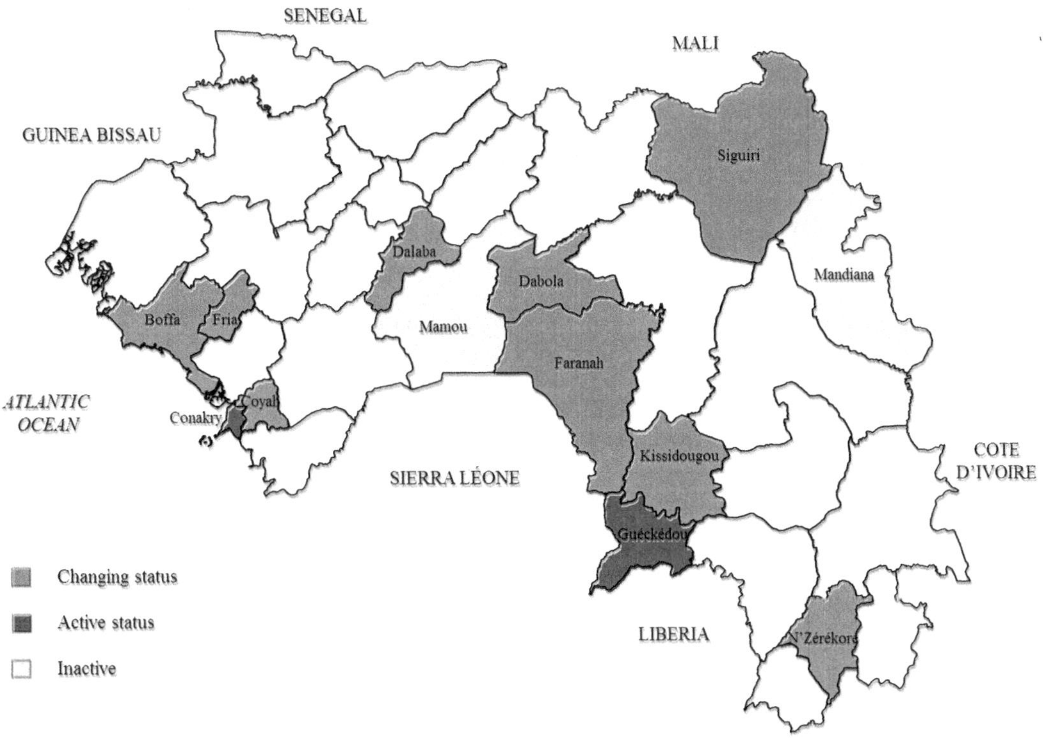

Fig. 19.1 Map of Guinea showing selected prefectures

Public facilities visited for data collection within selected prefectures and city districts included the central hospital (regional or district) and two health centers serving the nearby communities. Similarly, in Conakry, the medical centers from three city districts were selected, along with two health centers located in the areas around each of these medical centers. Private clinics serving the same catchment populations as the surveyed public facilities (within and outside of Conakry) were visited for interviews with health workers when possible. Health District Offices of selected prefectures and city districts were also visited for record review.

The record review included a number of maternal and child health services, as well as overall outpatient visits, and spanned the months from October 1, 2013 to December 31, 2014 (for a total of 15 months). Private clinics were not included in the record review due to the differences in record keeping. Brief, structured interviews were conducted with health-facility directors or managers and up to two providers of MNCH services. These interviews helped to document provider shortages/absenteeism, changes to service delivery, infection control practices, and provider safety concerns, among others. Field work training, pretesting, and data collection were conducted in collaboration with StatView International, a private research firm based in Conakry. Six data collection teams, each including a physician and three quality assurance teams, were deployed. EPI Info 7 was used for data entry. Consent was obtained prior to initiating the interviews from all facility directors and health care providers. Permission to undertake the survey was granted by the Guinea Ministry of Health and Public Hygiene (MOHPH). The activity was reviewed and received an exemption from the Institutional Review Board (IRB) at the University of North Carolina at Chapel Hill.

19.3 Analysis

Data on MNCH service delivery and staffing were collected by month and aggregated into quarters. For each indicator, the percent change in the median number of services recorded between Quarter 1 (October 1–December 31, 2013) and Quarter 5 (October 1–December 31, 2014) was calculated and disaggregated by type of public facility, i.e., hospital/city district medical center ("hospital" hereafter) or health center. The choice of the same annual quarters for comparison (i.e., the months of October through December one year apart) helped to account for seasonal variation in service provision. The Wilcoxon-signed rank test was used to test for statistically significant differences between the indicator medians calculated for the two quarters (Q1 and Q5). Trends over time using all 15 months of data were plotted for select indicators that had statistically significant differences in Quarter 1 and Quarter 5 median numbers. Information from the facility director and provider interviews was analyzed and is reported as univariate distributions (frequencies). Responses to open-ended questions were reviewed and categorized by content and frequency. Selected portions of this qualitative information are reported to add insight into interpretation of the quantitative results. Information from the health districts was collected for the period April 2014 through December 2014 and is presented as univariate statistics (Fig. 19.2).

19.4 Results

A total of 45 public facilities were visited for record review, including 16 hospitals and 29 health centers. Most facilities were classified as being in Ebola "Changing Status" areas (n = 26), while thirteen were in Ebola Active areas and six were in Ebola Inactive areas. The median number of services provided for selected indicators in the comparison time periods (October–December 2013

Fig. 19.2 Data collection team reviewing health services records at the regional hospital in Mamou, February 2015. Photograph by Mamadou Alimou Barry

and October–December 2014) are shown in Table 19.1. Findings show that there were declines across a number of MNCH services. The median number of outpatient visits experienced a statistically significant decline between Quarter 1 and Quarter 5 in surveyed hospitals (31%) and health centers (6%). Findings for maternal health indicators also suggest declines in routine services between the two time periods, though only one indicator, testing for HIV infection among pregnant women in surveyed hospitals, showed a significant decline of 51% between the two quarters. On the other hand, many indicators for child health services showed significant declines between the two time periods. The median number of Pentavalent 1 and 3 vaccinations given at health centers declined 18% and 32%, respectively. The numbers recorded for Pentavalent 1 and 3 vaccinations at surveyed hospitals also showed declines, but these were not statistically significant. The number of children-under-five seen for diarrhea and acute respiratory infection (ARI) showed a large decrease over the 1-year period in both hospitals (60% for diarrhea and 58% for ARI) and health centers (25% and 23%, respectively).

The trends varied by Ebola classification. A few of the statistically significant trends are disaggregated by Ebola classification and presented in Figs. 19.3, 19.4, and 19.5. Figure 19.3 shows that in surveyed hospitals located in Ebola Active areas, there was an increase in outpatient visits between Quarter 1 and Quarter 2, followed by steady declines throughout the outbreak period of 2014. Hospitals in the Ebola Inactive and Changing Status zones showed the largest declines in outpatient visits in the final Quarter of 2014, when the outbreak was at its peak.

As detailed in Fig. 19.4, declines in Pentavalent 3 vaccinations given at surveyed health centers occurred in all three Ebola zones. The decline in the number of Pentavalent 3 vaccinations at these health centers began at the time of Ebola case detection, after Quarter 2, and was especially evident in Ebola Active zones.

Figure 19.5 shows the trends for diarrheal cases in children under five in the surveyed hospitals. The trends indicate that the median number of children receiving care for diarrhea declined in hospitals in all three Ebola status zones. However, the smallest amount of change was noted in the Ebola Changing Status zones.

Our assessment also found that stockouts of key medications were fairly common throughout the study period. Almost one-half of all surveyed facilities reported at least one stockout of Oral

Table 19.1 Median number of services at 45 public health facilities between Quarter 1 (Oct.–Dec. 2013) and Quarter 5 (Oct.–Dec. 2014): Facility record data, Guinea, 2015

Indicator	Median number Quarter 1	Median number Quarter 5	% change	Missing
Outpatient visits				
Number of outpatient visits:				
Hospitals	1355	930	−31**	2
Health centers	1223	1147	−6**	1
Maternal health				
Number of pregnant women tested for HIV:				
Hospitals	112	55	−51*	7
Health centers	255	246	−4	15
Number of pregnant women seen for first antenatal care visit:				
Health centers	337	295	−13	1
Number of pregnant women seen for third antenatal care visit:				
Health centers	245	205	−16	2
Number of births:				
Hospitals	303	281	−7	0
Health centers	100	69	−31	4
Child health				
Number of pentavalent 1 vaccinations given:				
Hospitals	504	316	−37	13[a]
Health centers	259	212	−18**	3
Number of pentavalent 3 vaccinations given:				
Hospitals	353	320	−9	13[a]
Health centers	244	167	−32**	4
Number of cases of diarrhea in children under 5 years:				
Hospitals	34	14	−60**	0
Health centers	16	12	−25**	0
Number of cases presenting ARI for children under 5 years:				
Hospitals	98	41	−58**	2
Health centers	108	83	−23**	0

$^{**}p < 0.01$; $^{*}p < 0.05$

[a]This indicator was not originally intended to be collected at the hospital level, but the information was found at three sites and is included here for illustrative purposes

Rehydration Salts (ORS) for the treatment of diarrhea (49%); in 18% of facilities, the stockout(s) occurred only after the Ebola outbreak began, and not in the preceding months. Similarly, 40% of the surveyed facilities reported at least one stockout of Cotrimoxazole for treating ARI over the 15-month period, and in 20%, the stockout occurred only after the Ebola outbreak began. Overall, these findings suggest that stockouts of common medications are ongoing concerns and were a greater challenge for one in five of the surveyed facilities in the wake of the Ebola crisis (Fig. 19.6).

Information obtained from the district health information system showed that, as of April 2014, 11 of 177 facilities reported closures (6%) and 15 (8%) reported service suspension (Table 19.2). There were fewer reported absences of doctors and auxiliary nurses/midwives as compared to nurses. From April to December 2014, the surveyed facilities reported a total of 44 Ebola cases in health care workers, with 22 deaths; N'Zérékoré was most affected by deaths in health care workers. Large numbers of health care workers had been trained in Ebola infection prevention and control; 50 health workers from surveyed facilities had been transferred to work in an Ebola Treatment Unit. Overall, closures,

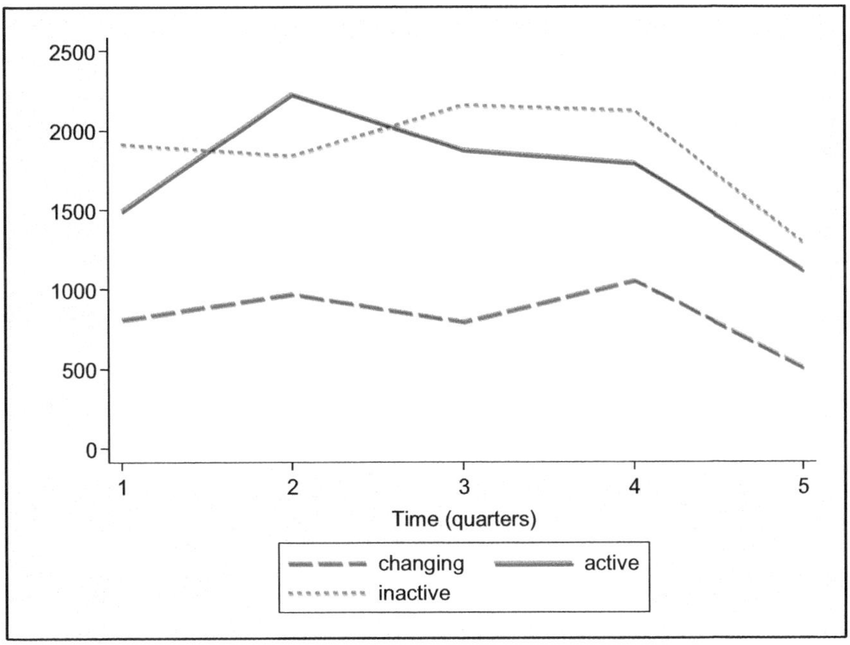

Fig. 19.3 Trend in median number of outpatient visits in hospitals by changing, active, and inactive status zone

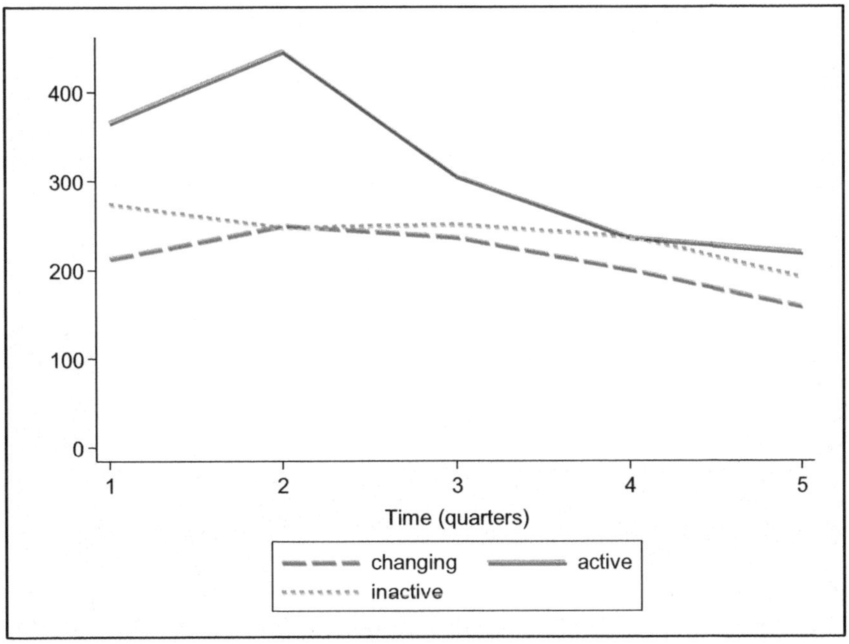

Fig. 19.4 Trend in median number of Pentavalent 3 vaccinations in health centers by changing, active, and inactive status zone

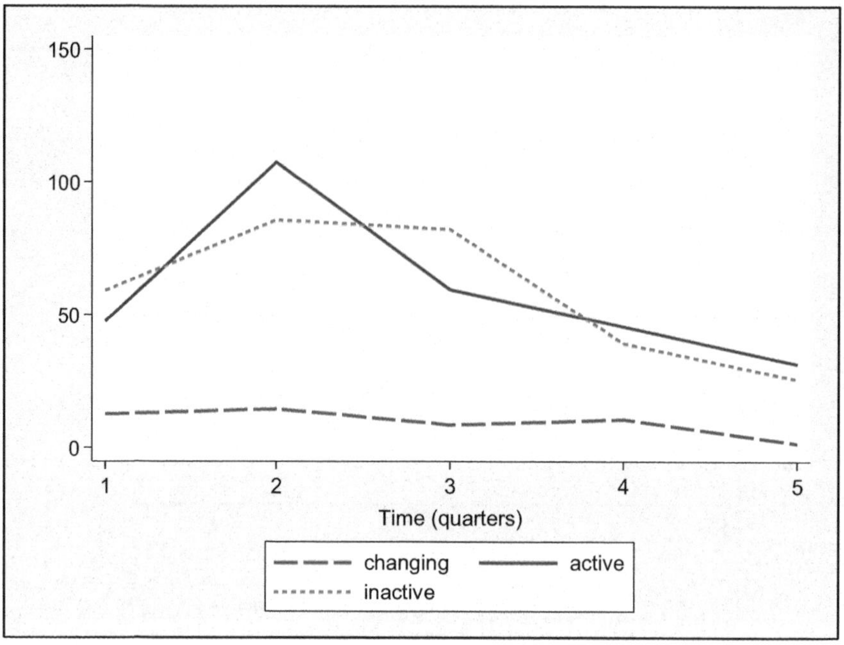

Fig. 19.5 Trend in median number of diarrheal cases in children under five in surveyed hospitals by changing, active, and inactive status zone

Fig. 19.6 Quiet day at the pediatric ward of Guéckédou Prefectoral Hospital, February 2015. Photograph by Jack Hazerjian

Table 19.2 Information from health records at 15 surveyed Health District Offices, April 2014 through December 2014, Guinea[a]

Indicator	Health districts (N = 15)	
	Facilities (N = 177)	Most affected health district(s)
Number of facilities reporting:		
Closures	11	N'Zérékoré
Service suspension due to the Ebola epidemic	15	N'Zérékoré
Absence of health care workers:[b]		
Doctors	2	Guéckédou
Nurses	13	N'Zérékoré
Auxiliary nurses/midwives	3	Guéckédou
Number of health care workers:		
Infected with Ebola	44	N'Zérékoré, Coyah, Matam
Died from Ebola	22	N'Zérékoré
Trained in infection prevention and control[c]	1293	–
Transferred to an Ebola Treatment Unit	50	Guéckédou, Siguiri

[a]Continuous monthly reporting by facilities. Individual facilities and workers may be counted more than once for the same indicator if included in separate monthly reports
[b]Missing (n = 1)
[c]Missing (n = 6)

service suspension, and provider absences appeared to have been localized, with N'Zérékoré having borne the brunt of these issues over the course of the epidemic in 2014.

Information collected at the Health District Offices also indicated that the reporting of data through the routine health system was not negatively affected by the Ebola epidemic. During this period, over 90% of the 177 facilities responsible for reporting information to the Health District Office consistently reported service data. The trend held true even for Health Districts categorized as Ebola Active, though the level of reporting was a few percentage points lower. This trend is in contrast to reports from Liberia, where in July 2014 only one half of health facilities were reporting data to the Health Ministry (Paye-Layleh and DiLorenzo 2014).

Additional information on the effects of Ebola on service delivery and utilization was obtained through interviews with 62 health directors/managers ("directors") and 117 providers from 64 public and private facilities. The main findings from interviews with health directors indicate that provider absences during the Ebola crisis varied from 8% to 21% at their respective facilities, depending on staff type. Only a small proportion of health directors reported that their facility had to reduce hours of operation (13%), close (15%), or suspend services (13%). Just over one quarter of directors (26%) reported that stockouts of medications or supplies increased during the Ebola crisis. Almost 20% of surveyed directors believed there had been an increase in complications due to delays in seeking care. While 76% reported a likelihood that community members had concerns about the safety of care at their facility, most also reported improved infection control practices at the facility: 68% of those interviewed reported having personally received training on Ebola infection control and 85% reported that other providers at the clinic received such training. Eighty-five percent of directors reported having a screening/triage process for Ebola case identification, and 92% reported the implementation of additional infection control measures. Furthermore, most facility directors reported access to the supplies needed to prevent infection transmission, such as gloves (87%) and disinfectant (82%). Access to personal protective equipment was reported by 69% of the directors interviewed.

Providers interviewed at the facilities included doctors, nurses, midwives, and other health care professionals. The majority reported learning of Ebola early in the outbreak period; only 15% reported learning of the outbreak after June 30, 2014. Seventy percent of providers reported that they had been trained in Ebola infection control and good practices for risk reduction. Nevertheless, the majority (80%) expressed concerns about the safety of service provision at their facility. Only a small percentage of surveyed providers reported that services had been suspended due to Ebola (services suspended included maternity care/delivery, small surgeries, nighttime consultations, and pediatric services). The vast majority of providers reported making changes to their service provision practices, including wearing gloves and other personal protective equipment; washing hands; taking patient temperatures; implementing safety protocols; and disinfecting the facilities. Just over one-quarter of the providers (28%) reported an increase in complications due to patient delays in obtaining services. Frequently cited complications included those in delivery/maternal care, severe anemia, and advanced malaria, especially among children (Fig. 19.7).

Health directors and service providers were asked a number of open-ended questions about negative reactions they experienced from family members, friends, or within their communities, in their role as health professionals during the Ebola crisis. Thirty-six percent of the MNCH service providers interviewed, and 46% of the health directors interviewed, reported experiencing negative reactions and provided the following examples of how those reactions were expressed or experienced:

"*We are considered carriers of the Ebola.*"
"*My daughter keeps away from me. My mother is suspicious.*"
"*The (local) vendor forbids me to enter his shop.*"
"*Our children are no longer accepted by their friends.*"
"*My neighbors don't trust me.*"
"*Rocks have been thrown into the hospital's courtyard.*"
"*My family told me to leave Guéckédou and not step foot in the hospital there.*"
"*My children have asked me to stop working.*"
"*I have been threatened to be burned by those in my community.*"
"*There was talk of destroying my clinic.*"
"*(Health) personnel are accused of spreading Ebola for money.*"

Fig. 19.7 Empty pediatric outpatient ward, Donka National Hospital, Conakry, February 2015. Photograph by Jack Hazerjian

19.5 Discussion and Conclusions

The first assessment of MNCH service delivery and utilization during the Ebola crisis in Guinea was based on data abstraction and brief structured interviews at a nonrandom selection of health care facilities and district offices in 12 prefectures and three city districts of Guinea from January to February 2015. Though the facility sample was not necessarily representative of all facilities, the assessment in Guinea revealed a number of important findings.

First, there was an overall decline in service utilization, as seen in the median numbers of outpatient visits in facilities participating in the assessment. The decline was especially notable in hospitals. This finding suggests that hospital service utilization may have declined more from Ebola-related stigma or misunderstandings than did service utilization at health centers. Findings from a study on the effect of Ebola on malaria treatment in Guinea, which used stratified random sampling, reported similar levels of declines in all-cause outpatient visits (11%), cases of fever (15%), and patients treated with oral (24%) and injectable (30%) antimalarial drugs (Plucinski et al. 2015). Also, child health services, such as vaccinations and treatment of common childhood illnesses, were especially affected by the Ebola epidemic. These findings suggest that parents may have been reluctant to bring children for health care services out of fear of contracting Ebola. The reluctance to seek care out of fear and stigma may have had significant health consequences (Strong and Schwartz 2016, 2019). Delays in treatment may be one of the factors, along with ineffective contact tracing and low quality of care despite investments in training and equipment, that Guinea suffered the highest case fatality rate of the three most affected countries; a rate that did not improve during the course of the outbreak. Twenty-eight percent of interviewed health providers in our study also reported an increase in complications (in delivery, maternal care, severe anemia, and advanced malaria among children) due to patient delays in obtaining services. This proportion would likely have been even higher if we considered only the reference health facility (Hospital), which saw the most cases of complications. The observed declines in service utilization during the epidemic phase may have led to an increase of disease complications and mortalities.

A recent systematic review of 22 studies found decreases in utilization of MNCH services, particularly in cesarean sections and facility-based deliveries, in Guinea, Liberia, and Sierra Leone (Brolin Ribacke et al. 2016). Likewise, there were declines in antenatal care, postnatal care, and family planning services, as well as declines in child health services such as vaccinations. Declines were also evidenced in HIV/AIDS and malaria services, general hospital admissions, and major surgeries. While few hard numbers were available, estimates from these various studies were that indirect (non-Ebola) morbidity and mortality substantially increased as a result of these declines in utilization, particularly for outcomes determined by the continuing availability of routine reproductive and MNCH services. This review also found that it was more the case that uptake/utilization of health services faltered, rather than their provision during the course of the outbreak. Again, a fear of nosocomial transmission and mistrust of authorities likely played a role, as did factors such as misinformation on whether facilities were open and offering services.

Health care providers reported a number of improved infection control behaviors as a result of the Ebola outbreak, including more frequent handwashing and the use of disinfectants. It is reasonable to conclude that these changes in health practices and the wide implementation of safety protocols and measures contributed to the prevention of infection at health facilities. However, though tremendous efforts were made to train managers and health providers in infection control and prevention at the district and health-facility levels, approximately 30% of staff at both levels had not received training by the time of this assessment. This gap made providers more vulnerable to the risk of infection and less able to ensure adequate and secure health services.

Our results suggest that the negative effects on service availability (such as reduced hours, closures, and service suspensions) were likely to be regional and/or facility-specific. While this indicates a weakening of the system, it does not point to system failure. A systematic review of 13 articles supplemented with expert interviews of key country personnel evaluated the link between the Ebola virus disease (EVD) outbreak and health systems in the three affected countries (Shoman et al. 2017). The review analyzed country health systems in terms of the World Health Organization (WHO) health system building blocks, which identify six building blocks for a well-functioning health system: (1) health workforce, (2) health financing, (3) information and research, (4) medical product and technologies, (5) leadership and governance, and (6) service delivery (WHO 2017). Findings from the systematic review corroborate several of the findings in our study. For example, the reviewed articles indicated that a lack of a skilled health workforce contributed to the spread of the epidemic, and that health care workers often lacked knowledge of basic infection and control measures. Communities often stigmatized and rejected health care workers, who experienced adverse psychological effects. Stigma toward health care workers also led patients to avoid seeking needed care due to fear of contracting EVD. The review also mentioned that some patients had to travel great distances over difficult routes in order to access what services were available, pointing to the need to locate treatment centers and diagnostic laboratories close to the sites of EVD transmission and more rapid deployment of health care services to these sites during epidemics. The review also noted a lack of epidemiologic surveillance data collection and analysis, resulting in delayed detection and identification of EVD. Compromised and interrupted service delivery meant that routine health services were not offered at adequate levels. Interviews with experts suggested that morbidity and mortality from common causes such as malaria, diarrhea and pneumonia (among infants), childbirth (from nonhospital births), and measles (due to interrupted vaccination services) far exceeded those from EVD. As suggested by our study, the review concluded that community-level factors were important to the inability to contain EVD, and that patients need to be placed at the center of health care systems in order to maintain high levels of service uptake. As the capabilities and responsiveness of the health systems grow through sustained investments, communities will gain trust in these systems and utilize the available services.

The data gathered and analyzed for our study suggest that the routine health information system (RHIS) was still functioning with regard to primary data collection and data reporting from the health-facility level up to the district level. This is an indication that at the time of the outbreak the RHIS was functional with the transmission of health information between health facilities and district offices, though it was not sensitive to emergency disease outbreaks. As a result, the RHIS did not trigger an alert in the system of the outbreak and rapid spread of Ebola. In the event of future outbreaks and epidemics, an integrated RHIS and community-level disease surveillance system is needed to prompt a timely response. To this end, since 2016, Guinea's Ministry of Health with the assistance of the USAID-funded MEASURE Evaluation project and other partners, such as RTI International, eHealth, and Jhpiego, have worked together to build an integrated data management system. The new platform uses the District Health Information Software 2 (DHIS2) and provides real-time access to the data at upper levels of the information system, namely at the regional and central levels, for validation, interpretation, and decision-making. In addition, a disease surveillance module was customized and is being added to the functional DHIS2 platform to help manage epidemic-prone diseases and facilitate rapid disease detection and notification. The system modifications are expected to expedite outbreak investigations and response.

Finally, there were a number of negative social consequences to being a health care worker during the Ebola epidemic. These consequences reflected the serious concerns for safety, epidemic control, and maintenance of family and social networks. Despite the many real concerns, health staff showed their commitment to their profession by continuing to work and provide services as best as possible. Such commitment to service is an encouraging foundation on which to continue to build a strong health system.

References

Barbiero, V. K. (2014). It's not Ebola...it's the systems. *Global Health: Science and Practice, 2*(4), 374–375.

Barden-O'Fallon, J., Barry, M. A., Brodish, P., & Hazerjian, J. (2015). Rapid assessment of Ebola-related implications for reproductive, maternal, newborn and child health service delivery and utilization in Guinea. *PLoS Currents Outbreaks*. https://doi.org/10.1371/currents.outbreaks.0b0ba06009dd091bc39ddb3c6d7b0826.

Barry, A., Barden-O'Fallon, J., Hazerjian, J., & Brodish, P. (2015). *Rapid assessment of Ebola-related implications for reproductive, maternal, newborn and child health service delivery and utilization in Guinea*. Chapel Hill: MEASURE Evaluation.

Bolkan, H. A., Bash-Taqi, D. A., Samai, M., Gerdin, M., & von Schreeb, J. (2014). Ebola and indirect effects on health service function in Sierra Leone. *PLOS Currents Outbreaks*. https://doi.org/10.1371/currents.outbreaks.0307d588df619f9c9447f8ead5b72b2d.

Brolin Ribacke, K. J., Saulnier, D. D., Eriksson, A., & von Schreeb, J. (2016). Effects of the West Africa Ebola virus disease on health-care utilization—A systematic review. *Frontiers in Public Health, 4*, 222. https://doi.org/10.3389/fpubh.2016.00222.

CBC News. (2014). *Ebola outbreak compared to wartime by Doctors Without Borders*. Retrieved October 4, 2017, from http://www.cbc.ca/news/health/ebola-outbreak-compared-to-wartime-by-doctors-withoutborders-1.2737367.

CDC. (2016). *Ebola outbreak in West Africa—Case counts*. Retrieved January 27, 2017, from https://www.cdc.gov/vhf/ebola/outbreaks/2014-west-africa/case-counts.html.

Delamou, A., Hammonds, R. M., Caluwaerts, S., Utz, B., & Delvaux, T. (2014). Ebola in Africa: Beyond epidemics, reproductive health in crisis. *The Lancet, 384*(9960), 2105. https://doi.org/10.1016/S0140-6736(14)62364-3.

ICF International. (2015). *DHS Guinea 2012 via STATcompiler*. Retrieved April 6, 2015, from http://www.statcompiler.com/.

Jhpiego. (2015). *See the data: The impact of Ebola in Guinea on maternal health and family planning services*. Retrieved October 8, 2017, from https://www.jhpiego.org/field-notes/see-the-data-the-impact-of-ebola-in-guinea-on-maternal-health-and-family-planning-services/.

Menendez, C., Lucas, A., Munguambe, K., & Langer, L. (2015). Ebola crisis: The unequal impact on women and children's health. *The Lancet, 3*(3), e130. https://doi.org/10.1016/S2214-109X(15)70009-4.

Paye-Layleh, J., & DiLorenzo, S. (2014). *Ebola hits health care access for other diseases*. Associated Press. Retrieved October 8, 2017, from https://www.usatoday.com/story/news/world/2014/11/04/ebola-care-other-diseases/18468781/.

Plucinski, M. M., Guilavogui, T., Sidikiba, S., Diakité, N., Diakité, S., Dioubaté, M., et al. (2015). Effect of the Ebola-virus-disease epidemic on malaria case management in Guinea, 2014: A cross-sectional survey of health facilities. *Lancet Infectious Diseases, 15*(9), 1017–1023. https://doi.org/10.1016/S1473-3099(15)00061-4.

Schlein, L. (2014). General health systems damaged by Ebola in West Africa. *Voice of America News*. Retrieved April 6, 2015, from http://www.voanews.com/content/general-health-systems-damaged-by-ebola-inwest-africa/2542540.html.

Shoman, H., Karafillakis, E., & Rawaf, S. (2017). The link between the West African Ebola outbreak and health systems in Guinea, Liberia and Sierra Leone: A systematic review. *Globalization and Health, 13*, 1. https://doi.org/10.1186/s12992-016-0224-2.

Strong, A., & Schwartz, D. A. (2016). Sociocultural aspects of risk to pregnant women during the 2013-2015 multinational Ebola virus outbreak in West Africa. *Health Care for Women International, 37*(8), 922–942. Retrieved October 5, 2017, from http://www.tandfonline.com/doi/full/10.1080/07399332.2016.1167896.

Takahashi, S., Metcalf, C. J. E., Ferrari, M. J., Moss, W. J., Truelove, S. A., Tatem, A. J., et al. (2015). Reduced vaccination and the risk of measles and other childhood infections post-Ebola. *Science, 347*(6227), 1240–1242. https://doi.org/10.1126/science.aaa3438.

WHO. (2014). *World health statistics 2014 report*. Global Health Observatory (GHO). Retrieved February 9, 2017, from http://www.who.int/gho/publications/world_health_statistics/2014/en/.

WHO. (2015a). *Ebola situation report—1 July 2015*. Retrieved July 6, 2015, from http://apps.who.int/ebola/current-situation/ebola-situationreport-1-july-2015.

WHO. (2015b). *Guinea*. Retrieved April 6, 2015, from http://www.who.int/countries/gin/en/.

WHO. (2016). *Health workforce—African health observatory*. Guinea Statistical Factsheet. Retrieved February 9, 2017, from http://www.aho.afro.who.int/profiles_information/images/3/3b/Guinea-Statistical_Factsheet.pdf.

WHO. (2017). *The WHO health systems framework*. Retrieved February 9, 2017, from http://www.wpro.who.int/health_services/health_systems_framework/en/.

Strong, A., & Schwartz, D. A. (2019). Effects of the West African Ebola epidemic on health care of pregnant women—Stigmatization with and without infection. In D. A. Schwartz, J. N. Anoko, & S.A. Abramowitz (Eds.), *Pregnant in the time of Ebola: Women and their children in the 2013-2015 West African Ebola epidemic*. New York: Springer. ISBN 978-3-319-97636-5.

Part IV

Sierra Leone

Nowhere to Go: The Challenges of Caring for Pregnant Women in Freetown During Sierra Leone's Ebola Virus Epidemic

20

Gillian Burkhardt, Elin Erland, and Patricia Kahn

20.1 Introduction

Médecins Sans Frontières (MSF)-Doctors Without Borders has been deeply involved in the fight against Ebola for decades and was the first organization to respond when the index case of what became the West Africa outbreak was initially identified in Gueckedou, Guinea, in early 2014. During the unprecedented epidemic that followed, MSF built and operated 11 Ebola Treatment Centers (ETCs) and treated nearly 5,100 patients in Guinea, Sierra Leone, and Liberia. One of these centers, which opened in Freetown, Sierra Leone, in January 2015, was the first of its kind: a Maternity Ebola Treatment Center (METC) designed and equipped to provide Basic Emergency Obstetric Care (BEmOC) and staffed with trained midwives, nurses, and obstetricians/gynecologists (OB/GYNs), although without surgical capacity.

In previous Ebola outbreaks, pregnant women were reported to have extremely high mortality rates (89–100%), although this figure was based on very limited data (Bebell and Riley 2015). However, emerging evidence during the West African epidemic suggested that pregnant women could survive Ebola infection at much higher rates. This evidence, along with the shear scope of the epidemic, was the rationale for MSF to undertake the METC initiative. Another factor was the hope that an METC could address medical providers' underlying fear of caring for pregnant women during the time of the Ebola outbreak (Erland and Dahl 2017)—fear which may have contributed to the decreased number of women delivering in facilities during the epidemic (Jones et al. 2016; Dynes et al. 2015; UNFPA 2015). These concerns were understandable given that many hospitals lacked the equipment, setup, and/or training to prevent transmission and provide safe care in an Ebola context and also lacked capacity to

G. Burkhardt
Médecins Sans Frontières (Operational Center Barcelona), Freetown, Sierra Leone

Department of Obstetrics and Gynecology, University of New Mexico, Albuquerque, NM, USA

E. Erland
Médecins Sans Frontières (Operational Center Barcelona), Freetown, Sierra Leone

Telemark Hospital, Skien, Norway

P. Kahn (✉)
Médecins Sans Frontières USA, New York, NY, USA
e-mail: Patricia.Kahn@newyork.msf.org

D. A. Schwartz et al. (eds.), *Pregnant in the Time of Ebola*, Global Maternal and Child Health,
https://doi.org/10.1007/978-3-319-97637-2_20

rapidly test pregnant women so that those without Ebola could be admitted as regular patients. The specific goals for the METC were, therefore, to provide safe, dedicated obstetrical services for pregnant women with confirmed Ebola virus disease (EVD) and to triage pregnant women with suspected EVD, thereby hopefully improving overall outcomes for these women. Ultimately, it was the process of triaging and caring for pregnant women (including management of obstetrical complications) while awaiting Ebola test results that proved to be the biggest challenge for MSF field teams.

Managing pregnancy in the midst of this epidemic was extremely complex. In addition to the reasons mentioned above, another complicating factor was that the symptoms of normal labor, which often include abdominal pain, vaginal bleeding, nausea, and vomiting, as well as classic danger signs such as fever and headaches, overlap with the case definition of EVD[1] (see Fig. 20.1). Consequently, differentiating between EVD and normal pregnancy, but especially Ebola and complications of pregnancy, became a vicious circle. Failure to diagnose EVD in a pregnant woman placed families, health workers, and communities at risk for exposure to highly infectious blood, amniotic fluid, and other bodily fluids during delivery. On the other hand, screening all pregnant women for EVD potentially placed them at risk by delaying access to obstetric care—which could be life-threatening in some cases, such as those requiring emergency cesarean sections.

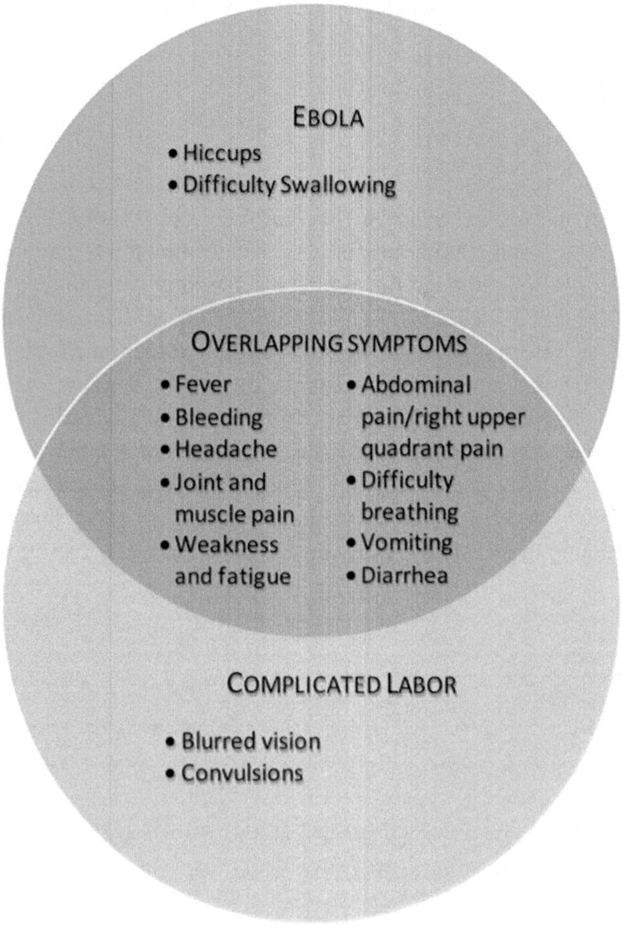

Fig. 20.1 Overlap between symptoms of complicated labor (danger signs) and the WHO case definition of Ebola virus disease

[1]WHO case definition for Ebola virus disease includes fever plus three of the following symptoms: headache, vomiting, anorexia/loss of appetite, diarrhea, lethargy, stomach pain, aching muscles or joints, difficulty swallowing, breathing difficulties, hiccup or inexplicable bleeding (WHO 2014a).

This chapter describes the experiences of the first two authors (GB, EE) as frontline medical workers (OB/GYN and midwife, respectively) at the MSF METC in Sierra Leone during the later phase of the outbreak. It also portrays how our team navigated the challenges of balancing the two sides of the 'safety versus rapid access to care' dilemma, especially in caring for pregnant women suspected of having Ebola. Lastly, through this lens, we address some broader consequences of the Ebola epidemic on maternal health in Sierra Leone.[2]

20.2 The Maternal Health Crisis in Sierra Leone and the "Four Delays"

When Ebola struck Sierra Leone, it made the country's extremely bad health care situation even worse. Maternal mortality statistics released by the World Health Organization (WHO) just prior to the outbreak showed that Sierra Leone was in very last place among all countries worldwide: 1100 maternal deaths per 100,000 live births—equivalent to a lifetime risk of 1 in 21 women dying during pregnancy or childbirth (WHO et al. 2014). Yet even these dismal figures represented a 52% improvement since 1990, reflecting the struggles of a country still rebuilding from its decade-long civil war that ended in 2002 and left the health care system with enormous gaps. A 2008 survey reported that the Sierra Leonean Ministry of Health had only 95 doctors, including only seven registered obstetricians, serving a population of over five million people (Kingham et al. 2009)—one of the lowest doctors-to-population ratios in the world (Milland and Bolkan 2015). Five years later, the Demographic Health Survey (Statistics Sierra Leone--SSL and ICF International, 2014) found that only 60% of pregnant women in Sierra Leone delivered with a skilled provider, that just over half (54%) delivered in a facility, and that home deliveries conducted with traditional birth attendants were a prominent feature of maternal health care in the country (Statistics Sierra Leone (SSL) and ICF International 2014). These findings highlighted the enormous unmet need for skilled delivery, and especially emergency obstetric services for the estimated 15% of pregnant women who have a complication during childbirth (WHO et al. 2009).

To help reduce mortality and provide care for this very underserved population, MSF had inaugurated its flagship Maternity Hospital, Gondoma Referral Hospital (GRC), in Bo District in 2008. This program tackled the three most common delays contributing to maternal mortality—delay in seeking care, in reaching care, and in receiving care (Thaddeus and Maine 1994)—by establishing a waiting home for pregnant women and providing 24/7 access to ambulances and Comprehensive Emergency Obstetric Services (CEmOC), which have the capacity to provide cesarean sections and blood transfusions on top of BEmOC services. In 2012, an MSF survey found that maternal mortality in Bo District had decreased by 61% since the program's launch, despite limited access to specialized care elsewhere in the district (Médecins Sans Frontières 2012).

The arrival of Ebola virus in Sierra Leone introduced a fourth delay for pregnant women: delays in accessing routine emergency obstetric care, because of new bottlenecks caused by the unprecedented outbreak. Due to the overlap between symptoms of Ebola and those of complications in labor, some women in labor presenting at health facilities had symptoms that could potentially indicate Ebola. Before these women could be admitted to a health care facility, the national protocol required them to have negative Ebola laboratory test results from two separate blood samples taken 48 hours apart, with the second sample drawn at least 72 hours after onset of symptoms. Women testing negative were then admitted to appropriate hospitals to manage the pregnancy, while those who tested positive were transferred to an ETC.

Many pregnant women were impacted by this new delay. The MSF program at GRC was also affected: once Ebola began spreading in Sierra Leone, MSF could not continue running the maternity unit safely, and therefore, made the extremely painful decision to close it down (Caluwaerts and Kahn 2019).

[2] All names of patients have been changed to protect patient privacy.

20.3 Understanding the Fourth Delay: Fatima's Story

Fatima, a 13-year-old girl pregnant with her first child, had been in labor for over two days in mid-March, 2015, when an ambulance brought her to the MSF METC in Kissy, Freetown. Her contractions had started at home, a rural village in the Port Loko district. After laboring for more than 24 hours without signs of imminent delivery, Fatima's relatives brought her to the Port Loko Government Hospital, a facility the community viewed with fear due to the stigma associated with hospitals during the Ebola epidemic. Fatima was admitted, but her labor failed to progress normally. Unfortunately, there was no doctor at the hospital who could perform the required cesarean delivery, so staff called for an ambulance to transport her to Freetown, over two hours away.

By the time the staff called Sierra Leone's Ebola Command Center, which dispatched all ambulances during the epidemic, Fatima had developed a fever. Because she was in labor and had symptoms of abdominal pain, vaginal bleeding, and nausea—symptoms also included in the case definition of EVD—the Command Center transferred her to the MSF METC in Kissy, where she was admitted to the labor unit designated for patients with suspected EVD. The MSF team believed that her symptoms stemmed not from Ebola, but from complications of a prolonged, obstructed labor and that she needed a cesarean section, which was not available in the METC. Nevertheless, staff collected a blood sample to test her for Ebola, administered antibiotics to treat her presumed chorioamnionitis (an intra-amniotic bacterial infection, and the more likely reason for her fever given her prolonged labor), and waited for the test results.

Six hours after arriving at the METC, and almost three days after her labor began, Fatima's blood results came back from the reference laboratory: negative for EVD. Because both her labor and the symptoms mimicking EVD had begun 72 hours ago, staff at the Princess Christian Maternity Hospital (PCMH) in Freetown, where she could obtain a cesarean section, agreed to accept Fatima with only one negative test instead of the standard two consecutive negative tests 48 hours apart.

After the METC's mandatory discharge shower in chlorinated water, Fatima was transferred to PCMH, where she waited another six hours for surgery before delivering a stillborn baby by cesarean section in the middle of the night. She remained at PCMH for treatment of her infection and was found to have a vesicovaginal fistula, a sequela of her prolonged obstructed labor. At the time, surgeons trained to repair such fistulas were no longer operating in Sierra Leone due to the Ebola outbreak. The staff at PCMH added Fatima's name to list for future fistula repair surgery and discharged her from the hospital with a preventable new medical condition that would seriously impact her life, unable yet again to access the surgical treatment she required.

In addition to personifying the impact of Ebola-related delays in receiving care, Fatima's experience also illustrates other factors that contribute to maternal morbidity and mortality in Sierra Leone. As a young girl she had undergone female genital mutilation (FGM),[3] which is associated with negative health effects including obstetric complications (Bjälkander et al. 2013). Furthermore, Fatima was only 13, although her pregnancy was not all that unusual in Sierra Leone: in 2013, the country's teenage pregnancy rate ranked among the ten highest in the world, with 28% of girls aged 15–19 years either pregnant or already having given birth at least once, and 10% of women becoming mothers by age 16 (Neal et al. 2012). There is ample evidence that the youngest girls (those under 16) have the highest health risks associated with adolescent pregnancy (Neal et al. 2012). One of these risks is obstructed labor, which in turn can lead to obstetric fistula.

[3]In Sierra Leone, the prevalence rate of FGM among women aged 15–49 years exceeds 90% (Yoder et al. 2013). FGM is common practice as part of the initiation ceremony of the Bondo society, a powerful all-women-led group, signifying that a girl has become a woman in her community.

Pregnancies in this age group are also associated with a higher rate of adverse neonatal outcomes (Conde-Agudelo et al. 2005). Sierra Leone has the world's third highest estimated perinatal mortality rate, with 30.8 intrapartum stillbirths and neonatal deaths on the first day of life for every 1000 deliveries (Save The Children 2014). Sadly, Fatima's child can be counted among these stark statistics.

20.4 Establishing the Maternity Ebola Treatment Center in Kissy, Freetown

While the outbreak was ongoing in Sierra Leone, there was little data available about its impact on access to care, including for pregnant women. One retrospective study published in 2016 compared periods of 10 months during the outbreak versus 12 months prior to its start (Jones et al. 2016) and found that the uptake of Antenatal Care (ANC), Postnatal Care (PNC), and facility-based deliveries all decreased significantly during the epidemic (by 18%, 22%, and 11%, respectively), while maternal deaths and stillbirths increased by 34% and 24%, respectively. Data from a modeling study (UNFPA 2015) estimated an average 22% increase in maternal deaths and 25% in newborn deaths for the period May 2014–April 2015.

MSF was acutely aware of the issues revealed in this study and that there were likely many women facing dire situations like Fatima's. Thus, in August and September 2014, MSF staff conducted an exploratory mission to identify the immediate maternal health needs, with a focus on Freetown. They found that services for pregnant women were generally available within the public sector in Freetown, but with two significant gaps. First, as seen in Fatima's case, there was a potentially life-threatening new delay in accessing obstetric care: testing laboring women for Ebola in a timely manner, especially at the height of the epidemic. By the time of the exploratory mission, many facilities were able to draw blood samples for Ebola testing, but insufficient laboratory capacity meant that it often took 24–48 hours for results to be returned—a critical time lapse for a woman in labor. Second, many pregnant women who met the EVD case definition and were awaiting Ebola test results were kept for days in "holding centers" that provided little or no care; others were sent directly to ETCs without prior testing at a holding center, even though many ETCs lacked procedures for managing pregnancy. It was, therefore, decided that the METC would focus on filling these gaps by providing a treatment center where pregnant women with either confirmed or suspected Ebola could obtain emergency obstetrical care (Fig. 20.2). Unfortunately, it could not address another critical gap: provision of cesarean deliveries, a key element of CEmOC services, since surgery of confirmed or suspected EVD patients was deemed too risky for both mothers and care providers.

The first step in creating the METC was the opening of a standard ETC in the Kissy Methodist Boys High School in Freetown, on January 8th, 2015 (In April, when schools reopened, the METC moved to the Police Training Station III in Hastings). In parallel, MSF updated its new guidelines on Ebola and pregnancy, based on the most recent experience at other MSF ETCs. These guidelines covered areas such as safe conduct of deliveries and called for minimizing invasive procedures, including episiotomies, vacuum deliveries, and manual removal of placentas, to protect staff and other patients (Caluwaerts et al. 2015; Caluwaerts and Kahn 2019). The guidelines also stipulated how to set up an ETC and manage patient flow in ways that meet the needs of pregnant women.

In keeping with these recommendations, the METC was designed with separate areas for confirmed and for suspected EVD patients, both with individual labor rooms (see Figs. 20.3 and 20.4).

Upon arrival at the METC, pregnant women with suspected EVD were triaged, isolated, and cared for in the non-confirmed area, and blood was drawn and sent for laboratory testing for Ebola.

Much thought also went into designing other aspects of the METC layout (see Fig. 20.5)—for example, deciding where to place the nursing station, where staff would don their PPE, how to monitor women in labor and allow staff to enter the high-risk area as quickly as possible in case of emergencies, and how to incorporate private rooms for deliveries. Additional nursing stations were placed

Fig. 20.2 Aerial view of the METC in Kissy, Freetown. Published with permission of © Doctors Without Borders/Médecins Sans Frontières 2018. All Rights Reserved

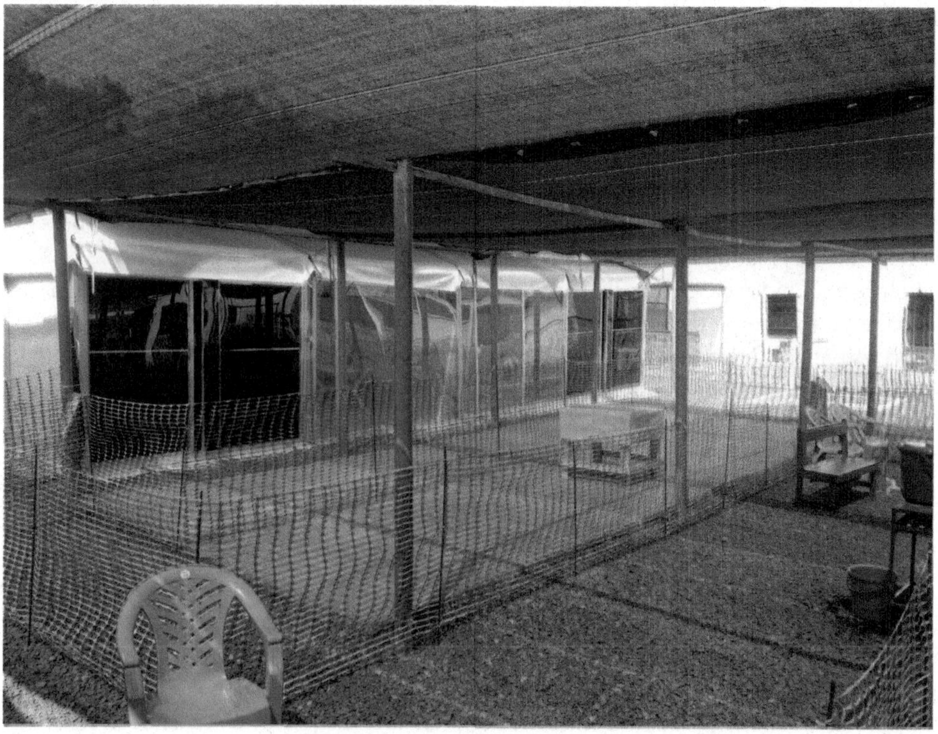

Fig. 20.3 Outside the delivery rooms for suspected cases of Ebola at the METC in Kissy, Freetown. Published with permission of © Doctors Without Borders/Médecins Sans Frontières 2018. All Rights Reserved

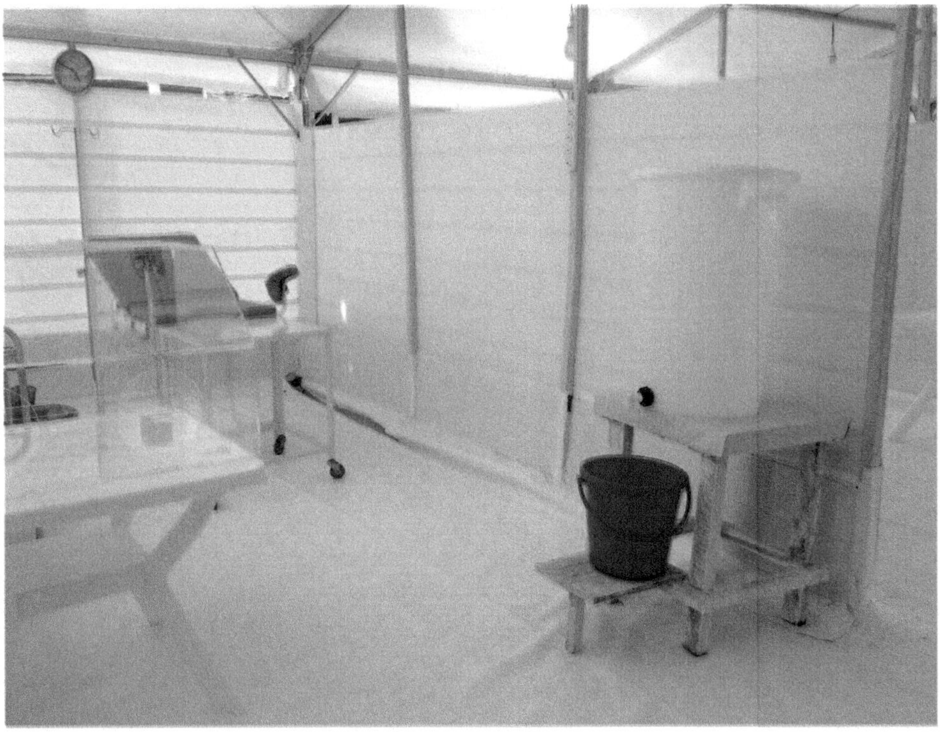

Fig. 20.4 A delivery room at the METC. Published with permission of © Doctors Without Borders/Médecins Sans Frontières 2018. All Rights Reserved

close to the delivery rooms and stocked with supplies and medications, such as misoprostol to control bleeding after delivery, antihypertensives to manage high blood pressure, and magnesium for rare events such as eclampsia. A neonatal resuscitation table was established for managing newborns requiring support at the time of delivery. Recognizing that providing hands-on nursing care to a pregnant woman throughout her labor would be practically impossible for staff, who would have to be wearing full PPE, a video system was installed to help monitor patients around the clock. With these adaptations, the MSF METC team of local and expatriate midwives was better prepared to manage obstetric emergencies.

The majority of obstetric deaths globally are caused by hemorrhage, sepsis, unsafe abortion, or hypertensive diseases of pregnancy (Say et al. 2014), which in their most severe form present with symptoms that can easily be confused with Ebola, as discussed and shown earlier (Fig. 20.1). However, taking probability into account, pregnant patients with these symptoms were more likely to be experiencing common complications of childbirth than to have EVD, as shown by a modeling study that calculated the incidence in Sierra Leone of EVD versus other pregnancy conditions (chorioamnionitis, HELLP syndrome, placenta abruption, septic abortion, bleeding placenta previa, ectopic pregnancy) and febrile illnesses (including thyroid, pyelonephritis, and Lassa fever) that mimic some Ebola symptoms (Deaver and Cohen 2015). This model suggested that only 1.5% of pregnant women with symptoms fitting the EVD case definition would actually have EVD, while most of the remaining 98.5% would have pregnancy-related conditions. But this analysis, as well as much of the evidence on managing pregnancy and Ebola, only became available after the epidemic. While the outbreak was ongoing, the lack of evidence left care teams to grapple with the reality that there were no good solutions for a pregnant woman who presented to the METC with complications.

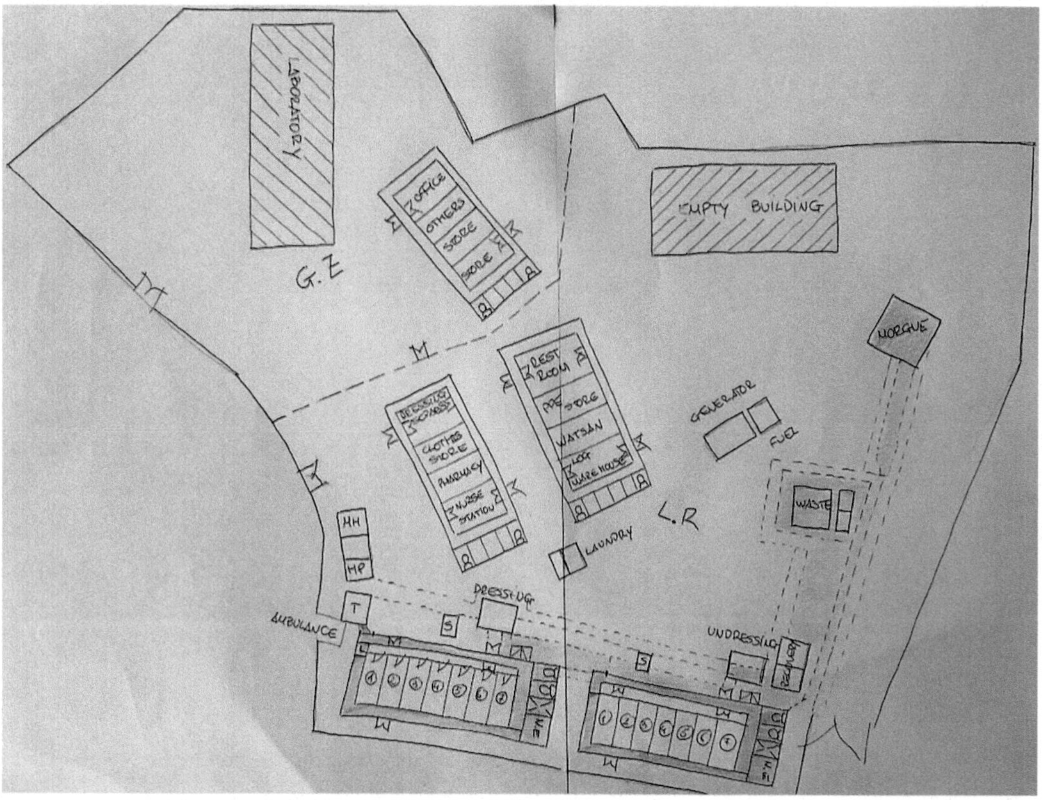

Fig. 20.5 Layout of METC in Hastings, Freetown (the second home of the METC). Published with permission of © Doctors Without Borders/Médecins Sans Frontières 2018. All Rights Reserved

20.5 Zainab and Mariama: A Tale of Two Eclamptics

Zainab was 21 years old and pregnant at term with her second pregnancy when she arrived at the METC in early April 2015. A routine ANC the previous week had found her antenatal course to be uncomplicated. The night before her arrival she began to have seizures. The family called 117, the emergency hotline established through the Command Center, but was told that all ambulances were busy and they should call a private vehicle. Zainab presented to the METC semiconscious and with a temperature of 37.9 °C, which increased to 38.1 °C over the next two hours. The staff initially wanted to transfer her to another facility, since their suspicion of Ebola was low despite her fever (which in rare cases can also accompany eclamptic seizures). But other facilities would presumably not accept her without a negative Ebola test, which takes time—and she needed to be stabilized quickly for presumed eclampsia. Zainab was, therefore, admitted to the METC and treated for eclampsia with magnesium sulfate (to prevent and control eclamptic seizures) and blood pressure medications—a difficult treatment under any circumstances since it requires frequent monitoring, but far more difficult for staff wearing PPE. She spontaneously went into labor on the afternoon of her arrival and very quickly delivered a healthy baby boy unattended, since the staff were unable to dress in PPE fast enough to assist. The placenta was delivered normally and Zainab was given oral misoprostol[4] to reduce her risk of postpartum hemorrhage. Because of her fever, two blood samples were take 48 hours apart and tested for Ebola virus; both samples returned negative. Over the next five days, her blood pressure improved and she was discharged with a happy and healthy infant.

Mariama was 16 years old with her first pregnancy and arrived at the METC by ambulance one evening in April 2015, referred from Command Central as a suspected Ebola patient. She had no known Ebola virus exposure in her community and was not on any contact list. According to her family, Mariama had started having convulsions a few hours earlier at home, and her only other prior symptoms had been swollen feet for the previous month and several days of a swollen face. Examination revealed a semiconscious woman with normal temperature, severely elevated blood pressure, and contractions. She had the classic symptoms and history of eclampsia and was in early labor. The Command Center sent her to the METC, and in keeping with Sierra Leone's policy, she was therefore admitted. Mariama was given standard treatment for eclampsia as well as medications to control her blood pressure, but because she was only semiconscious, she was combative and removed multiple intravenous solutions (IVs). Standard practice in an ETC is to avoid multiple IV placements in order to safeguard staff from potential needle injuries, but the staff repeatedly donned their PPE and replaced her IV so that she could receive the eclampsia treatment she urgently needed. The next morning, she delivered a healthy baby boy, but her placenta did not deliver, despite the use of standard medications to facilitate the last stage of labor. Multiple attempts to deliver the placenta were unsuccessful, and it remained firmly attached to the uterus. The medical team requested approval to transfer Mariama immediately to a surgical facility where the placenta could be removed in an operating theater (not available at the METC), as they feared her condition would otherwise continue to worsen. However, no facility would accept Mariama until she had two negative Ebola tests. So the METC team monitored her and provided supportive care while awaiting her test results. On the morning of her third day in the METC, her placenta still undelivered, she began to hemorrhage. She was given multiple rounds of medications to assist with the delivery of the placenta and to control the bleeding, all to no avail. She died later that morning. The first and second Ebola tests, the latter performed on the day she passed away, were both negative.

The MSF team did not believe either Zainab or Mariama had Ebola, nor did their symptoms meet the WHO criteria for Ebola. Nevertheless, without a negative Ebola test 72 hours after the onset of symptoms, there was nowhere to send them for treatment of eclampsia—a difficult condition to manage even in a non-Ebola context. A recent analysis of all pregnant women sent and admitted to the METC found that 9 of the 40 (22.5%) who proved to be Ebola-negative did not meet the WHO case definition for EVD (Lledo et al. 2016). While clearly there was an ongoing public health risk of potentially introducing Ebola into health facilities not equipped to manage EVD, the MSF medical team still felt caught in a system broken by Ebola and encumbered by the restrictions it imposed. It was especially devastating to see the consequences for patients, which included deaths such as Mariama's due to treatable conditions, or preventable but serious complications, such as Fatima's unrepaired obstetric fistula (described above).

The METC faced its share of problems. Within a short time of opening the doors for pregnant women, misconceptions about its role were spreading through the community. At times, pregnant women would arrive on foot to seek routine prenatal care; others would come by taxi, in labor, and without any signs or symptoms of Ebola, and were quickly referred to more appropriate health centers.

[4]WHO protocols recommend oxytocin IM for prevention of PPH. However, in the setting of an ETC, the preference is for misoprostol, which is also effective in preventing PPH and can be given orally.

After the cases of Zainab and Mariama, which occurred within a few days of each other, MSF staff found themselves with many questions. Were ETCs in fact properly equipped to handle not only normal but complicated deliveries? Were there certain invasive procedures, such as manual extraction of placenta or a manual vacuum aspiration (which may have saved Zainab's life) which *could* be safely done in this setting? Were the restrictions imposed by the METC policies against all invasive procedures, conceived at the height of the epidemic, still appropriate in its later stages? And would the staff feel comfortable and safe performing such interventions in the setting of Ebola?

In practice, the restrictions loosened somewhat as the months passed, the number of Ebola cases continued to decline, and the METC saw more patients like Zainab and Mariama. The MSF team of experienced midwives, nurses, and OB/GYNs had struggled to manage conditions such as diabetes, eclampsia, obstructed labor, a ruptured uterus, and even puerperal sepsis, in an METC designed to minimize interventions. Women presented to the METC late, both because they often did not know where to go and because they feared contracting Ebola in the hospital. In one tragic case, a woman presented to the METC with septic shock after a neglected breast abscess; her family told the METC health workers that fear of Ebola had kept her from seeking medical care earlier.[5] But with the number of new EVD cases dwindling, the METC field team, in consultation with MSF advisors at headquarters, gradually began to provide more and more invasive procedures inside the ETC, including manual removal of placentas and vacuum aspirations for retained placenta or incomplete abortions. Thus, as the epidemic evolved and declined, patient care inside the METC adapted and expanded the range of treatments provided.

20.6 Responding to Quarantines

Pregnant women admitted to the METC, or for that matter to any ETC, were by far not the only ones affected by the devastating epidemic. Another example of the epidemic's impact on maternal care in general comes from those caught up in a quarantine.

In June 2014, shortly after the first Ebola cases in Sierra Leone, the country's president declared a public health state of emergency in Kailahun and imposed limitations on public gatherings (WHO 2015a). In August, the outbreak was designated a national emergency, and Sierra Leone began to quarantine Ebola-affected households and some of the hardest-hit communities, often involving the military to enforce these quarantines. By November, the National Ebola Response Coordination (NERC) released Standard Operational Procedures subjecting the household of any Ebola contact to a mandatory 21-day quarantine. At best, quarantines can help stem the tide of infectious outbreaks by isolating persons exposed to a communicable disease. At worst, they restrict personal liberties and access to health care, food, and water, posing serious health, legal, and ethical concerns. Additionally, it is often vulnerable populations, poor and marginalized groups, and people with disabilities and other illnesses who are most at risk from quarantine policies.

Over time, the quarantine situation in Sierra Leone became dire and provoked ethical concerns within the global health community. An Oxfam report written during the first six months of the epidemic found several troubling aspects of Sierra Leone's household quarantines, in particular the slow or inadequate provisions of food, water, and other basic items. For example, in the Western Sierra Leone community of Leicester, "…community members reported that they only received food on the fourteenth day of the quarantine. They also complained of a lack of access to routine health care, including access to antenatal services for pregnant women" (Wahome 2014).

[5]Personal communication, Sonia Guinovart.

20.6.1 Kadiatu: Pregnant and in Quarantine

Kadiatu was 20 years old, pregnant with her second child, and in labor when she was brought from a quarantined household to MSF's METC in Kissy. There were rumors that she had assisted another quarantined woman in labor and delivery. Kadiatu's only risk factor was that she had been living in a quarantined home. Since the Command Centre had brought her to the METC, she was admitted to the area for suspected cases. As her delivery was imminent, the staff quickly dressed in PPE so they could assist her. Before the nursing staff could collect her blood sample, she delivered a healthy baby girl. Blood sampling was done immediately after delivery, but another confirmatory sample was required at least 48 hours later before she could be discharged. While awaiting this second test, both mother and infant remained in the suspect area, exposed to other potentially EVD-positive suspects. Health care workers conducted routine postpartum care, with the notable exception of advising the mother not to initiate breastfeeding in case she proved to be Ebola-positive. When the two tests came back negative, mother and baby were discharged home following the mandatory chlorinated shower, but followed up for the next 21 days as potential Ebola contacts.

Kadiatu's story illustrates the limited access to health facilities and delivery services for pregnant women in quarantine homes, as well as some of the perverse effects these broad quarantines had on pregnant women. Quarantines in Sierra Leone were applied to everyone living in a household compound where a case was detected. Households in Freetown often include multiple families and extended family members and may spread beyond just one physical home; subsequently, anyone in a quarantined household who became ill was then suspected of having Ebola, even if there had been no actual contact with the Ebola-confirmed patient. During the first half of 2015, when the country's EVD case load was rapidly decreasing and a semblance of normalcy was returning to Freetown, pregnant women in quarantine continued to be neglected—unable to access routine antenatal care and left without information on how and where they should deliver. Nor did quarantine policy address issues such as: Should pregnant women in quarantine households be restricted from traveling to clinics for ANC? Where was the safest place for pregnant women living in quarantine to deliver? Does being in a quarantine household automatically mean the woman must be considered as an Ebola suspect and deliver in an ETC? How many pregnant women were living in quarantine households?

Given these questions, in April 2015, the METC team, in collaboration with the NERC, conducted an informal survey that identified and then included eight pregnant women living in quarantine, as a way to help find some answers and understand the needs and perspectives of these women. Planning for the survey revealed that a formal system to identify pregnant women in quarantine homes was lacking; there were no complete lists, but rather multiple lists, often providing outdated information. Furthermore, because contact tracers, District Surveillance Officers (DSOs), NGO staff, and social mobilizers visiting the quarantine homes were not trained midwives or medical staff and did not ask specific questions related to pregnancy, many were unaware that pregnant women may have been living in the quarantine homes under their surveillance.

To find pregnant women in quarantined areas, the survey team, therefore, worked with the District Surveillance Officers (DSOs) in charge of the quarantine houses, and with contract tracers and other implementing agencies helping the NERC manage the quarantine homes. Each pregnant woman was then visited and interviewed by a Certified Nurse Midwife and a Health Promoter, both from MSF's METC staff and trained in Ebola contact precautions. Overall, eight women in five different wards, with pregnancies ranging in gestational ages from two to eight months, were interviewed. Four women self-identified as being in the first trimester, and two each were in the second trimester and third trimester.

Their responses to the survey, although representing just a very small sample, highlighted the complex issues of pregnancy while living under quarantine, particularly with respect to referrals and transport. Access to antenatal care was severely curtailed: only one interviewee said she had attended an Antenatal Care (ANC) visit while under quarantine. She was in her eighth month of pregnancy and had walked by foot to a private clinic for her previously scheduled appointment. The contact tracer responsible for her household was informed, but the clinic staff were unaware she was presenting from a quarantine house. All other women interviewed either said they were not allowed to leave the quarantined area or did not think to inquire about attending ANC visits. Women were asked where they would prefer to go if allowed, and most replied that they would prefer to go to clinics where they had a previously established relationship. Upon further probing, three women mentioned they would feel comfortable going to an ETC for ANC, one of whom specified that she would go to an ETC as long as her husband didn't know. Seven of the eight women said they would not want to use an ambulance to travel to clinics or ETCs for ANC visits.

The MSF survey also revealed that most women did not know the signs of pregnancy complications, nor did they have a clear plan for what to do in the event of a complication or if labor began while they were in quarantine. But they did have a preference to deliver at a clinic familiar to them rather than at an ETC. Concerns about delivering in an ETC were primarily related to stigmatization of family members; several women mentioned fear of going to the ETC because of the "false" swab tests.[6] One woman did not see the necessity to deliver in an ETC as she "knew she was not infected".

By the end of April 2015, the NERC, WHO, MSF, and other stakeholders began to make plans for managing quarantined pregnant women and acknowledged the limitations of contact tracers and DSOs, for whom ANC care was outside their area of expertise and training. With support from WHO, the Ministry of Health and Sanitation (MoHS) ultimately devised a strategy to proactively track pregnant case contacts nationally using two indicators contained in the daily contact tracing data reports: (1) number of pregnant contacts currently being followed; and (2) total number of pregnant contacts who have finished their 21 days of monitoring. Adapting WHO guidance on home-based ANC visits, the new strategy established a "Mobile Reproductive Health Team" (MRHT), comprising trained health workers, midwifes, and nurses, to provide basic ANC for pregnant women in quarantine homes. The strategy also included providing amenorrheic women with home urine pregnancy tests and called for the DSO and MRHT to monitor the pregnant women closely for complications, ensure the women had adequate nutrition and essential supplies (such as bednets and safe delivery kits), and help devise a plan in the event of any obstetrical emergency, including assuring a reliable means of transportation. Quarantined women in labor would be transferred via ambulance to an Ebola facility with a designated "mother-baby" unit, which at the time meant either the MSF METC or the Partners in Health (PIH) holding center at the Princess Christian Maternity Hospital (PCMH). Following delivery and discharge, mothers were urged to seek accommodation outside their quarantine home for 21 days, if possible, and were followed up as contacts.

This guidance was an improvement, even though it was devised near the end of the epidemic when the numbers of new cases and quarantined households were dwindling. However, for future outbreaks, many questions still remain on how to best manage pregnant women in quarantine households and on the very ethics of quarantines themselves.

20.7 Lessons Learned and Unanswered Questions

The West Africa outbreak laid bare the weaknesses in the health systems, not only of the most affected countries (Sierra Leone, Liberia, and Guinea), but also at the global level. Collectively, these three countries had a total of 28,616 confirmed, probable, and suspected cases of EVD and 11,310 deaths

[6] At this time in the epidemic, many rumors were being spread about false positive tests.

(WHO 2016a). Looking beyond this devastating death toll, only in retrospect has the global health community been able to assess the epidemic's broader impact, including its repercussions for maternal health.

In Sierra Leone, which reported 3,590 Ebola-confirmed or -probable deaths (WHO 2016b), several published reports have begun to document the reduction in uptake of essential maternal health-related services during the epidemic (Brolin Ribacke et al. 2016a, b; Delamou et al. 2017; Iyengar et al. 2015; Jones et al. 2016). Table 20.1 summarizes the available evidence so far, which demonstrates increased mortality for both mothers and infants, a decline in cesarean section rates (mostly due to the closure of private facilities), and a decline in ANC visits. Since maternal mortality had been decreasing in Sierra Leone in the 25 years prior to the outbreak, all indications are that Ebola has reversed these successes and put the country back years in these efforts.

The METC model was an attempt to help fill critical gaps in access to care that these disastrous outcomes reveal, although it opened only at the end of the outbreak when patient numbers were declining rapidly. During the last week of January 2015 when the METC officially opened its doors, there were 65 confirmed new cases of Ebola, compared with 248 in the first week of January, including 93 in Freetown (WHO 2015b). This was a stark contrast to the peak period in October and November 2014 when initial plans were being made to open an METC, and Sierra Leone had an average of nearly 500 new cases of EVD per week (WHO 2014b). In the end, the METC cared for 48 pregnant women, eight of whom were confirmed positive for Ebola while the remaining 40, who were suspected of having Ebola, ultimately tested negative. Of the women who tested negative, there were three maternal deaths in the METC and one woman who died after transfer to a CEmONC center; three of these four women needed potentially life-saving surgical care. Among the 40 women who were considered potentially infected, admitted to the METC, and tested negative for Ebola virus, 15 (37.5%) presented with an obstetric emergency and seven (17.5%) likely needed urgent referral that was ultimately delayed[7]; Fatima, Zainab, and Mariama were among these seven. However, despite these small numbers, the METC exemplified a novel and flexible strategy for managing pregnant women during a widespread Ebola outbreak and providing a place where they could receive at least some of the care they needed.

More broadly, while this epidemic provided many lessons on how to care for pregnant women with EVD, it also left many open questions, a few of which we consider here. The first is whether the case

Table 20.1 Impact of Ebola on maternal care services and outcomes in Sierra Leone

Service/indicator	Change[a]	Time interval	Citation
Cesarean deliveries	−20%	January 2014–May 2015	WHO (2015d)
Facility-based deliveries	−11%	May 2013–March 2015	Jones et al. (2016)
Antenatal visits	−18%	May 2013–March 2015	Jones et al. (2016)
Facility-based MMR[a]	+34%	May 2013–March 2015	Jones et al. (2016)
Overall MMR	+74%	May 2015	Evans et al. (2015)
Stillbirth rate	+24%	May 2013–March 2015	Jones et al. (2016)

[a]*Maternal mortality ratio—number of maternal deaths per 100,000 live births*

[7]Guinovart, unpublished.

definition of Ebola in pregnancy is adequate. During this epidemic, evidence emerged that cured pregnant women could still harbor Ebola virus in the placenta and amniotic fluid, where it could potentially be amplified (Baggi et al. 2014; Caluwaerts et al. 2015). In addition, one case report described an apparently undetected previous Ebola virus infection in an asymptomatic pregnant woman who tested negative at presentation to the clinic, but gave birth to a stillborn, Ebola-positive infant (Bower et al. 2016). Another report described a pregnant women who presented with symptoms of normal labor, but was found to have a high Ebola viral load, yet remained asymptomatic for EVD for three more days before becoming ill and dying of EVD soon after (Akerlund et al. 2015). These two cases (described more fully in Chap. 6 by S. Caluwaerts and P. Kahn in this book) provided important information about the natural history of EVD in pregnant women, while also adding to the stigma for women who were pregnant during this time.

As a result of the new knowledge on viral persistence and seemingly asymptomatic cases, WHO's September 2015 interim guidelines for EVD in pregnancy (WHO 2015c) called for a "higher level of suspicion" for Ebola infection in women with any of the following complications: spontaneous abortion, pre-labor rupture of membranes, preterm rupture of membranes, preterm labor/preterm birth, antepartum or postpartum hemorrhage, intrauterine fetal death, stillbirth, maternal death, or neonatal death. At first glance, this list may seem overly broad, but given the just-cited cases seen during this epidemic in which asymptomatic pregnant women could potentially have infected others during delivery, WHO's guidelines can be viewed as reasonable and necessary, especially when viewed through a public health lens, even in spite of the paucity of numbers and data on these types of cases. On the other hand, does such guidance mean even more women would be labeled as "Ebola suspects," probably furthering the stigma of Ebola as well as potentiate Ebola's negative impact on maternal health as a whole? As described in this chapter, many pregnant women were caught in this ambiguous case definition, labeled as "suspects", and consequently suffered often serious consequences. Looking ahead, it will be important to further refine the case definition for Ebola in pregnancy, to avoid having so many Ebola-negative pregnant women face life-threatening delays in receiving care.

A related, urgent need is to expedite the diagnosis of Ebola among pregnant women. Ultimately, just as with many other diseases, accessible and rapid point-of-care diagnosis could potentially avoid the delays many pregnant women encountered in trying to access obstetric care. Continued development of promising tests developed late during the outbreak (Coarsey et al. 2017), including the GenXpert® Ebola Assay (Raftery et al. 2018), should be pursued in these non-epidemic times.

Another unresolved question is whether a standalone METC was the right solution under the circumstances at the time, or would be the right solution were another major outbreak occur (a scenario that will hopefully never materialize). Or alternatively, should all ETCs be designed and equipped to manage pregnant women? Learning from the experiences of the 2014–2015 epidemic, one could envision a model in which all ETCs have the capacity to provide basic emergency obstetric care for both normal and complicated deliveries, including management of hypertensive disorders, provision of antibiotics and uterotonic drugs, resuscitation of newborns, manual and vacuum-assisted removal of retained placenta, and vacuum-assisted delivery of babies—the seven elements of BEmOC. In this model, alongside these facilities would be Comprehensive Emergency Obstetric ETCs that could manage complicated emergency obstetric care, i.e., those requiring blood transfusions or cesarean sections, for referrals.

If MSF and other medical care providers had incorporated pregnant women's special needs more broadly and earlier in the outbreak, perhaps this would have helped mitigate the impacts on maternal and neonatal outcomes, not only for Ebola-positive women, but also for women suspected of having Ebola and for those caught in the quarantine system. While Ebola-positive pregnant women had higher survival rates in this epidemic than in previous ones (see chapter in this book by Caluwaerts and Kahn), many noninfected pregnant women may have paid a heavy price for the lack of knowledge and capacity to manage pregnancy in Ebola contexts.

20.8 Epilogue

After Sierra Leone was declared Ebola-free, MSF reestablished maternal health programming in different regions of the country. For example, a collaborative program between MSF-Holland and Sierra Leone's MOHS supports the pediatric and maternity wards in Magburaka Government Hospital in Tonkolli district through the provision of human resources, training, medications, and equipment.

In the initial months after Sierra Leone was declared Ebola-free, the staff at Magburaka Hospital were faced with the challenge of balancing safety and fear. To maintain infection control but minimize impediments to care, they established an Ebola screening area and a Holding Unit at the hospital entrance. Before admission to the facility, all staff, patients, and visitors were screened for symptoms of Ebola and all deliveries were conducted in standard PPE for childbirth (face shield, gown, plastic apron, elbow length gloves, and rubber boots). Perhaps, one silver lining of the epidemic will be to reinforce these standard infection prevention practices, although it remains to be seen whether such measures will remain in place long-term, and if so, whether they will ultimately improve overall care.

Overcoming fear of EVD seems likely to be a slow process. Initially, Magburaka saw few deliveries, and anecdotal evidence from informal discussions with women at a birth registration facility close to the main hospital indicated that many had delivered at home.[8] This district also had a case of nosocomial transmission of EVD in a hospital early in the outbreak—a woman who had a cesarean section later developed symptoms and died of EVD. Forty-six contacts were reported, including 25 hospital staff, all of whom were quarantined in the maternity ward for 21 days (Dunn et al. 2015). Given such events, it will likely take time for the community to rebuild trust in the health care system.

Acknowledgments We would like to thank Patricia Lledo and Olimpia de la Rosa, who managed MSF-Spain's maternal health response during the epidemic and who provided us with tireless support both while working at the METC and in the writing of this chapter. Additional thanks also go to Sonia Guinovart and Severine Caluwaerts for their insightful comments to this chapter and to Michelle Olakkengil and Eleonora D'Amore for expert editorial assistance.

References

Akerlund, E., Prescott, J., & Tampellini, L. (2015). Shedding of Ebola virus in an asymptomatic pregnant woman. *New England Journal of Medicine, 372*(25), 2467–2469. Retrieved March 25, 2018, from http://www.nejm.org/doi/full/10.1056/NEJMc1503275.

Baggi, F. M., Taybi, A., Kurth, A., Van Herp, M., Di Caro, A., Wölfel, R., et al. (2014). Management of pregnant women infected with Ebola virus in a treatment centre in Guinea, June 2014. *Eurosurveillance, 19*(49). Retrieved March 25, 2018, from http://www.eurosurveillance.org/content/10.2807/1560-7917.ES2014.19.49.20983.

Bebell, L. M., & Riley, L. E. (2015). Ebola virus disease and Marburg disease in pregnancy: A review and management considerations for filovirus infection. *Obstetrics & Gynecology, 125*(6), 1293–1298. Retrieved March 24, 2018, from https://www.ncbi.nlm.nih.gov/pmc/articles/PMC4443859/.

Bjälkander, O., Grant, D., Berggren, V., Bathija, H., & Almroth, L. (2013). Female genital mutilation in Sierra Leone: Forms, reliability of reported status, and accuracy of related demographic and health survey questions. *Obstetrics and Gynecology International, 2013*, 680926. https://doi.org/10.1155/2013/680926. Retrieved March 24, 2018, from https://www.ncbi.nlm.nih.gov/pmc/articles/PMC3800578/.

Bower, H., Grass, J. E., Veltus, E., Brault, A., Campbell, S., Basile, A. J., et al. (2016). Delivery of an Ebola virus-positive stillborn infant in a rural community health center, Sierra Leone, 2015. *American Journal of Tropical Medicine and Hygiene, 94*(2), 417–419. https://doi.org/10.4269/ajtmh.15-0619.

Brolin Ribacke, K. J., Saulnier, D. D., Eriksson, A., & von Schreeb, J. (2016a). Effects of the West Africa Ebola virus disease on health-care utilization—A systematic review. *Frontiers in Public Health*. Retrieved March 24, 2018, from http://journal.frontiersin.org/article/10.3389/fpubh.2016.00222.

[8]Informal communications between Elin Erland and women at the Maternal and Child Health Post in Magburaka.

Brolin Ribacke, K., van Duinen, A., Nordenstedt, H., Höijer, J., Molnes, R., Froseth, T., et al. (2016b). The impact of the West Africa Ebola outbreak on obstetric health care in Sierra Leone. *PLoS One, 11*(2), e0150080. Retrieved March 25, 2018, from http://journals.plos.org/plosone/article?id=10.1371/journal.pone.0150080.

Caluwaerts, S., Fautsch, T., Lagrou, D., Moreau, M., Camara, A. M., Günther, S., et al. (2015). Dilemmas in managing pregnant women with Ebola: 2 case reports. *Clinical Infectious Diseases, 62*(7), 903–905. Retrieved March 25, 2018, from https://academic.oup.com/cid/article/62/7/903/2462754.

Caluwaerts, S. & Kahn, P. (2019). Ebola and pregnant women: Providing maternity care at Médecins Sans Frontières treatment centers. In D. A. Schwartz, J. N. Anoko, & S.A. Abramowitz (Eds.), *Pregnant in the time of Ebola: Women and their children in the 2013-2015 West African epidemic*. New York: Springer.

Coarsey, C. T., Esiobu, N., Narayanan, R., Pavlovic, M., Shafiee, H., & Asghar, W. (2017). Strategies in Ebola virus disease (EVD) diagnostics at the point of care. *Critical Reviews in Microbiology, 43*(6), 779–798.

Conde-Agudelo, A., Belizan, J. M., & Lammers, C. (2005). Maternal-perinatal morbidity and mortality associated with adolescent pregnancy in Latin America: Cross-sectional study. *American Journal of Obstetrics and Gynecology, 192*(2), 342–349. https://doi.org/10.1016/j.ajog.2004.10.593.

Delamou, A., El Ayadi, A. M., Sidibe, S., Delvaux, T., Camara, B. S., Sandouno, S. D., et al. (2017). Effect of Ebola virus disease on maternal and child health services in Guinea: A retrospective observational cohort study. *The Lancet Global Health, 5*(4), e448–e457. https://doi.org/10.1016/S2214-109X(17)30078-5.

Deaver, J. E., & Cohen, W. R. (2015). Ebola virus screening during pregnancy in West Africa: Unintended consequences. *Journal of Perinatal Medicine*. https://doi.org/10.1515/jpm-2015-0118.

Dunn, A. C., Walker, T. A., Redd, J., Sugerman, D., McFadden, J., Singh, T., et al. (2015). Nosocomial transmission of Ebola virus disease on pediatric and maternity wards: Bombali and Tonkolili, Sierra Leone, 2014. *American Journal of Infection Control, 44*(3), 269–272. https://doi.org/10.1016/j.ajic.2015.09.016.

Dynes, M. M., Miller, L., Sam, T., Vandi, M. A., & Tomezyk, B. (2015). Perceptions of the risk for Ebola and health facility use among health workers and pregnant and lactating women—Kenema District, Sierra Leone, September 2014. *MMWR Morbidity and Mortality Weekly Report, 63*(51), 1226–1227. Retrieved April 13, 2018, from https://www.cdc.gov/mmwr/preview/mmwrhtml/mm6351a3.htm.

Erland, E., & Dahl, B. (2017). Midwives' experiences of caring for pregnant women admitted to Ebola centres in Sierra Leone. *Midwifery, 55*, 23–28. https://doi.org/10.1016/j.midw.2017.08.005.

Evans, D. K., Goldstein, M., & Popova, A. (2015). Health-care worker mortality and the legacy of the Ebola epidemic. *The Lancet Global Health, 3*(8), e439–e440. Retrieved March 24, 2018, from http://www.thelancet.com/journals/langlo/article/PIIS2214-109X(15)00065-0/fulltext.

Iyengar, P., Kerber, K., Howe, C. J., & Dahn, B. (2015). Services for mothers and newborns during the Ebola outbreak in Liberia: The need for improvement in emergencies. *PLoS Currents Outbreaks, 7*, pii: ecurrents.outbreaks.4ba318308719ac86fbef91f8e56cb66f. Retrieved March 24, 2018, from http://currents.plos.org/outbreaks/article/services-for-mothers-and-newborns-during-the-ebola-outbreak-in-liberia-the-need-for-improvement-in-emergencies/.

Jones, S., Gopalakrishnan, S., Ameh, C., White, S., & van den Broek, N. (2016). "Women and babies are dying but not of Ebola": The effect of the Ebola virus epidemic on the availability, uptake and outcomes of maternal and newborn health services in Sierra Leone. *BMJ Global Health, 1*(3), e000065. https://doi.org/10.1136/bmjgh-2016-000065. Retrieved March 24, 2018, from https://www.ncbi.nlm.nih.gov/pmc/articles/PMC5321347/.

Kingham, T., Kamara, T., Cherian, M., Gosselin, R., Simkins, M., Meissner, C., et al. (2009). Quantifying surgical capacity in Sierra Leone: A guide for improving surgical care. *Archives of Surgery, 144*(2), 122–127. https://doi.org/10.1001/archsurg.2008.540.

Lledo, P., Guinovart, S., Bernasconi, A., Flevaud, L., Segers, N., & Saint-Sauveur, J. F. (2016). Impact of Ebola outbreak on pregnant Ebola negative-tested women [version 1]. *F1000Research*, (5), 840 (poster).

Médecins Sans Frontières. (2012, November). *Safe Delivery: Reducing maternal mortality in Sierra Leone and Burundi*. Retrieved March 25, 2018, from https://www.doctorswithoutborders.org/news-stories/special-report/safe-delivery-reducing-maternal-mortality-sierra-leone-and-burundi.

Milland, M., & Bolkan, H. A. (2015). Enhancing access to emergency obstetric care through surgical task shifting in Sierra Leone: Confrontation with Ebola during recovery from civil war. *Acta Obstetricia et Gynecologica Scandinavica, 94*(1), 5–7. Retrieved March 24, 2018, from https://obgyn.onlinelibrary.wiley.com/doi/full/10.1111/aogs.12540.

Neal, S., Matthews, Z., Frost, M., Fogstad, H., Camacho, A. V., & Laski, L. (2012). Childbearing in adolescents aged 12-15 years in low resource countries: A neglected issue. New estimates from demographic and household surveys in 42 countries: Childbearing in adolescents aged 12-15 years. *Acta Obstetricia et Gynecologica Scandinavica, 91*(9), 1114–1118. https://doi.org/10.1111/j.1600-0412.2012.01467.x.

Raftery, P., Condell, O., Wasunna, C., Kpaka, J., Zwizwai, R., Nuha, M., et al. (2018). Establishing Ebola Virus Disease (EVD) diagnostics using GeneXpert technology at a mobile laboratory in Liberia: Impact on outbreak response, case management and laboratory systems strengthening. *PLoS Neglected Tropical Diseases, 12*(1), e0006135. https://doi.org/10.1371/journal.pntd.0006135.

Save The Children. (2014). *Ending newborn deaths—Ensuring every baby survives*. Retrieved March 24, 2018, from http://www.savethechildren.org/atf/cf/%7B9def2ebe-10ae-432c-9bd0-df91d2eba74a%7D/ENDING-NEWBORN-DEATHS_2014.PDF.

Say, L., Chou, D., Gemmill, A., Tunçalp, Ö., Moller, A. B., Daniels, J., et al. (2014). Global causes of maternal death: A WHO systematic analysis. *The Lancet Global Health, 2*(6), e323–e333. https://doi.org/10.1016/S2214-109X(14)70227-X.

Statistics Sierra Leone--SSL and ICF International. (2014). *Sierra Leone Demographic Health Survey 2013*. Freetown, Sierra Leone: SSL and ICF International. Retrieved March 24, 2018, from http://dhsprogram.com/pubs/pdf/FR297/FR297.pdf.

Thaddeus, S., & Maine, D. (1994). Too far to walk: Maternal mortality in context. *Social Science & Medicine, 38*(8), 1091–1110.

UNFPA. (2015). *Rapid assessment of Ebola impact on reproductive health services and service seeking behaviour in Sierra Leone*. Retrieved March 23, 2018, from https://reliefweb.int/report/sierra-leone/rapid-assessment-ebola-impact-reproductive-health-services-and-service-seeking.

Wahome, W. M. (2014). *Quarantines in Sierra Leone: Putting people first in the Ebola crisis*. Oxfam. Retrieved March 23, 2018, from https://www.oxfam.org/sites/www.oxfam.org/files/file_attachments/quarantines_in_sierra_leone_-_putting_people_first_in_the_ebola_crisis.pdf.

WHO. (2014a). *Case definition recommendations for Ebola or Marburg virus diseases*. Retrieved March 24, 2018, from http://www.who.int/csr/resources/publications/ebola/ebola-case-definition-contact-en.pdf.

WHO. (2014b). *Ebola response roadmap situation report—19 November 2014*. Retrieved March 24, 2018, from http://apps.who.int/iris/bitstream/10665/144032/1/roadmapsitrep_19Nov14_eng.pdf?ua=1.

WHO. (2015a). *Ebola in Sierra Leone: A slow start to an outbreak that eventually outpaced all others*. WHO. Retrieved March 25, 2018, from http://www.who.int/csr/disease/ebola/one-year-report/sierra-leone/en/.

WHO. (2015b). *Ebola Situation Report. 7 January 2015*. Retrieved from http://apps.who.int/iris/bitstream/10665/147112/1/roadmapsitrep_7Jan2015_eng.pdf?ua=1&ua=1.

WHO. (2015c). *Ebola virus disease in pregnancy: Screening and management of Ebola cases, contacts, and survivors*. Retrieved from http://apps.who.int/iris/bitstream/10665/184163/1/WHO_EVD_HSE_PED_15.1_eng.pdf.

WHO. (2015d, April). *WHO statement on cesarean section rates: Executive summary*. Retrieved on March 25, 2018, from http://apps.who.int/iris/bitstream/handle/10665/161442/WHO_RHR_15.02_eng.pdf;jsessionid=62FF53FFE61EAAF3B136880E88FF72D0?sequence=1.

WHO. (2016a). *Ebola outbreak 2014-2015*. Retrieved from http://www.who.int/csr/disease/ebola/en/.

WHO. (2016b). *WHO Statement on end of Ebola flare-up in Sierra Leone*. Retrieved March 23, 2018, from http://www.who.int/mediacentre/news/statements/2016/end-flare-ebola-sierra-leone/en/.

WHO, UNFPA, UNICEF, & Mailman School of Public Health. (2009). *Monitoring emergency obstetric care: A handbook*. Retrieved March 24, 2018, from http://www.who.int/reproductivehealth/publications/monitoring/9789241547734/en/.

WHO, UNICEF, UNFPA, The World Bank, & The United Nations Population Division. (2014). *Trends in maternal mortality: 1990 to 2013. Estimates by WHO, UNICEP, UNFPA, The World Bank and the United Nations Population Division*. Retrieved March 23, 2018, from http://www.who.int/reproductivehealth/publications/monitoring/maternal-mortality-2013/en/.

Yoder, P. S., Wang, S., & Johansen, E. (2013). Estimates of female genital mutilation/cutting in 27 African countries and Yemen. *Studies in Family Planning, 44*(2), 189–204.

21 The Services and Sacrifices of the Ebola Epidemic's Frontline Healthcare Workers in Kenema District, Sierra Leone

Michelle M. Dynes, Laura Miller, Tamba Sam, Mohamad Alex Vandi, Barbara Tomczyk, and John T. Redd

21.1 Introduction

Sierra Leone is a geographically small country (71,740 km^2 total area) with a population of between six and seven million people (Central Intelligence Agency 2016). To most of the world, Sierra Leone is best known for its natural resources—particularly diamonds—and the 11-year civil war they fueled until 2002. Twelve years after the end of the war, a recovering but fragile Sierra Leone was ravaged and thrust into the global spotlight again, this time by the largest Ebola outbreak in history.

Once Ebola was detected, many global health experts expected that the outbreaks in West Africa would follow the course of previous outbreaks of Ebola and would be self-limited by geographic barriers. But this epidemic originated and spread at the intersection of three countries—Sierra Leone, Guinea, and Liberia—where tribal affiliation is more important than nationality. People move freely between these countries, often through informal border crossings, to conduct business, visit family,

M. M. Dynes (✉)
Division of Global Health Protection (at time of Ebola epidemic in West Africa), Centers for Disease Control and Prevention, Atlanta, GA, USA

Division of Reproductive Health, Centers for Disease Control and Prevention, Atlanta, GA, USA
e-mail: mdynes@cdc.gov

L. Miller
IRC Sierra Leone (at time of Ebola epidemic in West Africa), Freetown, Sierra Leone

International Rescue Committee, New York City, NY, USA

T. Sam
IRC Sierra Leone (at time of Ebola epidemic in West Africa), Freetown, Sierra Leone

M. A. Vandi
Kenema District Health Management Team, Sierra Leone Ministry of Health and Sanitation, Kenema, Sierra Leone

B. Tomczyk
Center for Global Health, Centers for Disease Control and Prevention, Atlanta, GA, USA

J. T. Redd
Center for Global Health, Division of Global Health Protection, Centers for Disease Control and Prevention, Atlanta, GA, USA

D. A. Schwartz et al. (eds.), *Pregnant in the Time of Ebola*, Global Maternal and Child Health,
https://doi.org/10.1007/978-3-319-97637-2_21

and participate in community gatherings, religious ceremonies, and major life events, such as weddings, births, and funerals. Furthermore, improved transportation infrastructure in Sierra Leone in recent years enabled greater ease of travel than ever before. Recent expansions of paved roads connected border regions to major cities. This included a new paved road that connected the district of Kailahun—where the country's epidemic began—to the rest of the country, just 6 months before the Ebola virus was confirmed.

Sierra Leone's first case of Ebola was linked to a funeral in Guinea, attended by dozens of people who were able to travel back to the four regions of the country and through all major cities. This patient had traveled from Kailahun to the city of Kenema to seek care at Kenema Government Hospital, home of the country's famous Lassa Fever Research Center. The hospital and its 250 beds served as the referral hospital for the Eastern Region, which includes the districts of Kailahun, Kenema, and Kono. Over the next 4 months, the grounds of the hospital would host an Ebola Treatment Unit (ETU) and the situation room of the district response (Figs. 21.1 and 21.2). It also became the site of the greatest number of documented health care worker deaths within a single health facility of the entire outbreak, with a case fatality ratio (CFR) of 69% (Senga et al. 2016). That is to say that 69% of the health care workers at Kenema Government Hospital who became infected with Ebola ultimately died of the disease.

During the outbreak in West Africa, health care workers became 21–32 times more likely to be infected with Ebola than people in the general population (World Health Organization [WHO] 2015a). Over 350 health care workers were infected in Sierra Leone, 92 of whom worked in Kenema District,

Fig. 21.1 Grounds of the Kenema Government Hospital in Sierra Leone looking toward the Ebola Treatment Unit (straight ahead) and Lassa Fever/Ebola Lab (on right). Photo courtesy of Joan Brunkard

Fig. 21.2 Burial team entering the morgue on the grounds of Kenema Government Hospital in Sierra Leone. Photo courtesy of Joan Brunkard

the site of one of the largest ever reported clusters of health care workers infected with Ebola (Senga et al. 2016). Moreover, at the end of October 2014, 12.9% of all confirmed Ebola patients in Kenema District were health care workers, the highest percentage across all districts in the country (Kilmarx et al. 2014). Nosocomial transmission, or health facility-acquired infection, had similarly characterized and propagated previous outbreaks, but previous outbreaks had been limited due to geographic isolation. Stopping the spread of those outbreaks had not needed to rely so critically on systems being able to simultaneously protect health care workers and deliver health services to patients within such a widespread scale of a highly lethal infectious disease. This type of response had not been attempted before.

As the outbreak progressed, some health care workers fled, but many continued to provide compassionate care for their patients until they became sick themselves. One health care worker in Sierra Leone described why they continued to work, despite their vulnerability to the virus:

> *"Well the risk is too much, but I have no choice because if we the health worker are also afraid and we relax saying that this sickness is dangerous...it will still continue. That is why we are still fighting it to the end."*—Kenema District Health Care Worker

The walls of Kenema Government Hospital remain scattered with pictures of health care workers who courageously gave their lives to help others.

This chapter is the story of the first few months of the outbreak in Sierra Leone in Kenema District and the challenges and successes of health care workers who continued to provide maternal, reproductive, and child health care, within the deadliest, longest, and most widespread epidemic of Ebola ever recorded. It is the story of health care workers and the lifesaving services they risked their lives to deliver.

21.2 Health System Prior to the Ebola Outbreak

Health services in Sierra Leone are delivered through a system of hospitals and peripheral health units (PHUs). The PHUs include a network of community health centers, community health posts, and maternal and child health posts that provide primary care. When Ebola hit in 2014, it struck a country with a health system that had improved since the end of a brutal, prolonged civil war, but was still working to overcome preventable deaths related to infectious diseases, such as diarrhea and malaria. National health plans had been established and per capita health expenditure had increased since the end of the war. The country had made substantial progress towards a number of the Millennium Development Goal (MDG) targets for health, based on the results of the two preceding Demographic and Health Surveys (DHS) conducted in 2008 and 2013. Between these years, notable gains were made across essential services: skilled birth attendance (42–60%), antenatal care visits (56–76%), postnatal care visits (56–72%), prevalence of modern contraceptive use (7–16%), complete vaccination coverage during the first year of life (31–58%), pneumonia treatment (46–72%), and diarrhea treatment (73–86%) (Statistics Sierra Leone [SSL] and ICF International 2014; SSL and ICF Macro 2009).

Despite encouraging gains in access to services, and declines in maternal and under-5 deaths, key mortality rates remained intractably high. The country remained one of the poorest and least developed countries in the world, ranking 181 out of 188 countries on the Human Development Index, with an average life expectancy at birth of just 50.9 years (United Nations Development Programme [UNDP] 2015). According to recent estimates, Sierra Leone had the highest maternal mortality ratio (MMR), which was estimated at 1360 deaths per 100,000 live births (WHO 2015b). It also had the fourth highest under-5 mortality rate (estimated at 114 deaths per 1000 live births) in the world (United Nations Inter-Agency Group for Child Mortality Estimation [UN IGME 2017).

The postwar health improvements may in large part be due to the introduction of the Free Healthcare Initiative (FHC-I) by the Government of Sierra Leone in 2010. The FHC-I represented a critical step towards the attainment of universal health coverage, providing free health care services for pregnant women, lactating mothers, and children under 5 years of age. Despite the FHC-I, many issues remained that affected both the supply of and demand for health services. Critical shortages of health care workers and an ineffective drug procurement and distribution system persisted. While investments were made to train more health care workers, the country continued to face a critical shortage, with only 2 skilled providers per 10,000 population, compared to the WHO-recommended 23 per 10,000 population (WHO 2014a). Findings from a recent evaluation of the FHC-I highlighted concerns from participants regarding the quality of care they received, poor health facility infrastructure, and the skills and attitudes of service providers (Witter et al. 2016). The health system remained highly dependent on foreign aid, with 70% of the annual health investments supported by international donors.

Important postwar maternal and reproductive health gains were made both in the Eastern Region of the country—the epicenter of the Ebola outbreak—and in Sierra Leone as a whole between 2008 and 2013 (Fig. 21.3). For example, the percentage of deliveries attended by a skilled birth attendant (SBA) in the Eastern Region increased from 50% in 2008 to 77% in 2013, with a similar increase in skilled birth attendance in Sierra Leone from 42% to 60% in the same period. Furthermore, modern contraceptive use in the region outpaced that of the national average with an increase from 5% to 17% between 2008 and 2013, compared to the national average increase from 7% to 16%.

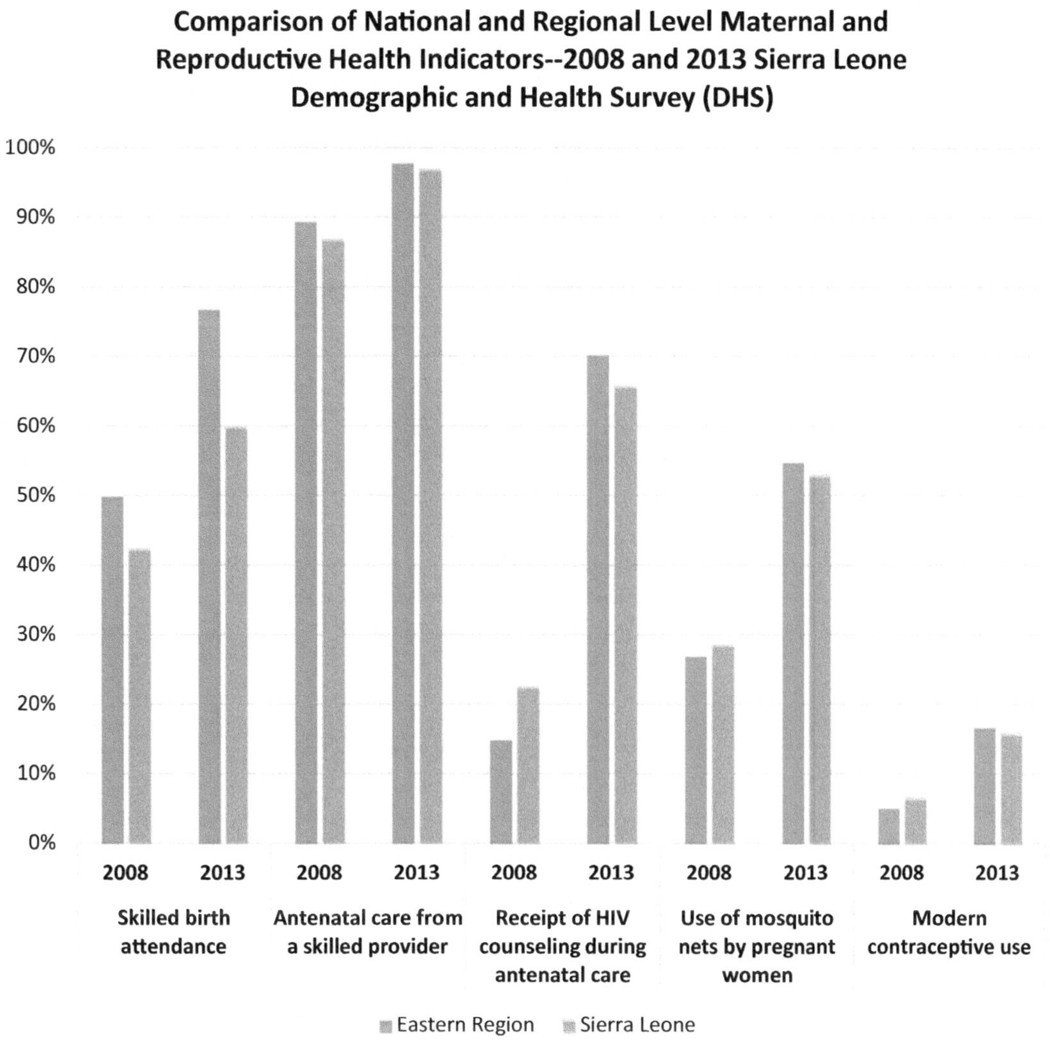

Fig. 21.3 Comparison of national- and regional-level maternal and reproductive health indicators—2008 and 2013 Sierra Leone DHS (Statistics Sierra Leone [SSL] and ICF International 2014; SSL and ICF Macro 2009)

Despite these major improvements in health care, the health system, as illustrated below, was unable to cope with an unanticipated, lethal, infectious epidemic of unprecedented proportions.

21.3 Protecting Access to Health Care during the Ebola Outbreak

At the onset of the Ebola outbreak—confirmed in May 2014—the Government of Sierra Leone allocated emergency funding from its limited national budgets to support Ebola response activities, but the resources were insufficient to meet the early needs of the response. In Kenema District, there was insufficient cash to fund even the most basic elements of the response. Case investigation teams lacked fuel, as did ambulances that needed to pick up sick patients. Ensuring a sufficient supply of paper for contact-tracing forms, printer cartridges, and mobile phone top-up were everyday concerns for early responders in Kenema. Both the hospital and PHUs lacked necessary diagnostic and waste

Fig. 21.4 Grounds of the Kenema Government Hospital in Sierra Leone looking toward the Ebola Triage Tent (on right) and the Triage Waiting Area Tent (on left). Photo courtesy of Joan Brunkard

management capacity. There were not enough body bags, gloves, personal protective equipment, or chlorine. The supply chain that supported the FHC-I had to reprioritize their efforts to focus on working the new supply chain for Ebola personal protective equipment.

Meanwhile, the health care facilities where patients were arriving were unprepared to screen, isolate, and confirm Ebola virus disease. No single symptom can indicate whether a patient has Ebola, and screening measures rely on whether an individual exhibits symptoms that are typical of the virus—but those same symptoms are similar to other common illnesses. The health system was soon overwhelmed. Health facilities ran out of beds for patients. Kenema Government Hospital quickly became filled with infected and uninfected patients confined together (Fig. 21.4). The staff tending to them often did not have the training and materials to protect themselves. In a focus group discussion with primary health care workers, one described their decision to continue working despite lacking personal protective equipment:

> "*We thought that if there was no protective gear, we would not work, but our boss told us to put that at the back of our mind, that we are soldiers, we should be ready to be at the war front.*"—Kenema District Health Care Worker

Within the first months, two deeply intertwined priorities emerged: one to fight Ebola and another to keep the health system functional for routine health services such as antenatal care and child vaccinations. Both relied on Sierra Leonean health care workers—but they needed to be protected. While protecting access to health services and protecting health care workers is important in any country, the people of Sierra Leone were particularly vulnerable to its erosion due to the country's high maternal and child mortality rates and the low density of health care workers. It was suspected that more people would die because of the lack of access to health services than from Ebola. Most of those deaths would be women and children.

Protecting health care workers emerged as a key to both interrupting transmission at health facilities and preventing the collapse of the health system. But ensuring strict adherence to measures intended to prevent infection in health care settings, also known as infection prevention and control (IPC), was as challenging as it was imperative. Many facilities at all levels of the health system across the district lacked basic isolation and waste management infrastructure, running water, electricity, and supplies required for IPC. Even prior to the epidemic, standard precautions were not widely understood, practiced, or promoted among health care workers. Significant investment in IPC training, supervision, infrastructure, and supplies was necessary, but planning for and implementation of such activities on a large scale—in the midst of an ongoing crisis—was particularly challenging.

21.4 Investing in IPC for Health Care Workers and their Patients

In August 2014, two-and-a-half months after the start of the epidemic, a small section of the Kenema Emergency Operations Center (created after the onset of the epidemic) launched the first effort in Sierra Leone to try to protect the public health system. The group was comprised of staff from the Kenema District Health Management Team (DHMT), International Rescue Committee (IRC), and WHO. With funding from the Swedish International Development Cooperation Agency (SIDA), the team designed a project with the aim of controlling the spread of Ebola within the PHUs in the district and linking community sensitization and referral protocols with community health workers. Under this pilot project, a rapid training program was provided by the IRC and DHMT to the clinical staff from 120 PHUs. Teams composed of two clinical staff members, one from IRC and one from the Ministry of Health and Sanitation, first filled a car in the city of Kenema with enough personal protective equipment supplies for 8–10 PHUs. They traveled to each assigned PHU with the objective of delivering personal protective equipment supplies and training health care workers within the health facility. The training included the basics of IPC and universal precautions, how to use the algorithm to screen patients for Ebola, what to do if the patient was suspected of having Ebola, and how health care workers could protect themselves while isolating and caring for patients.

Rapid health facility assessments conducted by the United States Centers for Disease Control and Prevention (CDC) and partners helped identify gaps in IPC in Sierra Leone early in the epidemic. The identified gaps included lack of IPC oversight, poor waste management procedures, lack of protocols for triage and isolation, lack or misuse of personal protective equipment, and inadequate standard infection control precautions. This work led to the development of partnerships with the WHO and other organizations and the establishment of a Ministry of Health-supported national task force to coordinate efforts to improve IPC in health facilities that did not have Ebola Treatment Units. Local and facility-level IPC specialists were trained to oversee and lead ongoing facility IPC improvements, including training health care workers and ensuring the availability of IPC supplies (Hageman et al. 2016).

Training on screening, IPC, and isolation was helpful, but not sufficient to prepare health care workers for the diverse set of situations and obstacles that they would face while combating the outbreak. The Ebola screening algorithm, or "decision tree," was used to decide whether a patient was a

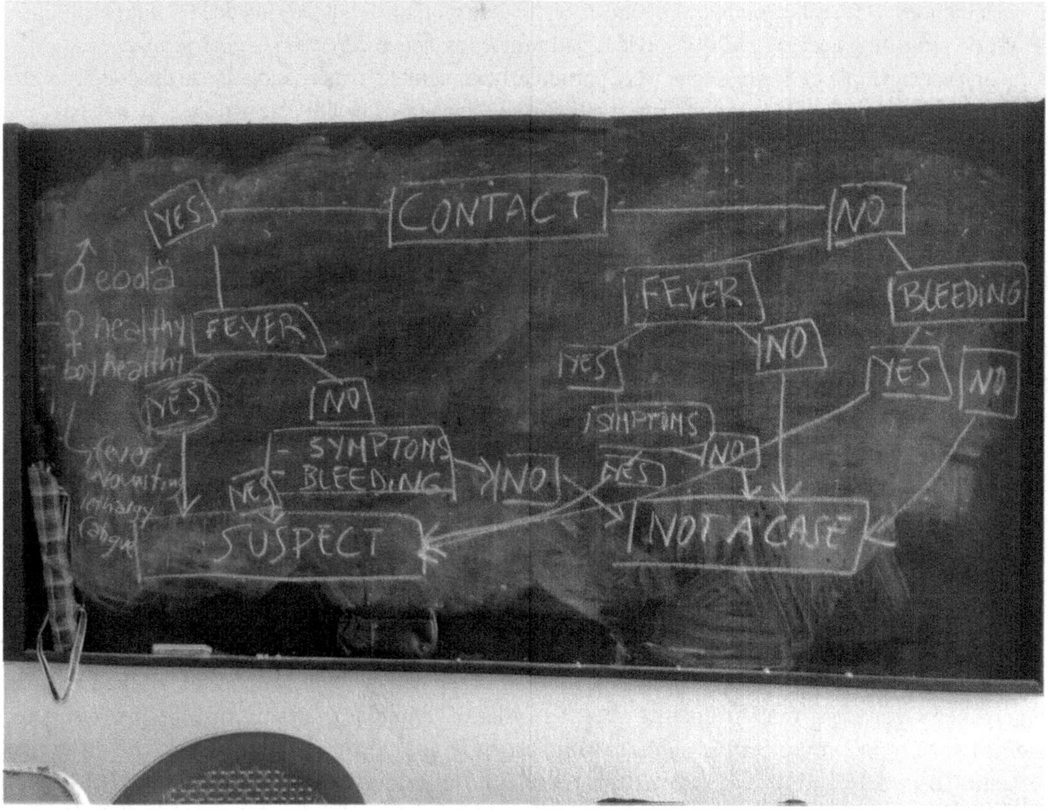

Fig. 21.5 Chalkboard showing the Ebola screening algorithm, or "decision tree"

suspected case or not (Fig. 21.5). The process of screening and use of the algorithm for decision-making was complex. During an Ebola outbreak, a suspected case of Ebola is defined through four potential pathways: (1) any person (alive or dead) suffering or having suffered from a sudden onset of high fever and having contact with a suspected, probable, or confirmed Ebola case or a dead or sick animal; (2) any person with sudden onset of high fever with at least three associated symptoms; (3) any person with inexplicable bleeding; or (4) any sudden, inexplicable death (WHO 2014b).

Many of the scenarios and challenges were as diverse as they were morally difficult (Table 21.1). Health care workers were faced with complex clinical decisions about how to adequately isolate patients when all of the patient rooms are occupied, or how to protect the babies of mothers who test positive for Ebola, yet still need the care of a mother. Moreover, health care works were faced with other tough decisions beyond the confines of the health workers' clinic day. For example, many of the health workers had to make the difficult decision to stay away from their own children during the epidemic, fearing putting their loved ones at risk for Ebola.

The emotional turbulence faced by health care workers during the epidemic was always present, as they encountered the breakdown of social connectedness and trust within and between health facilities, communities, and even their own families (McMahon et al. 2016). In addition to the IPC trainings, other efforts were made during this period to help safeguard the well-being of health care workers in the district. One such intervention was the construction of a second ETU in Kenema District by the International Federation of Red Cross and Red Crescent Societies (IFRC). The new IFRC ETU helped to reduce the burden of Ebola patients at Kenema Government Hospital where capacity was too often surpassed (Fig. 21.6). In addition, the psychosocial team at Kenema Government Hospital developed an onsite counseling center where health care workers and Ebola survivors could receive emotional support.

Table 21.1 Real-life examples of the numerous moral dilemmas that were faced by many health care workers during the Ebola epidemic

What is a health care worker to do...
• ...with suspected patients in a PHU that lacks an isolation ward when the ambulance will take 8 hours to arrive, and there is only one patient room occupied by a woman in the throes of delivery?
• ...with the baby of an Ebola positive patient who tests negative, yet still needs the love and care of a mother?
• ...with a suspected patient who must wait 2 days to get results back, facing either contact with others in the isolation ward, or risk dying outside of the ward if they don't receive early life-saving treatment? .
Beyond the confines of the health workers' clinic day, they were faced with other difficult decisions that were more personal. Should they...
• ...return home to provide care for their own children after a day of caring for an Ebola patient?
• ...continue to work in the ETU even after their family members forbade them from doing so?
• ...divulge that they see signs of Ebola in a colleague, risking either loss of a friend's trust or loss of his or her life? .

Fig. 21.6 International Federation for Red Cross and Red Crescent Societies (IFRC) Ebola Treatment Unit in Kenema, Sierra Leone

21.5 Mabel's Story

This story is an illustration of the challenges faced by individual health care workers and highlights the importance of support and training during times of crisis. Mabel Momoh is one of the health care workers who was trained as part of this IPC project in mid-August 2014 (Fig. 21.7). For 7 years, she had worked as a maternal and child health aide at the Konta Community Health Post in Kenema District. She was trained by IRC and DHMT staff on infection control, barrier nursing, and isolation

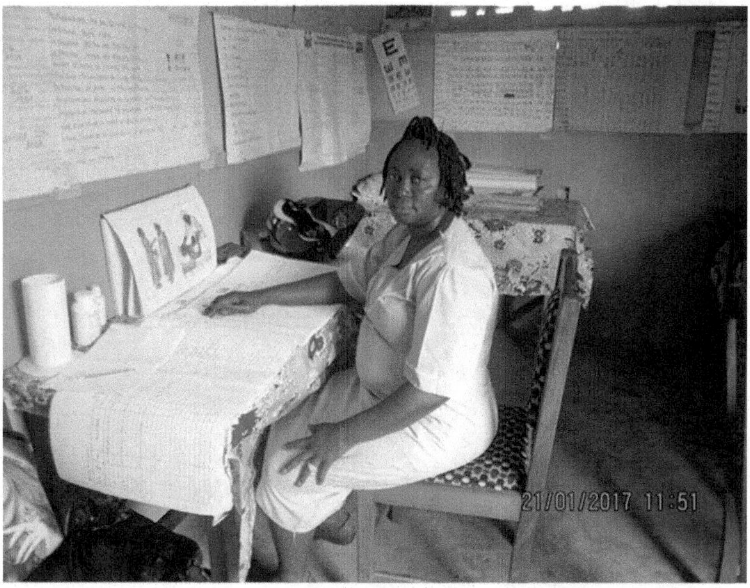

Fig. 21.7 Photograph of Mabel Momoh, a health care worker from Kenema, Sierra Leone

and referral of suspected Ebola patients. The team also provided her with essential supplies required to practice infection control and universal precautions, including chlorine, boots, gloves, and Veronica buckets (a bucket with a tap near the bottom that is perched above a basin for purpose of hand washing).

In early September, a few weeks after training, a woman visited the Konta Community Health Post for the first time. Mabel quickly learned that this woman was from Mano, a town under quarantine due to a high number of Ebola cases. The patient was both pregnant and bleeding. Mabel recognized that there was a very high chance that the woman had Ebola, so she tried to take an accurate history. The woman claimed to have only become sick in the last day and that her husband was away for work. Seeing the woman's symptoms and remembering her training, Mabel contacted a community health worker in Mano and discovered that the woman's husband had recently died of Ebola. Mabel immediately called the Alert Line and informed them of a suspected case of Ebola.

The woman was taken to the labor ward, and without touching her, Mabel assessed that the pregnancy was at 28 weeks gestation (7 months). During the assessment, the woman suddenly miscarried, expelling the fetus onto the bed. Wearing the universal gloves and apron that the IRC had provided, Mabel washed her hands in dilute chlorine solution, disposed of the fetus, and placed the wet linen and clothes into a bucket of chlorine solution. She soaked a piece of cloth in chlorine solution and cleaned the woman before washing her hands again, dressing the woman in new clothes and feeding her. The fetus was buried and the linen and clothes were burned, according to protocol.

A couple of hours later, the woman's sister came to Mabel and said that the woman was bleeding again. Mabel went to the woman, gave her oral rehydration sachets and malaria medicine, and the bleeding was controlled. Later that night, the woman's sister woke Mabel to tell her that the woman had died. Mabel called the Alert Line to inform them of the death.

Throughout the treatment, Mabel tried to educate the woman's sister on Ebola and guide her on appropriate infection prevention actions. Unfortunately, the sister did not believe it was Ebola that had killed her sister, a common response from community members early in the epidemic. The following morning, the woman's family from the village accused Mabel of lying and demanded that the corpse be given to them. Mabel refused to release the body, again adhering strictly to the protocol she learned. Instead, she took time to explain to the family about the Ebola virus and the dire situation facing Sierra Leone. She sent for the Health Development Committee Chairman and Town Chief, who also talked to the family. Eventually, the burial team from Kenema came to manage the corpse in order to prevent further spread of the virus. Before removing the body, the burial team took a swab of the body which later proved positive for Ebola. Through discussions with the woman's sister, Mabel created a list of all the people with whom the woman had come in contact. The next day, Mabel traveled to Mano with the surveillance team to gather more information. The community remained in denial, not accepting the information and telling the Chief that Mabel had killed the woman. Despite this resistance, Mabel stayed and talked with the community, further recognizing the importance of quarantining the house and completing the contact tracing. After 2 h of discussions, the community finally shared the location of the woman's home.

Following these events, Mabel and the two traditional birth attendants (TBAs) who had been supporting her throughout the incident were placed under 21 days observation for Ebola. After 21 days, they remained Ebola-free. Mabel's courage to continue providing care in the facility, her knowledge of Ebola risk factors, and quick implementation of IPC protocols likely contributed to reduced spread of Ebola both in the facility and among the community.

21.6 Community Perception of Health Services and Health Care Workers

The community fear and negative attitudes towards Mabel were not unique to the catchment area surrounding Konta Community Health Post. In August and September, 2014, a team from the CDC, the DHMT, and IRC collaborated to conduct focus group discussions among health care workers and pregnant and lactating women in Kenema (Dynes et al. 2015). The group discussions revealed favorable perceptions and experiences with the IPC trainings, in terms of increased use of routine services and perceptions of safety among health care workers. There was consensus among pregnant and lactating women that fear of contracting Ebola at health facilities contributed to reductions in use of health facilities during the epidemic. Early in the outbreak, there were rumors that health care workers were being paid for each patient that was referred to Kenema Government Hospital as a suspected Ebola case, and that health facility staff were injecting patients with Ebola or taking their blood for financial gain or magical power. All vehicles and/or foreigners that came into the community were thought to be bringing Ebola with them. It was reported that these rumors dissipated through time, but even in September 2014, some community members still feared coming to the health facility because of Ebola-related fears. In contrast, some health care workers noted that community members were coming to the health facility just for the purpose of washing their hands with the chlorinated water.

While fear was reduced among health care workers following IPC training, the risk and accompanying fear was never fully eliminated. One health care worker said, "*Fear – too much! Fear is always within us. You do not know where your patients have come from.*" Another health care worker said, "*If the health workers at the higher level [hospital] can get Ebola and die, then we can, too*!" Through training and access to IPC equipment and supplies, many health care workers were able to overcome their fear and remain in place so that women and children could continue to receive maternal and child health services that were integral in reducing preventable maternal and newborn deaths.

21.7 Impact of Ebola on Reproductive and Maternal Health Service Use and Provision

There is little doubt that the Ebola epidemic had a sudden and significant impact on maternal and reproductive health care services across Kenema District. One of the critical aspects of emergency obstetric and newborn care (EmONC) is the ability to conduct cesarean sections (C-sections) for high-risk pregnancies and labor complications. The total number of C-sections in Kenema district dropped

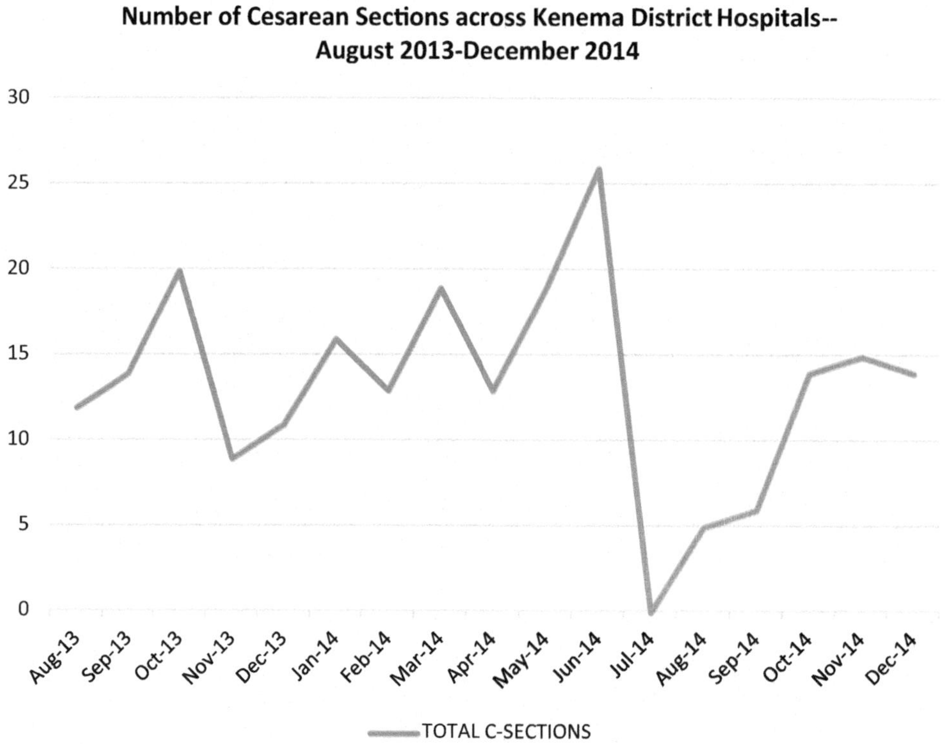

Fig. 21.8 Total number of cesarean sections during the early months of the Ebola outbreak in Kenema District (Unpublished data from the Sierra Leone Ministry of Health and Sanitation)

dramatically from 26 in June 2014 to zero in July of the same year (Fig. 21.8). Taking a slightly different view, the number of C-sections decreased from an average of 16 per month in the 5 months preceding the outbreak (January–May 2014) to an average of 10.2 per month during the 5 months of the outbreak (June–October 2014) (Sierra Leone Ministry of Health and Sanitation unpublished data 2014).

Routine outpatient pregnancy and family planning services also declined, with the most abrupt shifts in services occurring between May and July 2014 (Fig. 21.9). The total number of pregnant women who received at least four antenatal care visits across Kenema District decreased by 23% from May (1788) to July (1369), and the number of postnatal care visits within 48 h after birth decreased by 21% in the same period (1923 in May; 1512 in July). Moreover, Kenema health facilities distributed, on average, 534 fewer oral contraceptive pill cycles and 718 fewer contraceptive injections per month in the Ebola-outbreak period compared to the preceding 5 months of the pre-Ebola period (Sierra Leone Ministry of Health and Sanitation unpublished data 2014).

What might be even more striking than the initial reductions in provision of services was the swift return of services to near pre-Ebola levels. For example, the initial reduction in C-sections lasted only a short time, noting a steady climb back to baseline by October 2014. The average number of C-sections performed per month in Kenema District during the 5-month Ebola-recovery period (November 2014–March 2015) was 15.8, up from the average during the Ebola-outbreak period (June–October 2014) of 10.2 per month.

The distribution of family planning commodities also began increasing gradually during the Ebola-recovery period, though less steadily. For example, the average number of contraceptive injections given per month in Kenema District during the recovery period was 2263, up from the 1862 from

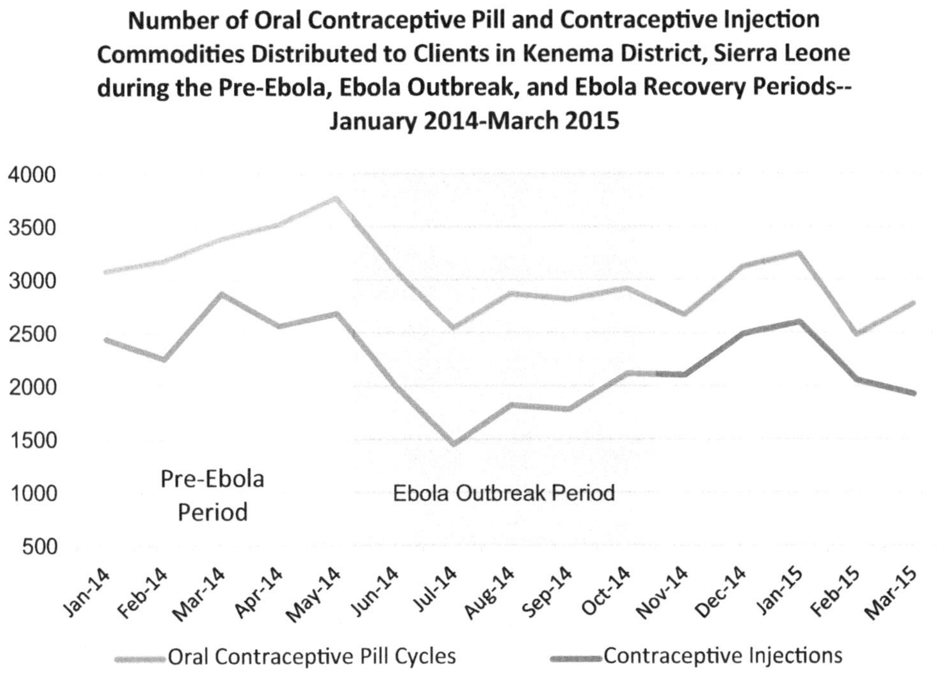

Fig. 21.9 Distribution of contraceptive commodities before, during, and after the Ebola outbreak in Kenema District (Unpublished data from the Sierra Leone Ministry of Health and Sanitation)

June–October in the first 5 months of the outbreak. The distribution of oral contraceptive pill commodities remained relatively constant in the recovery period with an average of 2883 cycles distributed per month (Sierra Leone Ministry of Health and Sanitation unpublished data 2014).

21.8 Conclusion about Health Care Workers and Health Services in Kenema

Sierra Leone, the Eastern Region of the country, and Kenema District have all made significant gains in health indicators since the end of the civil war in 2002 and subsequent introduction of the FHC-I in 2010. These major achievements highlighted the need to sustain gains and prevent loss of progress, despite the threat of Ebola.

Ebola tore at the fabric of communities across West Africa, destroying social connectedness and trust in health facilities, communities, and families. This experience was intensely and intimately felt by health care workers, who became frontline responders to the deadly outbreak. Over 350 health care workers across Sierra Leone became infected with Ebola—92 of them were from Kenema District (Senga et al. 2016). Despite the disproportionate risks faced by these health workers, only 4% of primary health care facilities were closed in Sierra Leone during the height of the outbreak (UNICEF 2014). The subsequent impact of serving on the frontline of the disease exposed health care workers to numerous threats beyond their physical health and well-being.

In focus group discussions in Kenema and Bo in December 2014, providers recounted Ebola's destruction of social connectedness and sense of trust within and across health facilities, communities, and families. They described feeling lonely, ostracized, unloved, afraid, saddened, and no longer respected. Many noted that family members stopped coming near them or talking to them due to fear. One person said that his family blamed him and other health workers for the outbreak. "*You, the health workers, currently we are afraid of you because you possess the disease*." He said that he felt as if he "*owns the Ebola*." A woman expressed experiencing loneliness after the workday ended, since she went home and had to sit alone and isolate herself from her family.

They also discussed restrictions on behaviors that enhance coping, including attending burials and engaging in physical touch (e.g., hugging, handshaking, sitting near, or eating with colleagues, patients, and family members). They described infection prevention measures as necessary but divisive because screening booths and protective equipment inhibited bonding or 'suffering with' patients. Health care workers felt that the goggles and masks protected but also dehumanized them in the face of patients and that 'thermometer guns' terrified community members. One provider used the phrase "*turning one's shoulder*" to describe how she felt while interacting with patients:

> "*I feel bad because I am a medical person, and this disease is preventing us from touching patients... I am an MCH [maternal child health] Aide and I always carry out deliveries and immunization...I must touch my patients*."—Kenema District Health Care Worker

The workers said they yearned for "*the way things were before Ebola*". They felt that they were "*not trusted*", "*not loved*", and "*not respected by the community*", that they had fewer friends, and that people wanted to "*keep a distance*" from them. Providers said they heard neighbors whispering things about them and were viewed with suspicion. One provider used the phrase "*killing my spirit*" to describe how the community's perception of her made her feel. She tried to address this problem by saying:

> "*If you are scared of me, it makes me feel bad. And what if I feel bad and get angry and decide not to go to the center [health care facility] again? What if all the health workers sit down and refrain from treating Ebola patients? Who will do that job?*
> *People have come from other countries to help us fight Ebola. If we sit in our own country and say we will not take part in that fight, how will the disease go?*"
> —Kenema District Health Care Worker

Ebola exposed, in extreme fashion, the complexity and grave consequences of neglecting the protection of health care workers at the forefront of any infectious disease outbreak—not only for *their* safety, but also for that of their patients and the world.

During the 2014–2015 Ebola epidemic in West Africa, primary health care workers were the first points of contact for most people who became infected with Ebola. Despite the risk of encountering an Ebola case, many continued providing routine health services. As a result, health care workers were substantially more likely to become infected with Ebola than the general adult population. Many primary health facilities and their staff courageously provided critical health services, even though they lacked access to the necessary supplies, training, or structures to protect themselves. Many facilities lacked the infrastructure for water and basic sanitation. Health facilities quickly became one of the leading sources of transmission. Health workers were vulnerable to infection, and many died as a result.

The loss of health workers gave a significant blow to an already weakened health system, both immediately and in the months that followed the epidemic, and will continue to impact health care delivery for years to come. Their loss was not only felt during the immediate Ebola outbreak, but also in the recovery period in long-term health care. Unfortunately, with the closure of medical training schools due to the Ebola outbreak, there will be a delay of over 1 year before a new cadre of health professionals will be certified and available to work in health facilities in Sierra Leone. The death of each health worker reduced access to health care for thousands of Sierra Leoneans.

Ultimately, efforts to support IPC in Kenema helped to save the lives of many health workers and stop transmission within health facilities. The lessons learned throughout this epidemic will be used to help protect health care workers in epidemics to come. The sacrifice that health care workers made was also seen through the continued use of maternal, reproductive, and child health services, which is quickly rebounding after the height of the outbreak in Kenema. The consequences of health care workers' critical service and sacrifice—and the vulnerabilities that left them exposed in the first place—should not go unrecognized nor unaddressed.

Disclaimer The findings and conclusions in this manuscript are those of the authors and do not necessarily represent the official position of the Centers for Disease Control and Prevention.

References

Central Intelligence Agency. (2016). *Sierra Leone. The World Factbook.* Retrieved January 3, 2017.

Dynes, M. M., Miller, L., Sam, T., Vandi, M. A., & Tomczyk, B. (2015). Perceptions of the risk for Ebola and health facility use among health workers and pregnant and lactating women—Kenema District, Sierra Leone, September 2014. *Morbidity and Mortality Weekly Report, 63*(51&52), 1226–1227. Retrieved May 16, 2018, from https://www.cdc.gov/mmwr/preview/mmwrhtml/mm6351a3.htm.

Hageman, J. C., Hazim, C., Wilson, K., Malpiedi, P., Gupta, N., Bennett, S., et al. (2016). Infection prevention and control for Ebola in health care settings—West Africa and United States. *MMWR Supplements, 65*(3), 50–56. Retrieved May 18, 2018, from https://www.ncbi.nlm.nih.gov/pubmed/27390018

Kilmarx, P. H., Clarke, K. R., Dietz, P. M., Hamel, M. J., Husain, F., McFadden, J. D., et al. (2014). Ebola virus disease in health care workers—Sierra Leone, 2014. *Morbidity and Mortality Weekly Report, 63*(49), 1168–1171. Retrieved May 17, 2018, from https://www.cdc.gov/mmwr/preview/mmwrhtml/mm6349a6.htm.

McMahon, S. A., Ho, L. S., Brown, H., Miller, L., Ansumana, R., & Kennedy, C. E. (2016). Healthcare providers on the frontlines: A qualitative investigation of the social and emotional impact of delivering health services during Sierra Leone's Ebola epidemic. *Health Policy & Planning, 31*(9):1232–1239. https://doi.org/10.1093/heapol/czw055. Retrieved May 16, 2018, from https://www.ncbi.nlm.nih.gov/pmc/articles/PMC5035780/.

Senga, M., Pringle, K., Ramsay, A., Brett-Major, D. M., Fowler, R. A., French, I., et al. (2016). Factors underlying Ebola viral infection among health workers, Kenema, Sierra Leone, 2014-2015. *Clinical Infectious Diseases, 63*(4), 454–459. https://doi.org/10.1093/cid/ciw327. Retrieved May 18, 2018, from https://www.ncbi.nlm.nih.gov/pmc/articles/PMC4967603/.

Sierra Leone Ministry of Health and Sanitation. (2014). *Unpublished data.*

Statistics Sierra Leone (SSL) & ICF International. (2014). *Sierra Leone Demographic and Health Survey 2013.* Freetown, Sierra Leone and Rockville, Maryland, USA: SSL and ICF International. Retrieved December 20, 2016, from http://dhsprogram.com/pubs/pdf/FR297/FR297.pdf.

Statistics Sierra Leone (SSL) & ICF Macro. (2009). *Sierra Leone Demographic and Health Survey 2008.* Calverton, MD: Statistics Sierra Leone (SSL) and ICF Macro. Retrieved December 20, 2016, from http://dhsprogram.com/pubs/pdf/FR225/FR225.pdf.

UNDP. (2015). *Overview: Human Development Report 2015 – Work for human development.* Retrieved May 18, 2018, from http://hdr.undp.org/sites/default/files/hdr15_standalone_overview_en.pdf.

UN IGME. (2017). *Levels and trends in child mortality: Report 2017. Estimates developed by the UN Inter-Agency Group for Child Mortality Estimation.* New York: United Nations Children's Fund.

UNICEF. (2014). *Sierra Leone Health Facility Survey 2014: Assessing the impact of the EVD outbreak on health systems in Sierra Leone.* Survey conducted 6–17 October 2014. Retrieved May 18, 2018, from https://www.unicef.org/emergencies/ebola/files/SL_Health_Facility_Survey_2014Dec3.pdf.

World Health Organization (WHO). (2014a). *Global atlas of the health workforce.* Geneva: World Health Organization. Retrieved from http://apps.who.int/globalatlas/default.asp.

WHO. (2014b). *Case definition recommendations for Ebola and Marburg virus diseases, as of 09 August 2014*. Retrieved May 19, 2018, from http://www.who.int/csr/resources/publications/ebola/ebola-case-definition-contact-en.pdf.

WHO. (2015a). *Health worker Ebola infections in Guinea, Liberia, and Sierra Leone: A preliminary report 21 May 2015*. Retrieved May 19, 2018, from http://apps.who.int/iris/bitstream/10665/171823/1/WHO_EVD_SDS_REPORT_2015.1_eng.pdf?ua=1&ua=1.

WHO, UNICEF, World Bank, & United Population Division. (2015b). *Trends in maternal mortality: 1990 to 2015*. Geneva: World Health Organization; 2015.

Witter, S., Brikci, N., Harris, T., Williams, R., Keen, S., Mujica, A., et al. (2016). *The Sierra Leone Free Health Care Initiative (FHCI): Process and effectiveness review*. Freetown: Health and Education Advice and Resource Team, Oxford Policy Management. Retrieved May 18, 2018, from http://eresearch.qmu.ac.uk/4358/1/eResearch%204358.pdf.

22 Taking Life 'Off Hold': Pregnancy and Family Formation During the Ebola Crisis in Freetown, Sierra Leone

Jonah Lipton

22.1 In Congo Town for the Celebration

It was the evening of May 17th, 2015, in Freetown, Sierra Leone. The state of emergency, declared on July 31st, 2015, had now been effective for almost a year. A weariness surrounding the Ebola crisis was taking hold in the capital. The city had witnessed a steep rise in cases in late 2014 and early 2015, when new clusters of cases were regularly popping up in different neighborhoods. This created some degree of panic, as it was hard to predict where and when a new case would emerge. By now, however, the rate of Ebola transmission was on the way down. Every evening, there was an announcement of the numbers of new cases countrywide, their location, and numbers of deaths. I had listened to the announcement together with my neighbors in the Congo Town area who had gathered for an ad-hoc 'bachelor's eve' celebration for one who was getting married the following day.

At that point, there had been several days without any new cases identified locally, but the daily announcement carried the bad news of a handful of new cases upcountry. The state of emergency—through which hosts of stringent regulations had been implemented, such as travel restrictions, curfews, and lockdowns—was set to be lifted after 42 days without any new cases, a measure established by the World Health Organisation (WHO), which amounted to double the 21-day incubation period of the Ebola virus. The clock had been reset.

The party was on a much smaller scale than it might normally have been. Gatherings of more than ten people were illegal under the state of emergency. By this point, the habitual bribing of the police had become a standard means of bypassing this law. The groom, however, like many people during the crisis, was particularly short of cash. Our neighbors had collected some money for a crate of locally brewed Guinness. One, who worked for the National Power Authority, used his influence to secure electricity for our area—never a surety and at times a rarity—so that we would have light and could dance to an RnB, Afrobeat, and Reggae playlist that my close friend James,[1] who's family compound I lived in, had put together. James, normally the first to dance and crack a joke, was uncharacteristically quiet and contemplative, sitting by himself on the veranda. As one of our other neighbors explained to me, "James is feeling the effects of Rachel being close to giving birth".

[1] All names are pseudonyms.

J. Lipton (✉)
Firoz Lalji Centre for Africa, London School of Economics, London, UK
e-mail: j.h.lipton@lse.ac.uk

D. A. Schwartz et al. (eds.), *Pregnant in the Time of Ebola*, Global Maternal and Child Health,
https://doi.org/10.1007/978-3-319-97637-2_22

22.2 Rachel and James

Rachel and James, a couple in their mid-20s, lived in the Congo Town neighborhood of Freetown (Figs. 22.1 and 22.2), where I conducted long-term ethnographic fieldwork between October 2013 to September 2014 and March 2015 to September 2015. Congo Town, built along the Congo Valley River, was initially settled by freed-slaves from Congo (Fyfe 1962), yet today its residents identify with numerous tribal groupings. Muslims and Christians inhabit the neighborhood in roughly equal proportion. The neighborhood is comprised of both established family compounds, as well as more temporary dwellings, which stand in close proximity to each other. As with many neighborhoods in Freetown, Congo Town is close-knit and socially dense, yet has a crowded and somewhat unstable quality, owing in part to a major boom in population in recent decades.

James and Rachel had been together for about 2 years, and Rachel had become pregnant in the summer of 2014, shortly after the declaration of the state of emergency. Rachel was studying business management and finance at college, but the college had been closed until further notice during the Ebola crisis. Rachel also ran several informal businesses, including an *esusu* (rotating credit association) among mostly female market traders, and a very small store selling household goods, that she had set up in partnership with James. James had worked for several years at a restaurant frequented by tourists and expatriates ("expats"), but had been out of work since September 2014, after their operations had been closed.

Fig. 22.1 View over Congo Town. Published with permission of © Jonah Lipton 2018. All Rights Reserved

Fig. 22.2 View over Congo Town. Published with permission of © Jonah Lipton 2018. All Rights Reserved

The neighbor was right. As James and I arrived back at his family compound, Rachel was sitting on the veranda of James's uncle and auntie, Samuel and Susan. The two of them were both standing close by, alongside two of Rachel's close female friends from the neighborhood. Rachel was in severe pain and her feet were very swollen, meaning that she could barely walk without assistance. A few hours before she had gone to a maternity clinic, run by a nongovernmental organization (NGO), where she had registered earlier in her pregnancy, but they told her that she was not close enough to delivering and sent her home. James suggested going instead to the Connaught Hospital—one of Freetown's main hospitals—but this idea was rejected. Rachel had attempted to enrol in its Maternity Ward several weeks earlier, but had not found it an easy process now that the capacities of state hospitals and medical staff were both severely overstretched during the epidemic, and hospitals were also deemed hotspots for Ebola transmission. Rachel and her friends instead left and headed down to Rachel's family compound, a few 100 m down a steep path into the valley, where her mother was preparing for a traditional *sara* (small ceremony) that aimed to bring about an auspicious delivery. They had decided to go to a local *mami*, an old lady and former nurse who was a midwife, who would deliver the baby at her home, using a combination of Western procedures and customary methods. James and Susan were particularly unsatisfied with this option. James told us, "*they do not have much faith in that family, Christians do not make sara*". Susan agreed: "*we wait for God to intervene*". Rachel's family were Muslim, although Rachel had started attending Church with James's sisters. I asked if God was intervening now, and they laughed. Samuel advised James against interfering further: "*this is women's business*". The tension and uncertainty weighed heavily on us all. It was exacerbated by the fact that this was Rachel's first birth. As Auntie Susan pointed out, "*if you have more experience you can give birth at home, but if you panic you may need a caesarean and they can't be trusted to do that*". James went to his pastor to pray privately with him, as is common for fathers before childbirth. On Susan's instruction, and against her husband's advice, we called a friend in the neighborhood, a taxi driver, who Susan and I drove with to the *mami*'s house on the other side of the valley, which at the time was experiencing a blackout. The *mami* was frail and seemed like she had

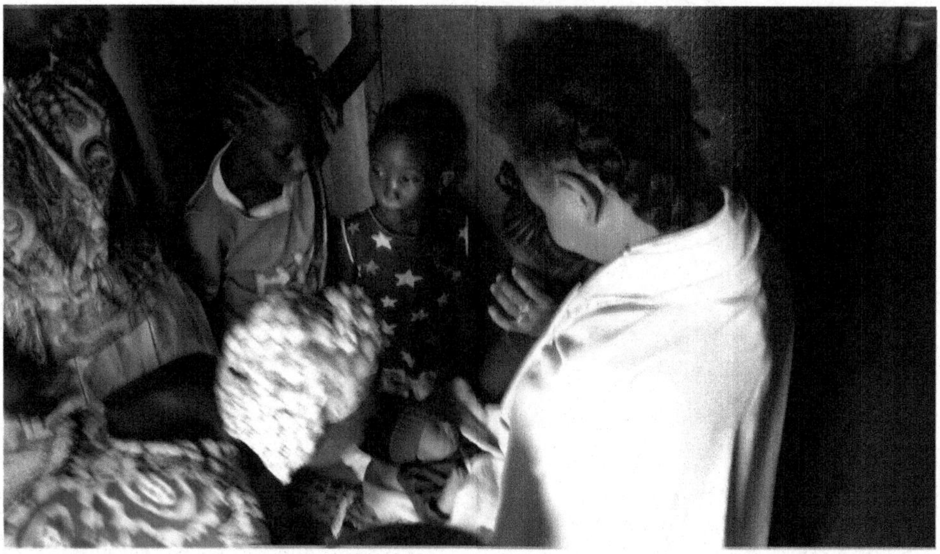

Fig. 22.3 Moses is introduced to some children from the neighborhood. Published with permission of © Jonah Lipton 2018. All Rights Reserved

been drinking that evening. From Rachel's swollen feet she had identified that she would require special treatment, and therefore, was prepared to refer her to the formal medical authorities. Susan fetched Rachel from inside, and we walked her back to the car, where they drove back to the clinic. Rachel was once again turned away, despite being in severe pain and feeling ready to deliver. The following morning a neighbor of Rachel's, who was married to James's adopted sister, took Rachel back to the clinic where she was finally accepted. James was still unhappy to be out of the loop.

Later in the day, James and I met Rachel at the gates of the clinic, carrying their newly born baby boy, who would be named Moses. Despite the fact that the clinic was publicized as free for patients, Rachel was forced to make an informal payment in order to be discharged, just as she had made a similar under-the-table payment in order to be registered.

We arrived back in the neighborhood with the baby, first stopping at James's family compound (Fig. 22.3), then down to Rachel's compound. The neighbors crowded around the baby telling the couple *tenki* (thank you). The baby was taken down to a room that Rachel and James had arranged for him and Rachel to stay in, near to her family. Despite her previous dismissive comment, Auntie Susan performed a *sara* of rubbing salt and leaf around the umbilical cord so that it would come off in 3 days, and the baby was washed. Food was served, prepared by Rachel's mother and neighbors.

The National Ebola Response Centre, which was comprised of national and international authorities involved in the large-scale humanitarian intervention, introduced a series of rules and regulations that defined the state of emergency. These rules and regulations threatened to put life 'on hold', as many of the operations of continuing social life were made more challenging than usual. Therefore, it was often during heightened moments of transition in the life-course, such as around birth or death, that 'Ebola', as the popular experience of living through the state of emergency, became particularly strongly felt (Lipton 2017). It was expected that eliminating the spread of Ebola was an ultimate and shared priority, and that when this objective was achieved, normal 'stable' life would continue.

This assumption, was, however, sharply opposed with those of many of the people who lived through the crisis. Rather than life being 'suspended,' it seemed that the continuity of biological and social life was as, if not more, pressing a concern than ever. Such a tension between continuity and stasis was particularly apparent surrounding bodily transformation, whether through illness or processes of the life-course. Pregnancy, as was evidenced in the events surrounding Rachel's pregnancy and Moses's birth, was a notable example of this. Rachel, as well as the loved ones around her, was experiencing a personal crisis of sorts, during which medical attention was required, yet in which certain established routines appeared harder to enact than normal. The state hospital that she had initially sought medical attention did not accept her, and the NGO clinic she had enrolled in was overstretched. More generally, many people could not afford the higher costs required to use medical facilities, while many, through loss of work, were more hard-up than usual. Public transport was not operating at night, and travel in and out of the city was more difficult than usual, all of which were obstacles to seeking formal healthcare. While the country was experiencing a huge increase of expenditures towards its healthcare system, this did not do much to build on its existing, limited capacity, and instead was largely directed towards Ebola. These included the funding of temporary Ebola holding and treatment centers, and ambulances dedicated for Ebola suspects and victims. A significant number of the country's doctors had died from the virus, and there were crackdowns on smaller clinics and pharmacies, where many people would typically go for treatment or medication.

On top of this, perhaps paradoxically, many were more resistant to using formal healthcare facilities than usual, which were seen as hotspots for catching the Ebola virus. Stories circulated of people being misdiagnosed as Ebola-positive and being put into isolation wards where they could not have contact with their family, or actually catching the virus from Ebola victims. Other rather sinister stories told of nurses actively injecting patients with the Ebola virus. These stories built on long-standing suspicions surrounding nurses as witchcraft practitioners, as well as theories that Ebola was being actively spread in order to secure continued foreign aid funding, which the elites in particular were siphoning off for personal use. Equally, drawing attention to one's illness during the time of Ebola risked stigmatization within the neighborhood, which was another reason for avoiding seeking medical attention (Strong and Schwartz 2016, 2019).

In this chapter, through the central case study of Rachel, James, and their new-born baby Moses, I examine attempts to start a family during the Ebola crisis. The chapter demonstrates how the processes of social and biological reproduction could not be put 'on hold' and appeared to be particularly pressing priorities, in part because the very networks and relationships that are established and maintained in the process of family formation are equally key to day-to-day survival during times of crisis. However, the challenges surrounding pregnancy, birth, and the social processes that accompany them were not necessarily radically different from some of the more enduring challenges that are faced in 'normal' circumstances in a context like Freetown, where material scarcity is widespread, medical infrastructure is weak and de-developed (Robinson and Pfeiffer 2015), birth outside of stable family structures is common, and a range of social beliefs and practices coexist. Responses to birth during the Ebola crisis, therefore, were informed by, and closely connected to, existing practices and responses.

22.3 Taking Life 'Off Hold'

Rachel's arguably untimely pregnancy coincided almost exactly with the declaration of the state of emergency, which became a marker of collective crisis for Sierra Leoneans; it signaled that anyone was at risk of catching the Ebola virus, and that the rules and regulations of the state of emergency were inescapable. However, from the perspective of the ground up, it often appeared that personal experiences of the Ebola crisis, rather than representing a singular period of upheaval, rupture, or danger, were marked more by a series of smaller, interconnected crises. While the Ebola outbreak was, on the one hand, a humanitarian crisis of global proportions, on the other, it often was felt along intimate dimensions.

In the case of James and Rachel, they both experienced significant financial strains that coincided with the pregnancy and the declaration of the state of emergency. Given the restrictions on travel in and out of the country, as well as daily curfews on business activity during the crisis, the restaurant and guesthouse that James worked in had closed its operations; James lost his job and received little financial compensation. Rachel experienced her own financial crisis as the *esusu* (rotating credit associations) that she ran dramatically crashed. One figure quoted was that she found herself in 9 million Leones (USD 1800) debt, as money she owed to her clients could not be paid back. One client threatened legal action, after which Rachel ran away for several days, not telling her family or James where she had gone. In addition, the small grocery store that Rachel and James ran struggled during this time, in part because the *luma*, the monthly produce markets on the outskirts of Freetown where Rachel would typically buy produce in bulk, were banned under the state of emergency regulations.

Rachel and James were unmarried when Rachel got pregnant. James in particular was keen to arrange a marriage, so that the birth of their child would mark the start of the establishment of an honorable family. However, this was no easy task, especially now that money was harder to come by. James attempted to *lay kola* for Rachel. This represents the 'traditional' or 'country' method of marriage, where a bride price and other symbolic and monetary gifts are given from the husband and his family to the bride and her family at an engagement ceremony. The initial meeting between the families, however, was not fruitful. James complained that Rachel's family was being unreasonable in asking for too much for the bride price, and not taking into account the circumstances of the crisis, or the ways that he had previously supported Rachel. Some of Rachel's friends were also critical. As one female friend put it, 'they should not treat her like meat for sale, that is old-fashioned thinking.' They equally did not offer to prepare food for the engagement ceremony, which is traditionally an obligation of the bride's family, as James is a 'stranger' coming to the household in need of being hosted. James called a family meeting (with his family) as to whether he should attempt to renegotiate, but his father advised him strongly not to make another offer as it would bring shame on the family. Rachel later told me that she had not wanted her family to make an affordable offer, as it would lead to even greater embarrassment as they would not have the resources to put together a marriage programme that would meet the widely held high expectations of the grandeur and scale considered appropriate for such an event. In this way, the Ebola crisis was widely not seen as an excuse to halt the pursuit of family formation.

By the time of Rachel's delivery, described above, James and Rachel remained unmarried and did not have plans in the pipeline to do so. This further contributed to the instability around Moses's birth and immediately after, as parental roles and responsibilities were not clearly defined. Perhaps for this very reason, Rachel and James organized a large-scale *pulnador* (baby-naming ceremony), which was performed 2 weeks after the birth. Traditionally, this ritual is performed 1 week after the birth, symbolizing the taking of the baby outside of the home and introducing them to the public. A name would be given to the baby, often administered by an elder or a religious leader. James and Rachel arranged a small ceremony a week after the birth, according to this tradition, administered by James's uncle Samuel. However, they spent an extra week making arrangements for a much bigger public ceremony. This was not only challenging because of Rachel and James's financial situation, but also legally challenging given the restrictions on gathering of over ten people under the state of emergency bylaws. James, through his connection with the local police station through a friend and neighbor, was able to secure a uniformed police presence at the ceremony, which gave it the necessary semblance of legality.

Hundreds of people were invited, including numerous significant figures, such as James's old boss at the restaurant and guesthouse, who made a financial contribution. James was relieved that the relationship had not died; and in fact, the *pulnador* had provided a means of keeping it alive. Rachel invited members of the Muslim youth group that she had been part of. Many of the guests were mutual friends and neighbors from the neighborhood. The programme took place in an open space behind James's family compound, where they erected a tarpaulin, under which the official proceedings took place. They rented chairs, which were carefully placed in neat rows around the small square. Fresh ginger ale was served to the guests while they were waiting for the ceremony to start. The ceremony was overtly Christian; proceedings were led by the pastor of a church that James had recently started attending. Hymns were sung, verses from the Bible read, and the pastor delivered a sermon. During the sermon, the pastor commented on how childhood should be taken more seriously in Sierra Leone, given that people are 'once an adult, but twice a child'. He noted how parents often start to neglect children after a few years. He also commented on the fact that many Sierra Leoneans had children out of wedlock, which was true of Rachel and James, but that God will judge the parents according to the way that they raise the child. This was a welcome and consolatory message. The sermon was followed by the official naming of the baby, who Rachel held in her arms as guests came and placed money in her lap, as was customary. Afterwards, food was served on disposable plates, prepared by James's sisters and stepmother. The food was the same type of food, often named 'white food', that would typically be served at wedding receptions: *jollof* rice (also called *benachin* in West African countries), noodles, prawn crackers, and balls of beef. All of these features of the ceremony—the food, the religious aspects, and the scale—were more reminiscent of a Christian wedding ceremony than a typical *pulnador*.

The fact that the baby-naming ceremony resembled a wedding was far from coincidental and appeared to be a very conscious substitute for Rachel and James as a means of forwarding a positive public image as they set about starting their own family. The emphasis put on the *pulandor* in adverse conditions demonstrated the ways that the lifecycle ritual and family formation after birth were prioritized during the Ebola crisis. Such concerns were, in many ways, more enduring and significant than the relatively temporary concerns of the Ebola crisis. While in the short term the Ebola outbreak represented a pressing danger and the state of emergency a major obstacle, the challenges of family formation were for many people much more ongoing, with lasting concerns and priorities.

In addition, as the *pulnador* demonstrated, such long-term concerns were in many ways inseparable from the day-to-day means of living life and getting through the crisis. The *pulnador* provided a powerful means to secure Rachel, James, and Moses's place into a social network and community of support; while James and Rachel had limited income during this time, securing these relationships was more important than ever, as it is these very networks and relationships that are reached to in times of need. This proved to be the case, as many of those invited to the programme became actively involved in raising Moses and in supporting Rachel and James through the crisis and beyond.

22.4 Exception or the Rule?

While the *pulandor* had many effective elements, Rachel and James remained in a long-standing position of uncertainty; one that would last for many years. They finally were married in December 2016, a little over a year after Sierra Leone had been declared Ebola-free. In the meantime, the baby moved around between Rachel's family home and James's. Rachel also moved between living with James's family, with her own family in the neighborhood, and staying with her sister on the outskirts of town. None of these options proved completely satisfactory; living with James was not deemed appropriate while formal arrangements of marriage had not been made between the families, but equally being apart was challenging.

The liminality that they seemed to be occupying was in some senses emblematic of the liminality that was being collectively experienced during the Ebola crisis; a period of uncertainty, in which many of normal operations of life were challenging to perform. However, while the Ebola crisis did seem to exacerbate their precarious position, it must also be acknowledged that such unstable characteristics of motherhood and childhood are, in Freetown and elsewhere on the continent, the rule rather than the exception.

Jennifer Johnson-Hanks's research in southern Cameroon among *Beti* people reveals the ways that "rather than a clear threshold into female adulthood, here motherhood is a loosely bounded, fluid status. Contrary both to folk intuition and to the assumptions of a life cycle framework, *Beti*[2] motherhood is not a stable status. *Beti* women who have borne children are not necessarily mothers, at least not all the time. Motherhood, instead, constitutes a temporary social status, an agent position that can be inhabited in specific forms of social action" (Johnson-Hanks 2002:865). Becoming a mother is not achieved through the biological process of giving birth, but rather through honorable family formation, which "demands mastery of very different spheres" (ibid 870-1), including education, biological motherhood, and employment. However, these different spheres do not follow one another in a linear manner, and at times, can play out against one another. Pursuing education might be a mark of seniority for a woman, but it might equally undermine her status as a mother or adult in the process as they become a 'school-girl'. Such observations resonate strongly with popular experiences and understandings of motherhood in Freetown, including Rachel's own experiences as a biological mother whose education has been stalled during the Ebola crisis.

In Sierra Leone, children are often not raised in single households, but rather move between several different households through the course of their childhood and are raised by range of actors within the home, village, or urban neighborhoods. This is often a result of shifting household circumstances: such as the death of a guardian, migrations, legal or informal disputes, and financial fluctuations. In many cases, children are born outside of a stable parental partnership. Negotiating the rights over and responsibilities of looking after the children is often a drawn-out process, which can last for years in some instances, often involving a range of family, religious, customary, and state authorities.

James, Rachel, and Moses's experiences, therefore, did not necessarily represent a huge deviation from the typically unstable processes of family formation in Freetown, where the necessary resources to establish households are extremely hard to come by, and there is a long history of the mixing social practices and traditions rather than a singular set of routines. Social practices surrounding the life-course, which have now become commonplace in Freetown, are seemingly reflective of this reality. For example, it is common for people to rely on multiple authorities and methods—biomedical, religious, and customary—throughout the processes of pregnancy and childbirth. This was evidenced in the events surrounding Rachel's birth, which involved various medical institutions—run by the state and an NGO—as well as the consulting of a pastor and a *mami* and the performance of *sara* in the household after the delivery in a clinic. Equally, the *pulnador* ritual was part of a process of building and strengthening of social networks, through which family formation takes place in an ongoing and protracted manner, rather than representing a singular moment of the establishment of a new family.

22.5 Conclusion

In this chapter, I have described the ways that a young couple attempted to start a family during the Ebola crisis in Freetown, Sierra Leone. Rachel became pregnant during the early stages of the state of emergency, delivered her baby during its height, and married and moved into a home with James a year after it was lifted. James, Rachel, and Moses's story demonstrates the ways that the onerous rules

[2]The *Beti* people are a Bantu people residing mostly in central Cameroon. They speak a dialect of the *Fang* language, termed *Banti* or *Edwondo*, that belongs to the Niger-Congo family of languages. The *Beti* people practice double exogamy.

and regulations of the state of emergency in effect were geared towards putting life 'on hold', as the elimination of the Ebola virus became a national and international priority. However, for those living through the crisis, the processes of biological and social reproduction were either unavoidable realities or priorities.

I have suggested that this can in part be explained by the fact that such processes were reflective of more enduring concerns than those of the Ebola virus; and that, in addition, the day-to-day business of getting through the crisis relied on support networks that life-course ritual and family formation were central to building and maintaining. However, rather than such responses being entirely novel, I have suggested that these strategies are in keeping, and reflective of, the ongoing difficulties and dangers surrounding reproduction in a context with a limited formal health infrastructure and widespread material scarcity.

AcknowledgementsThe research for this chapter was generously supported by the Horowitz Foundation for Social Policy, the Halperin Memorial Fund, the Alfred Gell Memorial Studentship, and the ESRC Centre for Public Authority and International Development grant: ES/P008038/1.

References

Fyfe, C. (1962). *A history of Sierra Leone*. Oxford: Oxford University Press.

Johnson-Hanks, J. (2002). On the limits of life stages in ethnography: Toward a theory of vital conjunctures. *American Anthropologist, 104*(3), 865–880.

Lipton, J. (2017). 'Black' and 'white' death: Burials in a time of Ebola in Freetown, Sierra Leone. *Journal of the Royal Anthropological Institute, 23*(4), 801. https://doi.org/10.1111/1467-9655.12696.

Robinson, J., & Pfeiffer, J. (2015, February 2). *The IMF's role in the Ebola outbreak: The long-term consequences of structural adjustment*. Retrieved October 7, 2017, from http://www.brettonwoodsproject.org/2015/02/imfs-role-ebola-outbreak/.

Strong, A., & Schwartz, D. A. (2016). Sociocultural aspects of risk to pregnant women during the 2013-2015 multinational Ebola virus outbreak. *Health Care for Women International, 37*(8), 922–942. https://doi.org/10.1080/07399332.2016.1167896.

Strong, A., & Schwartz, D. A. (2019). Effects of the West African Ebola epidemic on health care of pregnant women—Stigmatization with and without infection. In D. A. Schwartz, J. A. Anoko, & S.A. Abramowitz (Eds.), *Pregnant in the Time of Ebola: Women and their children in the 2013-2015 West African Ebola Epidemic*. New York: Springer.

23 Providing Care for Women and Children During the West African Ebola Epidemic: A Volunteer Physician's Experiences in Makeni, Sierra Leone

Emily Bayne

23.1 Ebola and Me

Prior to the onset of the Ebola virus outbreak, my background included a variety of clinical work in the United Kingdom, with a focus on Emergency Medicine. I also had some experience of working in such developing countries as Madagascar, Kenya, and Somalia and in remote areas on expeditions. Despite also completing the Diploma in Tropical Medicine and Hygiene (DTM&H) at the University of Liverpool, I had not done anything that could really have prepared me for working in an Ebola Treatment Centre (ETC).

When I was first approached about working in Sierra Leone with Ebola virus disease (EVD), I was asked to give a decision in principle based on whether I felt able to get involved. I didn't hesitate to say yes, although in reality I probably didn't fully consider the implications at first. I was initially asked in September 2014 at the height of the Ebola Outbreak. On 8 August, 2014, the Director-General of the World Health Organization had declared the Ebola outbreak in West Africa a Public Health Emergency of International Concern (PHEIC) (WHO 2014) This was a time when the media were sharing distressing scenes of patients dying in the streets due to a lack of medical provision, and for me, it was impossible not to feel motivated to help in any way I could.

As everyone involved in humanitarian work will know, funding, planning, and ultimately decision-making take time and it wasn't until the end of November 2014 that my deployment was confirmed.

There were many clinicians that left their regular jobs, families, and often countries to help during the Ebola crisis. It was a decision some thought was selfless, and others thought was madness, but one that I saw as ordinary. I became a doctor to help people and this was a country with a crisis that they were unable to manage alone, and they needed help. The people we forget when we make these decisions are often our loved ones; I didn't give mine a choice, I merely told them where I was going and why. Fortunately, they respected this decision, but I'm sure they worried about the possibility that I could become infected with the Ebola virus, and what the implications would be if this occurred. My parents and the rest of my family were very supportive and didn't really share any concerns they may

E. Bayne (✉)
National Health Service, Kendal, Cumbria, Great Britain

International Medical Corps, Bombali District, Sierra Leone

D. A. Schwartz et al. (eds.), *Pregnant in the Time of Ebola*, Global Maternal and Child Health,
https://doi.org/10.1007/978-3-319-97637-2_23

have had. I was also lucky to have a very supportive group of friends. Some of my colleagues found it difficult to understand why I felt I should go, especially when it meant leaving other colleagues to fill a gap in the rotation.

I was fortunate to undergo a week of 'Ebola Training' prior to being deployed, and it was during this period that the reality of what I might face started to hit home. As we were given a lesson about the various items of Personal Protective Equipment (PPE) we would need to wear, I had an impending feeling of claustrophobia. I was aware that I might get slightly different kit in the field, but it was very useful to have an opportunity to try on the basics, although with frost on the ground outside in England, the temperature was somewhat different than in Sierra Leone. My first-time dressing in the bulky equipment was clumsy and difficult, and the end result I can only describe as resembling a claustrophobic alien, but second time around I had started to get the hang of it and felt more comfortable with the process.

My training also focused on the importance of resilience and being able to look after yourself and fellow colleagues. It included cultural awareness sessions and the opportunity to learn a few words of commonly used language. I was also able to discuss a variety of clinical issues and explore my own motivations and concerns for deployment. Realistic expectations about what you are going to see and what you can achieve in a situation such as this are important.

The training was also an opportunity to gain an insight into how an Ebola Treatment Centre (ETC) may be set up, and to consider what may be important in terms of design and function.

At the end of the training, armed and ready with my new friendly thermometer, I set off on my long journey. I knew that I would have to spend the next 9 weeks measuring my temperature twice a day, as that was the protocol. Somehow it wasn't a fear of Ebola that I had, rather a fear of another more common, simple illness with any symptoms in common with Ebola. There was a moment of realization that a short bout of traveller's diarrhea could become a serious concern that I might have Ebola.

23.2 Makeni, Sierra Leone

My first real experience with the Ebola outbreak came on arrival at Lungi airport in Sierra Leone. It was 4 o'clock in the morning as I stepped off the plane to the smell of chlorine. I stood in the darkness waiting to wash my hands at a bucket tap a few feet away on the tarmac. A few minutes later, a man in a facemask and gloves held a temperature gun at my forehead and I was allowed to leave the airport. Despite the darkness and early hour, I could already feel the heat, a mere 26°C, and a stark reminder of the challenges I would face wearing PPE in the heat of the day.

I discovered that I was being deployed to work at an Ebola Treatment Centre in a place called Makeni. Makeni is the fifth largest city in Sierra Leone by population and is the capital of Bombali District (Figs. 23.1 and 23.2). Makeni had a population of 126,059 in the 2015 census and lies approximately 137 km (85 miles) east of Freetown (The Infolist n.d.). Within Makeni, there is a mixture of both Christianity and Islam, and people live happily side-by-side celebrating both of these religions. There are also a range of languages within Makeni and the Bombali district including Krio, Temne, and English. The infrastructure in Makeni is generally good, with a tarmac road in and out of the town, a large government hospital, and university. Stepping off the tarmac, the majority of roads are sand or dirt and getting around in vehicles can be a challenge, especially in rainy season. Locally, most journeys are made on motorbike, with many motorbike taxis readily available.

The international response to the Ebola Epidemic took some time. When I finally arrived in Makeni, the Centre there had not yet been finished. Located a few kilometres down a tarmac road just outside of the town (Fig. 23.3), it was still under construction. This felt frustrating, as I was keen to start seeing patients. The team I had been deployed with from the UK, a group of volunteers with a variety of clinical backgrounds, including nurses, doctors, health care assistants and paramedics, all felt the same frustration. We had met during our 1-week training in the UK prior to our deployment, but once

Fig. 23.1 The Bombali District in northern Sierra Leone. The Temne and Limba tribes are the predominant ethnic tribes, and the district is mostly Moslem with a large Christian minority. Available from: https://commons.wikimedia.org/wiki/Atlas_of_Sierra_Leone

out in Makeni we were teamed up with National Staff and other clinical staff employed directly by the nongovernmental organization (NGO). There was a wide range of nationalities among all of the staff including British, Kenyan, and American.

Initially, our tasks mostly involved logistics and planning; from the basics of finding equipment and unpacking it from the many pallets and boxes, to developing rota patterns for staff. I spent an afternoon in the blistering heat moving pallets and boxes in an attempt to get equipment and supplies where they most needed to be.

During these first few days, I was fortunate enough to visit a nearby ETC that was already up and running and it was only then that suddenly everything began to feel tangible. I had my temperature checked as I left my accommodation, and once again at a road check point and as I entered the ETC. I washed my hands in chlorine at the door and my feet were sprayed with chlorine. As I went in, I had my first glimpse of the reality of an ETC.

There was no air conditioning in any of the tents, and with an outside temperature of approximately 40 °C, the heat was stifling. There were some air conditioning units for the medical office, but they kept tripping the generator and so were not being used, and there was nothing in the wards. Despite the heat, the visit was extremely useful, as it provided an insight into how our ETC could run and an opportunity to discuss some of the teething problems that this centre had, and how we might avoid them.

Learning to use the PPE safely was a key skill, as one mistake opened the possibility of catching Ebola. So despite our training in the United Kingdom (UK), we had several dry run training sessions at the ETC. These were an opportunity both to practice dressing (donning) and undressing (doffing)

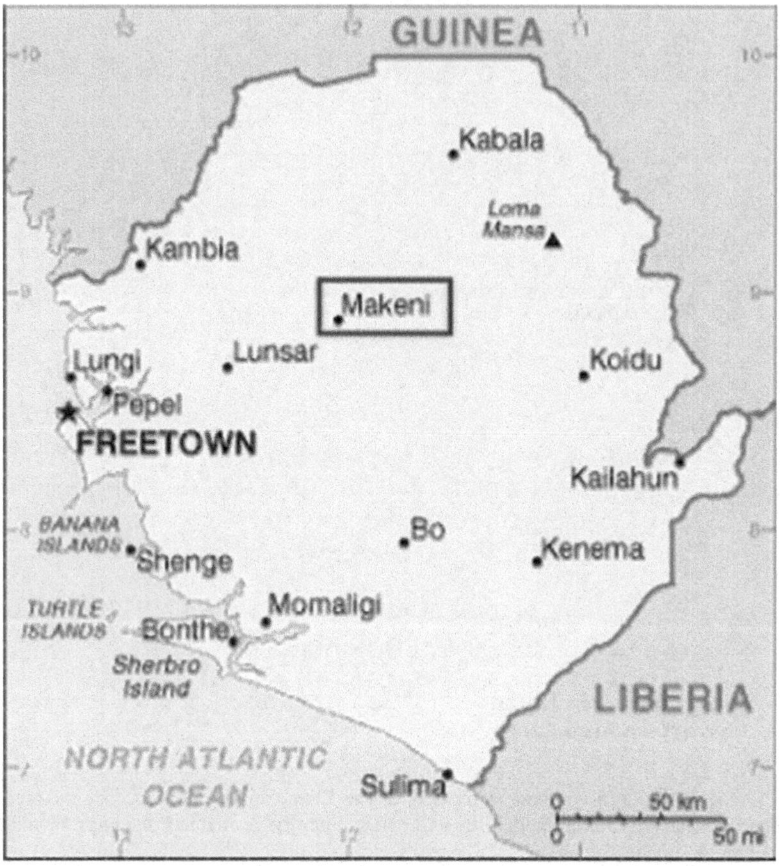

Fig. 23.2 Makeni, the capital of Bombali District in Sierra Leone. Available from: https://upload.wikimedia.org/wikipedia/commons/a/ab/Makeni_in_Sierra_Leone_CIA_map.jpg

safely and to acclimatize to the heat. The irony of calling it a dry run still makes me smile, given that you were completely soaked to the skin by the time you started to undress! The PPE varied slightly in different centres, but the PPE I had involved a pair of boots, a pair of gloves that reached up to your elbows, a plastic splash proof suit, a separate hood that covered all of your face except your eyes, a big heavy rubber apron, a pair of goggles, and then a second pair of long protective gloves (Fig. 23.4). This ensured there was absolutely no exposed skin. I can vividly remember peeling off the gloves for the first time and showering the floor with sweat and then realizing that my feet were sloshing in my boots and taking off one boot at a time and literally pouring sweat out on to the floor.

During my second dry run, I managed 45 min in my suit and I recall feeling really proud of myself. As I finished undressing, a press officer appeared and I got photographed very red-faced, with sweat pouring off me. I think it is difficult to describe the feeling of wearing PPE in this kind of heat, confronted with the hazard of Ebola. It is one of the hottest, most claustrophobic things you can possibly imagine and the knowledge that you have to undress extremely slowly and carefully and that there is no quick way out is difficult to deal with (Fig. 23.5). If you overheat, or panic, you still have to undress safely or you risk exposing yourself to the Ebola virus. Despite this, the PPE did have one benefit and that was touch. We had a strict no touch policy with colleagues outside of PPE, which meant you literally could not touch anybody at all. This was designed to help keep everyone safe, but a complete absence of human contact felt very lonely.

Fig. 23.3 Unpacking at the Ebola Treatment Centre (left) and the road to the ETC

Fig. 23.4 The author, Dr. Emily Bayne, donning her protective clothing

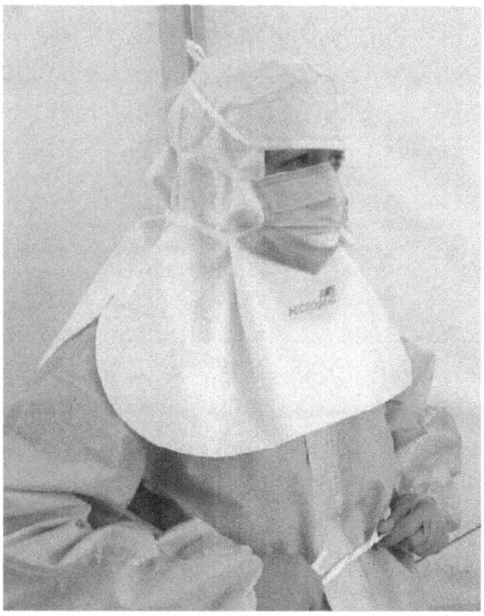

While we waited for the Centre to be completed, we had to look at the whole process of triage, admission, discharge, and death and consider each step and how it would work. There was a large group of national staff that we were working with, an interesting mix of teachers, nurses, and other workers who had found themselves without a job because of the Ebola outbreak. It felt frustratingly slow; with a lot of issues to sort out before it was safe to open to patients. There is so much involved in running an ETC—from the very important water and sanitation team (WASH), to the nurses and

Fig. 23.5 A sprayer assisting a clinical worker to safely remove their PPE

medical staff, the cooks, health monitors, site workers, finance, and many others. An enormous number of people needed to all be working efficiently as a team to ensure a safe working environment. In addition, we needed a guaranteed supply of water and different strengths of chlorine. On the last day before we opened to patients, we were still busy with training, making sure that everything was as safe as possible.

23.3 Pregnancy and Ebola

The first time I really considered the implications of pregnancy and Ebola was during one of our training sessions. At this stage, we didn't have any specialist midwives, and thus, we were working with the information and guidance that was available. It became evident that pregnancy and Ebola was a particularly sad and difficult situation, as I discovered that the previously natural process of giving birth is incredibly high risk to everyone, mother, baby, and any family member or health care worker involved.

Although not more likely to contract Ebola, limited evidence did suggest that pregnant women were likely to be at increased risk of severe illness and death when infected with Ebola virus (Mupapa et al. 1999). This was thought in part to be due to the increased risk of haemorrhagic complications, specifically vaginal and uterine bleeding associated with abortion or delivery (Jamieson et al. 2014).

It is worth noting that during delivery, or through procedures frequently performed during birth, there was felt to be a very serious risk of infection to health care personnel, other patients, and caregivers (Medecins Sans Frontieres 2014). As a result, we needed to carefully consider the likelihood of health care worker exposure to large amounts of blood and body fluids during labor and delivery, the overall physical condition of the patient, and the likelihood of neonatal survival. This had to be balanced with our very limited facilities and resources. With all of this in mind, it became clear that there was very little we could do to intervene in the face of any difficulties during pregnancy. Training focused on keeping staff safe, ensuring we only admitted women who definitely met the criteria and offering pain relief and comfort measures to women when needed. This particular training session really bought home the difficulties we were likely to face in caring for pregnant women.

23.4 Opening the Centre

Our staff were organized into teams with a rolling shift pattern. These teams were a mixture of UK volunteers, international staff, and national staff. We had one team leader to coordinate the shift (as well as taking their turn in the hazard zone) and then aimed for a mix of nursing and/or paramedic skills and doctors. Everybody took responsibility for staff safety and basic patient care, as entries to the red zone were often in groups of only two or three. The doctors were able to make treatment and prescribing decisions, but it was very much a team effort.

I was team leader for one of the teams and can recall the nerves that first day we opened (Fig. 23.6). The reality of standing in a stifling triage room, two staff in full PPE with hoods and masks, and the patient the other side of a glass divide attempting to communicate between everyone, often in an unfamiliar language; the aim of triaging the patient, deciding if they meet the criteria for possible Ebola infection, and whether to admit them, possibly to catch Ebola if they don't already have it. That first shift was understandably filled with a lot of uncertainty and new challenges.

As my first shift drew to a close, the wash team came to find me to explain that the young girl we had admitted earlier was dead. I was just about to handover her case to the next team and I didn't really know what to do. We kept entries to the hazard zone at an absolute minimum to keep the exposure risk as low as possible, so I couldn't really justify two staff getting into PPE just to confirm a possible death. After

Fig. 23.6 Rows of boots and aprons drying inside the Ebola Treatment Center

a discussion with the clinical lead from the next shift, we decided we would in fact go into the hazard zone together to see what had happened. On this occasion, I was lucky and the patient stood up and talked to us. I had to remind myself that sometimes patients might sleep on the floor in an attempt to keep cool. I also had to remind myself that we would have to learn that certifying a death could wait.

One of the first patients I admitted to the ETC was a 14-year-old girl. She was sadly convinced that the ETC was trying to poison her, and as a result, refused all medication, food, and drink. One of the key early issues with our Centre was managing the attitudes of the local people. There was a lack of education and people were worried that we were there to cause harm. There were many rumours that included Western people stealing organs and poisoning patients. The reality was that the majority of people admitted to hospital early in the outbreak were already dying, and therefore, went in and never came out. Safe disposal of Ebola victims was paramount and the bodies were bagged and buried in designated sites; a complete change to usual burial practices. Often family members were not able to see the dead and it was easy to see how the rumours had started. With no fluids, no food, and no medication, there was little I could do for this young girl. I watched, as the inevitable happened.

23.5 23-Day-Old Babies

Ebola was unremitting, choosing both children and adults as its victims. The ETC saw children of a variety of ages, but just before Christmas we admitted 23-day-old twins. During the triage process, I established that mum had died following development of a fever following a caesarean section. The triage process took a long time as I also triaged grandma who had brought the children to the ETC. I got lots of PPE baby cuddles as I established what had happened. The babies looked well, but I discovered that mum had tested positive for Ebola after her death. I had to heel prick both the twins and take blood swabs for Ebola, a difficult task in full PPE.

A parent or caregiver often attended triage with a child with symptoms, but was well themselves, as was the case of the grandmother of these twins. They did not require admission if they were well, just careful follow-up to ensure they didn't develop symptoms given their possible exposure. If they chose to be admitted with the child or children, in order to care for them, they risked further exposure to the virus and possible infection. It was a very difficult task to try and counsel these parents or guardians in triage and even more difficult to admit them knowing the potential risks. This was balanced with witnessing the heartbreak of a mother leaving her child alone for admission because she had other children to care for.

Once admitted, children were particularly tricky to manage in the ETC. A young child could not be left alone all day to care for his- or herself, and small babies needed to be fed regularly. Entries into the hazard zone only occurred at set times for specific tasks. In the heat of the day in Makeni, more than an hour in PPE was generally not possible and definitely not safe. This meant that, in reality, there was very little time for patient care and certainly not time to look after a frightened child. Admitting patients was primarily about protecting the community from further exposure to Ebola. Once admitted, our primary concern was for our own safety and after that came the needs of the patients. As a doctor, this involved a huge shift in perspective from the usual approach of prioritizing the needs of the patient and, when this was a child, it was especially difficult. That said, we scheduled more entries into the hazard zone in order to ensure babies were fed and children were looked after, but it was a careful balance of necessary tasks and their risks.

Putting an IV drip into a small child is difficult in a normal setting, but in poor light and wearing PPE with two sets of gloves it was even more challenging. If you managed to get it in, then you could only give a small fluid bolus before attending to other tasks and, by then, your time in the hazard zone would be up. Initially, we couldn't leave fluid connected to a child, the dangers of not being there to supervise the child or the other patients were too great. We did use nasogastric tubes to try and hydrate small children, but these frequently blocked and insertion could cause bleeding.

Fortunately, as time progressed, we were able to train up 'caregivers', survivors who were safely able to enter the hazard zone and provide care for lone children, but at first this was not an option.

The day after the twins were admitted, I was on a late shift and enjoying a later breakfast. The night shift arrived back at our accommodation and I knew straight away that both the twin babies had died overnight. Ebola is a cruel disease.

23.6 Playtime

Children of all ages were admitted over the first few days and we gradually developed ways to play and interact with them. Each ward had an outside area with plastic chairs and a see-through chicken wire fence. There was then a gap and another see-through chicken wire fence with hazard tape to keep staff a safe distance from the patients. I remember a 12-year-old boy sitting out on a plastic chair, just sitting quietly and not doing anything. Next minute, there was much laughter as one of the staff attempted to teach him to juggle with stones through the fence. I remember him giving up on the juggling and heading back inside to the ward, only to reappear a minute later clutching a radio we had put into the hazard zone. He managed to tune the radio into some Christmas songs and I found myself dancing around in my scrubs to Christmas songs with some of the other staff, separated from this small boy dancing on the other side of the fence. Surrounded by gloves, masks, and death, a child's smile was a powerful motivator.

23.7 Sadness and Joy

The lead nurse and Ebola survivor in my team didn't arrive for work one day and I discovered that it was because her husband had contracted Ebola. I knew that she had had the disease and survived, but I also knew that she had two young children at home. It was difficult to imagine what she was going through.

The 12-year-old boy that had been with us for a few days received a second negative Ebola test result and was ready to be discharged. He had to shower in chlorine and then with soap and water in a special shower that we nicknamed the 'Shower of Joy'. He came out to music and dancing with a great big grin. I arrived home from work that day to a freshly made mince pie, thanks to one of our lovely team and it almost started to feel like Christmas.

23.8 Christmas Cancelled in Sierra Leone

During December, I was able to follow snippets of the news from back home in the UK and I saw the headline 'Sierra Leone Forced to Cancel Christmas' (Malm 2014). This was not far from the reality as on 12 December, Sierra Leone banned all public festivities for Christmas or New Year because of the outbreak (BBC 2014). By that December in 2014, six districts and around half the total population of Sierra Leone were "locked down", under strict travel restrictions that prevented people from entering or leaving these districts without special permission (Rothe et al. 2015).

Christmas Eve fell on a night shift for me and I spent the morning wrapping Christmas presents, rather like I would have done at home. We planned an evening of carols by candlelight at our ETC for

Christmas Eve, and we hoped that the sound would drift across into the confirmed ward and provide some comfort to our patients. There were 6 deaths over the few days of Christmas, but two of our confirmed patients also started to test negative and provided some hope among the frustrations.

23.9 Incidents

Despite all the training, safety protocols, and precautions, there was still room for error. I was observing in the doffing area during a shift (Fig. 23.7), when there was an incident with the spraying of chlorine (Fig. 23.8). Everything had to be carefully sprayed before each item could be painstakingly removed, washing hands after each item. One of the nurses got chlorine across her facemask and the next minute had ripped of her goggles and mask and tripped and fell. The situation was understandably stressful, but protocols were followed and the situation was dealt with safely. The other members of the team were frightened and the rest of the shift was busy with admissions. It was an exhausting day and a reminder of how careful we needed to be.

Our first survivor had an enormous impact. The cycle was broken; it was clear that patients could come in, get better and go out. We started a 'survivor wall', a coloured handprint from each survivor, which could be seen by patients still in the confirmed ward (Fig. 23.9).

Fig. 23.7 A worker sits waiting for an opportunity to doff

Fig. 23.8 A worker stands ready to be sprayed with chlorine

23.10 The New Year

The New Year of 2015 arrived with two Ebola patients negative and ready for discharge. It was just the good news that we badly needed and the patients were received out of the Shower of Joy with the usual music and dancing. We also got the patients to put their brightly coloured handprints on our 'survivor wall' (Fig. 23.10). It was a lovely start after a difficult couple of days. Unfortunately, the rest of the shift was tinged with sadness as we had two deaths, and then a new patient that arrived for triage was found dead in the back of the ambulance.

Just into the New Year, an ITN[1] film crew arrived at the Centre wanting to film what was happening. I was surprised when they wanted to make a short piece about my role, as well as another member of our staff.[2]

It was a really busy shift, as I attempted to coordinate the discharge of four Ebola-negative patients; a 9-day-old baby, a 9-month-old baby with mum, and a 14-year-old boy. In the midst of this activity, four separate ambulances arrived with four new admissions. One of the admitted patients climbed out of the back of the ambulance and walked into triage looking very weak. We laid him down on a stretcher while we triaged another gentleman. A few minutes later, it became evident that the patient

[1]ITN, or Independent Television News, is a global news and content provider based in London, England.

[2]The video can be seen at: https://www.youtube.com/watch?v=hinXwk4wRb8

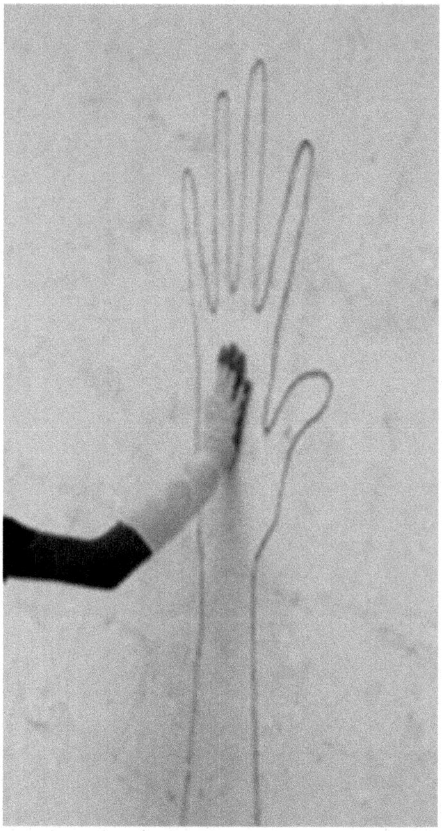

Fig. 23.9 Our first Survivor Wall handprint

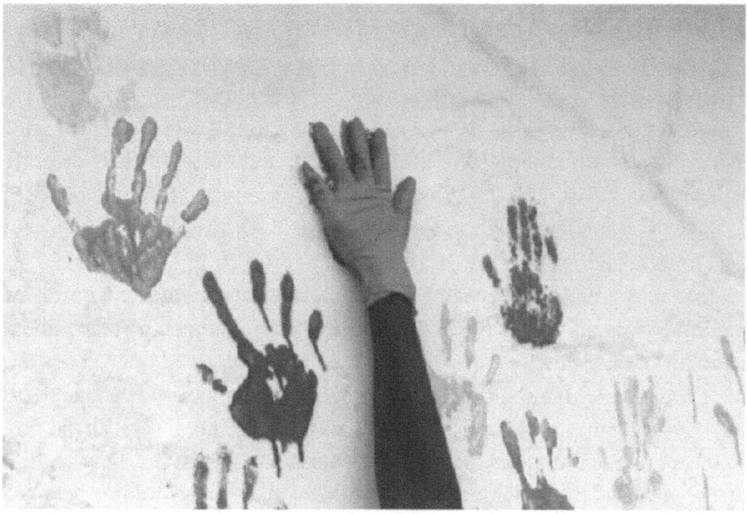

Fig. 23.10 The growing Survivor Wall

lying on the stretcher had died; he didn't even make it through triage. I took a swab from his mouth to test for Ebola and then the WASH team came to spray and bag the body ready for the morgue. At the same time, one of our doctors was a few feet away being interviewed for the ITV News.[3]

The next day, the film crew was back and planned to follow our lead nurse and myself for the shift. There were no admissions, but the burial team arrived to take the body of the gentleman who had died in triage the previous day. The family also arrived and wanted to see the body before it was taken for burial. It was unusual for a family to come and see the body because of the difficulties with communication, travel, and money, but it was important for us to facilitate this whenever it was possible. I explained the process to the family and asked for their permission for us to film what was going on. I was moved by their response to this request; they told me they were keen for the world to see their suffering and hoped that it would encourage others to come and help. The team transported the body from the morgue on to the back of the truck, we then opened up the body bag and the family came to pay their respects. The gentleman's wife was overcome with grief and it was a poignant moment. The film crew was at the fence asking me questions; I didn't want to let my emotions get the better of me, so I explained the importance to the community of allowing families to see their loved ones after death.

23.11 'Mary'

Over Christmas, we admitted a beautiful 5-month-old baby with her mother. Unfortunately, her father had died before admission and her mother was admitted, too—and later died in our confirmed ward. She had a grandmother, but after considering the risks, the grandmother felt she could not stay in the treatment centre to look after the child. We were lucky to find two Ebola survivors willing to take on shifts inside the hazard zone to look after the little girl. We were delighted when she tested negative for Ebola and we were able to discharge her to a monitoring Centre. However, she was readmitted due to fever and unfortunately then tested positive for Ebola.

There was no getting away from the emotion of watching a 5-month-old baby girl slowly get worse from a disease that we knew was likely to overwhelm her. She deteriorated gradually, and despite all our efforts overnight, she died the next day.

23.12 Orphans

As I came out of the hazard zone, I saw a 7-year-old little boy sat out by the fence. He was clearly in need of some entertainment. We grabbed some surgical gloves and started blowing them up and drawing faces on them. The little boys face lit up as we passed them over the fence for him to play with. We then had a rendition of head, shoulders, knees, and toes and, by the third attempt, he was laughing and joining in with the actions. This little boy was negative for Ebola, but both his parents were dead and we were waiting for an uncle to arrive from Freetown so he could be discharged.

One of the many problems we faced as we treated children was their family, or lack of it. Ebola often hit whole families and many children were orphaned as a result. For those children that did have some surviving relatives, it was often difficult to find out the details or get in contact with them. Our 7-year-old boy with an uncle in Freetown was one of the luckier ones.

By mid-January 2015, there were 8185 registered Ebola Orphans in Sierra Leone (UNICEF 2015).

[3]ITV News is a news programme on the British television network ITV, having the second-largest television news audience in the United Kingdom.

23.13 Convalescing Patients

Patients had to have two negative Ebola tests prior to discharge from the Centre. This meant that they could spend around 3 weeks in the Centre, and quite a lot of that time in our convalescent ward. This was designed for patients who were no longer having any 'wet' symptoms (diarrhoea or vomiting), but were still testing positive for Ebola virus on their blood tests. During convalescence, we tried to make time for some fun with the patients able to sit outside. Two of our convalescing patients put on a bit of show singing to each other and our staff and psychosocial team soon joined in with the singing and the drums and all the instruments came out. It was wonderful moments like these that were really special and I felt privileged to be a part of them.

23.14 A New Team

I initially had a deployment of 7 weeks and, at this point, I knew that there would be another group of international staff coming to the Centre to help the national staff continue triaging and treating patients. It felt uncomfortable giving up responsibility for what we had created together. There was a small overlap between teams to allow for handover and extra training.

Our last shift arrived and it felt like the last 7 weeks had flown by. I put on my PPE for the last time and took the new team leader into the hot zone. The shift finished with the whole group of volunteers arriving for a short handing over ceremony.

We managed to squeeze in a final get together for lunch before leaving for the UK. One of the team managed to organize the purchasing of a goat; which had been wandering around our 'garden' for a day. The goat got humanely dispatched and barbecued and was extremely tasty.

We set off at 9.30 in the evening of Saturday January 17th, 2015, and after repeated temperature checks, some long queues, and two flights, we arrived back in the United Kingdom (UK). We were roll called off the plane ahead of other passengers and escorted directly to Public Health England (PHE) for a health check, before we were allowed to leave. I was last out of the entire group for my health check and by the time I made it through, the baggage carousel had stopped and my bag was sitting on its own in the middle of the room. I escaped out through customs and most people had already left. It felt strange that our close-knit team had gone their separate ways and I didn't even say goodbye to many of them.

That first morning back in the UK, I woke up in a panic, not knowing where I was or what was going on. Eventually, my brain kicked in and I remembered that I was in London, I looked at my watch and discovered that it was only 5 AM! I tried to go back to sleep, but eventually gave in and got up at around 6 AM. I opened the PHE box that was given to me at the airport and discovered my biohazard kit, complete with gloves, scoops, and instructions, as well as a thermometer, crossing my fingers that I wouldn't need anything except the thermometer. Later on in the morning, I got a phone call from PHE explaining the monitoring process. I had to take my temperature twice a day and report in with PHE every morning. I was also advised about not sharing my toothbrush (I can't believe anyone does that in normal life?!) and not taking long journeys on public transport or staying in hotels. Although I was back in the UK, everything still felt somewhat surreal.

I managed to catch up with a few friends in London and life started to feel more normal. However, I will admit that talking about your experiences after a deployment such as this can be difficult, and it is common to focus on the lighter and happier moments with friends and family. In addition, measuring my temperature at 37.4° (when the cut off is 37.5°) induced a slight panic and I had to remind myself that everything was completely fine. The last 7 weeks of my life had been a true adventure, as well as an emotional roller coaster, with the opportunity to meet some inspiring people and forge friendships.

23.15 Returning to Sierra Leone

Even when I was preparing to leave, I had already begun to think about going back to Sierra Leone. I had some commitments at home until the middle of February, but following these I was keen to get back to the ETC. The NGO running the ETC had offered me a longer-term position starting at the beginning of March. This was a difficult decision, remembering the oppressive heat and the difficulties of PPE, but wanting to be able to continue to help. I was due to be Maid of Honor for a very close friend's wedding at the beginning of May and I was conscious to ensure that I would be able to return home for this. It was difficult to explain my decision to my friend and I knew she would be worried about my safety and about me getting home for the wedding. Explaining that I was more than prepared to take those risks was definitely a challenge, but I think our 20 years of friendship helped us both to accept my decision. Many other people asked me why I wanted to go back. This was a tricky question; as I started to explain how hot it was and how awful PPE was, I found them looking at me with a puzzled expression on their faces. If I'm honest, the last week before I set off on my journey back to Sierra Leone, I was asking myself the same question and struggling to find the answer.

Having undergone rigorous training prior to my initial deployment and having worked in the hazard zone for several weeks, I only needed some short refresher training on my return to the ETC. After my second shift back in the Ebola Treatment Centre, I knew exactly why I had chosen to go back. The warmth of the welcome I received from all the national staff that I worked with last time was truly incredible, and we discharged two survivors. Helping survivors to paint handprints on the wall and watching people leave was a wonderful feeling. Despite the exceptionally oppressive heat and hideousness of PPE, it was all worth it.

During my time back in the UK, the Ebola Treatment Centre had its busiest period and sadly there were many deaths. A cluster of cases from the same village, started by one unsafe body washing ceremony, led to the deaths of entire families. We had a gentleman; 'James' in our convalescent ward, who had lost his mother, wife, and two daughters to Ebola in our Centre and his brother was still in our confirmed ward.

Our Centre was completely full, and the staff worked round the clock doing their very best for the patients, but sadly watched many die, and tragically many of those were children. It was difficult to sit at home in the UK knowing all of this was going on and it was definitely another motivation for coming back to help. When I arrived back, it was evident that I had not been there for the hardest weeks and I did feel guilty. The experience had clearly impacted on the staff and they had become closer than ever.

There were several other members of staff that returned for further deployments over the coming months and it was lovely to see familiar faces.

23.16 A Different Kind of Burial

Since my return to the ETC, we had been caring for a 4-year-old little girl and her mum in our confirmed ward. Her father had died. Despite all our efforts, we watched this little girl becoming progressively more unwell, becoming less alert, and finally bleeding from her mouth before she died.

Her mother was understandably distraught, having also lost other family members, including her husband in our treatment Centre. Words cannot convey the suffering this lady had seen.

A few days later, I went to watch that little girl be buried, in among the rows and rows of others that this outbreak had taken. It was my first visit to the cemetery and I was struck by the enormity; so many mounds of earth, each with its own identification tag. It felt like a war cemetery and Ebola was certainly a battle.

Some family members, including the little girl's grandfather, arrived for the burial. He approached me and offered his thanks; saying that he knew it was God's decision to take her and that there was nothing we could have done. I was lost for words; there was nothing I could say to this gentleman.

The burial took but a couple of minutes; a small, white plastic container, lowered in among the earth. Her grandfather quietly murmured a prayer and then it was over (Fig. 23.12).

Another set of discharges, one is a 12-year-old boy, and somehow this moment helped to lessen the sadness of the days before. He came out smiling, and one of the nurses lifted him on to his shoulders so he could place his green handprint right in the middle at the top of survivor wall. He was going home to a village that had lost so many here to this disease.

The Centre was staffed 24 h a day and, despite the darkness and mosquitoes, we continued to care for patients. The coolness of the night was a blessing and often enabled us to spend longer in PPE and provide more care to our patients (Figs. 23.13 and 23.14). During one busy night shift, I was just about to get a cup of tea, when the radio crackled into life announcing the arrival of two ambulances. We got ready to receive and triage the patients. As they walked into triage, we realized that we knew them both. They had both been successfully treated for Ebola in our ETC and discharged a few weeks previously. Happily, they looked exceedingly well. A few questions later, we established that the lady had had one isolated episode of vomiting. She explained that she had drunk a large volume of oral rehydration solution a little too fast and that she was otherwise fine. This was certainly not a case of Ebola, but it demonstrated the ongoing issues and lack of education in the community. It was incredibly sad and frustrating, as both patients lost all their clothes and possessions that they entered with; they had to be burned (Fig. 23.11).

Finally, in April, our last two patients in convalescence were ready to be discharged. One of these survivors was a gentleman that most thought would die. He was incredibly sick and spent several days in our confirmed ward; during which time he watched many relatives and friends die. I was privileged to have listened to him talking through his experiences with one of our nurses. He said that during the dark hours alone in the confirmed ward he felt utterly hopeless, watching others die around him and he was desperate for death to release him from his suffering. His discharge was a very special moment.

23.17 Schools

During my stay in Sierra Leone, schools had been closed. This was part of attempts to reduce the spread of EVD. During April, I was lucky enough to see schools reopen in the Makeni district, although often there was a shortage of teachers. Children all wore smart school uniforms and it was lovely to see them out on their way to school.

With a day off from the ETC, I was privileged to spend a morning in a local school delivering some science lessons. Some of the international research staff had experiment equipment and we were able to run sessions including extracting DNA from bananas. The class was very large with a wide range of ages and abilities, but everyone was very keen and the morning was a great deal of fun.

Fig. 23.11 The incinerator burns away at the back of the Ebola Treatment Centre

Fig. 23.12 A row of newborn babies in among the dead at the local cemetery

Fig. 23.13 The hazard zone at night

Fig. 23.14 A glimpse inside the hazard zone

23.18 Personal Risk

I volunteered to go out to Sierra Leone and I volunteered to work with Ebola. I knew the risks; however, nothing prepares you for the feeling of dread when a colleague and friend working with you becomes unwell.

In April, my teammate and friend developed a fever and was unwell. She had a few symptoms and so I decided we should probably recheck her temperature. The thermometer beeped reading 38.6°; she met the criteria for a possible Ebola case.

I felt dreadful as I carefully backed away, explaining that she could not leave her room. I made a quick phone call and our experienced staff health team expertly handled the situation, but agreed that she needed to be tested for Ebola. I had to go back to her door and explain, at a distance, that she would have to have an Ebola test. I was sure she didn't have Ebola, but the thought of it hung in the air between us. The reality of staff in PPE and an ambulance journey to an ETC slowly sank in, as I was sitting on the other side of the door waiting.

The sprayer arrived and she was escorted into the ambulance. As it pulled away I still could not imagine what it was really like to be on the receiving end of care from people wearing suits and goggles. I was just wishing for her to come back safely and as quickly as possible.

She phoned me once she has arrived and all things considered seemed to be in good spirits. She was actually feeling a lot better and we joked about her new orange scrub uniform and green flip-flops that she had been given to wear. The clothes she took with her were burnt.

My birthday arrived, but with my teammate still in an ETC, I didn't really feel like celebrating. Then I heard the news that she definitely didn't have Ebola or malaria, just a simple gastrointestinal infection. I was so relieved and happy that she was returning safe and sound. The experience was a challenge for everyone, despite swift and supportive care. It was a reminder to all of us of the risks that we faced and a testament to the protocols and care that were in place if we became unwell for any reason. Little did I know I would be facing the same prospect myself just 1 month later.

23.19 Rainy Season

The ETC was a series of tents for the wards and offices, but we spent a lot of time outside going between them. With the ETC remaining open into the rainy season, there was work to be done to try and keep everything dry. It started to feel a bit like an arc, as wooden covered walkways sprung up all around the Centre. Overnight, there was an impressive thunderstorm, but this was just a taster of what it would be like when the rainy season properly arrived (Fig. 23.15).

Discharge day for our last confirmed Ebola patient arrived with the sun beating down. I was at the Centre when we had our first survivor discharge back in December, and I couldn't help hoping that this would be the last Ebola case the Centre would see. Most of the staff seemed to feel the same way and this discharge ceremony was particularly special. There was a lot of singing and dancing and, by the time we stopped, the sweat was pouring off us (Fig. 23.16).

Following this, the next week saw a substantial drop in numbers of patients and the Centre started to feel empty. This was fantastic news for the battle against Ebola, but I almost felt a bit redundant in my role. It did, however, give us an opportunity to focus on teaching and building capacity among the national staff.

Fig. 23.15 Preparing the Ebola Treatment Centre for the rainy season

Fig. 23.16 The Survivor Wall

23.20 The Reality of Being a Patient

Working at the Centre was at times tough and emotionally challenging and we did get the opportunity to have leave. I was really excited to be coming home for my friend's wedding, but just a few days before the flight I got sick. After the experience with my colleague I was assuming it was just an ordinary bug, and that I would get better in a few hours. Sadly, I didn't. After nearly 24 h of confinement, shut away in my room, my colleagues decided that I would need an Ebola test. This meant people in full PPE coming up to my room to get me and spraying chlorine behind me as I walked to the ambulance. I climbed in the back and endured a very bumpy and uncomfortable couple of hours to one of the ETCs. As I arrived I went through the triage process and was put into a room. The reality of being in a tent on your own, with nothing but a pair of scrubs and flip-flops is scary. When you realize you are having an allergic reaction to a medicine and you know that there is no one else in there, even more so. Someone has to go through the laborious process of putting on PPE before they can come in, which can take up to 20 min.

Fortunately, despite being unwell, my Ebola tests were negative and I was discharged. This experience gave me a tiny taste of what it was like being treated in an ETC; I didn't have Ebola and I was the only patient in my tent. Nothing like it was for patients with Ebola, watching family members die and not knowing if they themselves would live.

23.21 Sick, But Not Ebola

The number of patients in the ETC with Ebola steadily decreased, but the ETC remained active. There had been poor access to health care across the country even before the Ebola outbreak, but Ebola exacerbated the problems. With local health care provision poor and difficult to access without sufficient funds, the ETC, despite the risks, was a better option in the minds of some patients than no health care at all. This was a credit to how far we had come since opening, when patients were terrified of the ETC, but we were only set up to treat Ebola and malaria and to give broad spectrum antibiotics; patients that did not meet the triage criteria could not be admitted. In these circumstances, we did our best to give advice, medicines, and referral to the local services we had, but many of the patients could not pay for local services and had no other access to health care. There were times when I discharged patients knowing they were going home to die, not from Ebola, but more ordinary problems such as a stroke or liver failure.

If patients met the triage criteria, then they were admitted for the 48 h it took to get two Ebola negative tests. During this time, we did what we could to diagnose and treat them.

We admitted a 17-year-old girl who became very unwell. Her first Ebola test was negative, and I found myself inside the hazard zone in PPE with sweat pouring down my face, into my goggles and mask. One of the nurses and I were carefully cleaning and turning her. We placed her in the recovery position, desperately hoping that she wouldn't be sick when we left her. I reached up to hang some IV fluids and sweat poured down the inside of my suit and into my eyes, I couldn't see. I realized the cannula had stopped working, so we got a new one and I was fighting the sweat in my eyes as I tried to put it in. The tape got stuck to my gloves and I was over heating; I wrestled with it, trying not to rip my glove. We gave her IV fluids, but she looked desperately sick, and as we left, I was sure she would not survive the night.

One day we admitted a 2-year-old boy to the ETC and I donned my suit, knowing that this child was very sick. When I went in to the high-risk zone, I was confronted with a child who was breathing very noisily and very fast, foaming at the mouth and in a rigid posture. One look was enough for me to know that this child would likely die. I explained this carefully to grandma, who became understandably distressed. I felt that we should try everything we could for the child, so we desperately looked for a vein to cannulate. As my sweat dripped on to this child's tiny body I realized that it was going to be extremely difficult. In the end, I decided that intraosseous access would be our only hope (drilling a hole directly into the leg bone in order to give fluids and medicines). It took some time to assemble what we needed and explain the procedure to grandma. It seemed cruel, as I knew the likely outcome, but I felt like we had to try. The IO went in easily, but by then I was overheating, and my heart was racing. I gave the fluids and all the medicines that I could, at least knowing that the child was more comfortable. Then we explained to grandma that we had to leave, but that we would send someone else in shortly.

My shift finished at 8 pm and I finally left at 8.45 pm. One of the nurses was in charge on the night shift and I knew that I was likely to be called back to certify a tiny body in the near future. I was home for an hour when my phone rang. I got a lift back to work and got back into PPE. At least, the child was finally at peace; an example of one of many children to die before their fifth birthday in Sierra Leone and nothing to do with Ebola. Sadly, I will never know the exact cause of death of this little one, as patients that died within the ETC were buried according to the strict EVD protocols. There was no provision for investigation of other diseases or illnesses. If patients had two negative tests and could be discharged from the ETC, we did our best to arrange further investigation and treatment locally.

23.22 Health Promotion

I was fortunate to be able to spend a day out with the Health Promotion Team, another part of the work that was going on in Sierra Leone. There were several teams at work in the community alongside those focused on the ETC. These included Psychosocial Teams going out into the community to support survivors and communities affected by Ebola and Health Promotion Teams looking at improving education around basic health measures in communities such as hand washing. The Health Promotion Teams often had members of staff that came from a specific community working with that community and other local areas; this allowed good communication and acceptance by the local communities (Fig. 23.17).

Fig. 23.17 This is an example of a 'Tippy Tap' hand washing station. It was one of the many simple interventions and areas of education that the Health Promotion Team were involved with

Fig. 23.18 Village women making palm oil by hand

During the day, we visited a number of relatively remote communities and reviewed their water, sanitation, and health facilities. The communities had a health promotion officer, who was responsible for two or three communities and visited them each day. The tasks were varied and, during our visit, we constructed some simple taps, placing them next to the toilet and other areas. The team had already done work with the community on the importance of hand washing. I recall asking about the drinking water supply, only to be told that there was none. These people were drinking from the river and in one community I visited, from a swamp. I watched the women making palm oil by hand (Fig. 23.18), their children sitting nearby. One of the children needed a drink, her brother was sent to fill the cup from the river. She took a drink and promptly vomited down her front. The next thing she had diarrhea. It was no surprise to find infections like this were common among these children. Ebola had touched this community and people here had died. I was pleased to meet one young man, a survivor, who had started work as a caregiver at the ETC, helping patients.

23.23 Closing the Doors and Leaving

As we stopped seeing Ebola cases, eventually the number of patients coming in for triage also started to drop. Work concentrated on teaching and training, but things began to get quiet. I decided it was time for me to return home.

Fig. 23.19 Dr. Emily Bayne (the author) dressed in one of her parting gifts with her ETC colleagues

By now, my Sierra Leonean colleagues had started to feel like an extended family and my send-off left me feeling very emotional (Fig. 23.19).

When I left Sierra Leone, I went back to the UK and back to a job in the Emergency Department. The transition was difficult; how do you go from the ravages of Ebola and the poor resources of Sierra Leone to a busy UK Emergency Department on a Saturday night? I'm not sure I have the answer, even now (Fig. 23.20).

A few months later, I was extremely privileged to be invited to return to Sierra Leone again, to look at the work that was happening as the country prepared to reach 42 days Ebola-free and begin the decommissioning process. The United States Centers for Disease Control and Prevention (CDC) had reported a total of 14,124 Ebola cases in Sierra Leone during the outbreak and 3956 deaths (CDC 2014). Many of these were in the Bombali district where I had been working.

Returning to Sierra Leone was a real honor. I had witnessed all of the staff grow in confidence, with the sharing of knowledge and ongoing training during the time I had been in Sierra Leone and the opportunity to surprise those remaining at the Centre with a visit was truly wonderful. Some had already moved on to new jobs, or returned to old ones, in the case of some of the teachers. During the many months we had been open, there had been a few staff members admitted with infections such as malaria or GI infections, but fortunately at this particular ETC, there had been no cases of Ebola among the staff.

Fig. 23.20 '*Tenki for cam*"—Thank you for coming!

This visit was a culmination of what had been an incredible journey for me; from arriving at an unfinished ETC back in 2014, right through to celebrating the fact that it was no longer needed and watching it close its doors.

When I went back, the Centre was very quiet, with no patients. They were planning to close the doors for the last time and then begin the decommissioning process. I had been invited to make a short video clip exploring my feelings at returning to the ETC and the impending closure. This was an emotional time; the happiness to come so far in the fight against Ebola, tinged with sadness for those we had not managed to save.

Three days after my visit, on November 7, 2015, WHO declared Sierra Leone free of Ebola virus transmission after 42 days (two incubation periods) had passed since the last Ebola patient tested negative. There was a setback a few weeks later, with two new confirmed cases reported in January 2016, but on March 17, 2016, WHO declared the end of the flare-up (CDC 2014). Throughout the outbreak, the National Ebola Response Centre and the local District Ebola Response Centre had worked closely with a range of NGO partners, the ETC, the CDC, and the WHO; it was culmination of all of this work that enabled us to end this outbreak.

23.24 Recovery

The end of the Ebola crisis was really just the beginning for many of those affected. The disruption to education will have long-lasting impacts and many lost family members and friends. In addition, health infrastructure is poor in Sierra Leone, so equipping the country with the means to manage a possible infectious disease outbreak in the future is an enormous and ongoing challenge, but a hugely important one. It is really a matter of time before another infectious disease outbreak somewhere in the world and there will certainly be more sudden onset disasters such as the recent earthquake in Nepal. One of the key lessons we should take from this outbreak is the need to upskill and build capacity in resource-poor countries and increase their resilience for what may lie ahead. The life expectancy in Sierra Leone of women was just 50.7, and for men, just 49.7 from 2010 to 2015 (United Nations 2017).

23.25 Return to My Life and Career

This was an enormous challenge, but I felt ready to leave the ETC and the role I was in. As I hinted at earlier, the transition to a Saturday evening in a busy UK Emergency Department is a complex one. Colleagues imagined it would be easy after working in an ETC, but actually it was very challenging. I found myself anxious, with slightly rusty skills, lots of technology, and high patient expectation and demands. Fortunately, it was a very supportive department, with lovely colleagues and over a few months I was able to settle back in.

After a further year of Emergency Medicine, I decided to embark on training in General Practice (or Family Medicine) as I felt some of these skills would better equip me for working in resource-poor settings in the future.

My time in Sierra Leone has definitely changed me, but I am very keen to get involved with further humanitarian work and, if anything, feel more inspired following my experiences during the Ebola epidemic.

References

BBC. (2014). Ebola crisis: Sierra Leone bans Christmas celebrations. *BBC News Online*. Retrieved November 1, 2017, from http://www.bbc.co.uk/news/world-africa-30455248.

CDC (Centers for Disease Control and Prevention). (2014). *Ebola outbreak in West Africa—Case counts*. Retrieved November 1, 2017, from https://www.cdc.gov/vhf/ebola/outbreaks/2014-west-africa/case-counts.html.

Jamieson, D. J., Uyeki, T. M., Callaghan, W. M., Meaney-Delman, D., & Rasmussen, S. A. (2014). What obstetrician-gynecologist should know about Ebola: A perspective from the Centers for Disease Control and Prevention. *Obstetrics and Gynecology, 124*(5), 1005–1010. https://doi.org/10.1097/AOG.0000000000000533.

Malm, S. (2014). Sierra Leone forced to cancel Christmas over Ebola. *Mail Online*. Retrieved November 2, 2017, from http://www.dailymail.co.uk/news/article-2871790/Is-Ebola-s-biggest-victim-Sierra-Leone-forced-cancel-CHRISTMAS-struggles-cope-spiralling-caseload.html.

Medecins Sans Frontieres. (2014). Pregnant women in Ebola Treatment Centre. *Guidance paper*. Retrieved November 5, 2017, from https://www.rcog.org.uk/globalassets/documents/news/etc-preg-guidance-paper.pdf.

Mupapa, K., Mukundu, W., Bwaka, M. A., Kipasa, M., De Roo, A., Kuvula, K., et al. (1999). Ebola hemorrhagic fever and pregnancy. *Journal of Infectious Diseases, 179*(Suppl_1), S11–S12. https://doi.org/10.1086/514289.

Rothe, D., Gallinetti, J., Lagaay, M., & Campbell, C. (2015). Ebola: beyond the health emergency. Summary of research into the consequences of the Ebola outbreak for children and communities in Liberia and Sierra Leone. *Plan International. Report*. Retrieved November 4, 2017, from https://www.plan.ie/wp-content/uploads/2015/03/GLO-Ebola-Final-IO-Eng-Feb15.pdf.

The Infolist. (n.d.). Retrieved September 13, 2017, from http://theinfolist.com/php/SummaryGet.php?FindGo=Makeni.

UNICEF. (2015). *Sierra Leone Ebola Situation Report 29 January 2015*. Retrieved November 5, 2017, from https://www.unicef.org/appeals/files/UNICEF_Sierra_Leone_EVD_Weekly_SitRep_29_January_2015.pdf.

United Nations. (2017). *Country profile: Sierra Leone*. Retrieved September 13, 2017, from http://data.un.org/CountryProfile.aspx?crName=sierra%20leone.

WHO. (2014, August 8). *Statement on the 1st meeting of the IHR Emergency Committee on the 2014 Ebola outbreak in West Africa*. Retrieved November 5, 2017, from http://www.who.int/mediacentre/news/statements/2014/ebola-20140808/en/.

When the Patient Comes Third: Navigating Moral and Practical Dilemmas Amid Contexts of Pregnancy and Risk During the 2013–2015 Ebola Epidemic in Sierra Leone

24

Rebecca Henderson and Kristen McLean

24.1 Introduction

Mary is a Sierra Leonean nurse who worked on the maternity ward of a large urban hospital during the 2013–2015 Ebola epidemic. As we sat together in her small office behind the critical care unit, she narrated her experiences treating and caring for pregnant and laboring women. Even prior to the Ebola outbreak, the situation on the ward was dire. This is despite the Sierra Leonean Government's passage of the Free Health care Initiative in 2010 for pregnant women and children under five. "*The routine supply of even ordinary gloves we found at times difficult to get,*" Mary explained, "*and the late referral...And there would be no sutures, there would be no electricity...Yea, and again the doctors were demanding money—they were asking them to pay.*"

With the arrival of Ebola in Sierra Leone in early 2014, things quickly became much worse. "*Ambulances everywhere,*" she exclaimed. "*Kong, kong, kong, kong—everywhere!*" Her hospital experienced its first case in July. By August, nurses were beginning to flee their posts. Her medical superintendent promised additional pay to avoid staff shortages, though as she regretfully admitted, this money never materialized. Over the course of the next hour, Mary narrated the fear she faced every day as she went to work, and the stigma experienced even among her neighbors and close friends: "*When they realized that you are working at the Ebola center, people will just try to snub you, ignore you while you are passing.*" She also described the difficult ethical and moral situations she constantly encountered as she attempted to balance provision of care for her patients with the risk of infection to herself, her peers, and the other women on the ward. There was one particular case that stuck with her in which an Ebola-positive woman delivered stillborn twins, unattended and alone, in the hospital's isolation ward. "*At that time all the treatment centers were packed full with patients, so there was no space for her... we had nowhere to put her, so she was with us until she finally died.*" Explaining her predicament and rationalizing the

R. Henderson (✉)
University of Florida, Gainesville, FL, USA
e-mail: Rrhenderson@ufl.edu

K. McLean
Yale University, New Haven, CT, USA
e-mail: kristen.mclean@yale.edu

D. A. Schwartz et al. (eds.), *Pregnant in the Time of Ebola*, Global Maternal and Child Health,
https://doi.org/10.1007/978-3-319-97637-2_24

decision to keep the woman in isolation and to not to pursue life-saving treatment, Mary said, "*But it's because she alone can infect hundreds or thousands of people. Then who will be at fault? It will be our own fault.*"

Despite harrowing accounts such as these, Mary also described instances of strength and triumph during these trying times. She discussed the confidence gained by both her staff and the patients themselves, for example, when support finally arrived from international health care workers: "*They were the ones that motivated us... they gave us more zeal, they showed us how to handle some of the patients.*" She also described how she encouraged the other nurses when they feared infection, reminding them of their training and their efforts to take appropriate precautions: "*Didn't you put on your gloves? Didn't you wear PPE* [personal protective equipment]*? Didn't you protect yourself?*" By December, things had begun to settle down, and Mary was even acknowledged for her leadership with a gift from the hospital administration during their annual Christmas party. Though the Ebola outbreak took a great toll on her both personally and professionally, she expressed hope that if there was another epidemic—God forbid—they would be better equipped to handle it the next time around.

Mary's story is just one of many describing how health care workers—both domestic and international—navigated difficult practical and moral dilemmas in attending to vulnerable populations during the Ebola epidemic in Sierra Leone. While some existing literature addresses factors related specifically to pregnant and postpartum women during the epidemic—including issues of heightened risk of infection and the ethical and practical difficulties in making treatment decisions amidst medical uncertainty (Black 2015)—few studies have examined the subjective experiences of health care providers working with this particularly difficult-to-treat population (Erland and Dahl 2017; Jones et al. 2017; McMahon et al. 2016; O'Hara 2015). This vignette illustrates many of the themes we address throughout this chapter, which seek to capture the lived experiences of care provided by nurses, midwives, physicians, traditional birth attendants (TBAs), and others working directly with women who were pregnant, delivering, or postpartum. Additionally, this chapter assesses how medical providers managed personal risk in a context of extraordinary ethical, logistical, and clinical uncertainty, while operating under the omnipresent fear of contagion. We describe the difficult and ethically charged decisions these health care workers had to make on a daily basis—life-or-death decisions embedded in a constantly shifting calculus informed by temporal, personal, and contextual factors. By presenting these experiences, we hope to suggest fruitful avenues for further examination of maternal health issues in Sierra Leone that are relevant both to the current health care landscape and informing future epidemic disease outbreaks of this nature.

24.2 The Present Study

This chapter draws from 30 ethnographic interviews conducted between 2015 and 2017 as part of the Ebola 100 Project, an international, multi-partner research initiative that seeks to capture the experiences and stories of those who contributed to ending the West African Ebola outbreak of 2013–2015. The project focuses on individual experiences during the response, the knowledge economy that surrounded the response, and how local culture, context, and human capacity factors impacted the course of the epidemic. Its premise is that an historical and ethnographic approach will inform future global responses to epidemic and pandemic threats and will add to a growing literature on the West African Ebola outbreak.

The set of 30 semistructured interviews included in this analysis focused on capturing the experiences of health care providers working specifically in the context of maternal health in Sierra Leone during the Ebola crisis, either as part of international teams responding to the epidemic, or as employees or volunteers at nationally run hospitals or clinics. Interviewees included physicians, nurses, midwives, administrators, pharmacists, and TBAs working in affiliation with public and private hospitals and/or clinics in urban centers within two districts of the country, one in the East and one in the Western Area of Sierra Leone. Half of these interviews ($N = 15$) were conducted with local Sierra Leonean health care staff, the other half with expatriate medical and/or public health professionals. These interviews were conducted by the chapter coauthors in English or *Krio*,[1] depending on participant preference, and were done both in-person and online via Skype (the latter being reserved for expatriate health care workers who were no longer residing in Sierra Leone). Participants provided written consent prior to the start of their interviews, which lasted approximately 1 h each. Interviews were audio-recorded, transcribed, and translated into English if necessary. Data were then coded thematically and analyzed by the coauthors, with the use of QSR NVivo v.11 for assistance with data organization and retrieval. Ethical clearance for the study was obtained from the Sierra Leone Ethics and Scientific Review Committee, the Yale Human Subjects Committee, and the University of Florida Institutional Review Board for the Protection of Human Subjects. Findings from the study are described below, divided into two sections according to the timeline of the epidemic and stage of the response.

The first section of this chapter describes the "early response" to the Ebola crisis, defined as the period from winter 2013 through summer and early fall 2014 (WHO Ebola Response Team 2015). As this period occurred before the full international response, much of this data center on the experiences of Sierra Leonean health care workers employed mostly by government facilities, and a few international health care workers from agencies responding early, or already working in Sierra Leone, when the Ebola epidemic struck. The next section describes the period from mid-fall 2014 through the final cases of the epidemic in late 2015/early 2016, in which the international response played a substantial role. This section focuses on the experiences of volunteers and employees of nongovernmental organizations (NGOs) who arrived during this period and were directly involved in the treatment of pregnant, postpartum, and lactating women, as well as the experiences of Sierra Leonean staff in working with their new expatriate colleagues.

24.3 Mistrust and Maternal Health in the Early Epidemic

The emergence of Ebola was particularly traumatic for health care workers already caring for pregnant women in Sierra Leone. Their tenuous position and the neglected state of maternal health nationally that existed before the Ebola epidemic initially threatened both institutional and community support for these workers. At the time of the Ebola epidemic, Sierra Leone had one of the world's highest maternal and child mortality rates (WHO et al. 2014). Health care workers who cared for pregnant women immediately before the emergence of Ebola described a landscape of birthing options in which hospital delivery was only one possibility. Women could access some prenatal services at hospitals or local clinics known as PHUs (Peripheral Health Units) or could seek out traditional birth attendants (TBAs) or unofficial midwives who possessed varying skill levels. While home birth had been technically made illegal and all involved were potentially subject to fines, many women nevertheless continued to prefer to give birth at home with the help of family, community members, or TBAs. Hospitals and health clinics remained difficult to access for rural women, particularly in cases of emergency, and the reputation of hospitals and health clinics as places to safely give birth was tenuous at best. Women who desired a hospital birth often faced an uncomfortable, costly journey that

[1] *Krio* is spoken by 97% of the population of Sierra Leone, and although it is the lingua franca and de facto national language of the country, it has no official status. It is the native language of the *Krio* ethnic group of Sierra Leone, and its widespread use acts to unite the other ethnic groups in the country.

likely meant delivering alone, without the presence of family members. Furthermore, women feared delivery via cesarean section, which was not uncommon given trends of late presentation for obstetric emergency (Chu et al. 2012) and was rightly viewed within this context as a dangerous and painful treatment of last resort.

When the first cases of Ebola were reported, government-provided free maternal health care in Sierra Leone was still a relatively recent development, fraught with suspicion of corruption, low-quality care, and hidden costs (Maxmen 2013; Pieterse and Lodge 2015). In 2010, less than 4 years prior to the outbreak of Ebola, the Sierra Leonean government introduced the provision of free health care for both pregnant women and children under 5 years of age, widening the availability of antenatal care and hospital birth to women previously unable to afford it. As one hospital administrator explained, "*the expectations were high*" as officials touted free maternal health care as a major public health achievement. However, many pregnant women in Sierra Leone incurred costs and found limited care options discordant with their expectations of free health care. Women giving birth in "free" hospitals were sometimes charged for supplies or services not fully covered by the government plan, casting doubt on both the quality and availability of genuinely free government-provided health care as well as the integrity of health care workers themselves. Some women assumed that free maternal care included nonmedical goods and services and arrived at the hospital expecting free clothing for themselves or their babies. As one hospital administrator described, "*people started losing confidence, saying that the free health care is not working and all that.*" Compounding the problem of high expectations was a real shortage of drugs, supplies, and especially skilled health providers, particularly at smaller or more rural health centers. The number of physicians and midwives present at health care facilities was almost always insufficient to meet demand, and as a result, the majority of laboring women (even in hospitals and clinics) were attended to by nurses or Maternal Child Health Aids, all of whom had limited obstetric training. The confluence of these factors in the face of grandiose government promises meant a continuous loss of confidence in government-provided care.

With this erosion of confidence came corresponding erosions of trust, and in particular, a suspicion of corruption within the health care system. Even before the emergence of Ebola, health care workers described an environment in which pregnant women arriving at the hospital to deliver viewed them with suspicion. Nor were these suspicions unfounded, as "free" maternal care might exclude certain medicines or treatments or require a small payment to recoup the government's direct costs. While the policies driving these pricing decisions were clearly documented and aimed at making the free service sustainable, patients tended to view any charge from service labeled "free" as inherently duplicitous and avaricious and were often unable to shoulder the burden of these payments (Ponsar et al. 2011). One Sierra Leonean nurse described the situation before the epidemic:

> *But they don't understand. They complain, "government says free health care. We don't have to pay any money." That's where the arguments come from. They cuss the nurses, they cuss the doctors, they cuss the bosses. They say we want to take their money. They don't understand the difference. They say, "no everything is supposed to be free." They say, "we've come all the way from the village, we've come for the free health care." So we have to explain to them, we have to really explain for a long time before they understand. But some of them don't understand. They say they haven't come with money. Then they frown at us, they say we want to take their money.*

Furthermore, many health care workers caring for pregnant women during this time had tenuous relationships with the hospitals and health care systems that were ostensibly employing them. After finishing nursing school, for example, nurses in Sierra Leone are expected to "volunteer" within a government-run hospital, working long hours without pay in the hopes that they will eventually be awarded a paid position. During this time of unpaid service, they often depend on "gifts" from patients and their families, which incentivize better care and create an informal health care economy within an officially "free" system. One nurse described her payment history during her time as a nurse:

Yes, and I never got paid a salary. For two years I volunteered at the government hospital. [During the Ebola outbreak] they gave us a risk allowance. Yes they gave that every week. But when that money cut off, we got nothing again. And we, the ones who had been working on the maternity ward—it was free health care. We didn't ask patients for money, except like when we'd do delivery, because that's our culture here. If you give birth you must send a blessing for your child, even if it's a small thing they'd give us, then that small bit we would share amongst ourselves, and some days we are five or six on duty. We share that small thing. Sometimes one person gets 5000 [Le] (equivalent to less than $1 USD), and how long will that last? You won't be able to sustain your family.

For maternal health care providers, the expectation of free service deprived them of even this minimal remuneration or forced them into situations in which patients mistrusted them due to perceived corruption. The government, therefore, promoted the use of a "free" health care system that in reality contained multiple hidden costs—for medications, supplies, procedures, and care. In the face of these promises of free care, health care workers were perceived as corrupt, taking advantage of vulnerable patients for personal gain. These sentiments were further fueled by the nonnegligible amount of corruption: "*health care workers may really over-charge patients for materials that they're supposed to get,*" explained a Sierra Leonean pharmacist. This mix of system failures, inconsistent messages about the health care system, and actual and perceived corruption starkly undermined confidence in health care providers and institutions and set the stage for health-seeking behaviors during the Ebola outbreak.

During the epidemic's earliest days, information on Ebola within Sierra Leone was unreliable and traveled mostly by word of mouth. As a result, levels of suspicion and mistrust among the general public were extremely high (Yamanis et al. 2016). As such, antenatal visits and rates of hospital deliveries dropped precipitously (Jones et al. 2016). The sense of the epidemic being out of control, and potentially uncontrollable, was particularly felt early on by those already working in Sierra Leone's understaffed and undersupplied health care facilities, who were also among the first to become directly exposed to the virus. The mounting numbers of early health care worker deaths served both to heighten levels of stigma against those working in hospitals and health care facilities and to fuel a sense of panic. According to one community health worker speaking of her neighbors, "*they would say the government has given money to us, the Ebola workers, so that we will inject people with Ebola.*"

The stress of the early epidemic revealed the cracks already apparent in local trust of the health care system. The suspicions of corruption already felt by pregnant women accessing a supposedly "free" health care system prior to Ebola were exacerbated, and anyone who could potentially reap financial benefit from the sickness came under scrutiny. Rumors concerning corruption within the health care system abounded: that Ebola did not exist, that it had been created to attract international aid, that the chlorine used as disinfectant was the real cause of the sickness, and even that the government had simply purchased too many body bags and needed "the sickness" to use them all up. While some literature reflects the loss of trust in the health care system during the Ebola epidemic (Dhillon and Kelly 2015; Yamanis et al. 2016), less attention has been paid to health care workers' own traumatic experiences of this loss of trust, as they risked their lives to come to work only to face mounting estrangement from their communities. A midwife underscored the pervasive stigma faced by Ebola health care workers:

In my area the neighbors' children don't go to my house, they are afraid, because they tell them not to go to my house and that, "that woman is an Ebola nurse, don't you see, even her belt is red, it's human's blood." Yes, it affected me also in my personal life, because I remember, even my husband... ever since, when we lie in bed we never share a blanket, but before this Ebola we shared a blanket. He said, "you play with patients for the rest of the day over there. I am afraid, I am afraid of you."

The trauma of that stigma was undoubtedly exacerbated by the persistent suspicion that the community's fear was not entirely unfounded, and that the risk of spreading infection to one's loved ones was real.

Public discourse about the epidemic included disbelief about the official narrative. In searching for an answer, many blamed health care workers either for exaggerating the seriousness of the situation, or for creating the health crisis for their own benefit. Some health care workers described being the victims of outbreaks of violence, in the form of riots directed against health workers, and many reported being afraid to wear their uniforms in public. A hospital administrator described the situation:

> *They came out with all sorts of weapons saying that they were going to kill us. They wanted our heads! For the whole day—we were four doctors—for the whole day we were being kept in a secret location, being heavily guarded. And then they came to the hospital, they wanted to burn the hospital, this one. They came to the hospital three times with petrol and matches and all that.*

In an economy bolstered by international aid, some suspected that the crisis had been fabricated for financial gain, a suspicion fueled by the foreign aid that began to pour into the health sector following the epidemic's peak. As a Sierra Leonean physician recounted:

> *We the health workers are being perceived as killers, just because—in fact there was this notion that because the money for HIV and AIDS was getting finished that's why we, the health workers, have brought this new disease. It was only when the magnitude became great, and the health workers were dying, and through extensive health talks that we did and people started believing that indeed we had an outbreak.*

Containment and quarantine protocols related to treating and managing Ebola patients further exacerbated fear and lack of trust. Isolation practices raised suspicions about what was going on out of sight and also challenged cultural practices related to communal care for the sick.

Thus, maternity wards emptied as pregnant patients either avoided the hospital system or sought treatment only in cases of emergency (Jones et al. 2016; Ly et al. 2016). "*During the course of being transported from their homes to the hospital, they would just give birth in the ambulance,*" explained a local pharmacist, "*I witnessed so many women on arrival they would just give birth in the outpatient department.*" As a result of the high risks of contagion associated with contact through the labor and delivery process, the government cracked down on its absolute ban on home delivery, accompanied by high fines for anyone who delivered at home. The health care staff explained in detail how women and their families arrived in terror, and often against their will, and the difficulties they faced trying to calm their patients amidst their own fears. "*Some days if we called the doctor, they could become afraid that we would take them for that thing, for the Ebola,*" described one nurse. "*They would say 'please don't call the doctor, I beg you, don't call the doctor.'*"

As will be discussed further below, this hesitation in seeking care had critical consequences as expectant mothers often presented in late stages of labor, and as a result, suffered serious postpartum complications, conditions that frequently manifest with high fever and hemorrhage. Thus, many of these women immediately met the case definition for Ebola, triggering fears and protocols that further delayed treatment for even the most fundamental of problems. Especially during the beginning of the epidemic, pregnant women were sometimes ignored, abandoned, or placed in isolation in the midst of an obstetric emergency, leading to conflicting feelings of guilt and regret. As one pharmacist described:

> *Patients were just being abandoned because of symptoms which they presented that sort of mimicked the Ebola virus, like some of them were really abandoned and passed away at that time, and before the burial team came to take away the corpse, they would take a swab and go and do an Ebola test. When the test would come in and it happened that the patient was negative, then you really felt that guilt, that it's the fear that made this patient to have lost their life.*

Hospitals emptied not only of patients, but also of health care workers, as the dangers of coming to work multiplied along with the virus and health care worker deaths mounted. The already undersupplied maternity hospitals and clinics simply did not have the resources to protect their employees from the threat of contagion, and health care workers were left to manage their risk individually. Some avoided hospitals altogether, whereas others refused to perform certain duties that would put them at heightened risk of infection. One nurse described the way she navigated the risk posed to nurses by pregnant or laboring women at this time:

> *You know, a pregnant woman likes when her stomach is being touched so that she will feel good. They would say everywhere they go to join a clinic they don't touch their stomach, that's why they come here... the nurses were afraid to touch. So what I did, because at that time there was no supply during the outbreak.... Because there were no gloves, mask, cap, full dress, you know, for Ebola dressing. The government and NGOs were trying to make a system to be able to supply it. It was not there, so what I did was that I tried to protect myself.... what I did, those plastics that they sell in the market by yard, nylon, it's there by yard... I bought it and made it for myself as an apron. Since they don't touch, but I do touch, so I needed to protect myself, I should not wait for government, I should be doing something to protect myself, to save my own life. Because I am with my family, so I should not wait for 'papa' government. So, if gloves are not there, I will take out of my purse and buy. Then if I can't find gloves I will look for plastic nylon-bags that they sell. I will put my hands there and use, then I will use again my dettol [an antibacterial soap] and smear it all over my body.*

Even where supplies existed, no standardized protocol for the use of such equipment had yet been established, and basic principles of infection prevention control were not widely understood. One international health worker who helped to administer trainings in infection, prevention, and control (IPC) at this time described the level of knowledge they initially encountered:

> *What was missing from that is like a real good understanding of germ theory...It was just completely like such an abstract concept. It started to make more sense of why you'd have somebody who's maybe got three latex gloves for the day, who just uses one in the morning, one in the afternoon, and one in the evening. Because they aren't understanding that like with each patient they touch, that there's still pathogens on these gloves, and when they go to the next patient, that can pass around and all of these things, which was just, it was really kind of mind-blowing.*

Government messaging surrounding prevention control was widely misunderstood, leading to waste of the limited supplies available through misuse, and when these alleged protections seemed to fail, subsequent loss of confidence in infection control grew more generally (Yamanis et al. 2016). Without an adequate understanding of correct procedures for donning and doffing PPE, for example, nurses saw colleagues they believed to be protected die from the virus.

> *We lost our friends... trained nurses in maternity. So they said to quarantine the maternity ward. But if we quarantine maternity, what about the pregnant women? So they quarantined the female ward but not maternity because of the maternal care. If we don't treat them, one or two will die. Maternal deaths will go up. So like we the nurses, we risked our lives to go work there. Even our companions, a pregnant woman went to them in the night and infected them with Ebola. Not to say they weren't dressed, that wasn't the problem, but to remove it [the PPE], no one had taught us, it hadn't been too standard at that time.*

Health care workers were also very conscious that when they came to work and risked exposure to the virus, they also risked spreading it to their children and family members every time they returned home from the hospital. Several nurses described elaborate procedures of decontamination undertaken before entering their homes or caring for their children. Often these would be combined with strategies to avoid letting neighbors know that they worked in the hospital. Furthermore, as the epidemic grew, many health care workers were personally impacted by the loss of life. For example, one maternity nurse who continued to see patients throughout the epidemic mentioned only at the end of her interview, "*In this outbreak I lost most of my family members.*" Payment of nurses at government hospitals during this time also remained unclear, with some nurses receiving "hazard pay" along with informal incentives, but often without a clear payment structure. This further challenged their ability to balance and weigh risks to their health versus the ability to provide for their families. As one international health worker described:

We're asking people to do stuff that nobody in the U.S. would ever do. And then we're saying that we're going to pay them $300 a month to do it and we couldn't come up with that money...Most people aren't going to go in there and not get rewarded for it.

Staffing shortages became a larger issue as more and more patients presented with Ebola infection. The growing number of deaths of health care workers seemed to establish that staff were not adequately protected; it also forced exhausted nurses to leave their stations with no one to relieve them. A maternity nurse described the situation as follows:

In the whole maternity ward it was just me. I would go from the morning and I would be there until 6 o'clock. Some days I didn't have anyone to relieve me. I left the ward and went home by myself. Another time it was just me and the doctor working, myself alone as the nurse, I got a C-S [cesarean] case, the doctor would come do the operation, there were no nurses, it was just me and the theater staff. And we would all do the operation. But I was able to manage. That day I was at the hospital until 9 o'clock. I was waiting for someone to relieve me, but I didn't see anyone so I just left the patient with the doctor, and I went home ...we didn't have time to write up reports because just imagine the time if you have two nurses on duty, it's not easy. You're not going to be able to do all the work. Some days we weren't able to give medication.

Understaffing also led to a reprioritization of patient care in which even the most basic procedures could not be performed:

Some days we weren't able to check vital signs. If we gave medication, we wouldn't be able to check vital signs. Some days we weren't able to do dressings...like the attention would all be in the labor ward and we would forget about the ones in the postnatal ward, you see. It wasn't easy.

The landscape, therefore, was one of increasing chaos, fear, and disintegration within the health care system, which was further exacerbated by the illness and loss of Dr. Sheik Umar Khan.[2] The death of Dr. Khan highlighted the vulnerability of those working to combat the epidemic, revealing that even the most highly educated, knowledgeable, and careful clinicians could be vulnerable, thereby adding to the sense of the epidemic as uncontrollable. In some cases, Dr. Khan had provided the IPC training that many physicians and nurses had received, which represented their only means of protecting themselves against infection. This single event was highlighted as a turning point for the epidemic by health care workers at all levels within Sierra Leone. Thought one nurse at that time, "*Oh, if Dr. Khan died then what will happen to us?*" At this point, health care workers felt they were confronted with a stark choice: to abandon their patients, or to consciously risk their lives. As one physician described the dilemma:

The situation was difficult. It was turbulent. But, as I said, there was need for people, for Sierra Leoneans to display patriotism. There was a dire need for people to make the sacrifice. Let's not forget the Hippocratic oath that we took... I was brave enough together with my team. We risked our lives, we went in. We were at least able to have saved the lives of these pregnant women even though we lost most of the babies.

An international health worker who was in Sierra Leone throughout the epidemic explained that Dr. Kahn's passing also had an impact on high-level policy:

In Sierra Leone everything changed when Khan died. So when Dr. Khan died that was the real change in terms of realizing the severity of the problem and the government really taking it seriously, and that coincided with

[2]Sheik Umar Khan (1975–2014) was a Sierra Leonean physician and specialist in hemorrhagic fevers and who spearheaded the medical treatment of patients infected with Ebola virus during the outbreak. He was head of the Lassa fever programme at the Kenema Government Hospital and was a Consultant to the United Nations Mission in Sierra Leone (UNAMSIL). He became infected while treating patients and was admitted to the Ebola Treatment Center in Kailahun where he was treated by doctors from Médecins sans Frontières (MSF). He died from Ebola virus disease on July 29th, 2014, 1 week after being diagnosed.

WHO declaring the Ebola emergency a ...public health emergency of global concern... By early September we had money.

As such, the situation began to improve, and health care workers' experiences and perceptions about the risk of exposure shifted as the response itself gained momentum. These changes influenced decision-making regarding patient care practices and the management of self-care.

24.4 Managing Pregnancy Risk and Uncertainty During Ebola: The International Response

The international response gathered speed through the early fall of 2014 as it became increasingly clear that the epidemic was not likely to run its course or to be contained by existing measures. The arrival of large-scale international aid rapidly transformed the landscape of the Ebola response. International health workers, armed with IPC training, comparatively bountiful supplies of PPE, and decidedly mixed levels of familiarity with the environment, sought to deploy the structures of modern medicine to impose order on an environment of rising fear and chaos. Western health care workers entered a dystopic, terrifying landscape in which the normal social order had been suspended and trust in the rituals of modern medicine had broken down. The international response helped to organize systems of care, attempting to prevent Ebola-positive patients from entering hospitals, creating holding units for unconfirmed suspects, and organizing Ebola treatment units (ETUs). In instituting systems of triage, quarantine, screening, and infection prevention, the humanitarian workforce simultaneously sought to control the spread of infection and reimpose a sense of control as an expedient to their own clinical practice and safety.

These attempts to reassert the control and containment protocols dictated by modern medicine were challenged by imperfect medical knowledge of the disease itself. Since it took over 24 h to receive results of Ebola screenings during much of the epidemic (Black 2015), it was sometimes impossible to know who was infected, and even with the best attempts at infection control, a moment's inattention could lead to disastrous outcomes. This section explores how health care workers navigated this environment of extreme risk and uncertainty in settings radically different from most Western clinical spaces. It also explores the intended and unintended consequences of this navigation for patients and health care workers.

Managing the narrative of controlled risk was important for international staff working in Sierra Leone during the early stages of the epidemic. The international response at this time centered on containing the infection in affected countries, preventing the flight of infected individuals, and screening entrants to medical facilities. Simultaneously, many NGOs pulled their international staff from Sierra Leone, fueling a sense of abandonment and a perception that the international community doubted the possibility of safely managing the outbreak. For those international health workers who remained within Sierra Leone, the worry that the international community would treat the epidemic as "too dangerous" to risk sending international medical staff was a pressure that shaped health care decisions. The continued deployment of international health workers depended on a perception of safety, and maintaining this perception required an approach to care that prioritized the safety of international aid workers even over the safety of patients. If facilities were not deemed safe to work in, that is, safe for international health workers, they would be closed. For example, MSF's Gondama Referral Center (GRC) near Bo suspended services when "*MSF could not guarantee the extremely high quality of medical services needed to treat patients and protect MSF's staff at GRC and from Ebola infection*" (MSF 2014).

One international health care provider who was present very early on in the response commented that this idea caused individuals and organizations to be more cautious in making triage decisions:

> *But part of our fear was if we have a situation that we get infected, we have a situation where our national staff get infected, but also if our expatriates get infected, well, we felt that would be severely detrimental to our efforts to try and get international engagement in responding to this outbreak because... this was before anyone was committing to coming and the general response was one of paranoia. If you look back around July, August time the main response was stopping flights from these countries, stopping people from traveling. It was a lot more about 'what if we get this in America, what if we get this in the UK or we get it in Europe.' The focus was not on how do we control this in the countries where they're severely affected.*

Pressure to control the narrative of containment and manageability of risk was thus added to the pressure to contain the virus itself. The continued provision of concrete medical aid and the morale of both domestic and international medical staff depended on a successful narrative of containment, despite what one international worker described as an initial atmosphere that felt like "*World War Z.... that the world was coming to an end.*"

As the international response to Ebola began to gather momentum, the expatriate health workers and increasing amounts of health resources arriving in Sierra Leone began to reshape the medical landscape of fear and uncertainty. Ebola treatment units were set up to handle confirmed positive patients, while systems of triage and quarantined holding areas managed the medical uncertainty of unconfirmed but suspected cases. For international health workers who arrived in Sierra Leone during this time, training consisted of a staged transition from theoretical to real risk. Trainings attempted to mediate the uncertainty and risk of working in an ETU by providing step-wise exposure, allowing trainees to experience the ETU as "safe" through multiple simulations before interacting with an actual Ebola patient. As one international health worker recruited during this time described:

> *We had a lot of PowerPoints that people had created where they had an overview of what the ETU would look like and what kinds of things we would see, and what symptoms look like and what the red zone meant, lots of donning and doffing. When we got with the WHO training we were actually physically doing it. It was a little bit redundant but went over again like what is EVD, what the patients look like, what are the protocols, what are the medications. Long days. Interspersed in that was the training for donning and doffing. They actually had in the very last day, they had us don and then go into a mock ETU and walk around and had people acting as patients in different scenarios. We had to role play what we would do in these different situations. That was kind of cool. Then, they had survivors come in and talk to us in small groups, about what it was like for them, the things that they remember, the things that they liked or didn't like or whatever. Then we fanned out and went to our different sites... We went from the dry run to a live run. The next day ended up at one of the ETUs that our team was going to work in, and then we were partnered with somebody who had been there for a while... We had our actual red zone training there the first time through. We just watched, no hands on. Then, the second time around, we started doing actual patient care. I think it was really, really adequate.*

Trainings for expatriate health workers during this time, therefore, emphasized ideas of certainty, predictability of experience, and manageability of risk. The repeated exposure to the ideas of risk within the ETU prior to an actual confrontation with this risk helped these workers develop trust that the risk associated with Ebola was clinically manageable.

> *I remember very clearly Nancy and Karl*[3] *helping a woman sit up to drink who obviously had Ebola. They both said, "come here, help us hold her up." I remember very clearly putting my hand on her back and thinking, "oh, my God, I'm touching an Ebola patient." I survived and realized, "okay, I can actually do this." Then they helped me doff and I felt a lot more confident and continued to go through it in my head just like I did with CPR, when I learned CPR years ago or anything like that. Then after that I don't remember being terrified ever again after that. I just remember like, "okay, this is just a job, and this is what I'm going to do" and try to do it to the best of my ability.*

---

[3] All names used are pseudonyms.

Protective equipment and the correct usage of PPE became extremely important in the management of risk. Donning and doffing PPE took on an almost ritual significance, as every task had to be done correctly in the right order to avoid contamination (Pallister-Wilkins 2016). Some international workers described practices they knew rationally did not make a clinical difference, but which they continued to perform as a way of managing the extreme risk associated in operating inside the ETU. One physician, for example, described the following:

> *One of the last things that the medical director said [in our training] was, you know there's a million virions per milliliter of blood and you need ten to get infected. So, that really stuck with you when you were in there... I mean, the requirement was that you wear two gloves. I wore three and I call the top layer of gloves my Lindsay layer, my wife's name is Lindsay. So I would wear one extra pair of gloves because I didn't want to get this and surely don't want to bring it home... And they actually have had a major breach in the organization. We call breach when your suit broke, and somebody actually cut themselves in the unit and had to get flown home. He was fine, he ended up not having anything wrong with it but there was a high level of anxiety, you know, kind of when we got there. You know the thing is, you're working next to people that were not quite as meticulous as you are and that was part of the challenge...and actually while we were there, three of our Sierra Leonean colleagues contracted Ebola and one of them died.*

Part of the correct implementation of ICP procedure by necessity involves emphasis on disease containment and prevention of health care worker infection even at the expense of the welfare of the individual patient. Via perfectionism in doffing and donning protective gear, and meticulous adherence to recommended procedures, infection protocol was largely successful in preventing international health worker infections during this time. In fact, the management of risk was so successful that some Sierra Leonean health workers interviewed even suspected that international workers possessed a secret vaccine for Ebola. While the Sierra Leonean health workers who believed in a secret vaccine were in the minority, the rumor serves to illustrate the tremendous effect that this projected sense of safety and confidence had on staff who had worked for months terrified of becoming infected or of inadvertently infecting family. As mentioned above, many local health care workers experienced increased confidence given successful management of risk, stating as one nursed did that their expatriate colleagues imparted in them the "*zeal not to be afraid*."

Another major problem that challenged this sense of control during the epidemic was the management of pure clinical uncertainty, which operated and was acknowledged on many levels. Guidelines for the appropriate treatment of Ebola-infected pregnant and postpartum women were seen as uncertain, overly cautious, or based on inadequate data. As one clinician noted, "*we were just making up the rules as we went based on what we thought was safest for both the clinician and the patient...I think people were really nervous when it came to doing obstetric care.*" Unlike the heavily protected environment of the Western hospital where these international workers had generally spent the bulk of their practice, this new environment lacked clear heuristics for making decisions, putting the onus for making the "right" decision more squarely on the individual health worker. Further, the possibility for intervention was greatly limited by lack of the medical technologies familiar to most health care workers and by a lack of resources more generally. In the face of a disease as fast and painful in its course as Ebola, medical practitioners who were used to intervening aggressively on behalf of their patients were forced to accept that there was, in some cases, nothing they could do but watch. As one foreign physician put it:

> *I mean, I have been a [type of doctor] since the 80's. I have never watched a child die of severe disease without multiple interventions, but that's what I almost had to get used to in my first days in Sierra Leone. I have practiced actually in West Africa before and I recognize the fact that you can't intubate, you can't put central lines. You can't do all the basic things and so unfortunately there's only so much you know you can do in that environment and that's a really, really rough adjustment. I would say, among my colleagues that came from the states with me, that was one of the hardest things for us to do, was to just basically hold a patient while they died of disease and you're thinking to yourself, I know we could be doing more for this kid.*

Furthermore, the clinical course of Ebola was seen as particularly unpredictable. Patients would seem to improve over days only to abruptly take turns for the worse. The sickest and weakest patients would sometimes be the sole survivors of whole families. As one international physician described:

There were some mother of infant pairs where the infant died and the mother didn't, and there were some mother-infant pairs where they both died... Miraculously, I think of one mother and her six month old daughter who was running a fever of 104, and their eight year old daughter who went into renal failure, and the mother's brother. The mother and the brother were not that sick. The kids were real sick and they all recovered. It was miraculous to me...I felt like it was the most difficult disease to predict of any I'd ever seen. I think it's because I had so little data of what was going on. No lab tests really to speak of and all that. That thing was very challenging.

In addition to a lack of a clear treatment heuristic, a lack of treatment options, and a lack of clear or reliable signs of disease improvement, those treating Ebola patients faced concrete barriers to their ability to care for their patients. PPE was bulky, hot, and permitted severely limited vision, thus reducing the number of procedures that could be safely performed with it on. Furthermore, Sierra Leonean and international workers alike described the process of working in PPE to be utterly exhausting and described the real risk of passing out due to dehydration. However, in the context of extreme contagion, a moment's carelessness was an unacceptable risk, and the ritual of donning and doffing PPE had to be completed perfectly every time. Thus, each health care provider could only remain inside the "red zone" for a set period of time, after which they would have to come out safely, doff safely, and rest or rehydrate. These factors meant that concrete, hands-on patient care was severely limited. This combination of challenges led to a state of stress and uncertainty for medical providers, particularly those from Western settings where hospitals allow for better control over disease processes. The stakes of this uncertainty were extremely high. As a physician recounted:

It was really unsettling because there was an awful lot to what was going on that you wondered whether or not you're doing the right thing. There's a baby that died that haunts me. It was so hard to get a line into the baby. We couldn't get an IV, then we finally put an interosseous line in but our time was up, then I went out. Then the interosseous line, either the line didn't function or the father took it out because it was oozing, then we went back in and talked to the father. The next time I went in somebody else had put an IV in the baby and given this little baby a liter of fluid which clearly put the child into almost like heart failure, pulmonary edema. The kid should never have gotten that much fluid...In the end the baby went on to die and it was just tragic because this guy had lost his entire family to Ebola. He kept saying over and over again, "I have nothing, now I have nothing" and I can picture him. It was very, very unsettling to know that not only did we not know what we were doing but because of the difficulty controlling the clinical situation things happened that were beyond your control.

The process of triage using diagnostic guidelines was another important means of combatting disease uncertainty. In the early epidemic, health care workers reported a large degree of uncertainty both within the community and among health care practitioners about what Ebola actually looked like. This contributed to the sense of loss of control, as it was impossible to determine without testing whether any given patient was likely to be Ebola-positive. In a setting where serious diseases such as malaria, Lassa fever, tuberculosis, cholera, HIV, and others are endemic (WHO 2011) and confound the diagnostic differential, this problem is obviously compounded. The ambiguous nature of an Ebola presentation strongly contrasted not only with medical textbooks, but with the official narrative presented by government agencies during the early epidemic. Family members would bring sick relatives to the hospital with symptoms that did not resemble the official description of Ebola, only to be told they were positive, which then discouraged treatment seeking for unrelated illnesses, as any illness with any presentation could turn out to be Ebola.

As a result, as described above, early in the epidemic patients suspected for any reason of being positive were often avoided rather than swiftly quarantined. In addition to potentially spreading the infection to others, this delayed or wholly prevented care for a host of other conditions. Furthermore, a patient who presented with minimal symptoms would oftentimes progress while waiting without medical care from a patient who could reasonably be assumed not to have Ebola. One European physician explained the complexity of the situation:

> *The sort of partial training that health care workers were getting was leaving them in a situation where they would feel not confident about determining whether a patient was suspect or not. So they would kind of keep it to themselves if they felt something was wrong. And what that inevitably ended up doing was the patient would be sort of suspicious around Ebola but not categorizing them as being a suspect patient and managing them properly. They would just simply ignore them. And many, many times I walked into the outpatient department where a woman was in mid-labor with obstructed labor, and she'd be laying on the floor of the reception area, and people would step over her and continue walking. They didn't want to deal with that patient....You know there wasn't any suspicions to begin with but now that you've left her for so long, she's got so much worse because she's been unattended? And at that point we would have to isolate the majority of times, even if it was a normal pregnant mother suffering complications, who really needed to go to theatre, but now... none of the medical staff would accept her, they wouldn't take her. And she had to come to the ETU, where unfortunately they, you know, a lot of the time they died in isolation receiving very minimal pregnancy care because we would not get the test result back quick enough to get them back out and back into theatre.*

As a way of dealing with this level of clinical uncertainty, systems of triage were developed based on official clinical criteria or "case definition" that would serve to distinguish patients high-risk enough to warrant testing and quarantine, usually before the provision of any aggressive treatment. These criteria included combinations of symptoms as well as a patient's history of exposure. Health workers agreed that these systems did not provide a perfect guide, and many described the pressure to be overly cautious in their administration: "*we were very strict on case definition,*" described one American nurse, "*because we did not want to be responsible for having somebody with EVD come and deliver inside the hospital side by side with who knows how many patients, exposing who knows how many nurses. We just didn't want to be responsible for that.*"

Women who were pregnant or going into labor during the Ebola epidemic were particularly vulnerable to these systems of triage. By necessity, in these systems the welfare of the patient comes third—behind the safety of health workers and other patients—as clinicians lose the ability to distinguish the sick from the well. The ambiguity surrounding the status of pregnant women is especially problematic as, by their nature, complications of pregnancy and labor mimic the presentation of Ebola in many respects (Black 2015). Women presenting with common complications of pregnancy immediately met case definition for Ebola, a status which, in the absence of testing, was enough to quarantine rather than treat them. Methods for treating these conditions likewise carry an increased risk of transmission as they require exposure to bodily fluids.

According to infectious disease experts, the risks surrounding the treatment of a pregnant woman were compounded by their likely high disease burden, as viral loads in amniotic fluid and in products of conception could remain high long after the woman herself had cleared the virus (Black et al. 2015). In fact, it became known that a woman could have already cleared the virus, but products of conception could still contain the live virus and be infectious (Black et al. 2015). This was especially devastating in the context of a country that already had a very high rate of nonelective, medically necessary cesarean sections under normal conditions (Chu et al. 2012). These guidelines would have caused pregnant women who tested negative to be admitted to Ebola treatment units where they were at a high risk for contracting the virus by exposure. Furthermore, the rates of Ebola-related mortality among pregnant women were thought to be extremely high (Black et al. 2015)—as high as 90% according to one of our respondents—and this contributed to a sense among some that treatment for these women was futile. According to our interviews, even during the height of the Ebola response, there were very few facilities and/or specialists that provided care for women who were pregnant and suspected of having Ebola. One physician described his experience of risk, and of attempting to garner support to provide obstetric care for pregnant women with Ebola, by stating:

Before I [began working in the] maternity unit I was told by many, many, many people not to do it because I would die. Not just Sierra Leonean health care workers but expert health care workers very experienced and clinicians who had worked in previous Ebola outbreaks. All of them told me if I did it... I and my staff would most likely get infected and die. Everyone until that point in the Ebola outbreak who had delivered children was dying. And you know the, the idea simply that you know these patients are too dangerous or they're too difficult to manage and therefore we shouldn't try... and yes there were risks and there would always be risks in managing dangerous diseases, and with the background I come from... just because there is a risk doesn't mean that it's not a job that's worth doing and it's not a risk that's worth taking.

If this stigma was present within the international medical community, it was certainly present at much higher levels within the local health worker community. Pregnant women were known to be sources of contagion, and pregnant bodies were seen as a source of danger—described by some as "Ebola bombs," which compounded women's risk as they neared delivery. The approach used to treat was thus "very black and white," translating to huge delays in care.

As the epidemic continued, this gauntlet of uncertainty, fear of pregnant women, ethical dilemmas of triage, and extremely high stakes were constantly negotiated by health care workers attempting to balance the competing needs of personal safety, disease containment, and patient care. Clinical guidelines for screening were known to provide an overly sensitive heuristic, capturing for quarantine far more patients than likely had the disease. Yet these guidelines provided the only means of mediating clinical uncertainty for Western practitioners used to relatively high levels of certainty in practice. Gradually, however, even these black-and-white screening protocols became sites of negotiation, where the hard and fast rules of Western medical science gave way, allowing shades of gray and alternative local knowledge to seep in. For example, one American nurse described how their staff relied on a Sierra Leonean nurse to handle triage, teasing that she could "smell Ebola." Importantly, this process was a common dilemma across multiple treatment sites especially towards the end of the epidemic, where dedicated health care providers knew that even though the case definition for Ebola provided their only means for "knowing," in the moment of care, whether a woman had Ebola, they simultaneously recognized that these screening guidelines were overly sensitive and that the majority of the women fitting criteria in the setting of pregnancy would not test positive for Ebola. For health care workers whose extreme compassion for their patients caused them to risk their lives in caring for their patients, the option of sacrificing lives for added security became unacceptable. They, therefore, began to subjectivize these seemingly objective criteria, creating room for holistic "clinical judgment" and exceptions to rules.

Especially toward the end of the epidemic, this process of negotiation became an extremely important, if unspoken and unacknowledged, part of the process for how health care workers coped with ethical and moral dilemmas in triaging patients. One physician explained how, "*they were still checking the same boxes in the same forms, because they hadn't changed, but they were definitely—they were not checking the box of vaginal bleeding on hell of a lot of patients who were clearly having that, or the fever and symptoms boxes as well, if they really did not think that it was something concerning.*" Another international health worker described the process of deciding whether to admit a pregnant woman to an ETU or to a hospital as beginning with an objective set of screening questions, but ending up as a series of negotiating steps sometimes informed by "gut instinct:" "*If your gut says just admit the patient, you bet it's best to admit them than be wrong, then to let them go inside the hospital, and you be wrong...*" One nurse described the screening process as follows:

I do think though the algorithm was a really good one, there is a slight gray area when it came to vaginal bleeding... A lot of times you're like, okay, how much are you bleeding. You go to the back, you put on PPE, and you're like, okay, let me see your pad. Let me see how much you're bleeding, let me see what it looks like. You're trying to determine, okay, does this seem like a normal amount of bleeding? Then if it looks normal to you, you're trying to weigh, okay, it looks normal, does she have any other signs and symptoms that could make this possibly Ebola instead. Has she been exposed to anybody? You look at all of that and weigh all of that. It did seem like she had absolutely no exposure whatsoever. There's not been a single person in the family died. She's not been sick whatsoever. She's only been pregnant, and she's nine months pregnant. And her labor is starting. She's got a little bit of vaginal bleeding on her pad. Then you tell it to the nurse, and then you got to talk to a doctor. Because if a doctor refuses to accept her, which does happen, no matter how much you explain to them, they can refuse. If they refuse, you have to take them into the holding unit, because he's not going to take them. If he doesn't take them, they will go home and they will labor and have a baby at home, so you have to take them.

The process of managing risk during clinical care was, therefore, a back-and-forth negotiation between the material need for clinical certainty as a way of guaranteeing safety and control of the epidemic and the ethical need for room for subjectivity and doubt. Different health care workers recognized different levels of holistic clinical expertise, particularly in the care of pregnant women, where many international responders operating in this space had limited experience with maternal health, even without the added complication of Ebola. Different practitioners and different treatment sites dealt with balancing these issues in different ways, and as the epidemic drew to a close, many health care workers began to admit more uncertainty into their practice. Many of the health providers interviewed described this process of negotiating knowledge and uncertainty to be *the* crucial difficulty in caring for pregnant women during the epidemic. As one midwife described, "*I mean, we did the best that we could…you know, it's so hard to wrap your head around because at the same time, you let a patient go through to the ward and they end up being EVD positive and now you've put so many other people at risk, so it was very hard to kind of grapple with that…once you open it up to grey area it became really hard because then it was judgement calls.*"

24.5 Conclusions

As this chapter demonstrates, the health care providers who cared for pregnant and laboring women during the 2013–2015 Ebola epidemic in Sierra Leone were faced with constant practical and moral dilemmas regarding whether and when to intervene and how to prioritize and minimize risks both to themselves, the public, and the individual patients in their care. In capturing the lived experiences of nurses, midwives, physicians, and others working in maternal health at this time, this study fills an important gap in the literature with respect to the subjective experiences of caregiving amidst great uncertainty and fear, and at varying timeframes within the Ebola response. The Ebola epidemic represented a situation where true biomedical certainty and control were impossible to come by and personal risk of infection was at its peak. In order to navigate this landscape of risk and uncertainty, health care workers responding to Ebola deployed various strategies allowing them to mitigate their fears while maintaining flexibility to negotiate ethically challenging care decisions. In preparing for the next epidemic, in which we may once again reach the limits of medical knowledge, it is important to consider the ways in which the creation of future medical resources, such as case definitions or clinical guidelines, could alleviate the necessary tension between the mediation of risk and the need for flexibility and compassion. For example, lessons learned during this outbreak can be used to develop detailed protocols tailored specifically to obstetric and gynecological care that can be used to reduce stress and personal accountability in the making of critical care decisions amidst high levels of risk.

According to our findings, underlying weaknesses and mistrust in the existing health care system—especially disillusionment with regard to the 2010 Free Health care Initiative—contributed to fear and avoidance of health care facilities and health care workers themselves at the start of the Ebola epidemic, exacerbating an existing scenario of inadequate health care resources. The result was that many local health care providers were stigmatized for simply doing their jobs, compounding traumatic experiences of fear and loss as they risked their lives to help others while simultaneously suffering the effects of the disease. This included constant anxiety for themselves and their family, as well as the loss of coworkers, family members, and friends. Fear of contagion also meant that pregnant women presented at late stages of labor, often mimicking signs and symptoms of Ebola, and faced delays in receiving life-saving treatment.

Thus, as Sierra Leone considers health care strengthening in the post-Ebola climate, it must do more than fill these gaps in human capacity and material resource provision. To address health services corruption, for example, health care workers need to be provided a living wage—not be forced to "volunteer" their services. It is also important that more be done to support health care workers themselves in future epidemics of this nature, by providing additional clinical resources and guidance. It is also important that health care responders in similar situations receive adequate social support and self-care. In our study, local Sierra Leoneans and international temporarily deployed staff described subjectively different experiences navigating risk and moral uncertainty within the response; more research is needed to understand the nuanced differences and similarities faced by these groups, especially in relation to their respective needs as well as the dynamics influencing effective collaboration.

While the Ebola epidemic initially ravaged the country's health care system and eroded trust in its health care workforce, there have been several upsides to having overcome the disease. In addition to influxes of material resources and human capacity related to IPC measures, health care workers also spoke about a renewed sense of identity and pride in their roles. According to Mary, the nurse introduced at the beginning of the chapter, once the epidemic waned her peers and neighbors began to admire her, commending her for her bravery. It seems that for some, particularly Sierra Leonean care providers, their membership in a larger international health community contributed to a growing sense of pride and professional identity. These factors are also important for further exploration as relevant to debates around health care strengthening in post-epidemic environments.

References

Black, B. O. (2015). Obstetrics in the time of Ebola: Challenges and dilemmas in providing lifesaving care during a deadly epidemic. *BJOG: An International Journal of Obstetrics & Gynaecology, 122*(3), 284–286.

Black, B. O., Caluwaerts, S., & Achar, J. (2015). Ebola viral disease and pregnancy. *Obstetric Medicine, 8*(3), 108–113.

Chu, K., Cortier, H., Maldonado, F., Mashant, T., Ford, N., & Trelles, M. (2012). Cesarean section rates and indications in sub-Saharan Africa: A multi-country study from Medecins sans Frontieres. *PLoS One, 7*(9), e44484.

Dhillon, R. S., & Kelly, J. D. (2015). Community trust and the Ebola endgame. *New England Journal of Medicine, 373*(9), 787–789.

Erland, E., & Dahl, B. (2017). Midwives' experiences of caring for pregnant women admitted to Ebola centres in Sierra Leone. *Midwifery, 55*, 23–28.

Jones, S. A., Goplalakrishnan, S., Ameh, C. A., White, S., & Van Den Broek, N. S. (2016). 'Women and babies are dying but not of Ebola': The effect of the Ebola virus epidemic on the availability, uptake and outcomes of maternal and new-born health services in Sierra Leone. *BMJ Global Health, 1*, e000065. Retrieved December 8, 2017, from https://www.ncbi.nlm.nih.gov/pmc/articles/PMC5321347/.

Jones, S., Sam, B., Bull, F., Pieh, S. B., Lambert, J., Mgawadere, F., et al. (2017). "Even when you are afraid, you stay": Provision of maternity care during the Ebola Virus Epidemic: A qualitative study. *Midwifery, 52*, 19–26.

Ly, J., Sathananthan, V., Griffiths, T., Kanjee, Z., Kenny, A., Gordon, N., et al. (2016). Facility-based delivery during the Ebola virus disease epidemic in rural Liberia: Analysis from a cross-sectional, population-based household survey. *PLoS Medicine, 13*(8), e1002096. Retrieved December 7, 2017, from http://journals.plos.org/plosmedicine/article?id=10.1371/journal.pmed.1002096.

Maxmen, A. (2013). Sierra Leone's free health-care initiative: Work in progress. *The Lancet, 381*(9862), 191–192. Retrieved December 8, 2017, from http://www.thelancet.com/journals/lancet/article/PIIS0140-6736(13)60074-4/fulltext.

McMahon, S. A., Ho, L. S., Brown, H., Miller, L., Ansumara, R., & Kennedy, C. E. (2016). Healthcare providers on the frontlines: A qualitative investigation of the social and emotional impact of delivering health services during Sierra Leone's Ebola epidemic. *Health Policy and Planning, 31*, 1232–1239.

Médecins Sans Frontières. (2014, October 16). *Sierra Leone: MSF suspends emergency pediatric and maternal services in Gondama*. Retrieved December 7, 2017, from http://www.doctorswithoutborders.org/article/sierra-leone-msf-suspends-emergency-pediatric-and-maternal-services-gondama.

O'Hara, G. (2015). Working on the front line. *Clinical Medicine, 15*(4), 358–361.

Pallister-Wilkins, P. (2016). Personal protective equipment in the humanitarian governance of Ebola: Between individual patient care and global biosecurity. *Third World Quarterly, 37*(3), 507–523.

Pieterse, P., & Lodge, T. (2015). When free healthcare is not free. Corruption and mistrust in Sierra Leone's primary healthcare system immediately prior to the Ebola outbreak. *International Health, 7*(6), 400–404.

Ponsar, F., Tayler-Smith, K., Philips, M., Gerard, S., Van Herp, M., Reid, T., & Zachariah, R. (2011). No cash, no care: How user fees endanger health—Lessons learnt regarding financial barriers to healthcare services in Burundi, Sierra Leone, Democratic Republic of Congo, Chad, Haiti and Mali. *International Health, 3*(2), 91–100.

World Health Organization. (2011). *Health Situation Analysis in the African Region: Atlas of Health Statistics, 2011*. Geneva: World Health Organization.

WHO, UNICEF, UNFPA, The World Bank, & the United Nations Population Division. (2014). *Trends in maternal mortality: 1990 to 2013: Estimates by WHO, UNICEF, UNFPA, The World Bank and the United Nations Population Division*. Geneva: World Health Organization.

WHO Ebola Response Team. (2015). West African Ebola epidemic after one year—Slowing but not yet under control. *The New England Journal of Medicine, 372*(6), 584–587. Retrieved December 8, 2017, from http://www.nejm.org/doi/full/10.1056/NEJMc1414992#t=article.

Yamanis, T., Nolan, E., & Shepler, S. (2016). Fears and misperceptions of the Ebola response system during the 2014-2015 outbreak in Sierra Leone. *PLoS Neglected Tropical Diseases, 10*(10), e0005077. Retrieved December 8, 2017, from http://journals.plos.org/plosntds/article?id=10.1371/journal.pntd.0005077.

25 Preserving Maternal and Child Health Care in Sierra Leone During the Time of Ebola: The Experiences of Doctors with Africa

Giovanni Putoto, Francesco Di Gennaro, Alessandro Bertoldo, GianLuca Quaglio, and Damiano Pizzol

25.1 Introduction

The recent outbreak of Ebola virus disease (EVD) in West Africa resulted in very high rates of infection and deaths both among patients and health care workers. The direct effects of EVD in Sierra Leone include over 14,122 infected persons, resulting in 3955 deaths, among them 221 were health care workers (World Health Organization 2015). Prior to the outbreak and following over 10 years of civil war, the high level of poverty, demolished infrastructure and extremely weak state of the health care system in Sierra Leone were striking. For example, the density of physicians per 1000 population was 0.022 (in comparison Belgium has 2.9 doctors per 1000 population) (World Health Organization 2017). Sierra Leone was also among the poorest countries in the world, ranking 183 out of 187 countries on the United Nations Development Programme (UNDP) Human Development Report. The Ebola crisis exacerbated problems that had persisted for decades in the affected area.

One of the major consequences of the Ebola outbreak has been its impact on maternal health, since the priority was to stop disease transmission and to prevent Ebola spread. During delivery or miscarriage, health workers are exposed to serious risks, and many times, pregnant women were denied hospital car and turned away, thereby convincing other pregnant women to avoid prenatal visits and assisted delivery (Hayden 2015). Sierra Leone's neonatal mortality rate of 35 per 1000 live births was one of the highest in the world in 2015 (UNICEF 2015a), and the lifetime risk of maternal death is one of the world's highest at 1 in 21 (UNICEF 2015b). In addition, 7% of Sierra Leonean are underweight and only 54% are put to the breast within 1 hour of birth (Statistics Sierra Leone 2014). Maternal and neonatal health in Sierra

G. Putoto (✉) · D. Pizzol
Doctors with Africa CUAMM, Padova, Italy
e-mail: g.putoto@cuamm.org

F. Di Gennaro
Clinic of Infectious Diseases, University of Bari, Bari, Italy

A. Bertoldo
Zerouno Procreazione, Centro di Medicina, Venezia Mestre (VE), Italy

G. Quaglio
Directorate-General for Parliamentary Research Services, European Parliament, Brussels, Belgium

D. A. Schwartz et al. (eds.), *Pregnant in the Time of Ebola*, Global Maternal and Child Health,
https://doi.org/10.1007/978-3-319-97637-2_25

Leone has remained a significant public health problem, exacerbated by service interruptions during the 2013–2015 Ebola virus disease epidemic (Sharkey et al. 2017). In this challenging context, it was difficult to preserve the essential continuity of maternal and infant health care. Several authors reported a dramatic decline in pediatric and maternal medical admissions (Delamou et al. 2014; Médecins Sans Frontières 2014).

Doctors with Africa (DwA) CUAMM is the largest Italian nongovernmental organization (NGO) that deals with development cooperation in the health sector (Doctors with Africa CUAMM 2016). It was present in Sierra Leone from January 2012 in the Pujehun district, where it was involved with a project funded by the United Nations Children's' Fund (UNICEF). The strategy of the project centered on the reinforcement of service delivery at both the hospital and district levels and improving quality of obstetric and neonatal care. The project also addressed the burning issue of the scarcity of qualified health workers, with placement of additional human resources, increasing training, and creating a conductive clinical environment. Doctors with Africa was working in the field when the first Ebola case was reported in Sierra Leone on May 23, 2014, and the members decided to remain and work in the district. The present report describes the experiences of Doctors with Africa as an exemplar of the maintenance of the reproductive health services in time of Ebola.

25.2 Doctors with Africa in Sierra Leone

Doctors with Africa CUAMM was the first nongovernmental organization (NGO) focused on health care to be recognized by the Government of Italy and is now the leading Italian organization providing health services to vulnerable communities in Sub-Saharan Africa. With funding from UNICEF, it began its intervention in Pujehun in January 2012, aiming to strengthen the capacity of District Health System to address the needs of the population, especially the most vulnerable groups—pregnant women and under-five children, as well as improving the services provided by the Basic Emergency Obstetric and Newborn Care (BEmONC) and the Comprehensive Emergency Obstetric and Newborn Care (CEmONC) centers.

The strategy utilized involved support from the District Health Management Team through technical assistance in planning, monitoring, and evaluation, together with reinforcing service delivery at the Hospitals and the six BEmONC Centers at the District level, with special attention paid to improving quality on the areas of obstetric, neonatal, and child health care.

The project also addressed the important issue of scarcity of qualified human resources for health—this is crucial to guarantee basic services including the placement of additional human resources, in-service training, formal training, creating a conducive clinical environment, and increasing access to obstetric, neonatal, and child care.

Because the intervention was meant to develop and support local capacities, planning included the development of an exit strategy at the conclusion of the 3-year project. This included the formulation of an ad hoc Memorandum of Understanding (MoU) with the Ministry of Health so as to ensure that the District Health Management Team of Pujehun was prepared to take over from Doctors with Africa CUAMM.

The intervention was consistent with the National Strategic Plan of the Sierra Leone Ministry of Health 2010–2015 and built upon the efforts undertaken since 2004 by UNICEF Sierra Leone with the construction and turnover of four maternity complexes in Kenema, Koinadugu, Kono, and the more recent Pujehun District. The project also reflected CUAMM's Strategic Plan 2008–2015, which aimed at strengthening African health systems in terms of effectiveness, efficiency, and accessibility of services to all people, and especially to those who were the most vulnerable.

25.3 Pujehun District

The Pujehun District of Sierre Leone is located in the south, near the border with Liberia (Fig. 25.1). The district has a population of approximately 345,000. The population is primarily from the *Mende* ethnic group, and the district is made up of 12 chiefdoms and is predominantly Moslem. Its capital, Pujehun Town, is a very small underdeveloped rural town with approximately 30,000 people, located 330 km south-east of the national capital of Freetown. Connections within the district are made difficult because of the very poor status of the roads and the fact that the district is separated in two parts by a large river, the Moa river. Access to the health services across the river was possible, except during the raining season, by using a flat barge. Unfortunately, the barge sank in April 2012. Thus, supervision and referral are often made by a large journey through Bo and Kenema.

25.4 Pujehun Health Services

Health services in Pujehun are mostly provided by the Ministry of Health and Sanitation (MoHS), which operates on the primary health care model. Funds for running the district hospital and the primary health care services come through the local district council. The primary health system of Pujehun includes 73 Peripheral Health Units (PHUs), 12 of which are designated as Community Health Centers

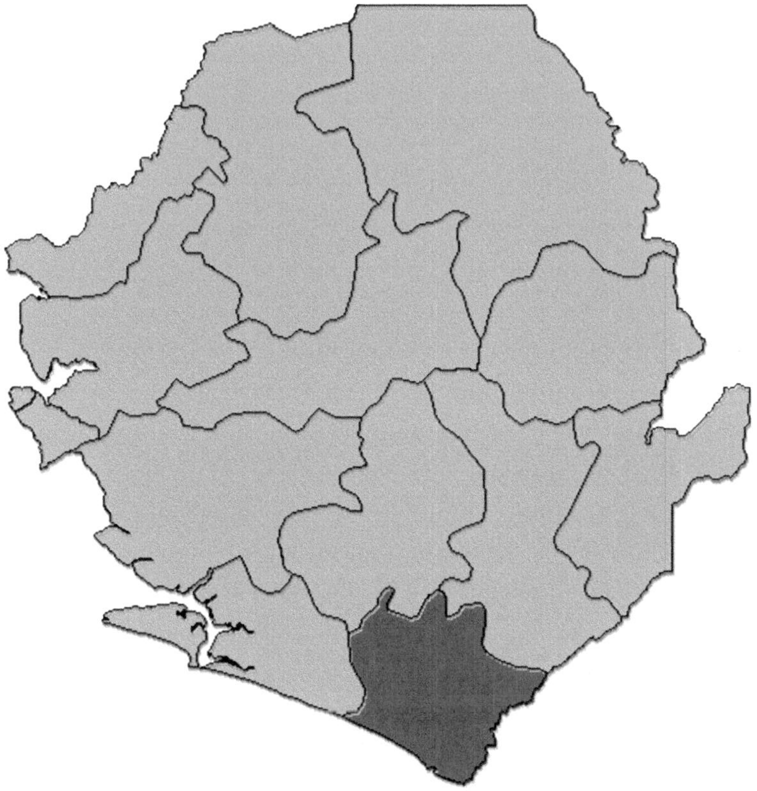

Fig. 25.1 The Pujehun District of Sierra Leone (red shading). Available from: https://en.wikipedia.org/wiki/File:Sierra_Leone_Pujehun.png

Fig. 25.2 Front side of Pujehun Maternity Hospital

(CHCs) and six of which have been selected to be upgraded to BEmONC centers. The secondary care consists of an 85-bed district hospital. With an uneven distribution of the available PHUs in the district, most of the population must travel long distances to reach the district hospital. This hospital is hosted in two different premises: the older structure that includes Male and Female Wards, Outpatient Clinics, Laboratory, dispensary, and store; and the new facility providing the health services covered by Free Health Care (FHC) including Maternity and Pediatric Wards (Fig. 25.2). Both the hospital and the Peripheral Health Units are seriously understaffed. The limited availability of health personnel in Pujehun district is due to a general shortage of professional staff in Sierra Leone; this is compounded by the fact that staff posted to remote facilities normally do not have accommodations, so that part of the staff that were posted to Pujehun preferred to leave their job there and request a new assignment.

The most important cadres in the delivery of PHU services are Community Health Officers (CHOs), who are designated to supervise the CHCs and the Maternal and Child Health (MCH) Aides who carry out most of the maternal and child health services given the shortage of nurses and midwives. In September 2012, three midwives were posted by the government to work in three of the BEmONC centers.

The hospital health care staff were very limited until March 2013. The Matron and the Medical Officer (MO) were often absent, resulting in poor organization of the hospital and limited health services provision, resulting in a new matron being posted in September 2012. Moreover, in 2012, there was no line item budget for the running cost of the new Maternity Complex.

Currently, at the Maternity Complex, there are 11 health staff in the Pediatric Ward and 16 in the Maternity Ward; 8 of them are not yet on the payroll. Other community-level personnel involved in delivery of services are traditional birth attendants (TBAs), trained to conduct normal deliveries at the PHU level and working under the supervision of MoH health staff. About 65% of the deliveries occur at the health facilities, while the remaining give birth at home assisted by relatives or TBAs (Fig. 25.3). An MCH Aides training school was set up in 2007 and graduated its first class of 36 students in 2009. In August 2013, 46 students have concluded the 2-year course,

Fig. 25.3 Women sitting with their babies to attend MCH clinic at Pujehun Maternity Hospital

but the percentage of students that graduate has been only 50%. The quality of the teaching is hampered by the low level of schooling of the trainees and the poor conditions of the premises, where there is only one room available to accommodate approximately 60 students. Beginning in 2010, UNICEF has funded an integrated case management program in Pujehun through Save the Children UK. To date, community mobilization has been performed in 10 out of the 12 targeted Chiefdoms, and a total of about 1000 community health workers have been selected. The programme is for case management of children with malaria, diarrhea, and respiratory infection. Moreover, World Vision is operating in five chiefdoms supporting the nutrition programme and the community treatment of under-five children with malnutrition.

25.4.1 Health Management Information System

Although a good District Health Information Software (DHIS) database to capture and analyze data exists, there is limited effort in data analysis at the district level. Many important data sets are lacking—basic data analysis including indicators for each facility catchment area, a data summary report for the district, comparisons among facilities, comparison with district/national targets, and comparison of data over time.

The hospital activities data are collected regularly in the Maternity Complex, but there is no certainty about the timeliness, completeness, and reliability regarding data concerning the main hospital wards. Data are, in fact, neither analyzed nor utilized by the planners at Pujehun District Government Hospital (PDGH.). The data collection is performed in the wards and departments according to the standard register of the Ministry of Health by the "In-charges", local staff specially trained to do data collection. A monthly summary is forwarded to the data clerk of the PHC department, who submits it to the Ministry of Health, without any data entry and analysis. It was apparent that some of the hospital indicators were being calculated, such as inpatients days, or not properly recorded, such as the occurrences and causes of death.

The utilization of the Pujehun Government Hospital is low—in 2013, the bed occupancy rate (BOR) was 25%, and the average length of stay (ALOS) was 4.3 days. In the Maternity Complex, the bed occupancy rate was slightly higher at 27%. In 2012 and 2013, the number of deliveries occurring in the Maternity Ward were 423 and 425, respectively, demonstrating an increase of 71.9% and 72.7% compared with 246 deliveries occurring in 2011. In 2012 and 2013, the number of cesarean section deliveries performed were 102 and 88, respectively. This represented an increase of 155% and 120% when compared with a total of 40 cesarean section deliveries occurring in 2011. In 2011, the cesarean section rate in Pujehun District was 0.2%, increased by 0.7% and 0.6% per annum in the 2 following years—this is contradistinction to the minimum recommended cesarean section delivery rate of 5%. The causes of operative delivery underutilization include poor accessibility due to economical and geographical barriers, low education and cultural barriers among the communities, drugs stock out, and poor training of hospital staff (Fig. 25.4).

Fig. 25.4 Pregnant women standing in front of Pujehun Maternity Hospital

25.4.2 Governance Following the Start of the Ebola Outbreak

In April 2014, just before Sierra Leone reported its first cases of Ebola virus disease in late May, collaboration with the local authorities of Pujehun District became instrumental in adopting important public health measures that would help to limit the epidemic—these included decisions to close the market; ban certain social activities, worship in churches and mosques; establish three check points in the main roads; restrict cross-border movements with Liberia; and prohibit the selling of bush meat and the transportation of deceased people (The Guardian 2015). During the outbreak, the district health management team was in charge for overall coordination of Ebola control measures and was supported by a number of national and international partners. The work was organized around the District Ebola Outbreak Control Plan which was drafted in July. Daily and weekly meetings were held with the District Health Management Team and the Hospital Management Team to address problems and find solutions. Training, supportive supervision, and supplies of basic equipment and drugs were provided to the peripheral health units throughout the epidemic, although sometimes with difficulty. Regular communication, exchange of data, procurement of available drugs, and protective equipments were made with the headquarters of the Ministry of Health.

25.4.3 Triage, Contact Tracing, and Quarantine

In August 2014, an initial triage system was set up at the entrance of the Maternity Complex. Mothers and children were first assessed for Ebola virus infection by trained health workers and senior staff in cases of suspicion, with a "no touch" policy in place. One woman with Ebola virus disease was identified in the Maternity Complex. During the outbreak in the district, 250 "contact tracing" personnel were trained, and 1222 contacts listed were performed in and around Dumagbe, Makpele Chiefdom, which was the main epidemic hotspot (Sierra Leone Ministry of Health and Sanitation 2016). The tracing team was very effective in terms of contacts' identification, listing, and follow-up, and there were no contact tracers that became infected. Around 1100 people were placed into quarantine for 21 days, including 11 hospital health workers as a result of violating safety procedures at the Ebola Holding Centre.

25.4.4 Isolation and Clinical Management

In order to help contain the , three were established, one in July 2014 within the old complex of with 6 beds, one in August at containing 12 beds, and the last one in November in with 8 beds. Overall, approximately 60 people were admitted in these EHCs, where they received supportive care and food (World Health Organization 2016). A sample of blood was taken from each patient and sent to the laboratories based in Kenema and Bo for Ebola virus testing—it took an average of 2 days to receive the result.

Patients found to be infected were immediately transferred to an Ebola Treatment Centre (ETC) in the neighboring districts. Initially, all positive cases were transferred to Kenema Government Hospital; later, they were transferred to an Ebola Treatment Centre run by the International Federation of Red Cross and Red Crescent Societies (IFRC) in Kenema District and an Ebola Treatment Centre run by Médecins Sans Frontières (MSF) in Bo District.

Safe burials teams were created, trained, and equipped. On average, 50 safe burials were carried out each week from August to December 2014, after which they decreased to 38.

25.4.5 Training, Provision of Personnel Protective Equipment, and Infection Prevention and Control

Training for Ebola virus disease was first provided to the health staff and personnel of the hospital and 75 health units beginning in April 2014. In August, further training was provided in Kenema for those people engaged in the EHCs, safe burial teams, laboratory, and drivers. Finally, on September 24th, health workers were trained in Ebola infection screening. All the PCHs were initially provided with a kit of protective equipment and few sets of personal protective equipment (PPE) to be used in dealing with suspected cases. Since the supply was irregular and scanty, only staff working in the EHC and burial teams constantly used PPE ensembles. Normal deliveries and cesarean sections were performed in the delivery room and operating theater using strict universal protection measures. A functional referral system was kept in place to deal with cases of maternal complications (Fig. 25.5). At the onset of the epidemic, the adoption of infection prevention and control (IPC) standards and procedures in the maternity wards was a difficult challenge that was compounded by lack of protective equipment, difficulty in changing staff behaviors, and inadequate hospital waste disposal. The improved supply of IPC barriers from mid-September, the provision of a dedicated IPC nurses, the setting up of a Hospital IPC Unit, and the upgrade of the clinical waste and disposal devices, including a new incinerator, contributed to improve the situation.

Fig. 25.5 Ambulance at Zimmi Primary Health Care Center to transport a pregnant mother

25.4.6 Human Resources

Management of human resources in the time of Ebola was of paramount importance. Fear, lack of motivation, and tensions between and within the national and international teams were the main problems that had to be confronted. The apex of the crisis was experienced during the time of sickness and, tragically, death of two infected health care workers. Performing continuous dialogues, formal and informal meetings, common decision-making processes on clinical and organizational issues, and alternating staff members were the major strategies used to help staff turnover and absenteeism and to defuse tension and fear. Of relevance was also the government monetary risk allowance system that was implemented in August. Despite being provided on irregular basis, which caused three health worker strikes, the government risk pay allowance proved to be an important driver for staff motivation and retention (Fig. 25.6).

Fig. 25.6 Midwife assisting a mother with her baby at Pujehun Maternity Complex

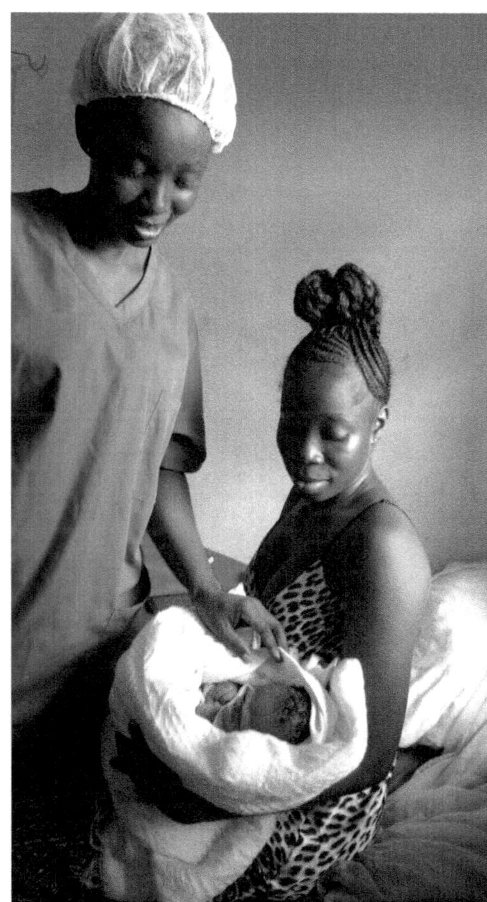

25.4.7 Community Sensitization

In order to gain trust and confidence among the population, a vast community sensitization initiative was designed and implemented through the outbreak. In all the 12 chiefdoms, traditional authorities were involved to help contain transmission of Ebola infections. Religious authorities, youth groups, and school leaders joined hands to help communities increase their awareness about the risks of Ebola infection and the importance of prevention measures. These communications and educational activities were disseminated using methods including radio messages, Ebola flip charts, flyers, posters, dramatic presentations, door to door messaging, and religious functions. The fact that communities were addressed in a culturally sensitive manner, particularly in relation to the suspension of traditional burial practices, may explain the widespread adherence of the people to the messages, as well as the absence of hostile reactions against health workers and contact tracers.

25.5 Data Collection and Analysis

25.5.1 Data Collection

To evaluate utilization of maternal and neonatal health services during the Ebola outbreak, both probable and confirmed cases of Ebola virus disease among pregnant women and infants were considered. The transmission chain was reconstructed through analysis of the registers of the two EHCs and contact tracing forms. In addition, health care workers involved in the management of the outbreak, surviving patients, and relatives of deceased patients were interviewed. Data on hospital admission and death/discharge were obtained from district health management team and registers of the two EHCs. Data on the utilization of maternal and neonatal health services were collected from the local district Health Management Information System, with the technical support of Doctors with Africa.

25.5.2 Data Analysis

In the statistical evaluation of the utilization of maternal and neonatal health services, data were analyzed using two-tailed student's t-test after acceptance of normal distribution with the Kolmogorov-Smirnov test. Following calculations, P values (two-sided) of less than 0.05 were considered to be statistically significant. All interventions by members of Doctors with Africa were approved by the Ministry of Health, Sierra Leone. The study was approved by the Sierra Leone Ethics and Scientific Review Committee. Patient records/information were anonymized and de-identified prior to analysis.

25.6 Results

Pujehun district first recorded a confirmed case of Ebola on August 8th, 2014. Eventually, this district suffered 42 deaths from 49 cases (31 confirmed and 18 probable) with an 85.7% case fatality rate (Ajelli et al. 2015). Of these, 19 (38.8%) were male and 30 (61.2%) were female. Fourteen cases were less than 15 years of age. Among all cases, three health care workers turned out to be positive (1 driver, 1 cleaner of EHC, one member of the burial team), two died. One case of Ebola infection was recorded in the Maternity Complex. The largest epidemic hotspot was located at Dumagbe village, in Makpele Chiefdom, east of Moa River. Pujehun has not had a recorded case since 26 November, 2014 and was declared the first Sierra Leone district Ebola-free in January 2015 (Sharkey et al. 2017). Admissions to the Pediatric Ward were 825 in 2013 and 750 in 2014, with an important decrease after end May 2014, when the first Ebola case was reported in Sierra Leone (Fig. 25.7). Admission to the Maternity Ward was 781 in 2013 and 716 in 2014, with an important decrease between June–August 2014 (Fig. 25.8). The total number of deliveries in Pujehun Hospital was 453 in 2013 and 460 in 2014, with a significant reduction between June–August 2014, before resuming a positive trend (Fig. 25.9). In particular, both natural and assisted deliveries decreased in 2014 compared to 2013 (323 vs. 350 and 13 vs. 20, respectively). The number of Cesarean sections in 2014 was 124, slightly more than that occurred in 2013. In Peripheral Health Units, the number of deliveries increased by 6.1%, going from 9657 deliveries in 2013 to 10,285 in 2014.

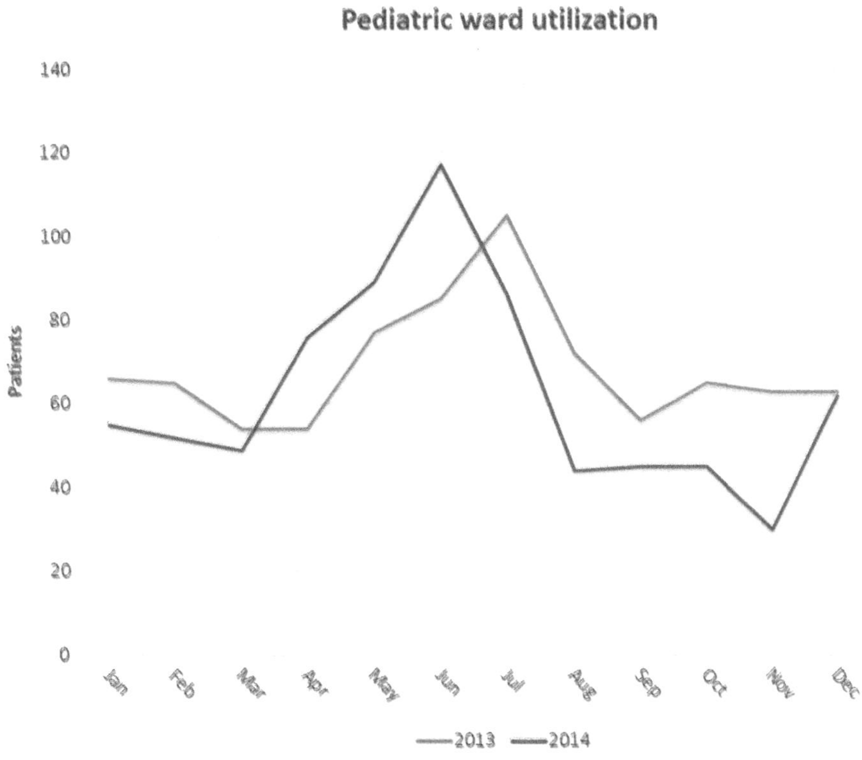

Fig. 25.7 Pediatric ward utilization at the Pujehun Hospital before and during Ebola outbreak

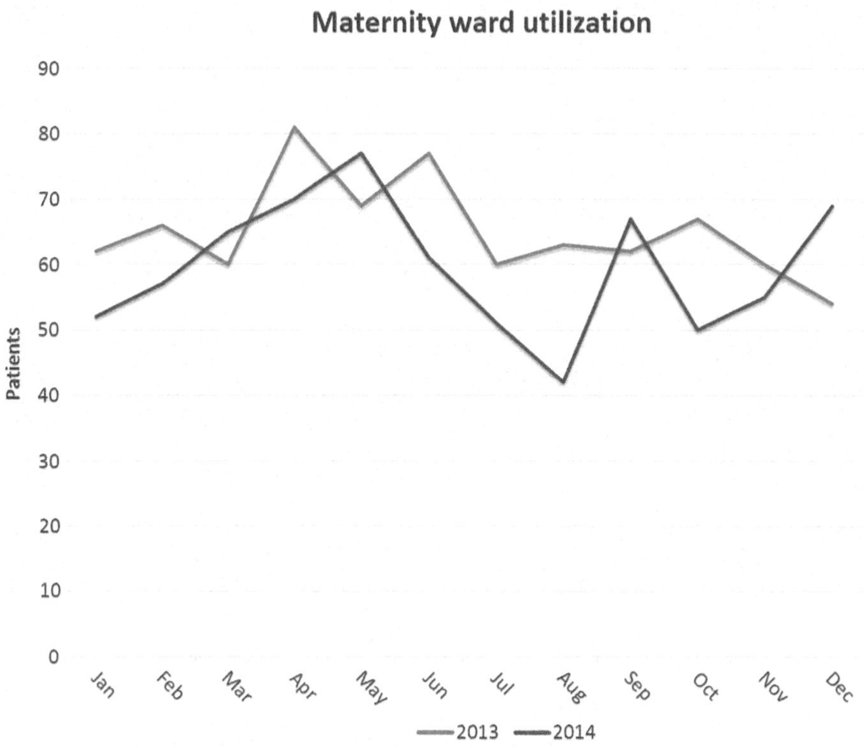

Fig. 25.8 Maternity ward utilization at the Pujehun Hospital before and during Ebola outbreak

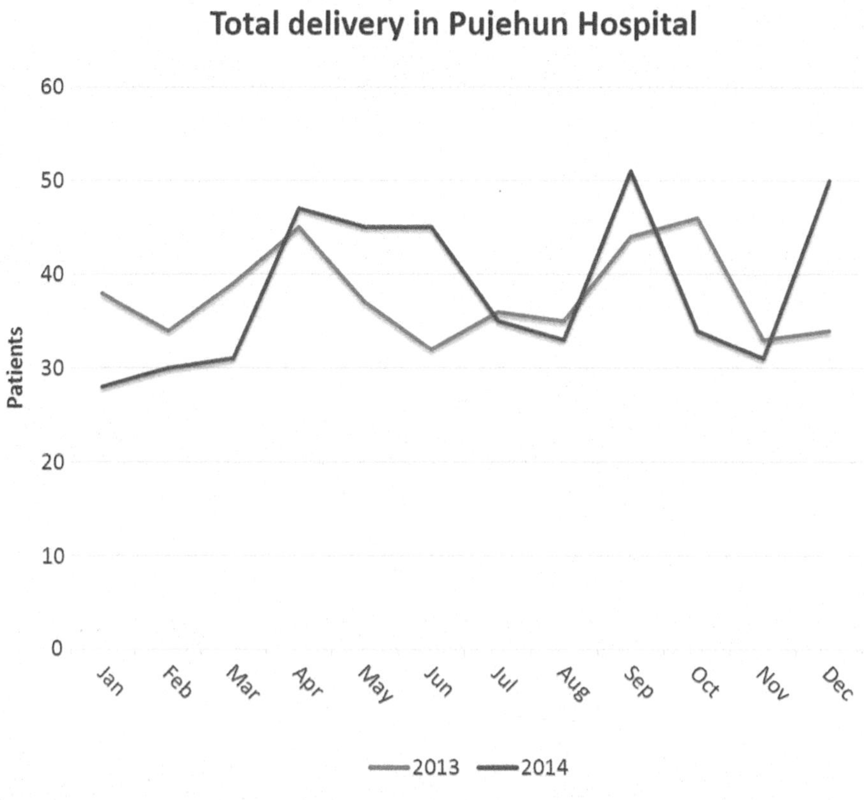

Fig. 25.9 Total deliveries at the Pujehun Hospital before and during Ebola outbreak

25.7 Discussion and Conclusions

In 2013, the maternal mortality ratio (MMR) was estimated to be 1100 per 100,000 live births for Sierra Leone, where the likelihood of dying as the result of being pregnant was among the highest in the world. Although this metric was extremely high just 1 year prior to the Ebola outbreak, the MMR had decreased from a high of 2400 in 1995 (The World Bank 2017). The Ebola outbreak has had many important effects on women and pregnancy in Sierra Leone. Firstly, pregnant women avoided going to clinics and hospitals where Ebola patients were concentrated for fear of becoming infected with Ebola, despite the need for antenatal and intrapartum care (Black et al. 2015; Strong and Schwartz 2016). How was it possible to achieve these results in time of the Ebola crisis? There may be different explanations of these results, some related to the health care services and some to other factors.

A recent study has documented the indirect effects of the Ebola outbreak on health facilities in Sierra Leone, showing a decline of 70% of hospital admission between May and October of 2014 (Bolkan et al. 2014). Moreover, effective containment of EVD in the Pujehun district was primarily ascribable to the intervention measures, including having an adequate number of EVD hospital beds and the effective detection of infected persons through contact investigation. These actions strongly depend on preparedness, population awareness, and compliance with intervention policies. Only one patient and two health care workers were infected in a hospital setting during the outbreak: this could reflect adequate health care system preparedness (Ajelli et al. 2015). As discussed above, to maintain maternal and neonatal services during the Ebola outbreak was challenging. The indirect negative effect of Ebola on reproductive health is primarily due to the fear of infection among potential patients, but also by health care workers. Most reproductive health interventions include handling blood or bodily fluids from patients whose Ebola infection status is often unknown. Appropriate and effective protection to prevent infections by health care workers is often inadequate, such as access to real-time diagnostic screening for Ebola infection. In addition, the initial diagnostic distinctions between Ebola infection and common complications of pregnancy can be difficult because bleeding and fever are common in pregnancy women in the affected areas. In addition to that, there is the chronic shortage of health personnel, which has been made even worse with the death of so many health care workers since the crisis began—approximately 30% of the medical staff in Sierra Leone who died from Ebola infections between April and September 2014 worked in women's and children's' health care (Hayden 2015). Data from Matam Maternity Hospital in Guinea show a substantial reduction in attendance in 2014, compared with 2013 (Delamou et al. 2014). A progressive decrease in pediatric and maternal admissions was also reported by Médecins Sans Frontières at the Godama Referral Center in Bo, the second largest city in Sierra Leone (Médecins Sans Frontières 2014); the Center is now closed because an effective infection control system could not be implemented. The geographical location is also important in establishing preventive measures for Ebola transmission and infection. First, Pujehun has one of the lowest population densities in the country: residents live in villages of less than 2000 people, and the administrative capital has a population of just over 30,000. Second, the rainy season, at its peak from August, played a significant role in making access very difficult, with no paved roads that are regularly passable. Third, the main hotspot of the epidemic in the surrounding area of Zimmi is separated from Pujehun town and the rest of the district by the Moa river. As a result, access to the hospital across the river was possible, except during the raining season, by using a flat barge. Unfortunately, the barge sank in April 2012. Thus, supervision and referrals were made mainly by boat.

The Ebola virus outbreak, devastating as it has been, should be an important lesson for clinicians, advocates, politicians, and researchers. In addition to jump-starting research efforts into vaccine development and novel methods of treatment, this most recent epidemic must motivate the world to continue to address gender inequities, because women and children are the most vulnerable parts of society. We should be strengthening health systems in deep and sustainable ways in low- and lower-middle-income countries through multisector support, public–private partnerships, and in partnership with communities (Sharkey et al. 2017).

In conclusion, in the Pujehun district the combination of health services factors and contextual (geographic and demographic) factors made it possible to contain transmission of the Ebola epidemic, while maintaining the overall functions of the maternal health care system. While there is no "magic bullet" that can address and solve the practical and ethical dilemmas posed by the outbreak of Ebola and its effects on reproductive health services, there are some lessons that can be learned. One is the importance of acting immediately; second is to adopt a systems approach involving evaluation of the entire situation; and third is to link emergency and development services to strengthen the resilience of the local health system.

Acknowledgments We are deeply grateful to all of the staff of Pujehun district hospital and PHUs, District Health Management, and the personnel of Doctors with Africa CUAMM who worked in Pujehun during the epidemic. Special thanks also to Dr. Agostine Kabano, Dr. Elamin Hayfa, and Dr. Yaron Wolman of UNICEF Sierra Leone for their continuous support and technical advice. We are in debt to Marco Ajelli, Stefano Merler, and Stefano Parlamento of the Fondazione Bruno Kessler for their accurate epidemiological investigation. Finally, we are extremely grateful to Sophie Mathewson for her comments and suggestions in the revision of the manuscript.

References

Ajelli, M., Parlamento, S., Bome, D., Kebbi, A., Atzori, A., Frasson, C., et al. (2015). The 2014 Ebola virus disease outbreak in Pujehun, Sierra Leone: Epidemiology and impact of interventions. *BMC Medicine, 13*, 281. Retrieved December 23, 2017, from https://bmcmedicine.biomedcentral.com/articles/10.1186/s12916-015-0524-z.

Black, B. O., Caluwaerts, S., & Achar, J. (2015). Ebola viral disease and pregnancy. *Obstetric Medicine, 8*(3), 108–113. Retrieved December 21, 2017, from https://www.ncbi.nlm.nih.gov/pmc/articles/PMC4582839/.

Bolkan, H. A., Bash-Taqi, D. A., Samai, M., Gerdin, M., & von Schreeb, J. (2014). Ebola and indirect effects on health service function in Sierra Leone. *PLoS Currents Outbreaks, 6*. https://doi.org/10.1371/currents.outbreaks.0307d588df619f9c9447f8ead5b72b2d.

Delamou, A., Hammonds, R. M., Caluwaerts, S., Utz, B., & Delvaux, T. (2014). Ebola in Africa: Beyond epidemics, reproductive health in crisis. *Lancet, 384*(9960), 2105. Retrieved December 27, 2017, from http://www.thelancet.com/journals/lancet/article/PIIS0140-6736(14)62364-3/fulltext.

Doctors with Africa CUAMM. (2016). *Doctors with Africa CUAMM Annual Report 2015*. Retrieved January 8, 2018, from http://www.mediciconlafrica.org/en/wpcontent/uploads/sites/2/2016/10/cuamm_ANNUALreport2015_INGLESE_ok_INTERObassaDISTILLATO.pdf.

Hayden, E. C. (2015). Maternal health: Ebola's lasting legacy. *Nature, 519*(7541), 24–26. Retrieved January 7, 2018, from http://www.nature.com/news/maternal-health-ebola-s-lasting-legacy-1.17036.

Médecins Sans Frontières. (2014). *MSF suspends emergency pediatric and maternal services in Gondama*. Retrieved December 19, 2017, from http://www.doctorswithoutborders.org/article/sierra-leone-msf-suspends-emergencypediatric-and-maternal-services-gondama.

Sharkey, A., Yansaneh, A., Bangura, P. S., Kabano, A., Brady, E., Yumkella, F., et al. (2017). Maternal and newborn care practices in Sierra Leone: A mixed methods study of four underserved districts. *Health Policy and Planning, 32*(2), 151–162.

Sierra Leone Ministry of Health and Sanitation. National Ebola Response Centre (NERC). (2016). *Ebola virus disease—situation report. Ministry of Health and Sanitation of the Republic of Sierra Leone*. 10 January 2015.

Statistics Sierra Leone. (2014). *Sierra Leone Demographic and Health Survey 2013*. Retrieved January 7, 2018, from https://dhsprogram.com/pubs/pdf/FR297/FR297.pdf.

Strong, A., & Schwartz, D. A. (2016). Sociocultural aspects of risk to pregnant women during the 2013-2015 multinational Ebola virus outbreak in West Africa. *Health Care for Women International, 37*(8), 922–942. https://doi.org/10.1080/07399332.2016.1167896.

The Guardian. (2015). *Sierra Leone declares first Ebola-free district*. Retrieved January 3, 2018, from http://www.theguardian.com/world/2015/jan/10/sierra-leone-first-ebola-free-district-who.
The World Bank. (2017). *Maternal mortality ratio*. Retrieved January 3, 2018, from http://data.worldbank.org/indicator/SH.STA.MMRT.
UNICEF. (2015a). Levels & trends in child mortality. *Report 2015*. Retrieved January 5, 2018, from http://www.childmortality.org/files_v21/download/IGME%20report%202015%20child%20mortality%20final.pdf.
UNICEF. (2015b). *Annual report*. Retrieved January 2, 2018, from https://www.unicef.org/publications/files/UNICEF_Annual_Report_2015_En.pdf.
World Health Organization. (2015). *Ebola response roadmap—situation report, March 2015*. Retrieved March 1, 2017, from http://www.who.int/csr/disease/ebola/situation-reports/en/.
World Health Organization. (2016). *Summary report of Ebola cases/deaths in Pujehun district 2014*. Geneva: World Health Organization.
World Health Organization. (2017). *Density of physicians*. Retrieved March 1, 2017, from http://www.who.int/gho/health_workforce/physicians_density/en/.

26 A Step in the Rights' Direction: Advocacy, Negotiation, and Money as Tools for Realising the Right to Education for Pregnant Girls in Sierra Leone During the Ebola Epidemic

Sinead Walsh and Emma Mulhern

The authors would like to dedicate this piece to the adolescent girls of Sierra Leone. The authors would also like to dedicate this to the Sierra Leoneans, within the Ministry of Education and at community level, who worked so tirelessly on this programme and believed that adolescent girls, pregnant or otherwise, deserved access to education.

26.1 The Situation of Adolescent Girls in Sierra Leone Before Ebola

On 25th May 2014, Sierra Leone announced its first confirmed case of Ebola virus disease (EVD). During the 18 months that followed, 14,122 Sierra Leoneans were to contract Ebola, and 3955 would lose their lives. The crisis would also have massive 'secondary impacts' on the health care system, the education system, and the economy[1]—reversing gains the country had made since the civil war. It would expose cracks in systems and exacerbate hardships for some already very vulnerable groups. One such group was adolescent girls.

To understand the impact of the Ebola epidemic on this cohort, it is important to understand what it was like to be an adolescent girl in Sierra Leone before Ebola struck in May 2014. Poverty and marginalisation pervaded many aspects of the lives of most teenage girls.

For illustrative purposes, let us imagine the life of a 15-year-old girl called Fatmata. Prior to the Ebola epidemic, there was a one in two chance that Fatmata would be illiterate; according to the Sierra

[1]This impact was compounded by the contraction in the iron ore market—"The iron ore price decline affected macro-financial stability and reversed the country's remarkable positive growth trajectory as economic growth declined from a buoyant 20.1% in 2013 to 4.6% in 2014 and thereafter contracted by 21.5% in 2015 according to the latest estimates" (African Development Bank Group 2016).

Sinead Walsh and Emma Mulhern worked for the Irish Embassy and the Irish Aid programme in Sierra Leone before, during, and after the Ebola crisis. Sinead was the Ambassador of Ireland to Sierra Leone and Liberia, based in Freetown from 2011 to 2016. Emma was Programme Advisor for Gender in 2014 and 2015.

S. Walsh (✉)
Department of Foreign Affairs and Trade, Irish Aid, Dublin, Ireland

E. Mulhern
Brighton, UK

D. A. Schwartz et al. (eds.), *Pregnant in the Time of Ebola*, Global Maternal and Child Health,
https://doi.org/10.1007/978-3-319-97637-2_26

Leone Demographic and Health Survey (DHS), only 52% of girls aged 15–24 are literate, compared to 70% of their male counterparts (Statistics Sierra Leone and ICF International 2014).

While there was quite a high likelihood that Fatmata had started school, it would have been touch and go as to whether she would still be in school at the age of 15. Only one in ten Sierra Leonean girls finished secondary school, and many dropped out in these early to mid-teenage years. A United Nations Children's Fund (UNICEF) study on out-of-school children highlighted that the primary reasons for dropping out of school were poverty, parental death, and teenage pregnancy (Coinco 2008).

There was also a significant chance that violence might be a part of Fatmata's life; 58% of women in Sierra Leone experience some form of violence during their lifetime (Statistics Sierra Leone and ICF International 2014). There was a high probability that Fatmata had undergone female genital mutilation (FGM); 90% of girls in Sierra Leone undergo FGM (Schwartz et al. 2019; UNICEF 2016). The United Nations Interagency Statement on the issue states that "[f]emale genital mutilation of any type has been recognised as a harmful practice and a violation of the human rights of girls and women" (UNAIDS et al. 2008).

There was also at least a one in four chance that Fatmata had been or would soon be pregnant. Before the epidemic, 28% of girls in Sierra Leone were pregnant as teenagers (Statistics Sierra Leone and ICF International 2014). There was also a high chance that Fatima would already be married; 48% of girls are married before the 18th birthday in Sierra Leone (UNFPA 2013).

In terms of what caused such a high likelihood of pregnancy for a girl like Fatmata, pre-Ebola, most studies conclude that it was a combination of factors, many related to poverty. It is important to note that poverty was and is a daily reality for most Sierra Leonean families. Sierra Leone ranks toward the bottom of all countries on the Human Development Index (HDI),[2] having a rank of 181 out of a total of 188 nations. In addition, 77.5% of the population of Sierra Leone are multidimensionally poor, taking into account health, education, and living standards (UNDP 2015a, b).

One important manifestation of poverty is educational attainment. The likelihood of Fatmata becoming pregnant as a teenager is related to her level of education; her probability of becoming pregnant could be as high as one in two depending on her education status. Fatmata only has a one in ten chance of finishing secondary school, compared with more than one in four of her male counterparts (UNDP 2015a, b). Sierra Leone's Demographic and Health Survey (DHS) finds that 46% of adolescent girls (aged 15–19) with no education have begun childbearing, compared with 22% of girls with secondary or higher level education. This is consistent with the United Nations Population Fund's (UNFPA 2013) assertion in the State of World Population 2013 report that:

> *Girls who remain in school longer are less likely to become pregnant. Education prepares girls for jobs and livelihoods, raises their self-esteem and their status in their households and communities, and gives them more say in decisions that affect their lives. Education also reduces the likelihood of child marriage and delays childbearing, leading eventually to healthier birth outcomes. Leaving school—because of pregnancy or any other reason—can jeopardize a girl's future economic prospects and exclude her from other opportunities in life.*

Unfortunately, even as the statistics predict that educating a girl like Fatmata would reduce the likelihood of her becoming pregnant as a teenager, community perceptions can be very different. In a finding that illustrates the vicious cycle at play, the aforementioned UNICEF research on out-of-school children in Sierra Leone highlighted that "the likelihood of a teenage pregnancy tends to pressure impoverished families into reconsidering whether or not sending a girl child to school is a good long-term investment" (Coinco 2008).

[2]The Human Development Index, or HDI, is a composite statistic developed by the United Nations that measures such variables as life expectancy, education, and per capita income indicators that can be useful in ranking countries in terms of human development. The aspect of health dimension is evaluated by life expectancy at birth, the educational dimension is assessed by the mean of years of schooling for adults (25 years of age and greater) and the expected years of schooling for children of school entering age, and the standard of living dimension is measured by gross national income per capita in the country.

Apart from her education level, what other factors would give Fatmata such a high chance of teenage pregnancy? As mentioned, she might already be married. She might be having sex with a boyfriend. She might be having transactional sex of some kind. In any of these cases, she may have limited access to accurate sexual and reproduction information and appropriate services. Lisa Denney of the Overseas Development Institute, in a review of teenage pregnancy reduction interventions in Sierra Leone, found that myths and misconceptions regarding contraception and its side effects were common among young people and adults alike (Denney et al. 2016).

The other possibility is that a girl like Fatmata might be coerced into having sex. Building upon research on teenage pregnancy in Sierra Leone, Denney notes that 'coerced sex' is widespread. Denney identifies transactional sex in exchange for basic needs and rape as two of the categories of sex that result in teenage pregnancy (Denney et al. 2016):

> *The reasons behind rape are distinct from those behind transactional sex and early sex with peer-age boyfriends. They relate to ideas about gender and power within society and can be exacerbated by impunity for offences... Most girls (and other respondents) did not seem to view all unwanted sex as rape, though a number of them said that they had been forced to have sex even after saying no to 'boyfriends' or other males. This type of coerced sex reportedly happens with boyfriends of all ages, teachers, men in the home and men/boys in the community.*

What's more, if an adolescent girl like Fatmata became pregnant pre-Ebola, she would have one of the highest probabilities in the world of dying from that pregnancy. To put it in perspective, if Fatmata started school in a class with 34 other girls, two of the class would ultimately die from pregnancy. With a maternal mortality ratio (MMR) of 1165 maternal deaths per 100,000 live births (Statistics Sierra Leone and ICF International 2014) and a lifetime maternal mortality risk of 1 in every 17 women (WHO et al. 2015), Sierra Leone has the highest level of maternal mortality in the world—teenage pregnancy is undoubtedly a significant contributor to this. The National Strategy for the Reduction of Teenage Pregnancy (2013–2015) highlights that adolescent girls account for 40% of all maternal deaths in the country (MICS 2010).

Prior to the advent of Ebola, the Government of Sierra Leone was aware of this dismal picture for adolescent girls like Fatmata and had put various measures in place. One significant example was the nationwide Free Healthcare Programme launched in 2010, a major objective of which was to reduce maternal mortality by providing free health care to pregnant and lactating women and girls.

In the education sector, the government, together with international partners, had managed to significantly improve enrolment figures. The 2013 DHS found that 71% of children of primary age were attending school; this represented an improvement since the 2008 DHS which estimated that only 62% of children of primary age were attending school. While attendance at secondary school remained low at 40%, according to the 2013 figures, this had dramatically improved since the 2008 DHS, which found that only 28% of children of secondary school-going age attended school.

The government also specifically recognised the problem of teenage pregnancy and launched the multi-sectorial *National Strategy for the Reduction of Teenage Pregnancy* in March 2013 with the tagline of 'Let Girls Be Girls and Not Mothers'. The president of Sierra Leone, Dr. Ernest Bai Koroma, himself launched the Strategy and mandated that the State House receive biannual progress reports on its implementation. A National Secretariat was established to implement the Strategy, housed in the Ministry of Health and Sanitation, and with a responsibility to convene relevant ministries and national and international organisation on this issue. The Strategy sought to create an enabling and empowering legal environment; improve access to quality sexual and reproductive health services for adolescents; provide age-appropriate comprehensive information to adolescents; and empower communities to prevent and respond to adolescent pregnancies. The Strategy brought partners together to work towards the ultimate aim of reducing teenage pregnancy in Sierra Leone. There was a real sense of optimism after this strategy was launched that the situation of teenage pregnancy would improve with the collective efforts of the government and many interested partners. Unfortunately, this optimism was short-lived.

26.2 How Ebola Impacted on the Vulnerability of Adolescent Girls

As discussed above, an adolescent girl like Fatmata in pre-Ebola Sierra Leone potentially faced many challenges including poverty, literacy and access to education, violence, and likelihood of teenage pregnancy. How did these challenges change during the Ebola epidemic? Unfortunately, the Ebola outbreak increased the vulnerability of teenage girls, making a bad situation worse. Let us take a look at each of the aspects we touched on above and see what role the Ebola virus epidemic played.

26.2.1 Access to Education[3]

After Ebola spread to Sierra Leone, the decision was made to close schools in order to reduce the possible risk of Ebola spreading among school children. This decision impacted 1.8 million children and adolescents and lasted 9 months in Sierra Leone—the longest of the school closures in the subregion (James 2015). We, at Irish Aid, were involved in the debates at the time as to whether closing schools was the right thing to do to curb the epidemic. On the one hand, restricting movement is always going to be important in an infectious disease outbreak and the concern was that children, particularly young children, might not be able to protect themselves by engaging in the 'no-touch' policy of Ebola avoidance that we all started to practice in the summer of 2014. On the other hand, some actors argued at the time that, in Sierra Leone, school children learning and bringing information home to their families had traditionally been an effective way to transmit health and hygiene messages.

We will, of course, never know the real impact of school closures on Ebola, as we can't know how things would have turned out had the schools stayed open. However, we do know about some of the other impacts of closing schools.

A joint advocacy brief by some key education actors, including UNICEF and Save the Children, painted a dire picture of the impact on education, noting that the "impact of prolonged school closures in a region with some of the lowest education indicators in the world is dire" (UNICEF et al. 2015). In particular, there are two ways in which the school closures during Ebola might have affected adolescent girls like Fatmata.

The obvious direct impact is that these girls were out of school for 9 months. In addition, and critically, adolescent girls' long-term likelihood of continuing or completing their education was thrown into jeopardy. To some extent, this applied to all children who were not in school for those 9 months. As the same education actors' report stated: "the longer a child stays out of school the less likely they are to return" (UNICEF et al. 2015).

However, as noted above, adolescent girls in Sierra Leone were already among the most likely children to drop out of school, particularly because of family pressures that they should earn money due to poverty, and because of pregnancy, which may have been linked to sexual violence.

And in a 'perfect storm', the Ebola epidemic made the three factors of poverty, pregnancy, and sexual violence more likely, as we will outline below.

26.2.2 Increased Poverty

By January 2015, the nongovernmental organisation (NGO), Street Child, had identified over 12,000 children who had lost their primary caregiver due to the Ebola epidemic (Street Child 2015). For the adolescent girls within this group, this meant more economic vulnerability.

[3]During the period of school closures, the government set up distance learning via radio and TV.

In addition, while many teenage girls around the country were not directly affected by Ebola infection occurring in their families, the country as a whole suffered economic hardship due to the general economic decline which occurred during Ebola, in part due to the emergency measures put in place by the government in an attempt to curtail the epidemic. For instance, affected districts were quarantined for months, leading to a major decline in trade; many people lost their jobs when all night-clubs and cinemas were closed; motorbike riders could not work after 7 PM, and hours of business of restaurants and supermarkets were reduced.

Transactional sex, already a survival strategy for some girls prior to the outbreak of Ebola, reportedly increased during the epidemic. For example, reports of 'sex for water' surfaced during the outbreak in which "men control access to water points and demand that women and girls (it is often girls who are sent to collect water) have sex with them in order to access water from community taps and wells" (Denney et al. 2015). More transactional sex among teenagers likely led to more pregnancy.

26.2.3 Increased Vulnerability to Violence

In early 2015, reports of increasing rates of violence against women and girls began to surface. These reports have since been confirmed by various quantitative and qualitative studies. A study by the United Nations Development Program (UNDP) in Eastern Sierra Leone (UNDP 2015a, b) concluded that:

> *"One of the social impacts of the crisis... has been heightened risk of sexual and gender based violence against women and girls. The findings of this assessment show that economic hardships caused by the crisis, as well as the by-products of the emergency regulations such as children being at home from school and limitations on movement, often translated into increases in cases of domestic violence and sexual abuse, especially of teenage girls and female spouses."*

Amnesty International, an international NGO, similarly highlighted that the school closures served to increase the vulnerability of adolescent girls in Sierra Leone, increasing their exposure to "sexual violence including abusive and exploitative relationships" (Amnesty International Publications 2015). Naturally, increased sexual violence will also lead to increased teenage pregnancy. A study by a group of international NGOs, capturing the views of over 1000 children in Sierra Leone, highlighted that "beatings—already common pre-Ebola—were now worse as a result of children being around the house more (due to the closure of schools)" (Risso-Gill and Finnegan 2015) and further found increased incidents of sexual violence perpetrated against girls:

> *"Sexual violence against girls was observed to have increased across all districts; and this was stated by both girls and boys. In nearly all of the FGDs [focus group discussions] where this issue was raised, children could relay a case of rape against a girl in their community, including attacks on girls in quarantine households. Data revealed that risk of rape was highest when girls went to collect water, travelled long distances to trade in other villages, or when using the bush to go to the toilet."*

26.2.4 Increased Teenage Pregnancy

The reported increases in transactional sex and sexual violence during Ebola meant that it was no surprise that we started getting reports of an increase in teenage pregnancy from our NGO partners around the country in late 2014. Naturally, other factors would also have played a role. Many people said that adolescents not being in school led to more consensual teenage sex. Furthermore, access to contraception for teenagers, already extremely low, would have been further compromised during the Ebola epidemic by the decreased operation of health clinics and NGO projects that might have provided contraceptives, compounded by a reluctance in attending health clinics for fear of contracting Ebola.

Fig. 26.1 Pregnancy reportedly increased during Ebola due to increased vulnerability of girls, resulting in increased transactional sex and sexual violence

A UNFPA survey found that over 18,000 girls had become pregnant during the Ebola outbreak (UNFPA 2016), but as Denney notes, "comparison is difficult as there were no precise figures before Ebola" (Denney and Gordon 2016); however, estimates of the percentage increase within communities ranged from 47% (Risso-Gill and Finnegan 2015) to 65% (UNDP 2015a, b) (Fig. 26.1).

26.3 The Ban

In early 2015, as response efforts saw a sustained 'bend in the curve' of the epidemic, plans began to shift to recovery and returning the country to some form of normality. An Early Recovery plan for 6–9 months, a part of the longer term National Ebola Recovery Strategy for Sierra Leone, was set in motion. The Early Recovery phase focused on four main areas—restoring basic health care and maintaining zero Ebola cases, returning children to school safely, social protection, and private sector recovery.[4]

Correspondingly, in February 2015, the preparation for the reopening of schools began, including the organisation of Basic Education Certificate Examination (BECE) exams for late March. The BECE, governed by the West African Examinations Council (WAEC), is a certificate of education in its own right, allowing students to transition from basic education to high school or secondary school. The representatives from the Ministry of Education, Science and Technology (hereinafter Ministry of Education) began to reiterate in public fora that pregnant girls were not allowed to sit for the upcoming exams, as established in 2010 in the form of a Cabinet Memo from the Ministry of Education. The ban on pregnant girls attending school was not formalised pre-Ebola, but had been a long-standing practice. It featured in the Truth and Reconciliation Commission (TRC) 2004 Recommendations after the civil war (1991–2002) in Sierra Leone. The TRC highlighted the practice as discriminatory and recommended that it be abolished (TRC 2004).

[4]The President's Recovery Priorities. (2015). Retrieved from http://www.presidentsrecoverypriorities.gov.sl/the-early-recovery Accessed 13 February 2017.

Fig. 26.2 Over the course of the programme, there were 14,500 girls enrolled

As the country prepared for the reopening of schools, on 2nd April 2015, the Ministry of Education, Science and Technology formalised the ban on pregnant girls attending school. In a press release, the government's position was justified with the following rationale:

> *[T]here are indications that if pregnant girls are allowed in schools, there is likelihood that many more girls will become pregnant. While such condition has the potential to negatively impact on their ability to concentrate and participate during lessons, it exposes them to ridicule by their colleagues and undermines the right ethical standards required in our educational institutional and in the process compromised the quality of education.*

Many national and international organisations in Sierra Leone objected to this announcement. Local women's groups and civil society organisations mobilised, NGOs issued a press release, as did the Human Rights Commission of Sierra Leone and the United Nations (UN).

A joint statement issued by the UN highlighted the negative human rights implications of the policy and concluded: "The United Nations in Sierra Leone stands ready to assist the Government in fulfilling the right to education for all its children, irrespective of their temporary status as a pregnant teenager."[5] However, the mass of debates that followed, both on radio and in the press, demonstrated that while many Sierra Leoneans felt it was unjust to ban pregnant girls from school, many others shared the government view that pregnant girls being in a classroom with nonpregnant girls would encourage the latter to become pregnant, and that pregnant girls in a classroom violated deeply held moral and religious beliefs.

As debates around the issue continued, it became clear that there were strongly held views on both sides and that polarisation was likely to continue. Meanwhile, at Irish Aid, we were inundated of reports indicating just how many girls would be affected by this policy. For the thousands of girls who had become pregnant during the Ebola outbreak, something needed to be done (Fig. 26.2).

[5]United Nations Country Team in Sierra Leone Statement on the Ban on Pregnant Girls Returning to Schools and Taking Exams. (2015) https://sl.one.un.org/2015/04/15/united-nations-country-team-in-sierra-leone-statement-on-the-ban-on-pregnant-girls-returning-to-schools-and-taking-exams-2/, Accessed 29 January 2017.

26.4 Negotiation and Action

Fortunately, we were not new to this issue. At Irish Aid, gender and women's rights programming had been one of our main areas of focus since the beginning of our engagement in Sierra Leone in the early 2000s. In the informal 'division of labour' that we had among donors and the UN agencies, Irish Aid acted as the lead on gender issues and convened the development partners working on gender issues in country. Quarterly and 'as-needed' meetings promoted coordination and joint action around gender programming and policy issues. We tried to make certain that the funding we gave to gender issues was cognisant of what others were doing or planning to do, and that we came together around issues of policy messaging to and support of government structures around gender.

Therefore, while Ireland was not significantly involved in the education sector at that time, through our leadership in the gender sector and our long-term work on the prevention of teenage pregnancy and sexual violence, we had been following the debate on the issue of pregnant girls' access to education and discussing internally what we should do. Understandably, policy discussions for the previous 8 months had been focussed on Ebola matters.

Fortunately, an opportunity soon arose to raise this with the government. The European Union organised a high-level conference on Ebola Recovery in Brussels, Belgium, in early March 2015 and our Ambassador was able to attend. At this conference, Ireland's Minister for State for International Development, Seán Sherlock, included the concern on access to education for pregnant girls in his address to the forum. Later that same day, the Minister and the Ambassador had a bilateral meeting with President Koroma and his team of Ministers on the issue.

There was no question that President Koroma was engaged in and concerned about the issue. Before the Ebola outbreak, the President's commitment to reducing teenage pregnancy was evident when he gave an impassioned speech to launch National Strategy for the Reduction of Teenage Pregnancy. In the foreword of the strategy, he noted that "girls hold the key a society without poverty. With the right skills and opportunities, they can invest in themselves now, and later, in their families." The President's concern about teenage pregnancy was once again evident during the Ebola crisis. His extensive social mobilisation tours around the country during Ebola had shown him the breadth and depth of the worsening problem of teenage pregnancy facing Sierra Leone.

The meeting in Brussels began a dialogue between the Irish and the Sierra Leonean government on the specific issue of how to ensure that girls, while pregnant, had access to education, while we all agreed that prevention of teenage pregnancy had to remain a high priority. After the conference in Brussels, the Government of Sierra Leone expressed willingness to act on the issue of access to education for pregnant girls and asked the Irish Embassy for some concrete proposals.

26.5 Coming Up with a Proposal

Access to education for pregnant girls is not an uncomplicated issue. Unfortunately, it is not simply a matter of a government saying "OK, you can remain in school". There are numerous complexities.

In the first place, pregnant girls have different health needs which need to be taken into account, as well as new educational needs on how to be a parent. In a country with high poverty rates like Sierra Leone, pregnant girls are likely to face serious economic challenges which affect their ability to attend school on regular basis. In addition, they will require a certain amount of time off from school, e.g. around the time that they give birth. Finally, and something we frequently heard, was the fact that pregnant girls may be stigmatised by their peers and/or the broader community if they attend school. In addition, some girls may already be alienated by their families and/or the father of the child due to the pregnancy. All of this can lead to an enormous amount of psychological stress. We needed to act quickly in order to take advantage of the positive momentum within government, but at the same time, the issue was very complicated. We engaged two groups of stakeholders to help us with this task.

Firstly, we approached a group of development partners which we knew would be supportive on this issue and which we thought would be in a position to help with nationwide solutions: DFID (the United Kingdom Department for International Development), UNICEF, and the United Nations Population Fund (UNFPA). As we expected, these organisations were extremely committed on the issue and happy to help. The willingness of the Government of Sierra Leone to engage at the highest level on pregnant girls' education was recognised by the development partners as a unique opportunity, arguably buoyed by the national efforts going into the Early Recovery plan. Crucially, these development partners were also willing to be flexible with their funds and their human resources, in order to work with government to find a viable solution.

This flexibility is a really important point, as this is not always the case; sometimes development partners are tied into preplanned programmes and do not have the flexibility to change plans, even when circumstances change on the ground. In this case, we had to go to the government and say: "Even though we don't know what the final solution on getting pregnant girls back into the education system will look like, we will support you in doing it." It was a leap of faith and we were fortunate that, both within Irish Aid and within these partner organisations, people were willing to take this leap.

The second group we approached for assistance was a mix of international, national, and local NGOs and individual activists concerned with issues around education and/or the rights of girls. While these organisations did not have the funds or scale to implement nationwide programmes, their field experience, ideas, and advice were indispensable to us as we tried to navigate this tricky area.

Technically, the government had not asked Irish Aid to consult others in coming up with the suggestions for getting pregnant girls back into the education system. But this teamwork approach had huge benefits. Between us, we had more ideas, more expertise, and more money.

For example, one of the most important functions of the NGO group was to help us to troubleshoot potential ideas, taking into account the contexts for pregnant girls in different parts of the country. For example, if we go with Option A, what might be the negative implications at the community level for girls? What about Option B? It is only the deep field experience of these NGOs and activists that could give us this reality check.

Another key contribution of both groups to this process was expertise in the education sector in Sierra Leone. In Irish Aid, we knew some sectors well, but education was not one of them. Therefore, these multidisciplinary conversations were extremely important, with some partners knowledgeable on teenage pregnancy, others knowledgeable about the school system, and others knowledgeable on other relevant areas, such as the reproductive health support that pregnant girls would need as part of their education.

Apart from these discussions with the other development partners, the NGOs and the activists, another way in which we came up with proposals for government was by conducting research into how other countries had dealt with pregnant girls in the education system. DFID gave important support here through an international research function that they have, and UNICEF reached out to their colleagues in other country offices to gain insights. What we found varied greatly by country, with no models that would fully suit the context in Sierra Leone, but we found models that gave us ideas about certain aspects of a potential system. It was also useful for us to be reminded what a complex issue this was, and that even developed countries struggled with this issue. For example, research in South Africa had highlighted the importance of supporting the access to health care and supporting the development of parental skills (Panday et al. 2009). Cash transfers were also recommended, but with the caveat that these must be designed in such a way as to not incentivise teenage pregnancy

(Panday et al. 2009). Research in the UK highlighted that, even when pregnant girls remained in school, there was a need to allow for the inevitable time that they would miss and help them to make up for this. Some local councils in the UK provide financial support to pregnant girls to ensure they can make the journey to school and one-to-one tuition for girls during maternity leave (Brighton and Hove Council 2014). Neighbouring Liberia had a system of 'afternoon schools' whereby pregnant girls attended the same schools but in the afternoon, whereas their colleagues attended in the mornings. In Mali, the focus is more on encouraging young mothers to return to the school system—being allowed to leave class for breastfeeding, free passes to go to health centres, being excused from physical education and toleration of late attendance (UNICEF 2015).

We also visited schools and the Ministry-run learning and vocational centres to gauge attitudes of teachers. The teachers we spoke with were worried that when girls left school due to pregnancy they would not come back. They were also willing to go that extra mile to ensure that pregnant girls continued their education. We heard of one teacher who sent notes and homework home to a student who had left due to pregnancy. In the learning centre, we saw girls as young as 14, with toddlers on their hips, learning sewing skills. The majority of the girls wanted to still be in formal education. These visits helped to inform our thinking on solutions and also strengthened our resolve.

26.6 The Proposal and the Government's Response

After this speedy process of consultation and research, we went back to the government. In our next meeting with the President, we emphasised the broad support of the development partners and NGOs to find a workable solution to this issue. We proposed certain principles that we felt were essential, after the consultation that we had done and given the information that we had gathered. We felt that proposing principles in the first instance, rather than a detailed plan, was appropriate, as the government obviously had the foremost expertise in delivering education to Sierra Leonean children, and therefore, they should lead on any detailed planning.

The most significant principle we proposed was the critical importance of formal education. While there were many voices saying that pregnant girls should exit formal education and learn practical skills such as tailoring, we stood firm that our interest was in supporting these girls to access formal education. We felt that we could not write off the formal education future of a girl simply because she was pregnant. If we did not find a way to keep her in the formal education system during her pregnancy, statistics clearly showed that she would be unlikely to return to the system later on. Fortunately, the President agreed strongly on this point. Gaining agreement that whatever programme we came up with for pregnant girls would be using the same official curriculum as other children was a critical shared starting point.

Having said that, another principle we proposed was that pregnant girls would need more than the usual curriculum given their special circumstances. We suggested that any final plan comprise a comprehensive package, including formal education, health information and services, and psychosocial support that would contribute to the safe delivery of her child and the girl's return to the school system after giving birth. The government was very supportive of this; the Ministry of Education, in particular, stressed the need for health education for these girls.

One particularly 'hot topic' was whether or not pregnant girls could simply be allowed to attend their regular schools, or if there needed to be a special separate programme. We expressed a preference for girls remaining in their existing schools. There were a few reasons for this. The first was from a rights-based perspective; we felt that becoming pregnant was not a reason for a girl to be banned from attending her school. In addition, there were cost considerations; the education budget for Sierra Leone was always challenged and we did not think that a separate programme for pregnant girls was sustainable in the long term. Finally, we were concerned to avoid possible stigmatisation of pregnant girls if they were educated separately. However, we acknowledged that this was a decision for government. We were aware that there were some vehement objections to pregnant girls being allowed in regular classes.

Ultimately, the government, while agreeing that there needed to be formal education for the girls as per the Ministry's curriculum and that the programme should be run by the Ministry, and while agreeing that pregnant girls should get additional services, did not agree that girls should be allowed to stay in their regular schools and instead decided that this programme would be physically located apart from the mainstream school system. They proposed the use of various vocational centres owned by the Ministry around the country for a bridging programme for pregnant girls only.

While this was not our ideal solution and left us with concerns, we discussed among the partners and agreed that this was still a significant step forward and would benefit thousands of pregnant girls in the short term. There was a sense of not letting perfect be the enemy of the good. The government made important points in the meetings about the need to bring communities across the country along with government policies, rather than moving too fast in areas which people felt violating their moral and religious beliefs. We agreed that the programme would include a significant social mobilisation component that would try to sensitise community members on the importance of pregnant girls finishing their education.

In practice, this programme would take place 3 days per week for 2 h in the afternoon, with 9 h of self-study per week which included radio lessons. In addition, the girls would have visits by midwives from the health service, referrals to health services, access to psychosocial support, and linkages to justice services if desired (Fig. 26.3).

26.7 Concerns and Criticism

This programme plan was not uncontroversial. The most vocal critic was Amnesty International, and the organisation wrote several reports and press releases on the issue. We worked closely with Amnesty International throughout this process, and in large part, sympathised with their concerns and sought any alternative ideas that they might suggest. There were four main concerns and dilemmas.

The first is mentioned above: the issue of rights. This issue brought the questions that used to keep us up at night! Does a physical separation of pregnant girls by banning them from regular school and allowing them only in this separate programme violate their right to education? Were we, by supporting this programme, actually bringing the girls' rights agenda backwards by accepting a substandard alternative? Amnesty International certainly thought so; they stressed that the ban on pregnant girls attending formal education was a violation of their right to education, which had been guaranteed to all children by the Government of Sierra Leone's ratification of various human rights treaties. As development partners, we were not sure. We understood that the government had to fulfill a child's right to education, but could choose to fulfill that right in different ways. We did not yet know if this alternative would be substandard or not. But we certainly wrestled with this issue.

Secondly, the issue of sustainability as mentioned above, which became an issue when the government decided that programme was to take place in separate centres. We believed that this was a costlier alternative for a budget-constrained government recovering from the Ebola outbreak. However, we felt that budgetary questions were really for government to decide upon. We could advise as development partners and obviously we have discretion over what we do and do not fund ourselves, but ultimately, we felt that decisions of resource allocation were for the government to make.

Thirdly, and this related to the 'rights' point, we were concerned about quality. Would 6 hours per week of a taught programme, together with additional radio programming and self-study, be sufficient

Fig. 26.3 The ban on pregnant girls being allowed to take State exams is still in existence and remains a serious hindrance to girls realising their right to education

for a quality education for these girls? At a glance, it would appear not, but the education experts advised us that, in effect, this was not significantly different from the hours other Sierra Leonean children had in school. The programme would leverage the investments made in the Emergency Radio Education Programme during the Ebola outbreak, building on its successes while trying to mitigate some of the challenges, for example through distributing complementary workbooks and radios where girls did not have access.

The fourth and final issue, as briefly mentioned above, was the ban on pregnant girls taking State exams, which had been in place since 2010. We, Amnesty International and others, strongly believed that this was deeply problematic for a girls' right to education. We lobbied that this new programme should include a change in this policy. The government agreed to consider this. However, rather than making this a precondition for our support for the programme, we development partners agreed that we would go ahead with the programme and continue to lobby on the exams point.

Ultimately, we at Irish Aid and in the broader development partner group with DFID, UNICEF, and UNFPA decided to proceed with the government's proposed 'bridging programme', despite our reservations as outlined above. We took a pragmatic approach, as we saw this programme as our best attempt at the time of promoting the right to education of pregnant girls. We recognised that a key barrier to the right of pregnant girls to education was public opinion, which would not change overnight. Public opinion in Sierra Leone, as evidenced by the attitudes of many people that we interacted with on this topic, from senior members of government to members of the public calling in radio shows, was favouring a ban on pregnant girls attending school. We understood that our lobbying on a complete removal of the ban on pregnant girls attending their regular schools was unlikely to succeed at that time. Therefore, we decided to go with the compromise of this 'bridging programme', while also prioritising social mobilisation activities that might positively impact public opinion in the long term.

Related to this social mobilisation point,[6] we saw this programme as an opportunity to change norms within government and at community level. The programme was important as it served as an acknowledgement by government, for the first time, that pregnant girls had a right to a formal education. What's more, in the President's speech to announce this programme, as part of the broader

[6]Other parties in-country were also involved in activities around the prevention of teenage pregnancy, of note were social mobilisation activities under the First Ladies Initiative. The First Lady, known for her work on this issue, continues to advocate for prevention of teenage pregnancy, increased community support for adolescent girls and a reduction in the number of girls dropping out of school through her Enhanced Inclusive Quality Education for all Girls project.

Ebola recovery package, he went further and said that he *expected* pregnant girls to continue their formal education. If we could help to normalise the fact, at community and at national levels, that pregnant girls were continuing with formal education, we might provide a basis upon which we could build further action and improvements. The Ebola crisis had given us a rare opportunity to make progress on a long-standing problem. It wasn't everything that we wanted, but we were keen to grasp it nevertheless.

26.8 How Did the Programme Work in Practice?

The Access to Education Programme for Pregnant Girls began in October 2015 and ran until August 2016. It was run by the Ministry of Education, the Ministry of Health and the Ministry of Social Welfare, Gender and Children's Affairs (MSWGCA), supported by UNICEF and UNFPA, with DFID and Irish Aid funding, respectively.

The physical locations for this programme, interestingly and due to a shortage of vocational or learning centres, were often the same schools where the girls' former classmates were being educated, but classes took place at different times and girls did not wear school uniforms. There were 303 schools and centres mobilised in total for the programme.

The teachers for this programme were the same teachers as in the regular school, but they received an additional stipend to their salary and training on how to implement the programme, with an emphasis on reducing stigma and discrimination towards the girls. The curriculum was exactly the same as regular schools at that time and comprised an 'accelerated' version of the usual curriculum which was rolled out across the education sector, which only included the core subjects of English, Integrated Science, Social Studies, and Mathematics, without the additional electives that students normally had. This was to make up for the academic time lost during the school closures. Class time was supplemented by 9 h of learning which was a combination of radio classes and self-study.

Over the course of the programme, there were 14,500 girls enrolled. This vastly exceeded our expectations—our initial estimate was for 3000 girls! Fortunately, the programme benefitted from some extremely committed Ministry of Education staff, who worked with UN partners to make adjustments for the far larger than expected numbers.

We had a great interest in data from this programme, in part because we hoped that data on a large number of girls being interested in continuing their education, and data on how much they could learn, would assist in our advocacy efforts with government to allow pregnant girls to take exams and to ensure that their education be a sustainable part of the education system. However, data capture in this programme proved to be very difficult. One reason for that was that pregnancy is obviously a temporary condition! While the findings of the evaluation of the programme are pending, at the time of writing, it was known that 9763 girls left the programme to reintegrate into regular school over the course of the programme. This was excellent, as one of the key objectives of the programme was to promote girls returning to the regular school system after they had given birth, providing a 'bridge' as it were. We had encouraged the government to make clear publicly as the programme was getting underway that girls were entitled to and encouraged to return to regular schools after childbirth, which they did.

In addition, other girls joined the programme months late, either because their awareness of the programme was delayed, or because they discovered they were pregnant after the official enrolment had ended. Fortunately, the Ministry of Education officials kept enrolment open for such eventualities. But this flow of girls through the programme did lead to challenges in terms of using data to monitor the programme beyond what is noted in this section.

In terms of qualitative data, we and the other development partners continuously conducted monitoring visits to assess the views of girls, teachers, and other stakeholders on the programme. One beneficiary of the programme told UNICEF (2015) that, thanks to the programme, she plans to return to school:

> *I attend the classes because I don't want to lag behind much, since I can't go to the normal school. I am very happy that this programme was brought here to help girls like me to continue learning even in this condition.*

Other external organisations conducted their own research on the programme which was useful. For instance, research from Amnesty International stated that girls are generally positive on the programme, but raise the issue of the more limited accelerated curriculum as mentioned above:

> *The majority of the girls Amnesty International spoke to had positive feedback about the bridging system. However, several girls stated that they would have preferred to stay on in their normal school if they had a choice. "[They] only taught us maths and English—but I was learning commercial studies in school and they did not teach us that," one girl told Amnesty International. "I would choose to go to my school if I had a choice as the school would give me results [exam grades]." When I give birth I will continue at my own school and the learning centre is only for when I am pregnant.*

As this quote also suggests, unfortunately and despite very significant lobbying on this point, we never managed to get a policy change on the issue of girls in this programme being able to take State exams. This continues to be a priority for development partners and NGOs.

An assessment of the programme commissioned by UNFPA (UNFPA et al. 2017), pending publication at the time of writing, highlights that:

> *The availability of such [education] services will undoubtedly support beneficiaries' integration back into the formal education system - more so than if the project had not been implemented. While there were some noted challenges with the education component (for instance, weak teacher attendance and lacking educational materials), the classes provided structure and access to continued education for the girls*

As mentioned, an additional output of the programme was that the girls would be provided with quality sexual and reproductive health information, services, and psychosocial support, as well as counselling and referral to justice and support services for survivors of violence. While the full data picture is not yet clear, 810 health care service providers were trained to conduct outreach services, creating a vital link between the schools and communities with the health facility. The aforementioned assessment of the programme concludes that while the psychosocial element of the programme was less effective:

> *A number of substantial gains were also recorded through the health intervention of the project. Beneficiaries demonstrated greater knowledge, access and practice of safe sex and contraceptive use. The project provided a suitable avenue for health information and access to sexual reproductive health and neonatal health care for the beneficiaries.*

For example, it has found that reported use of condom during last sexual intercourse increased from 10.7% of girls before the project to 23.8% afterwards (UNFPA et al. 2017).

Twenty-eight social workers were trained to provide psychosocial counselling to pregnant girls and new mothers. UNFPA estimates that a total of 4677 girls across the country were reached with psychosocial counselling services. During the period, a local NGO, Legal Access through Women Yearning for Equality, Rights and Social Justice (LAWYERS), provided approximately 326 girls with legal advice where cases of violence were identified.

26.9 What Surprised Us

Some issues we thought would be problems in the programme ended up being fine, whereas other problematic issues cropped up unexpectedly.

In the first category was the issue of stigma. Given the negative attitudes being expressed in the media about pregnant girls attending school, we were concerned that girls in the programme might face judgemental treatment by teachers and community members. We attempted to mitigate this risk with the social mobilisation component and by additional training for teachers.

These mitigating measures may have helped, but overall this did not end up emerging as a significant issue. On the contrary, our monitoring visits suggested that interactions of pregnant girls with teachers and project level staff in the ministries were overwhelmingly positive. On the whole, we encountered extremely dedicated, empathetic people, often going above and beyond to deliver the programme to girls. Community-level stigma also did not appear to be a significant issue. On the contrary, we found a lot of positive attitudes in communities to these girls being educated. In some cases, community members came together to offer childcare nearby, so the girls could focus while they were in class.

The higher than expected enrolment figures were also evidence that girls did not see the programme as stigmatising. On the contrary, some girls spoke about the mutual support and solidarity they received from being around others in the same situation. Amnesty International (2015) noted in a report that:

> *A few girls said they preferred [the programme] to their normal school due to the stigma faced. One girl told Amnesty International: "I was ashamed, in the normal school everybody would have laughed at me."*

Another issue that surprised us was that a large number of girls enrolled who had already given birth. In other words, these were girls who were officially entitled to go back to their regular schools, but chose to come to this programme instead. Monitoring visits suggested that, in some cases, it was because of the above-mentioned issue of girls feeling more supported and less stigmatised in this programme where other girls were in the same situation. In other cases, it seemed that girls appreciated the opportunity to have some 'refresher' education before returning to regular school.

The consequence of so many girls in the class having already given birth was that there were generally a lot of babies in the classroom! This led to requests from girls on our monitoring visits for a crèche facility. Sometimes girls reported that they had to bring their baby to class because they did not have family support. This was related to the other set of issues girls raised frequently which related to economic support. For instance, even though the centres and schools for the programme were widespread across the country, some girls still had to pay transportation depending on where they lived, and this was often a struggle. In other cases, girls struggled with the economic demands of being a mother, perhaps a single mother, in the context of high poverty rates, particularly post-Ebola. We had actually foreseen that poverty-related issues would emerge in some form, but they were difficult to deal with. In some cases, we were able to link girls with NGO programmes that might provide them some extra support, but the programme itself did not include economic support. One complexity here was that poverty rates are high across the board, within this programme and within regular schools. Would economic support be sustainable? Would it be seen to incentivise teenage pregnancy as some feared? These were difficult issues that we did not resolve during the programme period.

A final issue that we did not expect was that some girls who were not pregnant, and who had never had been pregnant, came to the programme, and even the occasional boy. These individuals had previously dropped out of school for various reasons and saw this programme as a potential bridge back to formal education. While this was outside the original plan, we saw it as positive that there was so much desire among Sierra Leonean children themselves to finish school. Observing this phenomenon led to discussions around a successor programme, Girls Access to Education (GATE).

GATE focuses on ensuring girls progress to and remain safely in secondary schools. Building on the bridge programme and assets built-up during that period, it also has a component on reintegrating out-of-school girls into the education system, including those who have dropped out due to teenage pregnancy. GATE targets community leaders and boys and men with awareness activities to promote the value of girls' education, as well as supporting a national campaign in an effort to challenge the entrenched social and cultural norms that perpetuate the discrimination which girls face in Sierra Leone. This 18-month programme began in November 2016 with DFID funding and technical support to the government from UNICEF. It aims to benefit over 120,000 adolescent girls. UNICEF reports that the component of GATE that targets pregnant adolescent girls has, once again, received an overwhelming response. With plans in place to support 8000 adolescent girls, over 17,000 have already registered, with more coming every day.

26.10 Conclusion and Lessons Learned

We believe that the Access to Education programme was a positive step in the right direction for the education of vulnerable girls in Sierra Leone. A large number of girls made use of the programme in its first year, and a good proportion of these have successfully transitioned back to the regular school system. While it is too early to say whether the programme has begun to influence societal norms around this issue, there were positive indications from Ministry staff and communities that there was a real appetite for these girls to be educated, even if some still felt that the education should be separate.

The ban on pregnant girls being allowed to take State exams is still in existence and remains a serious hindrance to girls realising their right to education.

It is extremely encouraging that GATE has begun as a successor programme, building on the lessons of the Access to Education programme. The wider scope of GATE is also welcome. However, we would emphasise, as we did earlier when GATE was being designed, that it is important that pregnant girls receive a special priority if resources are limited, as pregnant girls are the only group which is currently *required* to drop out of school.

26.10.1 Relationships Matter, and These Are Helped by Longevity

Irish Aid's relationships with key players in government helped to get the process of negotiating this programme started. Later during the design phase, positive relationships of UN staff with their ministerial counterparts were extremely useful. In both cases, relationships were strengthened by the longevity of international staff in-country. Negotiating on a controversial issue such as this requires trust and trust takes time to build. The high turnover in the aid sector often does not allow for the kinds of relationships needed to come up with creative ways forward such as this programme.

26.10.2 Flexibility Is Key

Another positive lesson was that all the four main actors (Irish Aid, DFID, UNFPA, and UNICEF) were ready and able to adjust plans and budget quickly to allow for this programme. What we did was also only possible because we were prepared to dialogue with government and compromise as necessary, and the government worked in the same spirit.

A DFID colleague noted that:

I think we found a middle ground which not only addressed the needs these girls had, but has also kept open the space for dialogue and started to build an evidence base about how to provide services for pregnant schoolgirls.[7]

26.10.3 If You Want to See It; Fund It (#1)

The budget and human resources of the Sierra Leone Government at the time of programme negotiation were extremely constrained by the Ebola crisis and other demands. Had we simply advocated on this topic without offering our technical and financial support, we would not have succeeded.

26.10.4 If You Want to See It; Fund It (#2)

Our efforts to engage the NGO community and actors such as the Human Rights Commission of Sierra Leone on a regular basis were positive in terms of getting useful advice on the programme design and ongoing feedback on implementation. However, we did not manage to reach our ultimate advocacy goals, such as lifting the ban on exams. While there was certainly strong advocacy by organisations and activists from time to time, the NGOs fed back to us that they struggled to have a systematic advocacy programme on this topic given the other demands that they faced, even though this topic was an avowed advocacy priority for most of the organisations. In retrospect, we could have funded NGOs to establish a more formalised advocacy programme on the advocacy goals of the programme, so that they could have had staff dedicated to this programme. Amnesty International continues to advocate on the issue, as do organisations based in Sierra Leone, which is welcome.

Acknowledgments We would like to acknowledge the work of Gibril Kargbo from Irish Aid on this programme. We are grateful to DFID, UNICEF, and UNFPA for their collaboration on this programme.

References

African Development Bank Group. (2016). *Sierra Leone economic outlook*. Retrieved February 13, 2017, from https://www.afdb.org/en/countries/west-africa/sierra-leone/sierra-leone-economic-outlook/.

Amnesty International Publications. (2015). *Shamed and blamed: Pregnant girls' rights at risk in Sierra Leone*. Retrieved February 13, 2017, from https://www.amnesty.org/download/Documents/AFR5126952015ENGLISH.PDF.

Brighton and Hove Council. (2014). *Policy: Ensuring a good education for children who cannot attend school because of health needs*.

Coinco, E. (2008). *The out of school children of Sierra Leone, executive summary*. UNICEF. Retrieved October 12, 2017, from http://allinschool.org/wp-content/uploads/2015/04/OOSC_Sierra-Leone.pdf.

Denney, L., & Gordon, R. (2016). *How to reduce teenage pregnancy in Sierra Leone, Briefing paper 18*. Retrieved February 13, 2017, from https://assets.publishing.service.gov.uk/media/5901d4f0ed915d06ac000288/BP18_How_to_reduce_teenage_pregnancy_in_Sierra_Leone.pdf.

Denney, L., Gordon, R., & Ibrahim, A. (2015). *Teenage pregnancy after Ebola in Sierra Leone: Mapping responses, gaps and ongoing challenges: Working Paper 39*. Retrieved February 13, 2017, from https://assets.publishing.service.gov.uk/media/57a0898ee5274a27b2000133/SLRC-WP39.pdf.

[7] Berry, C. (2017). Email. 8 February 2017.

Denney, L., Gordon, R., Kamara, A., & Lebby, P. (2016). *Change the context not the girls: Improving efforts to reduce teenage pregnancy in Sierra Leone*. Retrieved February 13, 2017, from http://www.securelivelihoods.org/resources_download.aspx?resourceid=402&documentid=528.

James, J. (2015). *In Sierra Leone, hopes of saving the school year*. UNICEF. Retrieved February 13, 2017, from https://www.unicef.org/infobycountry/sierraleone_81528.html.

Panday, S., Makiwane, M., Ranchod, C., & Letsoalo, T. (2009). *Teenage pregnancy in South Africa—With a specific focus on school-going learners*. Child, Youth, Family and Social Development, Human Sciences Research Council. Pretoria: Department of Basic Education. Retrieved February 14, 2017, from http://www.scirp.org/(S(351jmbntvnsjt1aadkposzje))/journal/PaperInformation.aspx?PaperID=70127.

Risso-Gill, I., & Finnegan, L. (2015). *Children's Ebola recovery assessment: Sierra Leone, Plan International, Save the Children and World Vision International*. Retrieved January 7, 2017, from http://www.alnap.org/resource/20472.

Schwartz, D. A. (2019). The Ebola epidemic halted female genital cutting in Sierra Leone: Temporarily. In D. A. Schwartz, J. N. Anoko, S.A. Abramowitz (Eds.), *Pregnant in the Time of Ebola: Women and Their Children in the 2013-2015 West African Epidemic*. New York: Springer. ISBN 978-3-319-97636-5.

Sierra Leone - Multiple Indicator Cluster Survey. (2010). *United Nations Children's Fund and Statistics Sierra Leone*. Retrieved from https://www.google.co.uk/search?q=mics+2010+sierra+leone&oq=MICS+2010+sie&aqs=chrome.1.69i57j0.4991j0j9&sourceid=chrome&ie=UTF-8.

Statistics Sierra Leone and ICF International. (2014). *Sierra Leone demographic and health survey 2013: Key findings*. Retrieved January 5, 2017, from https://dhsprogram.com/pubs/pdf/SR215/SR215.pdf.

Street Child. (2015). *The street child Ebola orphan report, January–February 2015*. Retrieved February 13, 2017, from https://www.street-child.co.uk/ebola-orphan-report/.

Truth and Reconciliation Commission. (2004). *Witness to truth: Report of the Sierra Leone Truth and Reconciliation Commission* (Vol. 2, Chapter 3, Report of the Sierra Leone TRC). Retrieved February 13, 2017, from http://www.sierraleonetrc.org/index.php/view-report-text-vol-2/item/volume-two-chapter-three?category_id=20.

UNAIDS, UNDP, UNECA, UNESCO, UNFPA, UNHCHR, UNHCR, UNICEF, UNIFEM, & WHO (2008). *Eliminating Female genital mutilation: An interagency statement*. Retrieved February 13, 2017, from http://apps.who.int/iris/bitstream/10665/43839/1/9789241596442_eng.pdf.

UNDP. (2015a). *Human Development Report 2015, briefing note for countries on the 2015 Human Development Report: Sierra Leone*. Retrieved February 13, 2017, from http://www.sl.undp.org/content/dam/sierraleone/docs/HDRs/Sierra%20Leone%20Explanatory%20Note.pdf?download.

UNDP. (2015b). *Assessing sexual and gender based violence during the Ebola crisis in Sierra Leone*. Retrieved February 13, 2017, from http://www.sl.undp.org/content/sierraleone/en/home/library/crisis_prevention_and_recovery/assessing-sexual-and-gender-based-violence-during-the-ebola-cris.html.

UNFPA. (2013). *State of world population 2013 motherhood in childhood: Facing the challenge of adolescent*. Retrieved January 27, 2017, from http://www.unfpa.org/publications/state-world-population-2013.

UNFPA. (2016). *Rapid assessment of pregnant adolescent girls in Sierra Leone*. Unpublished Report.

UNFPA, UNICEF, Government of Sierra Leone, & Nestbuilders International. (2017). *Pre and post-test providing services to the pregnant school girls project: Final report.*

UNICEF. (2015). *Best practices on educational support for pregnant girls and young mothers*. Unpublished Report.

UNICEF. (2016). *Female genital mutilation, a global concern*. Retrieved January 27, 2017, from https://www.unicef.org/media/files/FGMC_2016_brochure_final_UNICEF_SPREAD.pdf.

UNICEF, Global Education Cluster, INEE, & Save the Children (2015). *Safe access to learning, during and after the Ebola crisis*. Retrieved February 13, 2017, from http://educationcluster.net/wp-content/uploads/2015/02/EducationInEbola_JointAdvocayBrief.pdf.

WHO, UNICEF, UNFPA, World Bank Group, & United Nation Population Division. (2015). *Trends in maternal mortality: 1990 to 2015, estimates by WHO, UNICEF, UNFPA, World Bank Group and the United Nations Population Division, executive summary*. Retrieved February 13, 2017, from http://apps.who.int/iris/bitstream/10665/193994/1/WHO_RHR_15.23_eng.pdf.

Ebola Virus Disease Surveillance in Two High-Transmission Districts of Sierra Leone During the 2013–2015 Outbreak: Surveillance Methods, Implications for Maternal and Child Health, and Recommendations

27

Allison M. Connolly and Alyssa J. Young

27.1 Introduction

The West African Ebola epidemic in 2013–2015 triggered a massive response, not only from the affected countries, but also the international community, including governmental and United Nations agencies, universities, and nongovernmental organizations (NGOs). While key functions of the response such as safe burials and treatment for Ebola virus disease (EVD) have been well-described in academic literature and in the media, the function of case investigation in both communities and health care facilities has been less publicized. In communities, this work was conducted by surveillance officers, who were the first line of defense in controlling the outbreak as they were generally the response's initial encounter with ill or deceased patients. They were charged with making critical decisions, usually based on imperfect information and sometimes in communities which didn't welcome their investigations. Their effectiveness was dependent on their knowledge of EVD epidemiology and their investigative skills, in addition to the rapport and trust they could build up with the households affected by their investigations and with communities in general. In carrying out their responsibilities, they faced myriad challenges and decisions; some were directly related to Ebola, but many were not. Regardless, all of their interactions with a community had the potential to impact future efforts to control the outbreak in that community.

In this chapter, we will first provide an overview of the EVD surveillance system and how it articulated with other sectors of the response operation. Within the response, the surveillance sector consisted of five critical functions: case investigation, contact monitoring, quarantine, laboratory testing, and data management. Since we are seeking to describe the work on the front lines of the response to

A. M. Connolly (✉)
Palladium (Data, Informatics and Analytical Solutions), Washington, DC, USA
e-mail: allison.connolly@thepalladiumgroup.com

A. J. Young
Clinton Health Access Initiative (Malaria Analytics and Surveillance), Boston, MA, USA

D. A. Schwartz et al. (eds.), *Pregnant in the Time of Ebola*, Global Maternal and Child Health,
https://doi.org/10.1007/978-3-319-97637-2_27

identify and refer suspected cases for isolation, the focus of this chapter is primarily on the process and associated challenges of case investigation, which was the primary duty of surveillance officers. We will also touch upon the other functions of the response, both within and outside of the surveillance sector, insofar as they affected and supported the work of case investigation.

We hope to convey the comprehensive, complex, and dynamic nature of the role of surveillance officers during the EVD response in Sierra Leone. The challenges and consequences of conducting surveillance[1] in pregnant, postpartum, and breastfeeding women and their children will be specifically highlighted; identifying suspected EVD in pregnant and postpartum women and young children presented greater challenges than in the general population and increased their vulnerability, which was compounded by factors such as breakdowns in surveillance processes in the community, deficits in the implementation of quarantine, and lapses in infection prevention and control (IPC) in health care settings. A large cluster of Ebola cases which originated in a maternity ward and spread to several communities will be described to illustrate the impact of these shortcomings and the myriad challenges of caring for pregnant women and their infants in health care institutions, quarantined homes, and in the community during the crisis.

This chapter is written from our point of view as two international responders who worked in surveillance in two districts in Sierra Leone with some of the highest numbers of EVD cases: Bombali District and Port Loko District. Our work spanned from October 2014, during the height of the outbreak, through May 2015, when transmission had become sporadic and localized. Most of the text draws on the experience from the height of the outbreak in these districts, but we also present some issues which arose near its conclusion. Although surveillance processes and outbreak response should have been consistent across districts, we are aware through our own work that they were not. Therefore, not all processes and issues discussed here can be generalized to the rest of Sierra Leone, or to Guinea and Liberia.

Of the three countries most affected by the outbreak, Sierra Leone experienced the highest number of total cases (suspected, probable, and confirmed) at 24,207. However, only 41.1% of the total were confirmed cases ($n = 9957$) (Sarah D. Bennett, MD, MPH, U.S. Centers for Disease Control and Prevention, personal communication 2018). There were numerous reasons for the large percentage of cases that were classified as suspected or probable. They included lapses in the ability to identify, follow up and test ill persons as necessary, which persisted to varying degrees across districts until late 2014. These lapses were related to factors such as inadequate staff, transportation, bed capacity, and laboratory testing, which was required for a case to be confirmed. etc. Furthermore, the implementation of post-mortem EVD testing was also subject to limitations in the response's capacity to collect, transport and test the specimens, especially in its early phase. As a result of such conditions, the case counts should not be considered precise.

When considering the total number of cases, Port Loko was the district reporting the second highest number in Sierra Leone, accounting for approximately 13.5% of all nationally observed cases. In contrast, Bombali reported the seventh highest number of total cases among districts, representing 4.7% of the country's total. However, as shown in Table 27.1, the relative burden of cases in each district is greater when considering only confirmed cases. These two districts accounted for over a quarter of all confirmed cases in the country.

Port Loko borders Western Area, the district which reported the highest number of cases in Sierra Leone, both in terms of total cases and confirmed cases. Port Loko's proximity to the capital, Freetown, located in Western Area, and its situation on a major national highway, provides a direct line for

[1] As mentioned above, case investigation was just one of five functions within the surveillance sector. Nevertheless, the terms "case investigation" and "surveillance" were used interchangeably during the outbreak, while the other four functions were usually referred to by their given names. In this chapter, we have followed these naming practices.

Table 27.1 EVD cases in Port Loko and Bombali Districts, by case classification

	Suspected cases	Probable cases	Confirmed cases	Total cases	% of total cases nationally	National rank by district of total cases	% of confirmed cases nationally	National rank by district of confirmed cases
Port Loko	1068	565	1623	3256	13.5	2	16.3	2
Bombali	59	8	1066	1133	4.7	7	10.7	3
Sierra Leone total	12,106	2081	9940	24.1 27[a]	NA[b]	NA	NA	NA

Table based on data provided by Sarah D. Bennett, MD, MPH, U.S. Centers for Disease Control and Prevention, personal communication (2018)

[a]80 cases excluded due to missing data. [b]*NA* not applicable

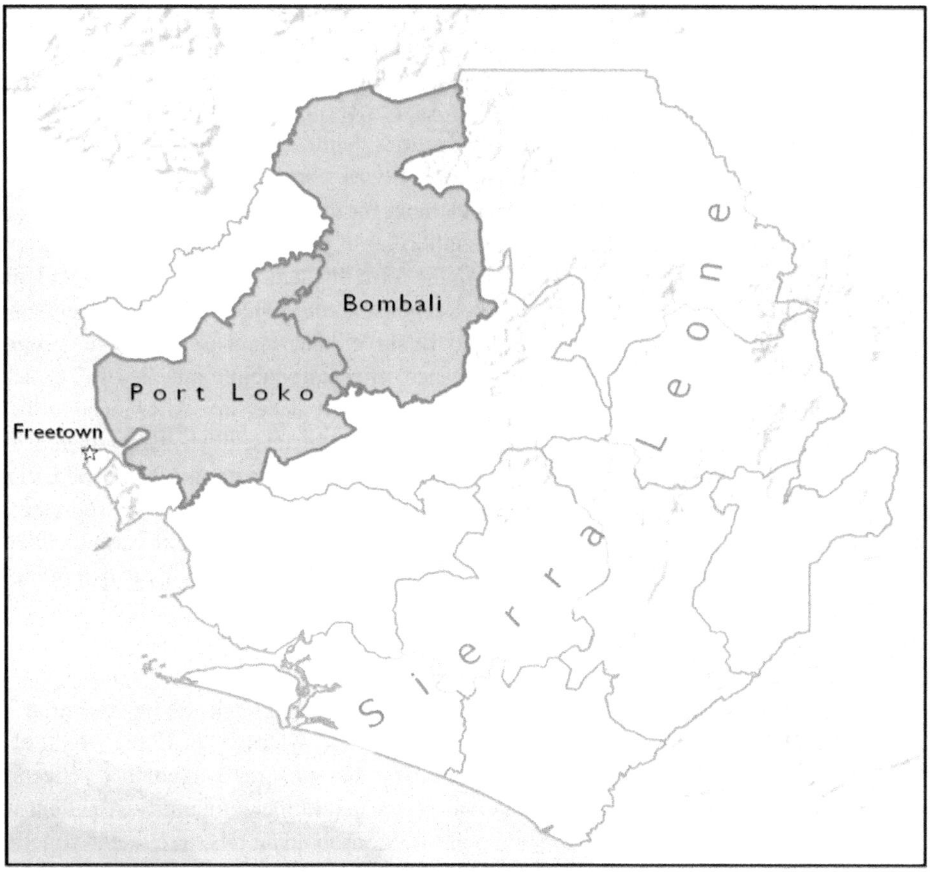

Fig. 27.1 Map of Sierra Leone highlighting Port Loko and Bombali Districts

transportation and migration from the capital to the eastern regions of the country. Bombali District is located to the northeast of Port Loko and is traversed by the same highway (Fig. 27.1). The population of each district is similar, with roughly 600,000 persons residing in each, often in rural, isolated villages. Nonetheless, both the country's fifth largest city—Makeni—and the main airport, are situated within the two districts.

In regard to confirmed cases, transmission in Port Loko was at its highest in the month of November 2014 ($n = 373$), whereas in Bombali newly-identified confirmed EVD cases were highest in September 2014 ($n = 306$) (Sarah D. Bennett, MD, MPH, U.S. Centers for Disease Control and Prevention, personal communication 2018). Bombali District's EVD case count declined sharply in December 2014, while this occurred in Port Loko about 6 weeks later (Yang et al. 2015). Eventually, there were often periods of time, sometimes weeks, in which no cases were reported. When cases were identified, they were typically limited to highly localized and isolated clusters. Bombali District experienced its last case of Ebola infection in September 2015. The final EVD case in Sierra Leone originated in Port Loko in January 2016, but was officially counted as a case in another district (World Health Organization 2016).

27.2 Community-Based Surveillance for Ebola Virus Disease

In August 2014, it became mandatory in Sierra Leone to report all deaths in the community to authorities regardless of presumed cause (Sierra Leone Ministry of Local Government and Rural Development, National Council of Paramount Chiefs 2014). It was also required to report anyone who was ill, although the messaging about whether to report all ill persons or only those with Ebola symptoms was less consistent across time and districts. People with injuries, pain, labor complications, and manifestations of chronic disease were routinely reported as there was a breakdown in the health care infrastructure such that it was often difficult to access treatment for these types of conditions.

A centralized alerting system was used to notify authorities of potential Ebola cases and to report all deaths in the community. This was operationalized in the form of a national Ebola hotline and district-based hotlines that were managed at District Ebola Response Centers (DERC). Information gathered through alerts called in to the hotlines was provided to the surveillance team assigned to the geographic area of the person reported as ill or deceased. Surveillance officers frequently travelled 1–2 h each way to respond to an alert, although as many as 8 h were occasionally necessary. After locating the individual, surveillance officers would conduct a case investigation, which ideally began with an interview with the patient or other household member to ask about symptoms and risk history for EVD. The demographic and symptom information that was collected is seen on page 1 of the shortened case investigation form implemented in late October 2014 (Fig. 27.2). Pregnancy status was not among the information collected on this or previous forms, and this has complicated the ability to determine EVD outcomes for pregnant women.

The standard WHO case definition was applied to classify an individual as a suspected case, or to rule out infection. If a person had a fever and at least three other EVD-related symptoms, he/she would be considered as meeting the definition of a suspected case[2] (World Health Organization 2014). An epidemiological link to a laboratory-confirmed case and one symptom of Ebola would classify someone as a probable case.[3] It is important to note that although some surveillance officers were health care professionals, no clinical assessment was performed in the community. A patient was to be classified solely according to the information he/she or a household member provided. Nonetheless, surveillance officers occasionally took into account obvious symptoms if the patient denied them (e.g., red eyes or extreme weakness).

[2] A person with unexplained bleeding was also classified as a suspected case. See "Case definition recommendations for Ebola or Marburg virus diseases 2014" for the complete WHO definition.

[3] The definition used for a probable case in these two districts was not among the recommended WHO case definitions.

IN THE PAST ONE (1) MONTH PRIOR TO SYMPTOM ONSET:

1. Did the patient have contact with a suspected or confirmed Ebola case in the one month before becoming ill?

☐ Yes ☐ No ☐ Unk

If yes, please complete one line of information for each sick source case:

Name of Source Case	Relation to Patient	Date of Last Contact (DD,MM,YYYY)	Village/Town	District	Was the person dead or alive ?
		__/__/__			☐ Alive ☐ Dead Date of Death: __/__/__ (DD,MM,YYYY)
		__/__/__			☐ Alive ☐ Dead Date of Death: __/__/__ (DD,MM,YYYY)

2. Did the patient attend a funeral in the one month before becoming ill? ☐ **Yes** ☐ **No** ☐ **Unk**

If yes, Name of deceased person	Relation to Patient	Date of Funeral (DD,MM,YYYY)	Village/Town	District	Did the patient participate? (carry or touch the body)?
		__/__/__			☐ Yes ☐ No

3. Did the patient travel outside their hometown or village/town before becoming ill? ☐Yes ☐No ☐ Unk

If yes, Village: __________ Chiefdom: __________

District: __________ Date(s): __/__/__ - __/__/__ (DD,MM,YYYY)

Case Report Form Completed by:

Name: __________ Phone: __________ E-mail: __________

Position: __________ District: __________ Health Facility: __________

Information provided by:

☐ Patient ☐ Proxy *If proxy*, Name: __________ Relation to patient: __________

Patient Outcome Information:

Please fill out this section at the time of patient recovery and discharge from the hospital OR patient death.

Date Outcome Information Completed: __/__/__(DD,MM,YYYY)

Final Status of the Patient: ☐ Alive/Recovered ☐ Dead

If the patient has recovered and been discharged from the hospital:

Hospital discharged from: __________ District: __________

Date of discharge from the hospital: /___/ (DD,MM,YYYY)

If the patient is dead:

Date of Death: __/__/__(DD,MM,YYYY)

Place of Death: ☐ Community ☐ Hospital __________ District: __________

Date of Funeral/Burial: __/__/__(DD,MM,YYYY)

Funeral conducted by: ☐ Family/community ☐ Outbreak burial team

Place of Funeral/Burial: Village: __________ Chiefdom: __________ District: __________

Case ID: __________

Fig. 27.2 Sierra Leone Ebola case investigation form (shortened version)

In the event that a sick individual met the suspected (or probable) EVD case definition, an ambulance was dispatched when one became available. Throughout most of the outbreak, the patient would be transferred first to one of the Ebola holding centers, operated by the Ministry of Health and Sanitation (MOHS), where he/she received testing for the virus and basic care. If the result was positive for EVD, the patient was transferred to an Ebola Treatment Center (ETC) once space became available. Until adequate ETC capacity existed locally, patients travelled up to 6 h via ambulance from the local holding center to an available ETC for treatment of their disease. The holding centers in Port Loko and Bombali closed between December 2014 and March 2015. During this time period, the patient management process gradually shifted so that suspected patients were transferred directly from the community to a local ETC. This became possible as there was adequate availability in these facilities to care for both suspected and confirmed patients due to the simultaneous decline in the number of cases and the opening of new ETCs.[4]

An ambulance was dispatched in occasional instances when a maternity patient with health care needs unrelated to EVD was identified during the case investigation; however, ongoing concerns about contamination of ambulances by patients with EVD, as well as their availability, made this a difficult prospect. Other patients who had been excluded as having EVD were sometimes referred to a non-Ebola health facility, but closure of Peripheral Health Units (PHU)[5] and restrictions on

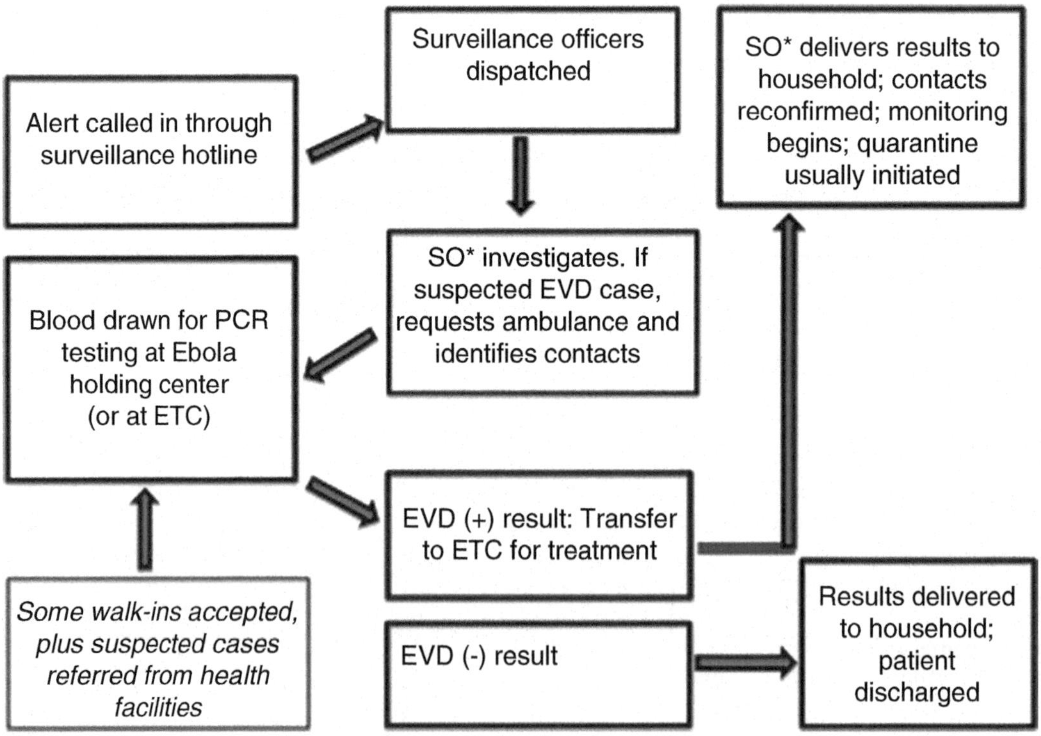

Fig. 27.3 Sick alert, referral, and case investigation process flow. (*SO=surveillance officers)

[4]After closing all of its holding centers, Bombali District reopened a holding center in February 2015 in response to a large Ebola cluster in the district.

[5]PHUs are designed as the delivery point for primary health care. There are three types of facilities: Community Health Centers; Community Health Posts; and Maternal Child Health Posts.

movement made accessing care difficult. See Fig. 27.3 for a breakdown of the sick alert, referral, and case investigation process flow.

When a death was reported, a burial team was dispatched to the location. A member of this team took an oral swab specimen from the deceased individual, irrespective of presumed cause of death. The surveillance team initiated an epidemiological investigation just as they did with ill patients, and later informed the household members of the laboratory test results.

When a suspected or probable case was identified in a living or deceased individual, a list of contacts was elicited. Frequently, this was also done for a deceased who did not meet the case definition. Upon receipt of an EVD-positive lab result, the household or compound[6] in which the confirmed case resided, in addition to his/her close contacts outside the household, were placed under contact monitoring, and usually, quarantine.

The description above provides a general overview of community-based surveillance during the outbreak. It is important to emphasize that surveillance work in the community was very complex, and the circumstances surrounding a reported illness or death that were revealed during an investigation were extraordinary in both their variety and, oftentimes, complexity. A common situation was difficulty or inability to find the person who was reported as ill or deceased. This was often due to abscondence or scant/incorrect information provided to the surveillance officers by the hotline. Perseverance and local knowledge and relationships were key to bringing such alerts to a resolution, if at all possible. Furthermore, there were usually some inconsistencies in the information gathered by surveillance officers in any type of investigation, so effective investigative work entailed triangulating the information obtained from the patient or family with that from neighbors, village elders, contacts who were named by the patient, and so on. In one of the districts in particular, there was a tendency for people to bypass the step of calling the hotline to report their illness and seek care by going directly to a holding center. Upon admission, some would falsely identify the community in which they had resided when their symptoms began, thus inhibiting the launch of an investigation in that location. The reason for this was usually to avoid the ramifications associated with a case investigation, especially the possible initiation of quarantine and monitoring by contact monitors.

27.3 Surveillance in Outpatient Facilities and Hospitals

Outpatient and inpatient facilities were required to conduct EVD screening on any person seeking entry. Among other requirements in the process, a health care provider was to ask about symptoms of Ebola and take each person's temperature using an infrared thermometer in a designated area. Inpatients were to be monitored for the development of Ebola symptoms during their stay. When someone who was suspected of having Ebola was identified during the screening process or within the facility, staff were responsible to notify the DERC. A surveillance team and an ambulance were dispatched, and the patient was transferred to an Ebola holding center. Postmortem laboratory testing for Ebola virus that was done throughout Sierra Leone was also a key element of surveillance in health care facilities.

Identification of contacts (contact tracing) of a suspected or confirmed case extended to both persons within the facility and in the community who may have had contact with the patient. At the facility, potential contacts would have included patients in nearby beds, visitors, health care workers, janitorial staff, laboratorians, and mortuary staff. The medical practices or procedures that may have facilitated nosocomial transmission were generally not covered in a lot of detail during the investiga-

[6]People commonly live in compounds, which are a group of nearby dwellings. They are often, although not always, occupied by members of an extended family. Residents of the compound often cook together and use the same latrine. When quarantine was initiated, the entire compound was usually quarantined, especially later in the outbreak. This chapter uses the terms household and compound synonymously.

tion unless the surveillance officers were accompanied by international staff with training in IPC or investigations of health care-associated infections.

In our experience, surveillance at health care facilities was less consistent than in the community, and there was a weakness in the link between health care facilities and the DERC. These issues are elaborated on later in the chapter. Nevertheless, the robustness of surveillance in health care facilities increased over time as the response provided more personnel and material resources for training, monitoring, and physical infrastructure.

27.4 Supplementary Activities Conducted by Surveillance Officers to Support Ebola Virus Disease Surveillance

In addition to their duties in case investigation, surveillance officers often tried to accommodate the social, practical, and communication concerns that occurred in households and communities under investigation for EVD. These unforeseen responsibilities tended to present themselves as petitions for food, water, and infant formula, as well as follow-up on the status of persons who had been referred for Ebola testing. The efforts by surveillance officers allowed them to build trust and rapport with communities, which were essential to conducting thorough epidemiological investigations. Communicating with village and higher-level chiefs about investigations and control measures in their area, as well as following up their concerns at the DERC, were also important.

27.4.1 Well-Being of Communities and Households Under Investigation and in Quarantine

Surveillance officers frequently visited quarantined homes, not only to assess the contact monitoring process, but also to inquire about the well-being of the inhabitants. This entailed checking if they were being treated acceptably by the community and anyone who may have been assigned to guard them, and if food had been delivered. A primary issue for households under quarantine was a lack of consistent and timely food delivery. Inadequate food and essential nutrients, along with inability to access prenatal care, put those pregnant women who were under quarantine specifically at risk.

In regard to maternal child health, one of the most common situations faced by investigators was a grave need for infant formula. This need often surfaced during a case investigation that had been initiated due to death during childbirth or indication of possible EVD infection in a postpartum woman. In the districts where we worked, the availability of formula for distribution when and where it was needed was a significant problem. Restricted movement of persons, coupled with extreme poverty, made infant formula inaccessible to many and likely resulted in negative outcomes for affected infants, although the extent was not monitored during the response. In one of the districts, a local volunteer tried to fill the gap by collecting funds and delivering formula to affected households. As formula became more accessible at the DERC, surveillance officers in Port Loko began carrying it in their vehicles and distributed it when a need was identified during an investigation.

27.4.2 Follow-Up on Individuals Who Had Been Referred for Ebola Virus Testing

Another common problem was the lack of communication with family members regarding the status of persons who had been referred to holding centers for testing, especially those who had been subsequently transferred to treatment centers outside the district. Although surveillance officers were charged with delivering the EVD laboratory test results to the family of a suspected case who had been referred, they were not responsible for keeping them informed of the patient's progress, due to competing priorities and lack of access to such information. Families and communities often had a feeling of alienation and despair from not knowing the whereabouts of relatives, if they were still alive, or the cause of their deaths. This jeopardized the future cooperation of a community, and officers usually attempted to obtain any information they could about people who had been referred.

27.4.3 Communication with Local Authorities

The involvement and cooperation of chiefs at all levels were vital for effective surveillance. A chief could often convince a reluctant household or neighborhood to cooperate during an investigation or help locate a missing individual. His involvement was also essential in ensuring that cases of illness or death were reported to authorities, and that traditional healing and burial rituals were eschewed. While there were many factors that led to the increased cooperation of chiefs over time, including stronger enforcement of the "Bye-Laws for the Prevention of Ebola and Other Diseases" established in August 2014 (Sierra Leone Ministry of Local Government and Rural Development, National Council of Paramount Chiefs 2014), surveillance officers' efforts to keep chiefs informed of the investigations in their area and plans to implement any control measures, as well as voicing their concerns to the DERC, engendered their cooperation.

The aforementioned responsibilities assumed by surveillance officers were not only time-consuming and complex, but also fell outside of their intended scope of work within the EVD response. Other actors within the response were charged with addressing social, material, and communication concerns, but lack of resources and organization hindered their ability to respond to such needs in a sustained and comprehensive way during the height of the outbreak. Surveillance officers' priorities were to respond to alerts and conduct case investigations, although it became evident as the outbreak evolved that their role was far more diverse. To their frustration, they were often unable to mitigate all issues raised due to the lack of authority or resources as well as weaknesses in communication protocols. Fortunately, the response gradually became more structured and organized. By the beginning of 2015, the availability of more human and physical resources somewhat alleviated the need for surveillance officers' involvement in the activities noted here.

27.5 Surveillance Challenges in Mothers and Their Children

The nonspecific nature of common Ebola symptoms, including fever, diarrhea, vomiting, headache, loss of appetite, and weakness, presented challenges for EVD surveillance. Symptoms similar to those associated with EVD are also associated with illnesses that are common in West Africa, notably malaria, as well as pregnancy and labor (Walker et al. 2015). Since a suspected EVD case in the community was defined primarily on the basis of reported fever and three other symptoms (for those without an epidemiological link to a confirmed case),[7] it was well known to staff working in both surveillance and at Ebola holding and treatment centers that many persons meeting the case definition would ultimately test negative for the virus. This was particularly true for pregnant women. Deaver and Cohen estimated that, in Sierra Leone, 98.5% of pregnant women identified as suspected cases of EVD by standard primary screening actually had other conditions, the large majority of which were related to complica-

[7] See "Case definition recommendations for Ebola or Marburg virus diseases 2014" for the complete WHO definition of a suspected case.

tions of pregnancy (Deaver and Cohen 2015). Investigation and referral of uninfected persons who met the EVD case definition, either as a result of an illness with symptoms similar to EVD, or in the case of pregnant women, having symptoms that also accompany pregnancy and labor, was detrimental to the individuals involved and put a strain on many facets of the response.

The breakdown in the primary health care infrastructure experienced during the outbreak had a substantial deleterious effect on the health outcomes of communities as treatment for non-EVD-related conditions was often inaccessible (Walker et al. 2015; Elston et al. 2015). For example, malaria treatment for children aged less than 5 years declined by almost 40% in Sierra Leone during the outbreak (Brolin Ribacke et al. 2016). Such lack of access to care provided an opportunity for falciparum malaria infections to rapidly progress, and even resulted in death. This was especially true for young children, who have less acquired immunity to the disease, and pregnant women, who are at risk of the complications that accompany placental infection (Schwartz 2013; Uneke 2007). Lack of primary health care services also meant that many pregnant women had infrequent or no access to other prenatal care during the epidemic (Walker et al. 2015). Additionally, difficulty in accessing delivery services meant that even women with complications often delivered in their homes.

27.5.1 Surveillance Challenges in Postpartum Women and Their Infants

In the situation where a woman had complications or otherwise became ill during the postpartum period and was classified as a suspected EVD case, the infant was in a quandary. Upon referral for Ebola virus testing, the mother may have had to choose whether to leave her (asymptomatic) infant in the community at risk of starvation, which was primarily due to the lack of infant formula or other replacement feeding. This possible eventuality had to be weighed against the risk of nosocomial EVD stemming from the lack of sufficient IPC protocols and practices in holding centers. Patients in both districts were referred to these facilities since the capacity for admitting unconfirmed cases to ETCs did not become available until after the outbreak had reached its peak.

On the other hand, if the mother had been infected with EVD by the time of giving birth, an additional consideration in the decision was the high likelihood that the infant would have also contracted the infection (Gomes et al. 2017). The very dependence of infants on others for all their needs put the caregivers of an infected infant at a high risk of infection once he/she developed symptoms (Check Hayden 2015).

Sometimes, families requested that surveillance officers advise whether the baby should be transferred to the Ebola holding center with his/her mother or remain in the home. The officers reported being very disconcerted when they encountered what they saw as an impossible choice: risk of death due to inadequate nutrition versus risk that an asymptomatic infant, who did not meet the criteria for referral, would acquire a nosocomial infection. Moreover, clear guidance from the response operation was not given to them. Fortunately, as the number of Ebola cases declined and more ETCs opened, suspected cases could be referred directly to the ETCs, as opposed to first spending time in a holding center. Since IPC protocols were closely followed in ETCs, the risk of nosocomial infection for the infant became much less likely. Moreover, formula was consistently available in ETCs, which addressed the risk of mother-to-child transmission from breastfeeding. Thus, the decision on whether the infant should accompany his mother to the facility or remain in the home was less complicated in the latter stages of the outbreak.

27.5.2 Surveillance Challenges in Other Breastfeeding Women and Their Babies

When a lactating woman who was past the postpartum period was identified as a suspected case, the decision regarding what to do with the baby was often more straightforward than in the preceding examples. First, an older baby's EVD status was more independent of its mother's status, as compared to a newborn (Bower et al. 2016). An infection in the mother did not guarantee transmission of the virus to the baby, despite exposure through breastmilk and other means. Furthermore, as the baby got older, its ability to survive in the community was less dependent on the availability of formula. Therefore, leaving an unweaned baby in the home was usually an easier decision than for a newborn, assuming that it did not meet the case definition of suspected EVD. Nonetheless, several competing factors sometimes arose, including the mother or even another family member insisting she take the baby with her. Another possibility was that there was no trusted individual available or willing to provide care for the baby, a situation often brought about by the way in which EVD tore through households and extended families, sometimes leaving them with few or no adults to care for children. A curious, ambulatory baby or toddler in a holding center without clinical justification was highly undesirable due to the inadequate IPC practices noted above. Furthermore, if the mother tested positive and was referred to a treatment center, an uninfected child would remain in the holding center until transportation was available to take him/her back to the community or someone came to retrieve the child. Timely discharge of uninfected patients from holding centers and return to their communities gradually improved in the two districts.

27.5.3 Surveillance Challenges in Gathering Information from and About Children

While it was expected that adults would answer questions on behalf of young children during an investigation, surveillance officers experienced challenges in obtaining information directly from older children due to the cultural norms of communication. The tendency was for adults to answer on behalf of children beyond the age where a child could accurately answer for himself. However, an adult was not necessarily aware of all symptoms or risks experienced by a child who had reached at least some level of independence in caring for himself and in his daily life (e.g., the risks incurred while playing or visiting with others in the village). In addition, there was a cultural expectation that a child would be submissive and not provide information that contradicted an adult. Effective surveillance required culturally-appropriate balancing of the need to respect authority figures in the household with the need to obtain accurate information about children in order to control the outbreak. One approach was to negotiate or deftly arrange a separation of the child from the adult. In one instance, a surveillance team was called to a PHU with the report of an ill girl about 8 years old. Her grandmother insisted the girl had just a few symptoms of EVD. The child concurred with these symptoms, which did not fulfill the case definition for Ebola. However, she appeared rather ill, and the pair had reported behaviors that were risks for EVD. After separating the child from her grandmother, she divulged additional symptoms that were consistent with suspected Ebola. The grandmother cooperated with the transport of her granddaughter to a holding center, where she was found to be infected.

Adding to the complexity of gathering information about children was that they were sometimes overlooked when adults provided information to investigators. This arose particularly when asking a suspected (adult) case about contact with other people. For instance, the interviewee might report that two outsiders had slept over in the house since the onset of her EVD symptoms. However, further probing would reveal that it was actually two *adults* that had slept over in the house, who were accompanied by one or more children. After this pattern of answering questions was detected, surveillance officers were trained to ask about children specifically, instead of assuming they had been accounted for in a similar manner as adults.

27.6 Weaknesses in Case Investigation and Related Functions of the Response and Their Effects on Pregnant, Postpartum, and Breastfeeding Mothers and Their Children

Deficiencies in case investigation including other functions of the response that it closely depended upon impacted outbreak control in these districts, as well as the welfare and health outcomes of mothers, their children and the rest of the population. We will highlight deficiencies which roughly can be divided into three categories: (1) irregularities in established surveillance processes or procedures; (2) lapses in infection prevention and control; and (3) failure to provide services in quarantined homes. In this section, we will focus on how these breakdowns affected mothers and their children.

27.6.1 Breakdown in Surveillance Procedures

Figure 27.3 illustrates that the intended point of entry to the Ebola response system for sick patients was via a call to the national or district hotline. This was to be followed by an epidemiological investigation conducted by surveillance officers, who would then determine whether the patient met the case definition and should be transferred to an Ebola facility via ambulance. However, in one of the districts where we worked, it was common for the DERC to dispatch an ambulance directly to the household, rather than wait for the assessment by surveillance officers. Another scenario in this district was that, rather than reporting an illness to the hotline, citizens would report them directly to ambulance staff, who were stationed with their vehicles around this district. In these circumstances, the ambulances usually arrived on the scene first. Ambulance workers, who were not trained in case investigation, would often transport the named individual irrespective of whether he/she met the case definition for Ebola. Indeed, some of the people who had called the hotline knew their symptoms were inconsistent with Ebola; they did so to comply with the requirement to report illnesses. In other instances, people were unaware that someone else had reported them as ill and found out only after an ambulance arrived. In situations such as these, where individuals were not expecting to be transported for EVD testing and surveillance officers were not involved in the investigation, the critical role of surveillance officers to determine, explain, and broker the appropriate next steps had been bypassed.

The practice of transporting people who did not meet the case definition to a holding center (or ETC) for EVD testing eventually led to a reluctance to report illnesses and to interact with the health care system overall. This coincided with a sharp decline in the number of sick alerts reported in Port Loko District. In the first 3 months of 2015, the number of sick alerts called in to the EVD hotline declined by approximately 65% in Port Loko District, to 1873 from the 5390 that were reported in the previous 3 months (Alpren et al. 2017). It is acknowledged that in addition to a mistrust of the EVD response system, there were other likely contributors to the decline such as "outbreak fatigue," and that sharp declines were seen elsewhere, notably, Western Area.

Deviation from the way surveillance procedures were intended to be carried out had a significant impact on young children in this district. As of April 2015, there was a surge in the number of deaths of young children reported through the EVD hotline in Port Loko District. Parents usually reported to surveillance officers that they had not taken their child for any medical treatment and would describe symptoms that were consistent with untreated malaria or another febrile illness. Follow-up with staff at health posts across the district revealed most had reopened, but that many parents feared that their children would be referred to an ETC and thus would not take the risk of seeking treatment or reporting illnesses to the hotline.

In response to the population's increasing disengagement with the surveillance system in Port Loko District, several steps were taken in both social mobilization and surveillance beginning in May 2015. Surveillance officers started active surveillance for Ebola at health posts (Young et al. 2016). They also conducted outreach to chiefs, other leaders, and to the community at large to encourage people to seek primary health care and report illnesses (Connolly et al. 2016). Priority was given to areas within the district that likely had the most significant underreporting, based on analysis of available data sources. The number of reports of both illness and death to the EVD hotline increased after these interventions (Tran et al. 2016).

27.6.2 Lapses in Infection Prevention and Control, Including Disease Surveillance, at Non-Ebola Health Care Facilities

At non-Ebola health facilities, there were significant and widespread issues with IPC (Kilmarx et al. 2014), which has disease surveillance as a core component. For such facilities in Bombali (Dunn et al. 2016) and Port Loko Districts, two of the substantial problems with surveillance were failure to screen entrants to the facility correctly or at all; and deficiencies in the monitoring of inpatients for EVD symptoms throughout their stay. Additionally, investigations were sometimes compromised by failure of the facility to report patients with EVD symptoms through the proper channels, as well as inadequate record keeping.

Surveillance officers reported, and our own experience revealed, serious lapses in screening at inpatient and outpatient facilities. At the end of 2014, it was still common to find that, at the time of the visit, no screener was in place, or basic elements of the process were missing. At hospitals, the large volume of patients and visitors, long opening hours, and the size of the facility required a level of coordination and oversight to the screening process that was difficult to implement and sustain.

For pregnant women, proper screening at entry to the facility was highly important. First, the overlap in EVD symptoms with normal pregnancy or complications of labor meant it was critical to precisely evaluate the woman's reported symptoms against the criteria for a suspected case. It is important to remember that for pregnant women, the goal of the screening was not only to avoid sending them to Ebola facilities unless it was warranted, but also to ensure that women who were genuine suspected EVD cases were excluded from the maternity ward, especially considering the amount of bodily fluids associated with delivery.

Ongoing monitoring of admitted patients for Ebola symptoms was essential since they may not have had symptoms of the disease upon entry to the facility, or there may have been a breakdown in the screening process. Retrospective identification of EVD through postmortem laboratory testing provided evidence that comprehensive monitoring of inpatients was not occurring. An investigation by Dunn et al. (2016) at a hospital in Bombali District also showed evidence of deficiencies in the ongoing monitoring for EVD. In February 2015, the MOHS issued guidance that all patients should be monitored at least three times daily (Government of Sierra Leone 2015).

Compounding the problems of surveillance in patients was the tendency for non-Ebola health facilities to overlook reporting cases of suspected EVD to the DERC. Commonly, investigators would become aware of the case only after the laboratory had sent test results to the DERC. Since the information received with the laboratory results did not indicate that the patient had been previously in a non-Ebola health care facility, the investigation at the facility, which should have commenced as soon as the patient was identified as a suspected case, was subject to further delay. Close coordination with the DERC was essential when any living or deceased patient was identified as a suspected or confirmed case; rapidly identifying their contacts within and outside the facility was essential so that proper follow-up could be made.

When a suspected or confirmed case of EVD was recognized, identification of contacts and collection of other epidemiological information such as a timeline of key events was hindered by deficiencies in employee and patient records. Examples included incomplete or misplaced employee log books, and patient charts that omitted information on symptoms, procedures that posed EVD risks (e.g., phlebotomy), and bed location. Moreover, in large facilities, surveillance officers often encountered an unclear managerial structure that made it difficult to identify the key persons to facilitate cooperation with the investigation.

A final challenge was that most surveillance officers in both districts did not have a background in case investigation prior to the outbreak. The training and technical assistance they received was oriented toward investigation in the field, where most cases were identified. In our experience, there was a shortage of international response personnel with expertise in surveillance, epidemiology, and IPC who were available to train and assist surveillance officers to conduct facility-based investigations. The time-consuming nature of these investigations and the obstacles noted here magnified the need for personnel resources.

27.6.3 Inadequate Services for Individuals in Quarantine

The use of household quarantine in Sierra Leone was widespread. It was to be guided by the Sierra Leone Emergency Management Program Standard Operating Procedure for Management of Quarantine, issued in October 2014 (Government of Sierra Leone 2014). The Standard Operating Procedure (SOP) called for the provision of water, food (including special foods for pregnant women, infants, and toddlers), bed nets, basic PPE for a caregiver, and a latrine disinfection kit. The problems with supplying sustenance have been extensively reported by the media and aid agencies working during the crisis (Westcott 2014; Wahome 2014). The issues with delivering sustenance were also widespread in the Bombali and Port Loko Districts (Maxmen 2015), while providing the other items mentioned in the SOP were beyond the capacity of the response for most of the outbreak.

Some of the difficulties with providing food and water arose from the vast numbers of persons in quarantine at any one time. People were frequently in quarantine for well over the 21-day incubation period for Ebola virus since each time someone in a quarantined home tested positive for EVD, the 21-day period started anew for the remaining inhabitants. Furthermore, there was a dearth of coordination among several organizations and the MOHS as to which would deliver the provisions. And, not all organizations supplied the full range of foodstuffs, which were to be delivered within 24 h—sometimes in far reaching areas—in accordance with the SOP. Surveillance officers frequently reported visiting households that were without food for 4 or more days after the initiation of quarantine or had an inadequate quantity or mix of food; our own experiences in the field were consistent with their reports.

As described earlier, people tended to live in compounds. In the event of a confirmed case, the entire compound was often quarantined, which typically contained between 15 and 25 people in these districts. The high fertility rate and frequency of polygamy meant that there were often many children and at least one pregnant or lactating woman in the compound, especially in rural areas.[8]

[8]The 2013 Sierra Leone Demographic and Health Survey showed that women in Bombali and Port Loko Districts had a total fertility rate of 4.4 and 5.3 children, respectively. The survey also found that over one-third of women in the country reported living in polygamous unions (Statistics Sierra Leone et al. 2014).

In addition to quarantine of individual houses/compounds, there was also village-level quarantine, when possibly hundreds of people were restricted to their homes or village. Although the SOP for Management of Quarantine stated otherwise, the amount of food distributed was often based on only those people who were to be monitored as direct contacts to an Ebola case. However, to plan for the nutritional needs of the entire community would have required taking a census. A system was also needed within the village to ensure that each person received the food that he/she required, especially vulnerable populations. According to our experience, these steps were not taken consistently, although procedures for supplying food to quarantined villages did improve slowly over time.

Health services were not officially provided to people in quarantine. This had a damaging effect on all segments of the quarantined population as issues ranging from injury to preexisting respiratory illness went untreated. Pregnant woman delivered their babies without a trained birth attendant, putting those in the home who assisted in the delivery at high risk of infection if the mother was infected with EVD.

The lack of a system to provide appropriate health services by trained staff resulted in myriad ad hoc steps to address health concerns. As the cluster investigation cited in the next section illustrates, one option was to escape quarantine in order to seek health care. Or, a nurse from the local health post may have attempted to fill the gap in services. The extent of this practice was unknown, although one of the authors was involved in two investigations in which a nurse had attempted to treat a quarantined individual who had EVD symptoms. One, a pregnant woman, and the second, a 9-year-old child, were each known contacts to EVD cases. These quarantined contacts had been treated for a few days before dying in their homes and tested positive for the infection after their deaths.

The provision of food and water was a stubborn problem that persisted until after the outbreak had crested in both districts. Its eventual improvement was an important development nonetheless, as quarantine was still widely used during the phase of low transmission and sporadic clusters.

27.7 Investigation of an EVD Cluster Linked to a Hospital Maternity Ward

The breakdowns described in case investigation and IPC procedures, as well as a lack of services in quarantined homes, impacted outbreak control directly and indirectly. The emergence of a large cluster of EVD cases in Bombali District from November 2014 to January 2015 provided an example of how such lapses impeded the identification of cases and contacts as well as control of the outbreak.

On November 28, 2014, a pregnant woman was admitted to the maternity ward of a local hospital. She came from a village in the midst of a large EVD outbreak in which 34 cases would eventually be confirmed, prompting a quarantine that would persist for approximately 8 weeks. The mother was a high-risk contact as she had washed the body of one of the four people infected with EVD that lived in her home. This mother (Mother 1) escaped the guarded village at night to seek health services. She was not known to have been screened at entry to the hospital and falsely identified herself as coming from a different village, which was not in quarantine. Eventually, she would be identified as the index case in this cluster.

Following the birth of her baby (Baby 1) the same night, the mother was moved from the delivery suite to the maternity ward, where she was placed next to a woman (Mother 2) who had not yet delivered. Mother 2 became frightened by the extent of Mother 1's hemorrhaging, which had been occurring since delivery and checked herself out of the hospital the following day.

Approximately one-half hour after Mother 1 delivered, another woman (Mother 3) gave birth in the same delivery suite and was assisted by the same nurse. The nurse, who was not known to have Ebola infection, did not fully change her personal protective equipment (PPE) between deliveries. Furthermore, it was reported that the delivery suite had not been cleaned between deliveries. Mother 3 and her infant, Baby 3, were discharged from the hospital on November 29, 2014.

Following the death of Mother 1 on November 30, 2014, Baby 1 was released to relatives in a village near the one experiencing the large outbreak. The postmortem laboratory sample taken from Mother 1 was positive for EVD. Nevertheless, the relationship of Baby 1 with Mother 1 was not established either through the hospital informing the DERC of Mother 1's suspicious death, or the epidemiological investigation that followed the positive laboratory result. Instead, surveillance teams only became aware of their relationship when several suspected cases of Ebola emerged in Baby 1's village. These came to light only after the clandestine burial of Baby 1 following his death on December 10, 2014. Six members of the baby's household were eventually confirmed as positive for EVD.

Mother 2, who had checked herself out of the hospital due to fear of Mother 1's bleeding in the next bed, went to a different hospital the next day (November 30, 2014). She delivered twins (Baby 2a and Baby 2b), and all three were subsequently released. Mother 2, who was not identified as a contact during the investigation of Mother 1, became ill 10 days later. Five days passed before an EVD lab sample was taken, and she died 3 days later (December 17, 2014). The twin infants were in the care of family until they were diagnosed with Ebola infection on December 19, 2014. Both children died 3 days later. A total of 7 cases emanated directly or indirectly from contact with the twins. No cases at the hospital where Mother 2 delivered were identified.

Mother 3, who had delivered immediately after Mother 1, also was not identified as a contact during the investigation of Mother 1. She became symptomatic for EVD and was readmitted to the same hospital on December 6, 2014. Similar to Mother 1, she was unknown to have been screened for Ebola symptoms upon arrival. Moreover, she remained in the facility for 3 days without being identified as a suspected case before her family removed her, attributing her illness to evil spirits. Mother 3 returned to her home, where she died the following day, and later tested positive for EVD via a postmortem specimen.

After Mother 3's second hospital discharge, the family elected to send her baby to the home of other relatives for care, where he died the following day. A total of 16 cases were identified as having had direct contact with the mother, the baby, or both. These included 9 people who had washed the body of the mother in a healing ceremony before her death. No cases were found at the hospital where she had been readmitted, despite her being symptomatic over the 3 days of her stay.

This cluster investigation provided a dramatic example of how various shortcomings in the response worked in synergy to facilitate direct and indirect EVD transmission from a pregnant mother to 35 people. It is possible that providing maternity care services to this pregnant mother while she was in quarantine could have avoided the start of this cluster. Subsequently, various lapses in IPC, including screening at entry, ongoing monitoring of inpatients, and improper use of PPE, appeared to have led to infections within the facility. A failure of communication regarding the death of Mother 1 and the discharge of her baby, in addition to obstacles in identifying other mothers who had been exposed in the facility, meant that two mothers and a total of four infants who should have been under contact monitoring were not. Furthermore, a breakdown in hospital surveillance during the second hospital stay for Mother 2 resulted in her remaining in the ward without being identified as a suspected case and transferred to a holding center. Thus, she was able to leave the hospital on her own accord, which provided the opportunity for relatives to perform a healing ritual before her death.

27.8 Recommendations

The 2013–2015 West African EVD Outbreak was unanticipated in its size, scope, and location. The health systems within these countries, as well as international health and humanitarian agencies, were inadequately prepared to manage the event. The lessons learned from this prolonged outbreak should be used to plan for future outbreaks of EVD and other epidemic-prone infectious diseases. We have selected a few key recommendations that would especially benefit mothers and children as well as disease surveillance during future outbreaks.

27.8.1 Health Facility-Based Surveillance

As highlighted in this chapter, EVD surveillance within health facilities during the outbreak was weak. One of our primary recommendations is to strengthen implementation of SOPs used for screening visitors and patients, monitoring inpatients, and referring suspected cases from PHUs and hospitals to holding centers and ETCs. This should be done through allocation of needed resources, training, and reinforcing supervision practices within facilities. It is also recommended to maintain accurate patient charts and records of staff shifts to effectively monitor IPC practices and to facilitate case investigations, including identifying potentially exposed patients and staff. SOPs should provide details on the type of information that must be reported to the outbreak response authorities and clearly define the surveillance-related roles and responsibilities of hospital staff and community-level surveillance officers.

27.8.2 Adherence to EVD Surveillance Procedures

The reluctance of persons to report sick and death alerts, intentional underreporting and/or masking of symptoms, and refraining from seeking primary health services were issues that resulted in part from fear of transfer to holding centers and ETCs. This fear appeared to be exacerbated if there was a tendency for ambulances to transfer persons prior to the involvement of surveillance officers—even if they did not meet EVD case definition. Such noncompliance with surveillance procedures could be avoided in the future by ensuring that SOPs related to surveillance and referral are clearly communicated to all sectors of the response, and that outbreak response authorities systematically monitor and address any breakdowns in these procedures.

27.8.3 Improvement of Basic Services for People Under Investigation for EVD and in Quarantine

Delayed food delivery and inadequate nutrition for those pregnant women and young children who were in quarantine resulted from poor coordination, limited resources, and weak resource mobilization. To help mitigate these problems, our recommendation is to refine the criteria for determining who should be placed under quarantine to reduce the number of persons needing access to quarantine-related goods and services. Additionally, development of a more efficient model of service delivery that is specific to an outbreak situation is recommended. Mass deployment of emergency food and health services for a prolonged period is more typically done in a common location; however, quarantine and other restrictions on movement require distribution for discrete periods of time at the household level, sometimes in isolated communities.

For pregnant women, access to specific health services during an outbreak that would improve both their well-being and disease surveillance should be considered. For instance, administration of sulfadoxine-pyrimethamine (SP) for intermittent preventative treatment in pregnancy (IPTp) to pregnant women under quarantine would reduce risk of malaria infection and help avoid unnecessary referral of pregnant women to holding centers and ETCs.

Finally, the measures taken to control the outbreak, including the function of disease surveillance, must be thoroughly analyzed for their effects on women and children. While the effects probably will not be identical in a future event, this outbreak highlighted that the potential separation of babies from their mothers, which arose when a mother was deemed a suspected EVD case, required the immediate availability of infant formula. The need for infant nutrition should be anticipated so that case investigators carry the appropriate products for distribution. This would eliminate the need for infants to be transferred to holding centers for the main purpose of continuing breastfeeding. It should also be anticipated that the outbreak itself will reduce the availability of obstetrical services for pregnant women. In order to provide safe deliveries for suspected and non-EVD-infected pregnant women, a strong reporting and referral network needs to be established rapidly to ensure transfer and admission, especially to specialized facilities.

Dedication and Acknowledgments This chapter is dedicated with respect and gratitude to the surveillance officers of Sierra Leone, who worked under arduous conditions on the front lines of this prolonged outbreak. For review of this draft, the authors would like to acknowledge Sorie Ibrahim Beareh Kamara, Salieu Jalloh (Bombali District), and Alhaji D. Kamara (Port Loko District), who worked as MOHS surveillance officers during the outbreak. The authors also acknowledge Indu B. Ahluwalia, MPH, PhD, of the U.S. Centers for Disease Control and Prevention, who collaborated with MOHS and WHO colleagues to conduct the cluster investigation described in this chapter.

References

Alpren, C., Jalloh, M., Kaiser, R., Diop, M., Kargbo, S., Castle, E., et al. (2017). The 117 call alert system in Sierra Leone: From rapid Ebola notification to routine death reporting. *BMJ Global Health, 2*, e000392. Retrieved May 1, 2018, from http://gh.bmj.com/content/2/3/e000392.

Bower, H., Johnson, S., Bangura, M., Kamara, A., Kamara, O., Mansaray, S., et al. (2016). Effects of mother's illness and breastfeeding on risk of Ebola virus disease in a cohort of very young children. *PLOS Neglected Tropical Diseases, 10*(4), e0004622. Retrieved May 1, 2018, from http://journals.plos.org/plosntds/article?id=10.1371/journal.pntd.0004622.

Brolin Ribacke, K., Saulnier, D., Eriksson, A., & von Schreeb, J. (2016). Effects of the West Africa Ebola virus disease on health-care utilization – A systematic review. *Frontiers in Public Health, 4*, 222. https://doi.org/10.3389/fpubh.2016.00222.

Check Hayden, E. (2015). Maternal health: Ebola's lasting legacy. *Nature, 519*, 24–26. https://doi.org/10.1038/519024a. Retrieved April 30, 2018, from http://www.nature.com/news/maternal-health-ebola-s-lasting-legacy-1.17036.

Connolly, A., Young, A., Mancuso, B., Hartley, M., Hoar, A., Kaur, G., et al. (2016). Surveillance strategies during low Ebola transmission in a district in Sierra Leone. *Online Journal of Public Health Informatics, 8*(1), e56. https://doi.org/10.5210/ojphi.v8i1.6470.

Deaver, J., & Cohen, W. (2015). Ebola virus screening during pregnancy in West Africa: Unintended consequences. *Journal of Perinatal Medicine, 43*(6), 649–651. https://doi.org/10.1515/jpm-2015-0118.

Dunn, A., Walker, T., Redd, J., Sugerman, D., McFadden, J., Singh, T., et al. (2016). Nosocomial transmission of Ebola virus disease on pediatric and maternity wards: Bombali and Tonkolili, Sierra Leone, 2014. *American Journal of Infection Control, 44*(3), 269–272. https://doi.org/10.1016/j.ajic.2015.09.016.

Elston, J., Moosa, A., Moses, F., Walker, G., Dotta, N., Waldman, R., et al. (2015). Impact of the Ebola outbreak on health systems and population health in Sierra Leone. *Journal of Public Health, 38*(4), 673–678. https://doi.org/10.1093/pubmed/fdv158. Retrieved April 30, 2018, from https://academic.oup.com/jpubhealth/article/38/4/673/2966926.

Gomes, M., de la Fuente-Núñez, V., Saxena, A., & Kuesel, A. (2017). Protected to death: Systematic exclusion of pregnant women from Ebola virus disease trials. Reproductive Health, 14(S3), 47–55. https://doi.org/10.1186/s12978-017-0430-2. Retrieved April 30, 2018, from https://www.ncbi.nlm.nih.gov/pmc/articles/PMC5751665/.

Government of Sierra Leone. (2014). *Sierra Leone emergency management program standard operating procedure for management of quarantine, Freetown, Sierra Leone, October 2014*.

Government of Sierra Leone. (2015). *Sierra Leone emergency management program standard operating procedures for the safe provision of hospital services during a haemorrhagic fever outbreak, with a focus on Ebola, Freetown, Sierra Leone, February 2015*.

Kilmarx, P., Clarke, K., Dietz, P., Hamel, M. J., Husain, F., McFadden, J. D., et al. (2014). Ebola virus disease in health care workers-Sierra Leone, 2014. *MMWR Morbidity and Mortality Weekly Report, 63*(49), 1168–1171. Retrieved April 25, 2018, from https://www.cdc.gov/mmwr/preview/mmwrhtml/mm6349a6.htm.

Maxmen, A. (2015). In Sierra Leone, quarantines without food threaten Ebola Response. *Aljazeera America*. Retrieved April 30, 2018, from http://america.aljazeera.com/articles/2015/2/19/in-sierra-leone-quarantined-ebola-survivors.html.

Ministry of Local Government and Rural Development [Sierra Leone], National Council of Paramount Chiefs. (2014). *Byelaws on the prevention of Ebola and other diseases*. Retrieved September 25, 2018 from https://www.humanitarianresponse.info/files/documents/files/by-laws.pdf

Schwartz, D. A. (2013). Challenges in improvement of perinatal health in developing nations: Role of perinatal pathology. *Archives of Pathology & Laboratory Medicine, 137*(6), 742–746. Retrieved May 1, 2018, from http://www.archivesofpathology.org/doi/full/10.5858/arpa.2012-0089-ED.

Sierra Leone Ministry of Local Government and Rural Development, & National Council of Paramount Chiefs. (2014). *Byelaws on the prevention of Ebola of Ebola and other diseases*.

Statistics Sierra Leone, Ministry of Health and Sanitation, & ICF International. (2014). *Sierra Leone demographic and health survey 2013*. Retrieved April 26, 2018, from https://dhsprogram.com/pubs/pdf/fr297/fr297.pdf.

Tran, A., Hoar, A., Young, A., Connolly, A., Hartley, M., Boland, S., et al. (2016). Strengthening community surveillance of Ebola virus disease in Sierra Leone. *Online Journal of Public Health Informatics, 8*(1), e164. https://doi.org/10.5210/ojphi.v8i1.6583.

Uneke, C. J. (2007). Impact of placental *Plasmodium falciparum* malaria on pregnancy and perinatal outcome in Sub-Saharan Africa. *Yale Journal of Biology and Medicine, 80*(2), 39–50. Retrieved May 1, 2018, from https://www.ncbi.nlm.nih.gov/pmc/articles/PMC2140183/.

Wahome, W. (2014). Quarantines in Sierra Leone: Putting people first in the Ebola crisis. *Oxfam*. Retrieved February 17, 2018, from https://oxfamilibrary.openrepository.com/oxfam/handle/10546/336992.

Walker, P., White, M., Griffin, J., Reynolds, A., Ferguson, N., & Ghani, A. (2015). Malaria morbidity and mortality in Ebola-affected countries caused by decreased health-care capacity, and the potential effect of mitigation strategies: A modelling analysis. *The Lancet Infectious Diseases, 15*(7), 825–832. https://doi.org/10.1016/s1473-3099(15)70124-6. Retrieved May 2, 2018, from https://www.thelancet.com/journals/laninf/article/PIIS1473-3099(15)70124-6/fulltext.

Westcott, L. (2014). Thousands break Sierra Leone Ebola quarantine for food. *Newsweek*. Retrieved May 2, 2018, from http://www.newsweek.com/thousands-break-sierra-leone-ebola-quarantine-food-282471.

World Health Organization. (2014). *Case definition recommendations for Ebola or Marburg virus diseases*. Retrieved February 10, 2018, from http://www.who.int/csr/resources/publications/ebola/case-definition/en//.

World Health Organization. (2016). *Ebola situation report - 20 January 2016*. Retrieved April 26, 2018, from http://apps.who.int/ebola/current-situation/ebola-situation-report-20-january-2016.

Yang, W., Zhang, W., Kargbo, D., Yang, R., Chen, Y., Chen, Z., et al. (2015). Transmission network of the 2014–2015 Ebola epidemic in Sierra Leone. *Journal of the Royal Society Interface, 12*(112), 20150536. https://doi.org/10.1098/rsif.2015.0536. Retrieved May 1, 2018, from http://rsif.royalsocietypublishing.org/content/12/112/20150536.

Young, A., Connolly, A., Hoar, A., Mancuso, B., Esplana, J. M., Kaur, G., et al. (2016). Use of peripheral health units in low-transmission Ebola virus disease surveillance. *Online Journal of Public Health Informatics, 8*(1), e177. https://doi.org/10.5210/ojphi.v8i1.6596. Retrieved May 2, 2018, from https://www.ncbi.nlm.nih.gov/pmc/articles/PMC4854600/.

28 Ebola and Accusation: How Gender and Stigmatization Prolonged the Epidemic in Sierra Leone

Olive Melissa Minor

28.1 We Know That Ebola Is Real

In March 2015, as this research took place, Sierra Leone marked an unhappy anniversary: 1 year since Ebola virus disease (EVD) had emerged within its borders. By the end of that month, the Government of Sierra Leone (GOSL) had confirmed 8545 cases of EVD and 3433 deaths. Through the Ministry of Defence, GOSL implemented a top-down Ebola response strategy made up of a centralized National Ebola Response Centre (NERC), headed by a CEO reporting to the President, and a system of District Ebola Response Centres (DERCs) led by district coordinators. Foreign and national military personnel directed the six-pillar Ebola response, organized into: (1) case management, (2) surveillance and quarantine, (3) safe burial, (4) logistics, (5) psychosocial support, and (6) social mobilization (Olu et al. 2016: 3).

In September 2014, Oxfam's humanitarian team began to implement Oxfam's three-pronged response to the outbreak, focusing on Treatment, Containment, and Prevention. Treatment meant partnering with medical actors to provide WaSH[1] services and materials for Ebola Treatment Units (ETUs), triage centers, and health clinics. Containment focused on community surveillance. In Sierra Leone, this involved house-to-house messaging and surveillance by Community Health Volunteers (CHVs), as well as contact tracing and Active Case Search (ACS). ACS meant carrying out door-to-door canvassing to locate people exhibiting active symptoms of possible Ebola infection and referring them for treatment. Prevention involved forming neighborhood Public Health Promotion (PHP) teams to increase community access to, and participation in, the uptake of Ebola prevention information, services, and materials (Adams et al. 2015).

By late October, however, humanitarian staff felt that they had hit a wall. Despite Oxfam's previous WaSH and social mobilization experiences with cholera, hepatitis E, and smaller Ebola virus out-

[1] WaSH is an acronym that stands for "water, sanitation and hygiene".

This chapter is partly derived from an article published in *Anthropology in Action* 24(2): 25–35, Berghan Books and the Association for Anthropology in Action, as Minor, O.M., Ebola and Accusation: Gender and Stigma in Sierra Leone's Ebola Response, https://doi.org/10.3167/aia.2017.240204.

O. M. Minor (✉)
International Rescue Committee, New York, NY, USA

Humanitarian Department, Oxfam, Oxford, Great Britain

D. A. Schwartz et al. (eds.), *Pregnant in the Time of Ebola*, Global Maternal and Child Health, https://doi.org/10.1007/978-3-319-97637-2_28

breaks, the scale of the 2014 EVD outbreak overwhelmed the templates developed in other contexts. Oxfam staff in Sierra Leone observed that most people in their areas of operation had moderate to high awareness of Ebola prevention and treatment information. They knew that "Ebola is real" and could list EVD symptoms, modes of transmission, and methods to prevent infection—yet they did not necessarily act in accordance with that information (Abramowitz et al. 2014: 1). Infections continued to ignite "hot spots" across the region. Oxfam staff realized that, in order to save lives, one-way messaging was not enough; their response needed deeper inquiry into the factors preventing people from taking the necessary precautions to avoid infection.

As an anthropologist embedded with Oxfam's humanitarian staff, I was originally tasked with identifying local-level barriers to community participation in the Ebola response, particularly among women. Although Ebola response activities initially overlooked gender as a potential factor in transmission of the Ebola virus (Harman 2016), later studies found that gender produced multiple vulnerabilities to EVD morbidity and mortality (Ministry of Social Welfare, Gender and Children's Affairs et al. 2014). A better understanding of gender relations in the context of the Ebola outbreak would ideally support a more inclusive and community-led Ebola response effort.

With a research assistant, I carried out interviews and focus groups with families under quarantine, EVD survivors, community and religious leaders, women's and youth groups, and health care providers in 17 communities affected by the Ebola epidemic. We also embedded with our PHP staff and volunteers as they carried out door-to-door messaging, contact tracing, and Active Case Search. This allowed us to engage demographically diverse household members in face-to-face conversations about Ebola virus prevention and treatment and to hear their perspectives. Cases in which CHVs located symptomatic persons and had to encourage them to seek assistance at an ETU offered critical insights into the individual, family, and community concerns that influenced treatment-seeking behavior.

This research quickly revealed that community-level Ebola risk factors, including gender inequalities, could not be understood in isolation from factors related to the broader Ebola response, including structural and power inequalities. Much of the disconnect between knowledge and action on EVD prevention advice resulted from severe shortages in the "staff, stuff, space, and systems" that Paul Farmer[2] (2014) had identified months before—not from cultural beliefs or practices. However, conversations with affected families and communities also revealed ongoing social and behavioral barriers to ending the epidemic. Many of these barriers—including avoidance of ambulances and ETCs, negative feelings towards burial teams, and resistance to government-led quarantine measures—ultimately rested on understandable fears of infection, social isolation, and death. Responses to these fears, however, meant individuals and families risked compounding the chances of further infection by evading detection of the virus; delaying biomedical treatment; avoiding quarantine measures; moving across districts to seek non-biomedical treatments; and burying those who died of EVD without wearing protective equipment.

On the national and international stage, much has been written about rumors and misinformation fuelling the 2014 Ebola epidemic. Often, these commentaries characterized Ebola-affected communities as irrationally "resistant" to public health advice and dismissed their concerns as ignorance. In conversations with affected people, however, I came to understand that this stigmatization and blame itself played a key role in the ongoing outbreak. Stigma and blame affected all levels of the Ebola response—from small-scale social interactions to large-scale government quarantine policies—and tended to be directed towards structurally disadvantaged individuals, families, and communities. These attitudes put pressure on gender roles in ways that compounded barriers to epidemic control, ultimately prolonging Sierra Leone's Ebola epidemic.

[2]See the chapter by Dr. Farmer and colleagues in this book, "*The Challenges of Pregnancy and Childbirth in Women Who Were Not Infected with Ebola Virus during the 2013-2015 West African Epidemic.*"

28.2 A Monster That Kills the Whole Family

This research with affected communities began by asking about local terms for EVD, to open up conversations about its impact; participants described the following terms for Ebola[3]:

Fitina (Temne): trauma, stress, God judges you
Bola Neh (Temne): suffering without the aid of the community
Bola-Bola (Limba): rolling in pain by yourself without the help of any person
Elbola (Kuranko): the sickness is in your hand; if you touch another person you will pass the sickness
Killbola (slang): it has no cure
Tawonah (Arabic): destruction
Boda Wuteh (Mende): a monster that kills the whole family
Sweh (Krio): fever that kills the whole family

Fitina, bola neh, and *bola-bola* point to the trauma EVD produces, in part because of the isolation and stigmatization of those affected. *Killbola, tawonah, boda wuteh,* and *sweh* highlight EVD's swift destructiveness. *Boda wuteh* and *sweh* also indicate an understanding of EVD as a disease that impacts whole families, not merely individual bodies. This supports findings from Liberia, where people used terms such as *Ju'pa* (Kpelle[4]), which means "to kill the whole family," and the darkly witty term "family visa" to the other side (Modarres et al. 2015). These conceptions of Ebola stand in stark contrast to the international community's public health approach, such as the 2014 Centers for Disease Control and Prevention (CDC) interim guidance, which centers on individuals exposed to Ebola infection without discussing family or collective actions (Calain and Poncin 2015).

After defining the terms, we asked community members to talk about EVD's etiology. We asked participants not just *how* Ebola spreads and causes sickness, but to discuss why some people became infected with EVD and others did not, and why some people died while others lived. Beyond biomedical etiology, these discussions offered existential explanations for the causes of misfortune (Bannister-Tyrrell et al. 2015):

1. God had "marked" or judged that person/family.
2. The affected people share blood, meaning a bloodline or blood group. Respondents explained that those who share the same blood were more likely to transmit infection to each other. Among respondents with biomedical education, this referred to ABO blood groups, but in most cases, "sharing blood" simply means two people are related.
3. The affected people are "dirty": careless about the cleanliness of themselves and their homes. Beyond the literal meanings of dirt, this evokes negative moral evaluations of individuals or groups judged to be dirty.

Similar to the local terms for Ebola, these discussions evoked fear and judgment of those affected by EVD (see also Berghs et al. 2014) and implicated families rather than merely individuals. As Richards

[3]Temne is a language of the Mel branch of the Niger–Congo language family that is spoken in Sierra Leone by approximately two million first-language speakers; Limba is a language spoken in Sierra Leone by the third largest ethnic group in the country and which forms its own unique branch of the Niger–Congo language family; Kurnko is a Mande language spoken by approximately 350,000 people in Sierra Leone and Guinea; Krio is the most widely-spoken language in Sierra Leone and was derived from liberated Creole slaves returned to Africa from North America.

[4]Kpelle is a Mande language of the Niger-Congo language family that is spoken by approximately one-half million people in Sierra Leone and three-quarters of a million people in Liberia.

et al. (2015) noted, a sick person will first turn to her family for help, as the most reliable source of assistance. This highlights the "need to understand Ebola risks from the perspective of family, and its notions of social obligations" (ibid.: 7).

Local understandings of EVD's etiology also involved the gendered division of labor in communities affected by the Ebola outbreak. During the epidemic, most communities held men responsible for security, and women responsible for family health—creating a division of public and domestic roles along gender lines. Affected communities expected women to maintain the personal health and environmental hygiene of their families, particularly children. Women, therefore, could not realistically follow EVD prevention advice to "avoid body contact," as their caregiving roles led to regular contact with the bodily fluids of children and other dependents. In our focus groups, women who had survived Ebola infection repeatedly described having fallen ill due to expectations of women's "sympathy" in attending to the sick and the dead—not due to denial, resistance, or hostility to EVD prevention activities. "This issue of denial is relating to emotional affection," said a female CHV in Western Rural District. "We believe in collective responsibility. When you are sick I will visit you, see you, and talk to you; you will say I love you. If I don't visit you, you will see me as an enemy." A woman who refused to care for the sick or the dead might be accused of such extreme moral failing that she could face allegations of witchcraft (Hunleth 2011: 210).

Counterintuitively, women's caregiving roles in Sierra Leone sometimes kept them from seeking health care. In Koinadugu District, "leaving the family home, even to give birth in a hospital, is viewed as a sign of failure to fulfil domestic responsibilities. Men do not carry out these chores in the absence of their wives and subsequently pressure them to remain at home," rather than seek medical assistance (Saez 2013: 10). Some men subjected their wives to violence and abuse based on these perceived failures (ibid). In the midst of the Ebola virus outbreak, widespread distrust of Sierra Leone's weak health care system—in which local health clinics already struggled to provide basic medicines or essential care—meant the burden of caregiving fell even more heavily on the shoulders of women. Women's unpaid care work underpins health care systems even in the best of times; in periods of crisis, women are expected to "absorb the burden of care through self-exploitation," leading to both direct and indirect damage to their health (Harman 2016: 525).

Amid failing health care systems, women also absorb the burden of stigma and blame when a family member falls sick or dies. Female Ebola survivors, in particular, encountered accusations that they had caused their own and others' suffering because they were "wicked" and "careless" in the face of illness. As a consequence, these survivors suffered social isolation, economic marginalization, accusations of witchcraft, expulsion from their communities, and violence. While male Ebola survivors reported stigma based on specific fears that they might transmit EVD via intercourse, female survivors more often reported a broad range of negative reactions, including eviction, family rejection, the loss of former community leadership roles, and violence.

Women in Ebola-affected communities, therefore, had to manage not only health and illness, but also relationships and identities—all within a context of structural violence and poverty, in which family and kinship networks remained crucial to survival (Richards et al. 2015). The Ebola outbreak disrupted these critical family and kinship networks. According to young men in Bombali District, for example, restrictions on movement severed ties with family members in Freetown, on whom relatives in the village depended for access to markets, cash, and commodities.

These disruptions left women and girls increasingly vulnerable, a fact underscored by an apparent 65% rise in rates of teen pregnancy during the outbreak (Rissa-Gill and Finnegan 2015; UNDP and Irish Aid 2015). Public commentary on this trend sought to place the blame on girls: with schools closed, some commentators argued, girls engaged in unprotected sex out of boredom and lack of moral bearing (Margai 2014). "In Koinadugu," joked a young male respondent, "the nights are long." Women's rights organizations, on the other hand, placed the blame on men, arguing that the disruption of families and communities during the Ebola epidemic had led to a rise in sexual violence against teen girls (Ministry of Social Welfare, Gender and Children's Affairs et al. 2014). Women and girls themselves pointed to a third factor: as the Ebola outbreak derailed women's livelihoods, destroyed families, and disrupted girls' hopes for education, some turned to romantic relationships with men to provide financial assistance. In Koinadugu District, for example, young women reported that a local bylaw stated that if a man impregnated a schoolgirl, he assumed responsibility for paying her school fees after she gave birth. Bearing children thus created new kinship ties that might offer a small measure of support in unstable circumstances.

The price of caregiving, stigma, and vulnerability prevented many women from assuming public roles, including leadership roles in the Ebola response. Female community health workers, for example, encountered accusations that they had infected community members with EVD. In one striking instance, neighbors threatened and threw stones at a Bombali nurse, who had cared for an elderly midwife as she died of EVD. The elderly woman had unwittingly become infected with Ebola after helping an EVD-positive woman in a neighboring village give birth. The pregnant woman, her child, and the midwife all subsequently died. Relatives of the elderly woman blamed the nurse for the dozens of additional deaths that followed, because she had called the burial team to handle the midwife's body. The midwife's family believed that the burial team and the nurse had brought the Ebola virus into their community. This narrative highlights a chain of vulnerability, infection, death, and blame that followed these women simply for performing their expected caregiving roles.

The limitations imposed by stigma and blame compounded preexisting restrictions on women's public participation in some areas. In Bombali, female community health workers told me that men generally expect them to stay at home. They sit behind the men in community meetings, and the men may "chase away" their female relatives (see also Carter et al. 2017). Women's groups in Bombali complained of exclusion from community meetings and decision-making. A Koinadugu District councilwoman said that, traditionally, women could not act as council members of chiefs, because they were not allowed to speak in front of men. "We are not supported in other [community] decisions—only if we have a program on 'women's issues.' In the traditional court women are not encouraged at all to talk. You will only see the woman who has her case. By law we are supposed to have elderly women who are supposed to represent us as women."

In areas where women faced strict limitations on community decision-making and public participation, all-male teams of surveillance officers, contact tracers, social mobilizers, and burial workers—often recruited from outside the communities in which they worked—assumed responsibility for the Ebola response. In Koinadugu, an all-male surveillance team theorized that women did not participate in the response because they are "shy" and find the work "too hard." In contrast, young women in Koinadugu pointed to restrictions on women's movement that made it difficult to join social mobilization activities: a woman seen moving from village to village, sometimes on motorbikes with men after dark, might risk accusations of bad behavior or promiscuity.

At the same time, all-male Ebola response teams could not adequately address the social and behavioral causes of ongoing outbreaks. If anthropologists had communicated one lesson from the Ebola epidemic thus far, it was that community members rarely trusted outsiders with sensitive personal information (Minor 2014; Richards et al. 2015). In order to be effective, Ebola response activities had to engage directly with women who could act as envoys in their communities, raising awareness among other women who represent the first line of response to illness in families (Minor 2014; Niang 2015). In the absence of women's direct involvement, men did not necessarily pass along critical information on Ebola prevention and treatment, thereby reinforcing cycles of infection, stigma, and blame (Minor 2014).

28.3 An Empty Bag Cannot Stand

As anthropologists, however, we do not just limit our observations to "the community" or "the other." Anthropology requires reflexivity: an awareness of how one's own cultural background and social position affect the research process and the analysis of data. Extending that reflexivity to organizations, anthropologists who study humanitarian aid do not merely study beneficiary communities, but extend their inquiries into the relationship *between* humanitarian actors and the communities in which they work. Ideally, a humanitarian organization will work to understand not only the perceptions and beliefs of beneficiaries, but will also interrogate the assumptions, beliefs, and categories that inform their own ideas about what kind of assistance communities need, and how it should be delivered.

In the context of the Ebola epidemic, stigma and blame did not operate only in community-level social interactions. In some ways, organizations involved in the Ebola response inadvertently placed blame on Ebola-affected communities as well, by relying on ethnocentric perceptions of Sierra Leonean cultures to account for the perceived lack of women's leadership or involvement in the response. The division of labor into men's public roles (such as community surveillance, task forces, and checkpoints) and women's primarily domestic roles (such as domestic surveillance and caretaking) cannot be viewed simply as expressions of patriarchal and unequal gender relations. They are also a way of organizing micro-social efforts towards disease containment in the absence of adequate health, infrastructural, and material support systems. As anthropologists in Liberia noted, community members relied on these efforts as "necessary but less desirable than a well-supported health systems-based response [...] involving considerable individual, social, and public health costs, including heightened vulnerability to infection" (Abramowitz et al. 2015). These efforts reflect a response to neglect and abandonment (ibid.).

Abramowitz et al.'s research in Liberia resonated with what I observed in Sierra Leone. While community members readily answered my questions about gender roles in disease containment, every single interview and focus groups emphasized the more pertinent lack of health clinics and schools in their areas. They called attention to the desperate need for water, transportation, and communication networks and pointed out the lack of basic first aid and hygiene supplies—including disinfectant, protective clothing, buckets, gloves, oral rehydration salts, and even soap (see also Richards et al. 2015: 13). The implicit message seemed clear: how can community members—and women in particular—have meaningful choices for participation in epidemic response in the absence of functioning *systems*—of communication, water, roads, health care, or schools?

When humanitarian organizations only look to household- or community-level explanations for women's limited participation, they risk engaging in what development scholars term "gender-scapegoating," in which "the suffering endured by women in the world's poorest communities is blamed on local patriarchal values rather than fully acknowledging the extent to which women's vulnerabilities (like those of men) are driven by broader global structures of impoverishment and inequality" (Diggins and Mills 2015: 2). Several anthropologists remarked that conversations blaming "non-modern" gender relations as a root cause of EVD deaths in Sierra Leone were "difficult to correlate with our experience of communities in which members of both sexes have held positions of religious and political authority, and have been active participants in public economic life, for as long as anyone can remember" (ibid.).

In my observations, the degree to which women participated in local power structures varied among communities, depending on the input of individual chiefs and women leaders, such as Mami Queens.[5] A female chief in Western Area District told me that

[5]The Mami Queen (or Mammy Queen) is a woman assigned to handle all of the activities of the women, but not the men, within their district.

> "when the outbreak started, [women leaders] held a meeting in our community and we suggested the idea of not washing bodies, frequent hand-washing, and we should call 117 to come and collect the deceased—and the men supported us in it. We came up with the idea before the organisations came to our aid."

In Koinadugu District, a councilwoman remarked that, due to prior Oxfam Sierra Leone projects supporting women's political participation, women's leadership had significantly increased in the past 10 years.

During the Ebola outbreak, however, gender- and culture-scapegoating led some humanitarian organizations to stigmatize and overlook "traditional" social structures that could have been mobilized to help stem the epidemic. In the early days of the outbreak, Sierra Leone's government banned local herbalists and healers from practicing traditional medicine and blamed traditional healers for driving the spread of infections. As a result, humanitarian organizations marginalized traditional healers for much of the outbreak. Within Oxfam, the pace of the epidemic hadn't allowed staff the time or bandwidth to deepen their inquiry into local health-seeking behaviors. As a result, we had initially overlooked networks of informal health care providers, including home-based nurses and pharmacists, midwives, traditional healers, and petty traders who supplied basic medicines to villages. These informal health care providers constituted a critical part of Sierra Leone's health systems and represented an important gap in early messaging, ACS, and referral strategies. Humanitarian organizations belatedly enlisted the help of local health care providers and traditional healers in efforts to report infections and provide families with EVD prevention materials and information. This created a significant missed opportunity in engaging the participation of "traditional" social structures in epidemic control.

Similarly, humanitarian agencies tended to regard Sierra Leone's gendered initiation societies as barriers to overcome, rather than potential partners in epidemic response. The government of Sierra Leone banned Bondo[6] societies from carrying out initiation ceremonies that included circumcision for the duration of the Ebola epidemic, due to concerns over infection (Schwartz 2019). Humanitarian organizations likewise viewed "secret societies" through the narrow lens of female circumcision—a view that portrays Bondo societies as barbaric and feeds ethnocentric tropes of African cultural practices as primitive or backwards. This view obscures the critical roles Bondo societies play in providing a forum for women to wield power as part of a broader collective. "Secret societies" allow women ways to control their time and movement, as men generally cannot question women's participation in Bondo activities (Saez 2013: 45). Bondo societies also provide a critical vehicle for women to relay information and organize responses to illness or death. As with traditional healers, ethnocentric biases initially prevented humanitarian actors from mobilizing community-based structures that could have provided support for epidemic response and recovery.

Moreover, blaming vulnerability on "local culture" harms the vulnerable, when it leads to "empowerment" approaches that ignore the very real consequences of material deprivation. For example, a basic but key barrier to both women's and men's participation in Ebola response activities was lack of payment. During the transition from emergency response to recovery, the government and humanitarian organizations reduced or ended the cash incentives they had been providing to community health volunteers, and began to rely on the donated labor of already impoverished men and women to carry out their priorities (Maes 2014). For women in particular, the expectation that they contribute unpaid labor to Ebola recovery activities compounded the double burden of domestic and farm labor that many women already shouldered.

[6]The Bondo society (also known as Sande and Bundu) is an all-female secret society in West African countries that initiates girls into adulthood through a number of rituals that include female genital cutting. Following their initiation, usually in a secluded area, girls are given Bondo society names in place of their birth names, pledged to secrecy, instructed in domestic skills, ritually washed and returned to their communities as marriageable adults.

Policy and program decision-makers seemed to share an unspoken assumption that women would continue to underpin the functioning of Sierra Leone's health system through their unpaid and supposedly elastic work (Harman 2016: 536). Meanwhile, men in Sierra Leone faced ongoing pressures to provide their families with cash and commodities within a radically disrupted economy and conditions of mass unemployment. At this point, volunteering for an international humanitarian organization may have felt more exploitative than empowering; non-participation may have reflected both men's and women's attempts to triage labor in a context of widespread poverty and volunteer burnout. As a female volunteer in Western District pointed out, "An empty bag cannot stand."

28.4 The Logic of War

Finally, the national-level Ebola response also participated in placing stigma and blame on affected families and communities. The government of Sierra Leone took a military approach to tackling the virus, emphasizing the more "masculine" aspects of disease containment—such as heavy-handed bylaws, quarantines, bans, and surveillance—over more "feminine" approaches, such as psychosocial support and social mobilization (Niang 2015). The National Ebola Response Centre (NERC) and District Ebola Response Centres (DERCs) tended to not only overlook the differential impact of EVD on men and women, but also reproduced gender hierarchies through the heavily masculinized and militarized spaces of decision-making and control (Harman 2016: 535).

Stigmatization and blame became ingrained in the tone of the Ebola response, through what anthropologist Cheikh Niang termed the "logic of war" (Niang 2015). National bylaws levied fines and imposed prison terms on families accused of hosting strangers, hiding sick relatives, or carrying out secret burials. This strategy implicitly characterized those who fell sick as ignorant, transgressive, and untrustworthy—deflecting responsibility for the belated and messy response from the government and international institutions and placing blame on affected families and communities (see also Calain and Poncin 2015: 127). International media outlets blamed "traditional rituals," "hostile" communities, and "confused beliefs" for the ongoing epidemic, rather than the ravaged health systems, lack of communication, and counter-productive quarantine policies that Sierra Leoneans observed in their communities (New York Times 2015). These approaches seemed to justify the degrading treatment of affected families and communities through tacit claims that they brought the virus on themselves.

Yet a different view emerged when anthropologists asked *why* families and communities might seem "resistant" to Ebola treatment and prevention advice. Conversations with community members suggested that the militarized and stigmatizing tone of the government's Ebola response led to violations of the basic rights and dignity of affected families and communities. For example, community members I interviewed raised critical questions about the confidentiality of personal information taken by surveillance teams. Some families described "ruthless" behavior by ambulance teams who treated the sick "like criminals" (Ministry of Social Welfare, Gender and Children's Affairs et al. 2014: 17). Others recounted horrifying stories of family members who were taken away and never heard from again, their medical records lost in the chaos, their bodies cremated and disposed of without the family's knowledge. Instead of "safe and dignified" burials, some grieving families witnessed burial teams—who feared infection themselves—handling bodies "like garbage" in burials that seemed like "little more than sanitary disposal" of a loved one (Richards et al. 2015; see also James et al. 2015). Security forces placed families under months-long quarantines without first ensuring access to food, latrines, or water. Nor could quarantined families carry out the economic activities necessary to survival: farmers' crops died; laborers could not work; traders could not access markets. Even after the quarantines lifted, some market women found that neighbors would no longer buy from them.

Faced with the prospect of suffering and dying alone, while their families endured quarantine, social exclusion, and economic devastation, some reacted with understandable disbelief and opposition (Calain and Poncin 2015: 127). The fear of stigma and blame thus prompted counter-productive behaviors: in some cases, individuals began to avoid quarantine by concealing sickness, fleeing their homes when they became symptomatic, or creating false documents stating that they had tested negative for Ebola at an ETU.

Affected families also interrogated the issue of hosting "strangers." Sierra Leonean households maintain a level of fluidity, with relatives periodically moving among households and villages. During the epidemic, however, national bylaws defined siblings, cousins, or parents who stayed in another village for a period of time as "strangers" and banned them from visiting relatives. Yet other strangers—including contact tracers, surveillance teams, and aid workers—seemed to have unrestricted access, without providing community members any check on whether they had come into contact with EVD. In other words, total strangers involved in the Ebola response moved freely through affected communities, while literally pathologizing the everyday movements of families in those same communities (Benton 2014).

Women who had married outside of their home villages, based on patrilocal marriage traditions, faced particular vulnerabilities due to restrictions on their mobility. Women who marry into another village may also be defined as a "stranger" (Richards et al. 2015: 7). In cases of domestic violence or conflict with the community, women I interviewed said that they would seek shelter in their home villages, until relatives and elders helped to resolve the problem. With entire villages under quarantine, and movement between districts highly restricted, many women could not access their usual sources of support.

Other studies documented additional vulnerabilities for women produced by the militarized EVD response: rising rates of poverty and food insecurity due to the loss of farms and trade; pregnant women forced to give birth under quarantine without medical support; and an increase in gender-based violence, exploitation, and rape; including quarantine guards coercing women and girls to exchange sex for access to aid (Ministry of Social Welfare, Gender and Children's Affairs et al. 2014; UNDP and Irish Aid 2015; Chavez 2015). In a number of ways, the military response to the EVD epidemic reinforced the pressures of women's domestic caregiving roles and contributed to cycles of infection, illness, and blame.

Few studies investigated the ways in which the EVD response affected men. The majority of "gender" assessments focused exclusively on women and girls, based on implicit assumptions that gender only produces vulnerabilities for these groups. Yet men in Sierra Leone also experienced particular types of vulnerability due to gender. The cumulative national data in Sierra Leone, for example, showed that infection rates among women were only slightly higher than those of men. Subsequent data also showed that, once infected, men faced a higher mortality rate than women. Some humanitarian staff took the relative parity of male and female infection rates to mean that gender was not a relevant factor in EVD morbidity and mortality. However, qualitative interviews with survivors, affected families, and health care providers revealed that the gendered division of labor in Sierra Leone put men and women at risk of Ebola infection in different ways. Women faced high risks as household caregivers, whereas men faced risks of infection due to their roles in transporting the sick. In several cases, the exposure of male motorcycle taxi drivers resulted in new chains of Ebola transmission in previously unexposed villages. An analysis that only considered women's vulnerability, or one that did not consider gender at all, risked ignoring these critical causal pathways, resulting in a response incapable of preventing further transmission of the virus. Lessons learned from the Ebola outbreak should challenge the burden of proof to demonstrate that women are disproportionately diseased or dying before concerns about gender can be heard (see Carter et al. 2017).

In interviews, men also emphasized the impact of loss, lack of freedom, isolation, demoralization, and profound grief. In Farekoro, a Koinadugu District village devastated by Ebola, a young man told me that "*everyone* is traumatized because their relatives, friends, and loved ones died in the cold arms of Ebola." In Bombali, a group of young men discussed the anger and helplessness they felt as they watched the epidemic progress. They could see the ways in which "Ebola money" and power flowed among international institutions and governments—yet none of it reached them in time to save their loved ones. The men acknowledged that feelings of fear, betrayal, and frustration drove the sporadic violence reported in Sierra Leone: for example, a group of boys in their area threw stones at a burial team for handling a body in a manner they found careless. Police arrested the boys' parents in order to force cooperation; when the boys came forward, a court sentenced them to 18 months in prison.

Anger and bouts of violence fed discourses blaming the ongoing outbreak on "hostile" or "resistant" communities. However, young men in my interviews pointed out that "resistance" was in reaction to the NERC's "logic of war," an approach that affected communities often experienced as an attack on their people, rather than the virus. As one young man reported, "They have come with guns to threaten us, and when you are diagnosed to have Ebola, they arrest you. That alone makes you to be depressed, and not for the disease but of the forces surrounding the patient. The entire family is looked at negatively" (James et al. 2015: 2). This stigmatizing approach compounded pressures on both men and women in ways that ultimately provoked responses that reinforced the epidemic.

However, the international community also cannot fully blame the NERC's response, without accounting for the ways in which global structures of inequality, exploitation, and impoverishment fed the Ebola outbreak. As Paul Farmer observed, "The three countries most afflicted by Ebola are those with some of the lowest public investment in healthcare and public health in Africa. They have been wracked by war, and by extractive industries, which have never failed to turn a profit" (2014: 38). Even before Ebola emerged, international financial policies that compelled disinvestment in public health and social services had severely eroded Sierra Leone's infrastructure and ability to meet basic needs. The acute lack of "staff, stuff, space and systems" that Farmer observed seriously hindered efforts to contain the Ebola outbreak. Critiques that place blame on Sierra Leone or stigmatize its people should consider this broader context.

References

Abramowitz, S. A., McKune, S. L., Fallah, M., Monger, J., Tehoungue, K., & Omidian, P. A. (2014). The opposite of denial: Social learning at the onset of the Ebola emergency in Liberia. *Ebola Response Anthropology Platform*. Retrieved April 10, 2018, from http://www.ebola-anthropology.net/case_studies/the-opposite-of-denial-social-learning-at-the-onset-of-the-ebola-emergency-in-liberia/.

Abramowitz, S. A., McLean, K. A., McKune, S. L., Bardosh, K. L., Fallah, M., Monger, J., et al. (2015). Community-centered responses to Ebola in urban Liberia: The view from below. *PLoS Neglected Tropical Diseases, 9*(5), e0003706. https://doi.org/10.1371/journal.pntd.0003706.

Adams, J., Lloyd, J., & Miller, C. (2015). *The Oxfam Ebola response in Liberia and Sierra Leone: An evaluation report for the disasters emergency committee*. Oxford: Oxfam GB. Retrieved April 9, 2018, from https://policy-practice.oxfam.org.uk/publications/the-oxfam-ebola-response-in-liberia-and-sierra-leone-an-evaluation-report-for-t-560602.

Bannister-Tyrrell, M., Gryseels, C., Delamou, A., D'Alessandro, U., van Griensven, J., Grietens, K. P., et al. (2015). Blood as medicine: Social meanings of blood and the success of Ebola trials. *The Lancet, 385*(9966), 420. https://doi.org/10.1016/S0140-6736(14)62392-8. Retrieved April 9, 2018, from http://www.thelancet.com/pdfs/journals/lancet/PIIS0140-6736(14)62392-8.pdf.

Benton, A. (2014). On Ebola and the pathological movements of others. *Ethnographic Emergency*. Retrieved April 8, 2018, from http://ethnography911.org/2014/07/28/on-ebola-and-the-pathological-movements-of-others/.

Berghs, M., Chandler, C., Fairhead, J., Ferme, M., Huff, A., Kelly, A., et al. (2014). Stigma and Ebola: An anthropological approach to understanding and addressing stigma operationally in the Ebola response. *Research for Health in Humanitarian Crises (R2HC) briefing note, Ebola Response Anthropology Platform*. Retrieved April 7, 2018, from http://www.ebola-anthropology.net/wp-content/uploads/2014/12/Stigma-and-Ebola-policy-brief-Ebola-Anthropology-Response-Platform.pdf.

Calain, P., & Poncin, M. (2015). Reaching out to Ebola victims: Coercion, persuasion or an appeal for self-sacrifice? *Social Science & Medicine, 147*, 126–133.

Carter, S. E., Dietrich, L. M., & Minor, O. M. (2017). Mainstreaming gender in WASH: Lessons learned from Oxfam's experience of Ebola. *Gender and Development, 25*, 205–220. Retrieved April 9, 2018, from https://www.tandfonline.com/doi/full/10.1080/13552074.2017.1339473.

Chavez, D. (2015). The socioeconomic impacts of Ebola in Sierra Leone. *The World Bank*. Retrieved November 9, 2018, from http://www.worldbank.org/en/topic/poverty/publication/socio-economic-impacts-ebola-sierra-leone.

Diggins, J., & Mills, E. (2015), The pathology of inequality: Gender and Ebola in West Africa. *Institute of Development Studies*. Retrieved April 8, 2018, from https://www.ids.ac.uk/publication/the-pathology-of-inequality-gender-and-ebola-in-west-africa.

Farmer, P. (2014). Diary. *London Review of Books, 36*(20), 38–39. Retrieved November 9, 2018, from http://www.lrb.co.uk/v36/n20/paul-farmer/diary.

Harman, S. (2016). Ebola, gender and conspicuously invisible women in global health governance. *Third World Quarterly, 37*(3), 524–541.

Hunleth, J. (2011). Being closer: Children and caregiving in the time of TB and HIV in Lusaka, Zambia (PhD dissertation, Northwestern University, Evanston, Illinois).

James, S. B., Mokuwa, E., & Richards, P. (2015). *Interviews on Ebola response, Bo, 15–17 December 2014, Njala University SMAC program and DFID Freetown*. Unpublished manuscript.

Maes, K. (2014). Volunteers are not paid because they are priceless: Community health worker capacities and values in an AIDS treatment intervention in urban Ethiopia. *Medical Anthropology Quarterly, 29*(1), 97–115.

Margai, M. (2014). Ebola, teenage pregnancy and education. *Sierra Leone Matters*.

Ministry of Social Welfare, Gender and Children's Affairs, UN Women Sierra Leone, Oxfam Sierra Leone, & Statistics Sierra Leone. (2014). *Report of the multisector impact assessment of gender dimensions of Ebola virus disease (EVD) in Sierra Leone*. Unpublished manuscript.

Minor, O. M. (2014). *Community perceptions of Ebola response efforts in Liberia: Montserrado and Nimba Counties, Oxfam Great Britain*. Retrieved April 8, 2018, from https://www.academia.edu/10354601/Community_perceptions_of_Ebola_response_efforts_in_Liberia_Montserrado_and_Nimba_Counties.

Modarres, N., Babalola, S., Figueroa, M. E., Wohlgemuth, L., Berman, A., Tsang, S., et al. (2015). *Community perspectives about Ebola in Bong, Lofa, and Montserrado Counties of Liberia: Results of a qualitative study. Health Communication Capacity Collaborative, Johns Hopkins Center for Communication Programs, Resource Center for Community Empowerment and Integrated Development*. Unpublished manuscript.

New York Times. (2015). Getting Ebola to zero. *New York Times*. Retrieved April 8, 2018, from http://www.nytimes.com/2015/03/09/opinion/getting-ebola-to-zero.html?_r=1.

Niang, C. I. (2015). *Anthropological study of the resistances to Ebola response interventions in Guinea. WHO Mission*. Unpublished manuscript.

Olu, O. O., Lamunu, M., Chimbaru, A., Adegboyega, A., Conteh, I., Nsenga, N., et al. (2016). Incident management systems are essential for effective coordination of large disease outbreaks: Perspectives from the coordination of the Ebola outbreak response in Sierra Leone. *Frontiers in Public Health, 4*, 254. Retrieved April 8, 2018, from https://www.ncbi.nlm.nih.gov/pmc/articles/PMC5117105/.

Richards, P., Amara, J., Ferme, M. C., Kamara, P., Mokuwa, E., Sheriff, A. I., et al. (2015). Social pathways for Ebola virus disease in rural Sierra Leone, and some implications for containment. *PLoS Neglected Tropical Diseases, 9*(4), e0003567. https://doi.org/10.1371/journal.pntd.0003567. Retrieved April 10, 2018, from https://www.ncbi.nlm.nih.gov/pmc/articles/PMC4401769/.

Rissa-Gill, I., & Finnegan, L. (2015). Children's Ebola recovery assessment: Sierra Leone. *Save the Children, World Vision International, Plan International and UNICEF Report*. Retrieved April 10, 2018, from https://www.savethechildren.org/content/dam/global/reports/emergency-humanitarian-response/ebola-rec-sierraleone.pdf.

Saez, A. M. (2013). *Accessibility strategy for the health care system of the District of Koinadugu, Sierra Leone. Medicos del Mundo*. Unpublished manuscript.

Schwartz, D. A. (2019). The Ebola epidemic halted female genital cutting in Sierra Leone – Temporarily. In D. A. Schwartz, J. N. Anoko, & S. A. Abramowitz (Eds.), *Pregnant in the time of Ebola: Women and their children in the 2013-2015 West African epidemic*. New York: Springer.

United Nations Development Program (UNDP), & Irish Aid. (2015). *Assessing sexual and gender based violence during the Ebola crisis in Sierra Leone*. Retrieved April 10, 2018, from https://docs.google.com/viewer?url=http%3A%2F%2Fwww.sl.undp.org%2Fcontent%2Fdam%2Fsierraleone%2Fdocs%2FEbola%2520Docs.%2Fundp.sle.ebola.SGBV.doc.

29 Ebola in Rural Sierra Leone: Its Effect on the Childhood Malnutrition Programme in Tonkolili District

Mohamed Hajidu Kamara

29.1 Introduction

Sierra Leone, a low-income country in West Africa with a population of 7.1 million (Statistics Sierra Leone 2016), has some of the most severe child malnutrition and mortality indicators in the world (Statistics Sierra Leone 2014). It is divided into four regions with 16 districts, of which Tonkolili is the most centrally located and is in the Northern region (Fig. 29.1). Greater than one-half of the population of Sierra Leone – over 3.5 million persons – are food insecure, and lack access to a sufficient quantity of safe and nutritious food. Of these, greater than 600,000 are considered to be severely food insecure, a number that has increased by sixty percent since 2010 (World Food Programme 2016).

The prevalence of severe acute malnutrition (SAM) and moderate acute malnutrition (MAM) in Sierra Leone is 4.0% and 9.3%, respectively, using the weight-for-height (WFH) ratio and 5.6% and 16.4% using the weight-for-age (WFA) ratio (Government of Sierra Leone 2012). The childhood mortality rate in Sierra Leone is very high at 156/1000 live births (Statistics Sierra Leone 2014). Before the Ebola virus outbreak began, the Ministry of Health and Sanitation (MoHS) had prioritized addressing childhood malnutrition (Government of Sierra Leone 2012). In 2011, it set a target to reduce SAM prevalence to 0.2% by the year 2016 (Government of Sierra Leone 2012). The strategies that were planned to achieve this goal included facility-based screening and treatment of all children who presented to health facilities, as well as community-based mobilization and screening with referral for facility-based treatment as required (Government of Sierra Leone 2012).

In Tonkolili District with a population of 530,776 in 2015 (Statistics Sierra Leone 2016), health care is provided by 107 health facilities consisting of 104 peripheral health units (PHUs), one private and two government hospitals, and deployment of 964 community health workers (CHWs) (Tonkolili District Health Management Team 2013). The prevalences of SAM and MAM in the district in 2011, prior to the Ebola epidemic, were respectively 1.9% and 5.5% using the weight-for-height (WFH) ratio, and 3.5% and 14.8% using WFA (Government of Sierra Leone 2012).

Beginning in late 2013, West Africa experienced the largest outbreak of Ebola virus disease (EVD) in history (Centers for Diseases Control and Prevention 2016). In Sierra Leone, at least 13,992 people became infected with EVD, of whom 3955 died (Ministry of Health and Sanitation 2016). The outbreak strained the country's already fragile health system, with consequences for health care delivery

M. H. Kamara (✉)
Ministry of Health, Freetown, Sierra Leone

D. A. Schwartz et al. (eds.), *Pregnant in the Time of Ebola*, Global Maternal and Child Health,
https://doi.org/10.1007/978-3-319-97637-2_29

Fig. 29.1 Map of Sierra Leone showing the Districts. The District of Tonkolili is in the central region of the country. Available from Wikipedia at: https://en.wikipedia.org/wiki/File:Sierra_Leone_Districts.png

reported across the health sector (Elston et al. 2015). The Ebola outbreak started in Tonkolili on the 26th of June 2014, and by 1st November 2016, when the national outbreak was declared to be over, there had been 456 documented cases (Ministry of Health and Sanitation 2016). By March 2015, EVD cases in Tonkolili had declined to zero, with a few sporadic cases reported in August 2015 and again in January 2016 (Tonkolili District Health Management Team 2016).

The Ebola virus epidemic may have had an impact on the malnutrition program in Sierra Leone, as has been reported for other services in EVD-affected countries (Elston et al. 2015, 2017; Government of Sierra Leone 2012; Lori et al. 2015). During the outbreak in Sierra Leone, the consequences to public health were dramatic—there were 221 health care workers who lost their lives (Ministry of Health and Sanitation 2016), funds for other programs were diverted to the EVD response, and some facilities underwent closure or reduced their services (Elston et al. 2015). There were reports of community members and health care workers being afraid to interact due to perceived risk of infection, and both healthy and ill pregnant women were frightened to seek medical care (Dynes et al. 2015; Strong and Schwartz 2016, 2019).

Because there were no data published describing the effects of the EVD epidemic on the malnutrition program, we sought to examine the association between the periods before, during, and after the outbreak of service delivery within the Tonkolili District malnutrition management program. Our objectives are to estimate (1) the number of children screened for malnutrition; (2) the number and proportion of children diagnosed with SAM and MAM among those screened; and (3) the number of children diagnosed with SAM who completed inhospital and outpatient treatment before, during, and after the outbreak.

29.2 Malnutrition Programme

The nutrition program of the MoHS includes facility- and community-based screening. In the facility-based program, all children aged 6–59 months coming to the facility for any reason should be screened for malnutrition. Fixed facility-based screening occurs during visits to a PHU or hospital, and 'outreach' screening occurs during visits to mobile clinics held in catchment areas of a given PHU. In both settings, children are screened using WFH (or length) ratio, weight-for-age (WFA) ratio, and mid-upper-arm circumference (MUAC).

Community screening includes social mobilization and sensitization by community workers on signs and symptoms of malnutrition, along with services available to prevent and treat malnutrition. Households and communities alert the CHWs of any child they perceive as malnourished. The CHW then conducts screening in the household using MUAC. If the criteria for diagnosing MAM or SAM are met, the child is referred to the nearest facility for further assessment and treatment.

MAM and SAM are defined as WFH or WFA ratios that are respectively <2 or <3 standard deviations (SD) below the mean using the World Health Organization (WHO) standard growth Chart (WHO 2009). Alternatively, a MUAC of <115 mm with or without medical complications also indicates SAM, while a MUAC of between 115 and 125 mm is indicative of MAM (WHO 2009). Children with SAM and medical complications are treated in an inpatient stabilization treatment program (STP), while those with SAM but without medical complications are treated in the outpatient treatment program (OTP), as per WHO guidelines (WHO 2009). The two government hospitals provide inpatient treatment for SAM, while 48 of the PHUs provide outpatient treatment for SAM without medical complications.

29.3 Tonkolili District, the People, and the Study Population

29.3.1 Tonkolili District

The Tonkolkili District lies in the North Central region of Sierra Leone. It comprises 11 chiefdoms, each with their own capital, and most of the inhabitants are Muslim. The predominant ethnic group in the Tonkolili District are the *Temne* people, who also make up the largest ethnic group in Sierra Leone at 35% of the population (slightly greater than the *Mende* people at 31%). The Tonkolilli District is significantly impacted by food insecurity largely due to widespread poverty which is exacerbated by low productivity, gender inequality, poor road and market accessibility, food price fluctuation and a lack of income generation diversification. Although most households are reliant on agriculture to obtain their livelihoods, only 4 percent grow sufficient rice to meet their annual needs, and rice production has been decreasing (World Food Programme 2016). Approximately 74% of households in the district are considered to be affected by moderate to severe food insecurity, and almost 5% of women have acute malnutrition (ReliefWeb 2015).

29.3.2 The Study Population

This study targeted all children <5 years of age residing in the Tonkolili district who were screened at either a health facility or at an outreach post during our period of study.

On a monthly basis, data from all treatment and screening facilities were gathered and sent to the Monitoring and Evaluation (M&E) Officer at district level, who collated and entered information in the District Health Information System (DHIS) software. These aggregate data were extracted from the DHIS and entered into epidata software where validation and analysis were done.

Variables captured were devoid of any traceable demographics and all ethical principles as to obtaining consent from District Health Management Team (DHMT) and ethical committee were followed.

29.4 Results

Data on malnutrition screening was completed for the 28 months of the study with the exception of those for SAM treatment admission and discharge. Six of eleven months in pre-EVD period, 8 of 11 months during EVD, and 1 of 6 months post-EVD were missing SAM admission data. Data on SAM treatment discharge from STP were missing for respectively 1, 1, and 2 months for pre-, during-, and post-outbreak periods. The findings of the study are described below.

29.4.1 Malnutrition Screening

The number of children screened per month varied between 8319 and 24,243. Most of these (78–96%) were screened in facilities, where the average monthly number of children screened decreased from 16,805 prior to the Ebola outbreak to 13,510 during the outbreak, and then subsequently returned to pre-outbreak levels. The decline in facility-based screening during the EVD epidemic was driven largely by the outreach sites, where average screening fell by 39% from pre-outbreak levels, and which had yet to recover at the time of the study. Facility-based screening at fixed sites increased after March, 2015.

29.4.2 Diagnosis of Severe and Moderate Acute Malnutrition

There were inconsistent trends and nonsignificant differences in the proportion of screened children with MAM or SAM during the outbreak compared with pre-outbreak levels. However, the proportion of screened children diagnosed with MAM by MUAC increased between the pre-outbreak and post-outbreak in the community screenings (3.6–8.2%) and in the health facilities (5.1–7.9%). MAM prevalence based on the other measures also increased between the pre- and post-outbreak periods. The prevalence of SAM using all measures was higher after the outbreak compared with the pre-outbreak period. For example, 3.5% of screened children were diagnosed with SAM using WFA after the outbreak when compared with the pre-outbreak periods, although the prevalence started increasing in March 2015.

29.4.3 Severe Acute Malnutrition Treatment Completion

Overall, one child completed SAM treatment for every four children diagnosed. The average number of children with SAM who were treated and discharged from STP per month declined from 67 pre-outbreak to 42 during the outbreak. There was a spike in STP discharges in March 2015. In comparison to the pre-outbreak period, STP discharges declined to an average of 8 per month post-outbreak. OTP discharges remain stable, at an average of 138, between pre-outbreak and the outbreak periods, but increased to 295 post-outbreak compared with the pre-outbreak level.

29.5 Discussions and Recommendations

We examined the effects of the Ebola outbreak on pediatric malnutrition screening, diagnosis, and management in Sierra Leone. We found that malnutrition screening declined during the outbreak, possibly because of the low turnout of patients at the health facilities, interruption in the access to care, and the fear of health care workers on EVD transmission (Elston et al. 2015; Brolin Ribacke et al. 2016). The diagnosis of SAM and MAM increased during the post-outbreak period. Across all the study periods, the ratio of SAM diagnosis to treatment remained well above one.

Decline in facility-based screening for malnutrition during the outbreak was consistent with patterns observed in services of maternal health, malaria, and hospital admissions across the three most affected countries (Bolkan et al. 2014; Brolin Ribacke et al. 2016; Elston et al. 2015, 2017; Lori et al. 2015). Such declines were partly due to community perceptions and fear of contacting EVD at health facilities (Dynes et al. 2015). As the number of EVD cases in the Tonkolili District fell from 12 to less than 5 cases per week by January 2015, and then ultimately to zero cases by March 2015, facility-based screening began to increase (Tonkolili District Health Management Team 2016). The decline in outreach screening by health facilities after the outbreak that occurred in the Tonkolili District was likely due to the system-wide reduction in outreach services, which had yet to return to pre-outbreak levels at the time of the study. However, the outbreak strategies employed by the MoHS for CHWs to provide caregivers with tape measures prevented a collapse in community screening, but did not make up for the shortfall (Centers for Disease Control and Prevention 2015). The timing and pattern of the decline, followed by the surge in facility-based (fixed) and community screening post-outbreak, suggests a major impact of the outbreak on screening and early health system recovery.

Discrepancies in diagnoses between the MUAC and the WFA/WFH ratios are common, with an estimated 40% agreement for SAM diagnoses between MUAC and WFH ratio (WHO 2009). Our findings are consistent with other studies that report a lower prevalence of MAM and SAM using the MUAC (Grellety and Golden 2016). MUAC increases with age, and the use of non-standardized measurements means that older children diagnosed with SAM using the WFH ratio may not be diagnosed with SAM using the MUAC (Myatt et al. 2009). Because those children diagnosed with MAM and SAM across the different measures may represent different subgroups of children, we restricted the comparison of MAM and SAM prevalence between the study periods to comparisons by measurement type.

The increase in prevalence of MAM and SAM that was diagnosed among those screened is concerning. The post-outbreak prevalence of SAM, measured using the WFA ratio, mirrors the 2010 estimates of 3.5% in Tonkolili from survey data and is well above the national goal of <0.2% (Government of Sierra Leone 2012). The increased prevalence of SAM among those children who were screened post-outbreak began after the number of EVD cases in Tonkolili District declined to zero cases per week, and screening had increased. One potential reason for this includes the possibility of underdiagnosis by caregivers who were provided with MUAC tape measurements during the outbreak, leading to delayed detection once CHWs recommenced taking MUAC measurements.

The outbreak may also have increased the proportion of children with acute malnutrition. During the outbreak, there was a decline in agricultural food production and growing food insecurity (Action Against Hunger 2015; Thomas et al. 2014). Greater than 12,000 children (>800 in Tonkolili District alone) (Street Child 2015) lost one or both parents or caregivers to EVD, and consequently, were highly vulnerable to increased poverty and malnutrition. Finally, the risk of malnutrition rose with the lack of early facility-based access to under-five services, including immunization, leading to an increased risk of preventable childhood illness, particularly measles (Suk et al. 2016).

Our findings suggest an unmet need in reducing the prevalence of malnutrition, with limited access to, and/or uptake of, appropriate management and therapeutic feeding. The low level of SAM treatment completion suggests a preexisting health system gap that remained stable during the outbreak, but worsened post-outbreak for STP and improved for OTP. The spike in STP and OTP treatments in March 2015 may reflect a surge in the reporting of treatments from previous months rather than a surge in treatments completed in March 2015, when the number of EVD cases fell to zero. We are unable, however, to verify whether this was the case; therefore, there is a need for further exploration of this issue. OTP may have increased post-outbreak, as it requires fewer resources, while STP requires facilities to be open, staffed, and stocked with supplies for inpatient treatment. OTP may also have increased in response to the larger number of SAM diagnoses, many of which may have been uncomplicated if they represented new diagnoses.

The strengths of this study include the use of population-level programmatic data. The conduct and reporting of the study adhered to the strengthening of the Reporting of Observational Studies in Epidemiology (STROBE) guidelines and sound ethics principles (Edington et al. 2012; von Elm et al. 2007), and the study was conducted to address a health system priority for the Sierra Leone MoHS. Study limitations include the use of secondary data without validation against the original paper records. The data were extracted as monthly district-wide totals from the DHIS database, and there was a large amount of missing data on treatment outcomes for SAM, reflecting the nature of report submission and data entry into the DHIS. Because individual patient-level outcomes were not available, and we used discharge from treatment as a proxy for successful treatment completion, our measure of SAM treatment outcome may not accurately reflect treatment uptake and effectiveness.

We believe that our study has important implications for Sierra Leone's health system. Prior to the Ebola outbreak, the MoHS had been committed to improving child health through the elimination of fees incurred by caregivers for health services provided to children under 5 years of age through the Free Health Care Initiative (FHIC). During the current recovery period, programs targeting food security, nutrition, and the loss of parents/guardians/caretakers must be considered by those implementers and donors looking to support pediatric services. Strengthening the program to close the gap between SAM diagnosis and treatment is needed to address this ongoing unmet need in child health.

In conclusion, the EVD outbreak in a rural district in Sierra Leone was associated with decreased malnutrition screening during the outbreak and an increased prevalence of acute malnutrition in the post-outbreak period. It is imperative that national policies on nutrition programs be strengthened and implemented to address the shortfalls created by the Ebola virus outbreak and to preempt such effects in future outbreaks.

References

Action Against Hunger, U. K. (2015). *Ebola in Liberia: Impact on food security and livelihoods*. London, UK: Action Against Hunger.

Bolkan, H. A., Bash-Taqi, D. A., Samai, M., Gerdin, M., & von Schreeb, J. (2014). Ebola and indirect effect on health services functions in Sierra Leone. *PLOS Currents Outbreaks, 6*. Retrieved November 7, 2017, from http://currents.plos.org/outbreaks/article/ebola-and-indirect-effects-on-health-service-function-in-sierra-leone/.

Brolin Ribacke, K. J., Saulnier, D. D., Eriksson, A., & von Schreeb, J. (2016). Effects of the West Africa Ebola virus disease on health-care utilization—A systematic review. *Frontiers in Public Health, 4*, 222. https://doi.org/10.3389/fpubh.2016.00222.

Centers for Disease Control and Prevention. (2015). *Recommendations for managing and preventing cases of malaria in areas with Ebola*. Atlanta, GA, USA.

Centers for Disease Control and Prevention. (2016). *2014–2016 Ebola outbreak in West Africa*. Retrieved November 7, 2017, from http://www.cdc.gov/vhf/ebola/outbreaks/2014-west-africa/index.html.

Dynes, M., Tamba, S., Vandi, M. A., & Tomczyk, B. (2015). Perceptions of the risk for Ebola and health facility use among health workers and pregnant and lactating women—Kenema district, Sierra Leone, September 2014. *Morbidity and Mortality Weekly Report, 63*(52), 1226–1227. Retrieved November 8, 2017, from https://www.cdc.gov/mmwr/preview/mmwrhtml/mm6351a3.htm.

Edington, M., Enarson, D., Zachariah, R., Reid, T., Satyanarayana, S., Bissell, K., et al. (2012). Why ethics is indispensable for good-quality operational research. *Public Health Action, 2*(1), 21–22. https://doi.org/10.5588/pha.12.0001.

Elston, J. W., Moosa, A. J., Moses, F., Walker, G., Dotta, N., Waldman, R. J., et al. (2015). Impact of the Ebola outbreak on health systems and population health in Sierra Leone. *Journal of Public Health, 38*(4), 673–678. Retrieved March 10, 2018, from https://academic.oup.com/jpubhealth/article/38/4/673/2966926.

Elston, J.W., Cartwright, C., Ndumbi, P., Wright, J. (2017). The health impact of the 2014-15 Ebola outbreak. *Public Health, 143*, 60–70.

Government of Sierra Leone. (2012). *Sierra Leone food and nutrition security policy implementation plan 2012–2016.* Freetown, Sierra Leone: Government of Sierra Leone.

Grellety, E., & Golden, M. H. (2016). Weight-for-height and mid-upper-arm circumference should be used independently to diagnose acute malnutrition: policy implications. *BMC Nutrition, 2*, 10. Retrieved March 10, 2018, from https://bmcnutr.biomedcentral.com/articles/10.1186/s40795-016-0049-7.

Lori, J. R., Rominski, S. D., Perosky, J. E., Munro, M. L., Williams, G., Bell, S. A., et al. (2015). A case series study on the effect of Ebola on facility-based deliveries in rural Liberia. *BMC Pregnancy and Childbirth, 15*, 254. https://doi.org/10.1186/s12884-015-0694-x.

Ministry of Health and Sanitation. (2016). *National Ebola Response Centre (NERC), Sierra Leone.* Retrieved November 7, 2017, from https://www.humanitarianresponse.info/en/operations/sierra-leone/office/national-ebola-response-centre-nerc-freetown-sierra-leone.

Myatt, M., Duffield, A., Seal, A., & Pasteur, F. (2009). The effect of body shape on weight-for-height and mid-upper arm circumference-based case definitions of acute malnutrition in Ethiopian children. *Annals of Human Biology, 36*(1), 5–20. https://doi.org/10.1080/03014460802471205.

ReliefWeb. (2015). Sierra Leone: Tonkolili District Profile (3 December 2015). Retrieved September 25, 2018 from, https://reliefweb.int/report/sierraleone/sierra-leone-tonkolili-district-profile-3-december-2015

Statistics Sierra Leone. (2014). *Sierra Leone Demographic and Health Survey, 2013.* Freetown, Sierra Leone: Ministry of Health and Sanitation.

Statistics Sierra Leone. (2016). *Sierra Leone 2015 population and housing census. Provisional results, March 2016.* Freetown, Sierra Leone: Ministry of Health and Sanitation. Retrieved February, 2017, from https://www.statistics.sl/wp-content/uploads/2016/06/2015-Census-Provisional-Result.pdf.

Street Child. (2015). *The street child: Ebola orphan report.* Retrieved November 8, 2017, from https://www.street-child.co.uk/ebola-orphan-report/.

Strong, A., & Schwartz, D. A. (2019). Effects of the West African Ebola epidemic on health care of pregnant women—Stigmatization with and without infection. In D. A. Schwartz, J. N. Anoko, & S.A. Abramowitz (Eds.), *Pregnant in the Time of Ebola: Women and their children in the 2013-2015 West African Ebola Epidemic.* New York: Springer.

Strong, A., & Schwartz, D. A. (2016). Sociocultural aspects of risk to pregnant women during the 2013–2015 multinational Ebola virus outbreak in West Africa. *Health Care for Women International, 37*(8), 922–942. https://doi.org/10.1080/07399332.2016.1167896.

Suk, J. E., Paez Jimenez, A., Kourouma, M., Derrough, T., Baldé, M., Honomou, P., et al. (2016). Post-Ebola measles outbreak in Lola, Guinea, January–June 2015. *Emerging Infectious Diseases, 22*(6), 1106–1108. Retrieved November 9, 2017, from https://www.ncbi.nlm.nih.gov/pmc/articles/PMC4880080/.

Thomas, A. C., Nkunzimana, T., Perez Hoyos, A., & Kayitakire, F. (2014). *Impact of the West African Ebola virus disease outbreak on food security.* Ispra, Italy: Institute for Environment and Sustainability, European Commission Joint Research Centre.

Tonkolili District Health Management Team. (2013). *District Community Health Worker Database, July 2013.* Tonkolili, Sierra Leone: Tonkolili District Health Management Team.

Tonkolili District Health Management Team. (2016). *Tonkolili District Ebola Surveillance Report Database February 2016.* Tonkolili, Sierra Leone: Tonkolili District Health Management Team.

von Elm, E., Altman, D. G., Egger, M., Pocock, S. J., Gøtzsche, P. C., Vandenbroucke, J. P., et al. (2007). Strengthening the Reporting of Observational Studies in Epidemiology (STROBE) statement: Guidelines for reporting observational studies. *British Medical Journal, 335*(7624), 806–808.

WHO. (2009). *WHO child growth standards and the identification of severe acute malnutrition in infants and children. A joint statement.* Retrieved November 8, 2017, from http://www.who.int/nutrition/publications/severemalnutrition/9789241598163/en/.

World Food Programme. (2016). *New report highlights fragile state of food security in Sierra Leone after Ebola outbreak.* Retrieved September 26, 2018 from, https://www.wfp.org/news/news-release/new-report-highlights-fragile-state-food-security-sierra-leone-after-ebola-outbrea

The Ebola Epidemic Halted Female Genital Cutting in Sierra Leone: Temporarily

30

David A. Schwartz

30.1 Introduction

Female genital cutting (FGC), also referred to as female genital mutilation (FGM) or female circumcision, is a procedure that involves the removal of parts or all of the entire external female genital organs, or other injuries to the external female genital organs, for nonmedical reasons (WHO 2018). The procedure is generally performed by nonphysicians, including elderly people within the community (typically, but not exclusively, women), traditional birth attendants or healers, barbers, members of secret societies (see Bondo Society below), herbalists, or even a close relative. In most areas where the practice occurs, FGC is still carried out by traditional circumcisers, who often play other central roles in communities, such as attending childbirths. In from 1 in 4 to 1 in 5 girls undergoing FGC, depending on the country, the procedure is performed by a medical professional, a process termed the "medicalization" of FGC (UNFPA 2018). The communities where it is practiced cite religious, social, psychosexual, aesthetic, religious, socioeconomic, or other cultural reasons for continuing this practice. Among some cultural groups it is performed as a "cleansing" ritual for girls and women or as a ritual circumcision that is analagous to that performed on boys and men; it may aso be performed to reduce female sexual desire, thus diminishing hypersexuality and promiscuity and ensuring fidelity (Gruenbaum 2000). The majority of girls and women who have had FGC reside in Sub-Saharan Africa and the Middle East (Fig. 30.1), but it is also practiced among certain ethnic groups in Asia, Europe, and elsewhere. Estimates are that from 140 up to 200 million girls and women throughout the world have been subjected to FGC—and the numbers are increasing (Bjälkander et al. 2013; UNFPA 2018; WHO 2018). At its present rate, it is estimated that 68 million girls will be cut between 2015 and 2030 in the 25 countries where FGC is routinely performed (UNFPA 2018); there are three million girls considered to be at risk for having FGC annually (WHO 2018).

FGC has no medical benefit—it only causes harm. The potential health consequences of FGC are significant and are dependent on several variables: the type of FGC procedure (see below), the skill of the practitioner, the level of hygiene of the location and sterility of the instruments, and the age and physical condition of the female infant, girl, or woman on whom it is performed as well as their degree of resistance to the procedure. The procedures are typically performed without anesthesia or adminis-

D. A. Schwartz (✉)
Department of Pathology, Medical College of Georgia, Augusta University, Augusta, GA, USA

D. A. Schwartz et al. (eds.), *Pregnant in the Time of Ebola*, Global Maternal and Child Health,
https://doi.org/10.1007/978-3-319-97637-2_30

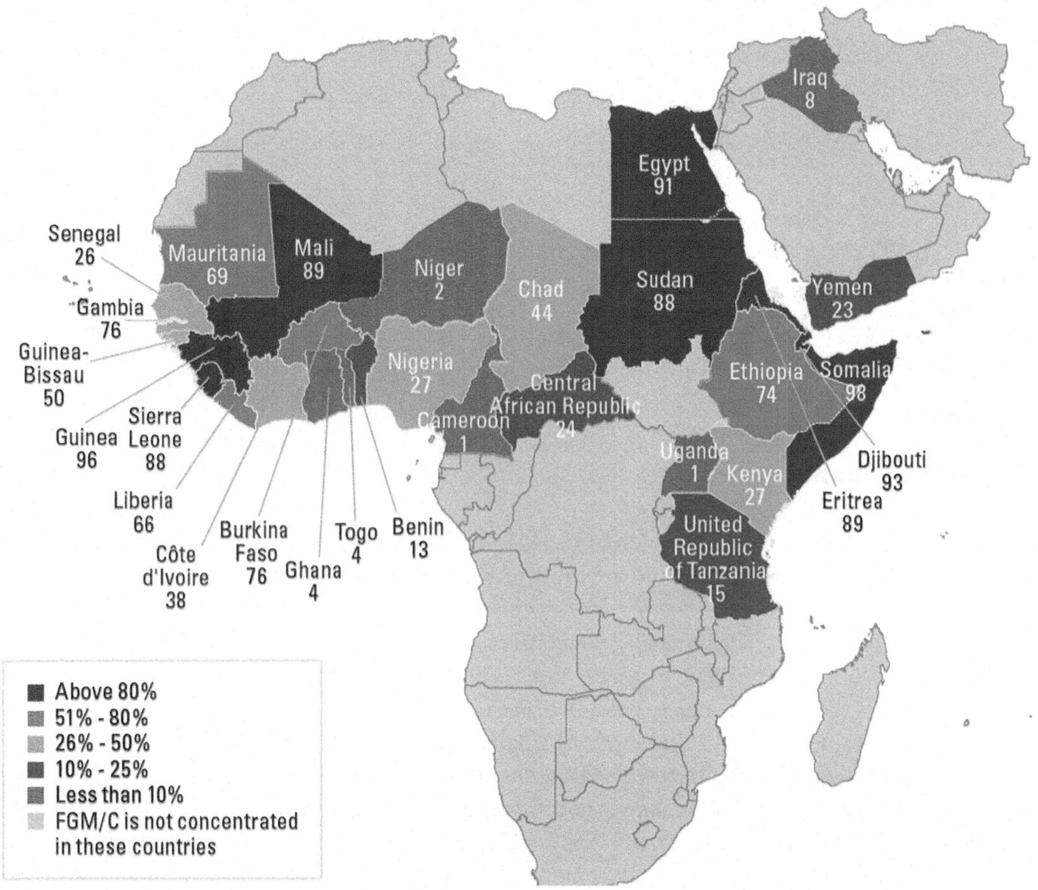

Fig. 30.1 Prevalence of female genital cutting in Africa. The three countries most affected by the West Africa Ebola epidemic, Sierra Leone, Guinea, and Liberia, all have high rates of female genital cutting. Photograph courtesy of UNICEF

tration of pain killing drugs. Some of the most frequent post-procedural complications include bleeding, swelling, infection and abscess formation, severe pain, shock, dermoid cysts, delayed healing, urinary obstruction and recurrent urinary tract infections, dyspareunia, infertility, and labial fusion. An increased risk of HIV transmission and tetanus is also associated with FGC, resulting from the unhygienic performance of the procedure and non-sterile instruments. Pregnancy-related complications also occur (US Department of Health & Human Services 2017; WHO et al. 2006)—these include increased risk for cesarean section, obstructed labor, prolonged labor, rupture of the uterus, obstetric fistula, postpartum hemorrhage, prolonged hospitalization, and need for episiotomy. Infants can be affected, as FGC of the mother is associated with low birth weight, infant resuscitation, and even perinatal death (WHO et al. 2006). It is difficult to estimate the number of girls and women that die as a direct or indirect result of having undergone FGC because records are not maintained, and deaths are not ascribed to this procedure; however, there is no doubt that they do occur. Interestingly, the majority of countries with a high prevalence of FGC also have high maternal mortality ratios (MMRs) and high numbers of pregnancy complications. Two high-FGC-prevalence countries, both in Africa, are among the four countries globally with the highest numbers of maternal deaths. Five of the high-prevalence FGC countries have MMRs of 550 per 100,000 live births and greater (UNFPA 2018).

Psychosocial problems are also more prevalent in girls and women who have had FGC, including depression, somatic disturbance, anxiety, and posttraumatic stress disorder (Behrendt and Moritz 2005; Vloeberghs et al. 2012). Genital cutting/mutilation is often performed on very young children, preadolescent, and teenage girls without their consent.

FGC is internationally recognized as a violation of the basic human rights of girls and women (WHO 2008). According to a number of international public health agencies (WHO 2008) including OHCHR, UNAIDS, UNDP, UNECA, UNESCO, UNFPA, UNHCR, UNICEF, UNIFEM, and WHO,

> *"Seen from a human rights perspective, the practice reflects deep-rooted inequality between the sexes, and constitutes an extreme form of discrimination against women. Female genital mutilation is nearly always carried out on minors and is therefore a violation of the rights of the child. The practice also violates the rights to health, security and physical integrity of the person, the right to be free from torture and cruel, inhuman or degrading treatment, and the right to life when the procedure results in death."*

30.2 Types of Female Genital Cutting

Female genital cutting is a general term that encompasses related but technically differing procedures performed on the external genitalia. There are four types recognized by the World Health Organization (WHO 2018) that include:

30.2.1 Type I: Clitoridectomy

This procedure involves the partial or total removal of the clitoris. In some cases, the prepuce, or clitoral hood, is also removed.

30.2.2 Type II: Excision

In this procedure, there is either partial or total removal of the clitoris and the labia minora. The labia majora may or may not also be excised. The amount of tissue that is removed varies widely from community to community.

30.2.3 Type III: Infibulation

Infibulation, also termed Pharoanic circumcision, refers to the procedure in which the vaginal opening is reduced by removing all or parts of the external genitalia (including clitoris, labia minora, and labia majora), and sewing, pinning, or otherwise causing the remaining tissue to fuse together during the healing process. This procedure involves narrowing of the vaginal opening through the creation of a covering seal. The seal is formed by cutting and sewing over the outer labia, with or without removal of the clitoris or inner labia.

30.2.4 Type IV

Includes all other harmful procedures to the female genitalia for nonmedical purposes. These may include cauterization, pricking, piercing, incising, or scraping, or placing caustic substances in the vagina.

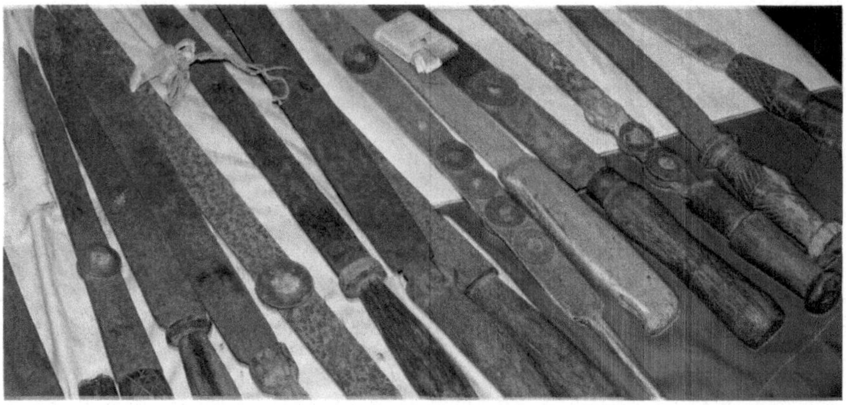

Fig. 30.2 Ceremonial knives used in FGM/C by members of the Bondo society in Sierra Leone. Photo courtesy of Bryna Hallam and the United Nations

Types I and II are the most prevalent forms of female genital cutting, although the prevalence will vary from region to region of the world where this practice occurs. Type III FGC, or infibulation, occurs in approximately 10% of cases, being most frequent in the Northern Sudan, Somalia and Djibouti. Type III FGC can cause total vaginal obstruction, resulting in the accumulation of menstrual flow materials in the vagina and uterus. It also results in a physical barrier to sexual intercourse, and if pregnant, vaginal childbirth. As a result, girls or women who have undergone infibulation must be cut open by the "circumciser" or husband to permit sexual intercourse on the first night of marriage. At the time of childbirth, many girls and women must be cut open again because the vagina is too narrow to permit passage of the infant (Fig. 30.2).

30.3 The Bondo Society

The Bondo (Bondu) Society is a woman's secret society that is active in the West African countries of Guinea, Liberia, Sierra Leone, and the Ivory Coast. Also known as the Sante Society, it consists of ethnically diverse groups of exclusively female initiates, transcending national, ethnic, and linguistic boundaries, which act as the repository for a wide range of gendered knowledge that is shared by the membership. The society is very old—at a very minimum it has existed for many hundreds of years. In the late 1600s, a description of the society was published by a physician and geographer from Holland, Olfert Dapper, who described the "Sandy" society as it existed from a first-hand account in 1628 from the Cape Mount region of Liberia (Dapper 1668). Among other roles, the Bondo society is a cultural organization that plays a major role in the lives of its membership, providing a source of power, solidarity, and support for girls and women and establishing a passage for their transition into adulthood. The Bondo society is also important in the sociopolitical landscape, where it has established deep roots for many hundreds of years in every village and town in Sierra Leone. The *sowei* is typically an older women who is a senior member of the Bondo society. *Soweis* yield enormous political power in Sierra Leone, especially in rural communities, and can command the support of female voters. As a result, it can be dangerous to speak out against the *soweis* and Bondo society, and activists have put their own well-being at risk to do so.

As part of the initiation rite into the Bondo society, young girls undergo female genital cutting that is highly ritualized. The girls are initiated in groups, often in the dry season, and initiation usually occurs in the Bondo Bush, which is a private enclosure that is constructed in a clearing up to several

Fig. 30.3 A historical photograph illustrating a group of Bondo (Bundu) female dancers all wearing necklaces of beads which are filled with medicines. The card reads "The dancers all wore fetishes peculiar to the order, each having special significance. These consisted of several ropes of cane cut into beads and rows of seeds which has been bored and filled with Bundu 'medicine'." Halftone after a photograph by T.J. Alldridge. Photograph courtesy of the Wellcome Images, Photograph No. V0015968

kilometers from their village. Each ethnic group in Sierra Leone, including the Fulah, Limba, Mende, Temne, Loko, and Susu, has its own Bondo Bush that is run by a *sowei*, who is also the traditional exciser or circumciser. Genital cutting is the first act that is performed during the initiation procedure. After the cutting, the girls are traditionally kept in the Bondo house in the forest for a period of many weeks to months while they heal. While there, they are instructed on such tasks of adult womanhood as sexual roles, household skills, farming, dancing, medicine, and how to get along with their future husband's family. The *soweis* are reimbursed for this instruction in food or, if available, cash. The fee is expensive—from 700,000 Leones in rural areas up to 1.6 million Leones (USD $208) in the capital of Freetown (Jones 2018), and is typically paid by the girl's father or her prospective husband, as a girl may not marry before initiation (Mgbako 2010). If the girl is found to not be a virgin at the time of initiation, the *soweis* may raise the price for circumcision. The girls are given new names, or Bondo names, signifying their emergence into adulthood at the conclusion of their initiation and their status, or rank, in the ritual hierarchy. They undergo a ritual washing and emerge from the forest to return to the community in new clothes as marriageable adults. Following genital cutting and the other initiation rites, the girls are permitted to return to the Bondo without obtaining permission from their husbands. Because membership in the Bondo society equates with women's power in Sierra Leone, members have a higher standing in Sierra Leonean society that do noninitiated women (Bjälkander et al. 2012, 2013; Bosire 2012; The Guardian 2015) (Figs. 30.3 and 30.4).

30.4 Female Genital Cutting in Sierra Leone Prior to Ebola

The history of female genital mutilation remains unclear, but it predates the origins of Christianity and Islam. In Sierra Leone, FGC is associated with the Bondo (Sante) Society and has been practiced regularly for many hundreds of years, and likely much longer. Prior to the Ebola virus epidemic,

Fig. 30.4 Sande society initiates in 1912, Sierra Leone. Photograph from C. H. Firmin.—Customs of the World

Sierra Leone had one of the highest rates of female genital cutting in the world. Although prevalence data have varied, studies have reported that from 88% to 89.6% of girls and women have undergone the procedure (Statistics Sierra Leone et al. 2013), with some sources reporting greater than 90% prevalence (Bjälkander et al. 2013; Yoder et al. 2004). In rural areas of Sierra Leone, rates of FGC have been reported to be as high as 94.3% (28 Too Many 2014). In fact, Sierra Leone is one of the only countries in Africa where the prevalence rate has exceeded 90% for the age group 15–49 years, and together with Guinea, has the highest prevalence rate in southwestern Africa (Yoder et al. 2013). There is no national law in Sierra Leone making the practice of FGC unlawful. Among women who have undergone FGC, 40% were circumcised between the ages of 10 and 14 years; 17% underwent circumcision during childhood between the ages of 1 and 4 years, and additional 13% were circumcised between 5 and 9 years of age. Children and women who are Moslem are more likely to undergo FGC (93%) as compared with girls and women of other religions (Statistics Sierra Leone et al. 2013).

This high prevalence should not be surprising based upon knowledge of the demographics of FGC. FGC is usually initiated before girls reach the age of 15 years and may even be conducted shortly after birth, on infants and toddlers. The total number of girls affected by FGC is a function of its overall prevalence or intensity in a society and the total number of girls at risk for the procedure; in other words, the age distribution of the female population. Most countries with highly prevalent FGC have large proportions of young adolescents and children from 0 to14 years of age; in 2015, Sierra Leone had 40.6% of its female population below the age of 15 years (UNFPA 2015). Countries with high prevalence of FGC also have high fertility rates (UNFPA 2015). In Sierra Leone, the fertility rate was 4.92 in 2012 and prior to the Ebola epidemic (Index Mundi n.d.), while the adolescent birth rate (number of births to women ages 15–19 per 1000 women in that age group per year) is one of the highest in the world at 126 (UNFPA 2017). A third metric that is highly associated with countries having high prevalence of FGC is the maternal mortality ratio. Prior to the Ebola epidemic, Sierra Leone had the highest MMR in the world—1165 per 100,000 live births (Mamaye 2014).

Sierra Leone was a signer of the Maputo protocol adopted by the African Union on July 11th, 2003 at its second summit in Maputo, Mozambique, which became effective on 25 November, 2005. This instrument, formally known as "The Protocol to the African Charter on Human and Peoples' Rights on the Rights of Women in Africa," guaranteed comprehensive rights to women in Africa including Article 5 providing an end to female genital mutilation (Scoop Independent News 2006).

On June 7th, 2007, the Sierra Leone Parliament passed the National Child Rights Bill. The passage of this bill provided a framework for ensuring adequate standards of care for all children in Sierra Leone. Specifically, it provided protection against such harmful traditional practices that affected children including female genital mutilation (UNICEF 2007). This bill was passed despite the fact that the Paramount Chiefs from all 149 chiefdoms in Sierra Leone had insisted that a ban on FGM be removed from the bill (Kargbo 2015).

The civil war that was waged from 1991 to 2002 left the country devastated, with much of the country's infrastructure destroyed and tens of thousands of people displaced. In the post-war reconstruction period, reduction of FGC was not a national priority. In the years leading up to the Ebola epidemic, there were efforts to reduce FGC in Sierra Leone and neighboring countries by various agencies. In 2008, United Nation's agencies issued a joint statement, *Eliminating Female Genital Mutilation: An Interagency Statement*, setting out the elements to uphold the rights of girls and women and calling on states, international and national organizations, civil society, and communities to develop, strengthen, and support specific and concrete actions directed towards ending female genital mutilation (WHO 2008). In 2011, with the assistance of Amnesty International, community members in northwest Sierra Leone signed a Memorandum of Understanding, which banned female genital mutilation for girls under 18 years of age. This agreement also stated that anyone above that age must supply their consent before the procedure could be carried out (Goldberg 2015). In a 2012 study conducted by 37 organizations working for the abolition of FGC in Sierra Leone, it was found that, as a result of the political sensitivity surrounding the topic of FGC, there was a notable lack of governmental commitment to approaching the issue, and that NGOs (nongovernmental organizations) were the main advocates of its abolishment (28 Too Many 2014). Despite some successes, however, FGC continued to be practiced with high intensity in Sierra Leone up to the national crisis caused by the Ebola virus epidemic in 2014.

30.5 The Ebola Epidemic Reduced Female Genital Cutting in Sierra Leone, But Now It's Back

As the Ebola epidemic spread through Sierra Leone in 2014, the government placed a temporary ban of female genital cutting in early November to prevent further transmission of the disease (Devries 2014). Because FGC is typically performed by *soweis* under non-hygienic circumstances and using non-sterile instruments and bandages, it was feared that the infection would be transmitted to girls being circumcised at the time of FGC from contaminated instruments and other materials, or even to the *soweis* themselves. Although there was still no law in Sierra Leone prohibiting FGC, the ban imposed a fine of 500,000 Leones (about USD $115) for performing the procedure. This ban on female genital cutting was enforced by Sierra Leone's 14 districts, although there was reportedly some variation in the amount of the penalties. According to Owolabi Bjälkander, a UNICEF consultant in Sierra Leone, "*The fact that some communities have chosen not to practice FGM during the crisis together with evidence on the effect and impact these bans have had on FGM could be valuable for developing interventions for total abandonment of the practice,*" (Devries 2014; Goldberg 2015). There were conflicting reports of whether any *soweis* had died of Ebola infection from performing FGC. A statement from Ann-Marie Caulker, founder of an NGO aimed at ending FGM, said that at least nine *soweis* had died from Ebola infection. However, the president of the National Council of Soweis in Sierra Leone, Mammy Koloneh, denied that any *soweis* had died from Ebola and wished to keep FGC continuing, as it has been part of the culture of Sierra Leone for many generations. Mammy Koloneh said that the loss of business has affected every one of approximately 2000 *soweis* across the country. "*Nobody is considering us, to help us, to give us rice or anything, since this Ebola started,*" she said (Devries 2014). Both Ann-Marie Caulker and Mammy Koloneh agreed that, as a result of the epidemic and governmental ban, the practice of FGC had come to a near halt. The Ministry of Health and Sanitation of Sierra Leone stated that they did not have statistics on whether FGC had been halted, or whether any *soweis* had died. And Dhuwarakha Sriram, a child protection specialist for UNICEF, stated that she would be surprised if the practice had stopped completely, but it appears to have been drastically reduced (Devries 2014).

The Ebola epidemic also induced the government of Sierra Leone to ratify the Maputo Protocol on July 2, 2015. Sierra Leone was the last of 37 African countries to ratify the Protocol, nearly 12 years after the country had initially signed the document in 2003 (Milton 2017). The temporary governmental ban of FGC also mobilized anti-FGC activists to capitalize on the halt to end the practice permanently. The chief medical officer at Sierra Leone's Ministry of Health, Dr. Brima Kargbo, said the Ministry was conducting outreach campaigns to educate traditional healers and *soweis* on the dangers of performing FGM during the Ebola crisis. The ban appeared to be working—according to unofficial reports, the prevalence of FGC had dramatically decreased and *soweis* were refusing to perform the procedure due to the risk of acquiring Ebola via blood and body fluids (Devries 2014; Merlan 2015). Local chiefs partnered with nurses and inspected rural provinces, explaining the legal and medical implications of genital cutting during the Ebola epidemic, and how the virus is transmitted through bodily fluids. By the end of the epidemic, it was generally agreed that the ban had been very effective in Sierra Leone and had resulted in the most dramatic decline in history of the rates of FGC (Acland 2016).

Following the World Health Organization's announcement on November 7th 2015 that the Ebola epidemic was ended in Sierra Leone, the President, Ernest Bai Koroma, made a televised speech in which he talked about a new start for his nation. "*This warrants that traditional practices which have a negative impact on health, and which were discontinued during the outbreak, should not be returned to,*" he said (Acland 2016). Women's rights activists, NGOs, and United Nations agencies were hopeful that, once seeing that FGC could be curtailed if necessary, the trend would continue following cessation of the Ebola epidemic. However, the end of the epidemic caused a return to "business as usual" by the country's *soweis*. One positive thing that emerged following the epidemic was that the shroud of secrecy surrounding FGC, the Bondo society, and the *soweis* and their political powers had been lifted during the ban. According to Zihalirwa Nalwage, a child protection chief at UNICEF, punitive measures would not be useful in eliminating FGC practices and would only drive the procedure underground, endangering the women and children who it is practiced upon. He stated, "*Prosecution itself would not be enough as they'd only continue to do it in hiding,*" and "*We need to work on a strategy which upholds tradition and gives the soweis an important cultural role in society, but one which doesn't involve cutting.*" Nalwage continued, "*I think it will take 10 to 20 years before the country sees a dramatic, long-term drop in the practice—but statistics are slowly declining and we're moving in the right direction*" (Acland 2016). However, as the practice of FGC returned to a post-Ebola Sierra Leone, in August 2016 a 19-year-old girl, Fatmata Turay, died in the northern town of Makeni soon after undergoing FGC and leading to the arrest of several *soweis* (Guilbert 2016). In another recent case, police reported that children had been rescued from a disused house in Magburaka, in the north of Sierra Leone, after undergoing an initiation ceremony there (The Guardian 2016).

Anti-FGC activists and campaigners have continued their battle against this deeply embedded cultural practice, attempting to garner support for national legislation to protect children and women. However, in a statement made in 2017, UN Women stated, "*But now 3 years on, the practice has returned even though the ban is still in place*" "*While gender inequality, myths, and cultural beliefs are at the root of the practice, for many rural women, FGM is also a matter of livelihood*" (Devi 2018). A statement made in January 2018 from Maya Moiwo Kaikai, the Minister of Local Government and Rural Development stated, "*FGM is a difficult issue for the government because it's very embedded into the hierarchy of the country. There is a strong constitution in favor of it given the high percentage of women who have gone through it.*" The Minister also stated that "*there has been an increase in forceful societal initiations*" (Devi 2018).

It was announced in February 2018 that the government of Sierra Leone has banned female genital cutting until March 31st, after its elections—this in order to prevent candidates from buying votes by paying for cutting ceremonies (Peyton 2018). According to anti-FGM campaigner Rugiatu Neneh Turay, formerly the Deputy Minister of Social Welfare, Gender and Children's Affairs, "*So many politicians use initiation into secret society during campaigns to gain votes, especially those of women.*" Most activists do not believe that this ban represents progress in eliminating FGC. "*This is just a political move*," Felister Gitonga of Equality Now told the Thomson Reuters Foundation. "*It has nothing to do with protecting the rights of women*" (Peyton 2018).

Unfortunately, following the Ebola outbreak in Sierra Leone, there are few services or individuals currently available to assist a girl or woman fleeing from her village or community to avoid the dangers of FGC. To make matters worse, following the destructive impact of the Ebola epidemic on the healthcare infrastructure and numbers of skilled healthcare workers and physicians in Sierra Leone, the availability of medical assistance to treat the multitude of health consequences of FGC, including recurrent genitourinary tract infections, sepsis, shock, pain, hemorrhage, and infertility, remains limited. Because there is no existing national legislation that forbids FGC in either Sierra Leone or neighboring Liberia, and Guinea also shares a border, there is an opportunity for families and cutters to move across national borders to perform the procedure and escape repercussions. There is no data available to even estimate the numbers of girls and women who are taken across borders to be cut (28 Too Many 2018). The world must watch and wait to see what progress, if any, is made to curtail the practice of FGC in Sierra Leone.

References

28 Too Many. (2014). *Country profile. FGM in Sierra Leone June 2014*. Retrieved April 13, 2018, http://www.refworld.org/pdfid/54bce6334.pdf.

Acland, O. (2016). Ebola ended FGM in Sierra Leone, but now it's back. *Vice News*. Retrieved April 15, 2018, from https://news.vice.com/article/ebola-ended-fgm-in-sierra-leone-but-now-its-back.

Behrendt, A., & Moritz, S. (2005). Posttraumatic stress disorder and memory problems after female genital mutilation. *American Journal of Psychiatry, 162*, 1000–1002.

Bjälkander, O., Bangura, L., Leigh, B., Berggren, V., Bergström, S., & Lars Almroth, A. (2012). Health complications of female genital mutilation in Sierra Leone. *International Journal of Women's Health, 4*, 321–331. Retrieved April 30, 2018, from https://www.ncbi.nlm.nih.gov/pmc/articles/PMC3410700/.

Bjälkander, O., Grant, D. S., Berggren, V., Bathija, H., & Almroth, L. (2013). Female genital mutilation in Sierra Leone: Forms, reliability of reported status, and accuracy of related demographic and health survey questions. *Obstetrics and Gynecology International, 2013*, 1–14. https://doi.org/10.1155/2013/680926.

Bosire, O. T. (2012). *The Bondo secret society: Female circumcision and the Sierra Leonean state* (PhD thesis, University of Glasgow).

Dapper, O. (1668). *Naukeurige Beschrijvinge der Afrikaensche Gewesten van Egypten, Barbaryen, Libyen, Biledulgerid, Guinea, Ethiopiën, Abyssine*. Amsterdam: J. van Meurs.

Devi, S. (2018). FGM in Sierra Leone. *Lancet, 391*(10119), 415. Retrieved April 20, 2018, from https://www.thelancet.com/journals/lancet/article/PIIS0140-6736(18)30189-2/fulltext.

Devries, N. (2014). Ebola fears bring female genital mutilation to near halt in Sierra Leone. *Aljazeera America*. Retrieved April 28, 2018, from http://america.aljazeera.com/articles/2014/12/4/sierra-leone-fgmebola.html.

Goldberg, E. (2015). Ebola could put a stop to female genital mutilation in Sierra Leone. *Huffington Post*. Retrieved May 3, 2018, from https://www.huffingtonpost.com/2015/01/16/female-genital-mutilation-sierra-leone_n_6481054.html.

Gruenbaum, E. (2000). *The Female Circumcision Controversy. An Anthropological Perspective*. Philadelphia: University of Pennsylvania Press.

Guilbert, K. (2016). Sierra Leone urged to ban FGM after death of teenage girl. *Reuters*. Retrieved April 15, 2018, from https://www.reuters.com/article/us-leone-fgm/sierra-leone-urged-to-ban-fgm-after-death-of-teenage-girl-idUSKCN10T15V.

Index Mundi. (n.d.). *Sierra Leone—Fertility rate, total (births per woman)*. Retrieved April 14, 2018, from https://www.indexmundi.com/facts/sierra-leone/indicator/SP.DYN.TFRT.IN.

Jones, S. F. (2018). Can Sierra Leone's female secret societies be allies in the fight against female genital mutilation? 50.50. *Open Democracy*. Retrieved May 1, 2018, from https://www.opendemocracy.net/5050/shanna-f-jones/sierra-leone-female-secret-societies-allies-fight-against-fgm.

Kargbo, M. A. (2015). FGM: Now that Ebola is over. *Cocorioko*. Retrieved April 30, 2018, from http://cocorioko.net/fgm-now-that-ebola-is-over/.

Mamaye. (2014). Facts and figures. The Sierra Leone Demographic and Health Survey 2013.

Merlan, A. (2015). *Ebola is temporarily halting female genital mutilation in Sierra Leone*. Retrieved April 19, 2018, from https://jezebel.com/ebola-is-temporarily-halting-female-genital-mutilation-1679493487.

Mgbako, C. (2010). Penetrating the silence in Sierra Leone: A blueprint for the eradication of female genital mutilation. *Harvard Human Rights Journal, 23*(1), 110–140.

Milton, B. (2017). Sierra Leone News: Salone hosts Mapotu Protocol… government pledges to support to women's empowerment. *Awoko*. Retrieved May 1, 2018, from https://awoko.org/2017/03/23/sierra-leone-news-salone-hosts-mapotu-protocol-government-pledges-to-support-to-womens-empowerment/.

Peyton, N. (2018). Seeking fair elections, Sierra Leone bans FGM during campaign season. *Reuters*. Retrieved May 5, 2018, from https://www.reuters.com/article/us-leone-women-fgm/seeking-fair-elections-sierra-leone-bans-fgm-during-campaign-season-idUSKBN1FP2CB.

Scoop Independent News. (2006). *UNICEF: Toward ending female genital mutilation*. Retrieved April 25, 2018, from http://www.scoop.co.nz/stories/WO0602/S00099.htm.

Statistics Sierra Leone (SSL), Ministry of Health and Sanitation, & ICF International. (2013). Sierra Leone Demographic and Health Survey 2013. Rockville: Statistics Sierra Leone (SSL) and ICF International.

The Guardian. (2015). Sierra Leone's secret FGM societies spread silent fear and sleepless nights. *The Guardian*. Retrieved April 15, 2018, from https://www.theguardian.com/global-development/2015/aug/24/sierra-leone-female-genital-mutilation-soweis-secret-societies-fear.

The Guardian. (2016). Captured and cut: FGM returns to Sierra Leone despite official ban. *The Guardian*. Retrieved April 12, 2018, https://www.theguardian.com/global-development/2016/sep/29/female-genital-mutilation-returns-sierra-leone-official-ban.

28 Too Many. (2018). Sierra Leone. The Law and FGM. September 2018. Retrieved November 22, 2018, from https://www.28toomany.org/static/media/uploads/Law%20Reports/sierra_leone_law_report_v1_(september_2018).pdf

UNFPA. (2015). Demographic perspectives on female genital mutilation. *United Nations Population Fund*. Retrieved April 15, 2018, from https://www.unfpa.org/sites/default/files/pub-pdf/1027123_UN_Demograhics_v3%20%281%29.pdf.

UNFPA. (2017). To secure a better future, teens in Sierra Leone look to family planning. *United Nations Population Fund*. Retrieved April 12, 2018, from http://sierraleone.unfpa.org/en/news/secure-better-future-teens-sierra-leone-look-family-planning.

UNFPA. (2018). Female genital mutilation (FGM) frequently asked questions. *United Nations Population Fund*. Retrieved April 15, 2018, from https://www.unfpa.org/resources/female-genital-mutilation-fgm-frequently-asked-questions#.

UNICEF. (2007). Sierra Leone approves the National Child Rights Bill. *UNICEF*. Retrieved May 3, 2018, from https://www.unicef.org/media/media_39951.html.

US Department of Health & Human Services. (2017). Female genital cutting. *Womenshealth.gov*. Retrieved April 13, 2018, from https://www.womenshealth.gov/a-z-topics/female-genital-cutting.

Vloeberghs, E., van der Kwaak, A., Knipscheer, J., & van den Muijsenbergh, M. (2012). Coping and chronic psychosocial consequences of female genital mutilation in the Netherlands. *Ethnicity & Health, 17*, 677–695.

WHO. (2008). *Eliminating female genital mutilation. An interagency statement—OHCHR, UNAIDS, UNDP, UNECA, UNESCO, UNFPA, UNHCR, UNICEF, UNIFEM, WHO*. Retrieved April 10, 2018, from http://apps.who.int/iris/bitstream/handle/10665/43839/9789241596442_eng.pdf;jsessionid=87022C9002FEE85E1B3A681FA3D5EB0B?sequence=1.

WHO. (2018). Female genital mutilation. *World Health Organization*. Retrieved April 19, 2018, from http://www.who.int/mediacentre/factsheets/fs241/en/.

WHO, Banks, E., Meirik, O., Farley, T., Akande, O., Bathija, H., et al. (2006). Female genital mutilation and obstetric outcome: WHO collaborative prospective study in six African countries. *Lancet, 367*(9525), 1835–1841.

Yoder, P. S., Abderrahim, N., & Zhuzhuni, A. (2004). *DHS Comparative Reports no. 7*. Female genital cutting in the demographic and health surveys: a critical and comparative analysis. Measure DHS+, Calverton. Retrieved May 1, 2018, from http://www.measuredhs.com/pubs/pdf/CR7/CR7.pdf.

Yoder, P. S., Wang, S., & Johansen, E. (2013). Estimates of female genital mutilation/cutting in 27 African countries and Yemen. *Studies in Family Planning, 44*(2), 189–204.

Index

D. A. Schwartz et al. (eds.), *Pregnant in the Time of Ebola*, Global Maternal and Child Health,
https://doi.org/10.1007/978-3-319-97637-2

J

K

L

P

Printed by Printforce, the Netherlands